International Symposium on Nanoscience and Quantum Physics 2011

(nanoPHYS'11)

Journal of Physics: Conference Series Volume 302

Tokyo, Japan
26-28 January 2011

Editors:

Susumu Saito
Hidekazu Tanaka
Takashi Nakamura
Masaaki Nakamura

ISBN: 978-1-61839-046-2
ISSN: 1742-6588

Printed from e-media with permission by:

Curran Associates, Inc.
57 Morehouse Lane
Red Hook, NY 12571

Some format issues inherent in the e-media version may also appear in this print version.

Printed by Curran Associates, Inc. (2011)

For permission requests, please contact the Institute of Physics
at the address below.

Institute of Physics
Dirac House, Temple Back
Bristol BS1 6BE UK

Phone: 44 1 17 929 7481
Fax: 44 1 17 920 0979

techtracking@iop.org

Additional copies of this publication are available from:

Curran Associates, Inc.
57 Morehouse Lane
Red Hook, NY 12571 USA
Phone: 845-758-0400
Fax: 845-758-2634
Email: curran@proceedings.com
Web: www.proceedings.com

TABLE OF CONTENTS

NOVEL NANOSTRUCTURE (NANOWIRES, NANOTUBES, SPIN-RELATED STRUCTURE ETC.)

NOVEL TRANSPORT AND ELECTRONIC PROPERTIES (GRAPHENE, TOPOLOGICAL INSULATORS, COHERENT CONTROL ETC.)

ELECTRONIC AND OPTICAL PROPERTIES OF NANOSTRUCTURE

FUNDAMENTAL PHYSICS AND NEW CONCEPT IN QUANTUM PHYSICS

QUANTUM PHYSICS – QUANTUM INFORMATION

QUANTUM PHYSICS – NUCLEAR AND HADRON PHYSICS

QUANTUM PHYSICS – ASTROPHYSICS ETC.

PREFACE

Quantum physics has developed modern views of nature for more than a century. In addition to this traditional role, quantum physics has acquired new significance in the 21st century as the field responsible for driving and supporting nanoscience research, which will have even greater importance in the future because nanoscience will be the academic foundation for new technologies. The Department of Physics, Tokyo Institute of Technology, are now conducting a "Nanoscience and Quantum Physics" project (Physics G-COE project) supported by the Global Center of Excellence Program of the Ministry of Education, Culture, Sports, Science and Technology of Japan (MEXT) in order to promote research and education in these important academic fields. The International Symposium on Nanoscience and Quantum Physics, held in Tokyo, Japan, 26–28 January 2011 (nanoPHYS'11) was organized by the Physics G-COE project of the Tokyo Institute of Technology to provide an international forum for the open exchange of topical information and for stimulating discussion on novel concepts and future prospects of nanoscience and quantum physics. There were a total of 118 papers including 34 invited papers. This nanoPHYS'11 is the forth symposium of this kind organized by the Tokyo Institute of Technology.

Topics focused on in the symposium included:

- Category 1: Novel nanostructure (Nanowires, Nanotubes, Spin-related structure, etc.)
- Category 2: Novel transport and electronic properties (Graphene, Topological insulators, Coherent control, etc.)
- Category 3: Electronic and optical properties of nanostructure
- Category 4: Fundamental physics and new concept in quantum physics
- Category 5: Quantum Physics – Quantum information
- Category 6: Quantum Physics – Nuclear and Hadron Physics
- Category 7: Quantum Physics – Astrophysics, etc.

All the papers submitted to this issue have been reviewed under a stringent refereeing process, according to the normal rules of this Journal. The editors are grateful to all the authors, the referees, and all the individuals involved in the symposium organization, in particular, all the committee members and secretaries who helped to make this symposium so successful. The organizing committee would like to take this opportunity to thank the invited speakers, the session chairs, and all the attendees for their contribution to the symposium.

Susumu Saito, Hidekazu Tanaka, Takashi Nakamura and Masaaki Nakamura, *Editors*

A Carbon Nanotube-based NEMS Parametric Amplifier for Enhanced Radio Wave Detection and Electronic Signal Amplification

B J Alemán[1,2,3], A Sussman[1,2,3], W Mickelson[2] and A Zettl[1,2,3]

[1] Physics Department, University of California, Berkeley, CA 94720, USA

[2] Center of Integrated Nanomechanical Systems, University of California, Berkeley, CA 94720, USA

[3] Material Sciences Division, Lawrence Berkeley National Laboratory, Berkeley, CA 94720, USA

Corresponding author: azettl@berkeley.edu

Abstract. We propose a scheme for a parametric amplifier based on a single suspended carbon nanotube field-emitter. This novel electromechanical nanotube device acts as a phase-sensitive, variable-gain, band-pass-filtering amplifier for electronic signal processing and, at the same time, can operate as a variable-sensitivity, tuneable detector and transducer of radio frequency electromagnetic waves. The amplifier can exhibit infinite gain at pumping voltages much less than 10 Volts. Additionally, the amplifier's low overhead power consumption (10-1000 nW) make it exceptionally attractive for ultra-low-power applications.

1. Introduction

The phenomenon of parametric amplification, such as occurs in a child's swing, has been studied since the early 19th century[1,2] and occurs in oscillating systems in which a parameter (*i.e.* resonance frequency, spring constant, degree of dissipation, *etc.*) is modulated at a sub-multiple of twice the resonance frequency, $2\omega_0/n$, to produce amplification in the response of the system. Parametric amplifiers have generated sustained interest because their operational noise can approach the quantum-limit[3,4] and several implementations of phenomenon have been demonstrated in electronic systems using varactors[5] and Josephson junctions[6], optical systems using nonlinear materials[7], and in electromechanical systems[8]. In this paper, we explore the parametric amplification of mechanical oscillations in a suspended carbon nanotube resonator. We show that by operating the device in the limit of strong coupling between mechanical vibrations and field emission tunnelling current, the device behaves as a tuneable, high-gain, phase-sensitive amplifier of AC current and voltage signals. Additionally, the same device can be used as a tuneable, highly-sensitive detector of radio frequency (RF) electromagnetic radiation or as an RF transmitter.

Suspended carbon nanotube field emitters have been explored as electron sources [9, 10, 11] and Fowler-Nordheim diodes[12], and have shown exceptional promise as resonators[13] in nanoelectromechanical systems (NEMS) with applications to radio wave detection[14], atomic-resolution inertial mass sensing[15], and self-sustained RF oscillators[16]. Apart from be able to actuate

Published under licence by IOP Publishing Ltd

carbon nanotube field-emission devices with electromagnetic fields, they offer a unique way to read out vibrational information because the field-emission current is strongly coupled to the position of the nanotube relative to the counter-electrode. This coupling occurs because the field emission current is dominated by the local field at the nanotube tip which can vary substantially as the tube position changes relative to the counter-electrode[14, 16]. Therefore, the vibrational dynamics of the tube can be ascertained from a direct monitoring the field emission current.

We propose a scheme for parametric amplification using a single suspended carbon nanotube field emitter that can operate as a current or voltage amplifier or as an electromagnetic wave transducer, with a direct field emission current output signal. A three-terminal prototype of such a device is shown in Figure 1 (a); the lithographically fabricated device consists of a singly-clamped multi-walled carbon nanotube, a counter (drain) electrode, and a pumping electrode to generate parametric modulation. We begin by describing the nature of the parametric modulation in our system and then explore the device's theoretical performance as an amplifier and transducer.

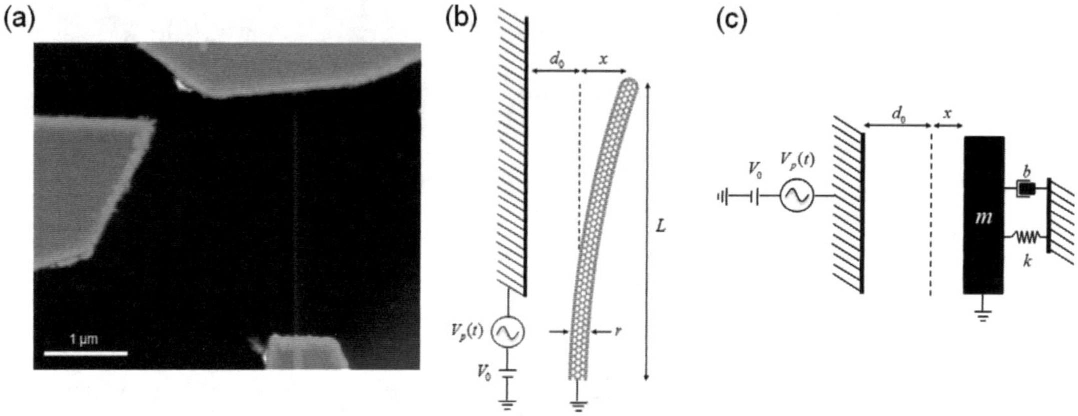

Figure 1: Prototype for a carbon nanotube parametric amplifier. (a) A false-color scanning electron micrograph of a lithographically-defined multiwalled carbon nanotube device. The source and drain electrodes (upper and lower in blue) maintain and measure the field emission current and the side pumping electrode (left in red) generates the modulation of the effective elastic spring constant of the nanotube. (b) A schematic of the device highlighting the interaction with the pump electrode and the basic electrical circuit wiring. (c) A schematic of the equivalent electromechanical system for the pump-nanotube interaction shown in (b). The pump electrode-nanotube interaction is modelled as a position variable capacitor of mass m connected to a rigid support by a spring of spring constant k and dissipation b.

2. Results and Discussion

We model our singly-clamped carbon nanotube device as a capacitively-coupled, driven-damped simple harmonic oscillator (Figure 1 (b) and (c)). The electromechanical coupling in the presence of a time-varying potential V(t), leads to a Hooke's Law force term with a time-dependent spring constant in the total electrostatic force, $F_E = -\frac{1}{2}\frac{dC}{dx}V^2(t)$, which, in turn, is responsible for the amplified response of the nanotube vibrations in the presence of an external driving force.

The form of the modulated spring constant is obtained from the capacitance of the nanotube, which we approximate with the classical expression for the capacitance of a metallic cylinder of radius r and length L a distance $d = d_0 + x$ from a conducting plane[17],

$$C(x) = \frac{2\pi\varepsilon_0}{\log\left(\frac{4d_0(1+x/d_0)}{r}\right)} L, \tag{1}$$

where x is the deflection of the nanotube tip, d_0 is the equilibrium tip distance in the absence of any forces, and $\varepsilon_0 = 8.85\times10^{-12}\ C^2N^{-1}m^{-2}$ is the vacuum permittivity. Although we are using x as the tip deflection, Eq. (1) agrees well (within 18 %) with finite-element simulations performed on a nanotube bending under the appropriate electrostatic load[18]. Some error cancellation is expected since we have neglected the capacitive contribution of the highly charged nanotube tip[16]. Preserving the first three terms of the Taylor expansion of Eq. (1) for $x < d_0$ and applying a voltage $V(t) = V_0 + V_p \sin(2\omega_0 t)$ to the pump electrode we obtain

$$F_E = -\frac{1}{2}\left(\frac{dC(0)}{dx} + \frac{d^2C(0)}{dx^2}x\right)(V_0^2 + 2V_0V_p\sin(2\omega_0 t) + V_p^2\sin^2(2\omega_0 t)) \tag{2}$$

Retaining only the terms involving products of x with V_0^2 or $\sin(2\omega_0 t)$ (neglected terms result in static deflection or produce off-resonant driving forces), and including dissipative, elastic, and driving force terms, the equation of motion of the nanotube becomes

$$m\frac{d^2x}{dt^2} + m\frac{\omega_0}{Q}\frac{dx}{dt} + [k + \Delta k\sin(2\omega_0 t)]x = F\cos(\omega_0 t + \phi) \tag{3}$$

where m is the nanotube mass, Q is the resonator quality factor, $\omega_0 = \omega_0(k)$ is the mechanical resonance frequency, $k = k_0 + \frac{1}{2}\frac{d^2C(0)}{dx^2}V_0^2$, and $\Delta k = \frac{d^2C(0)}{dx^2}V_0V_p$.

We use a standard result[19] from elastic beam theory for the spring constant, $k_0 = \frac{3\pi E r^4}{4L^3}$, with a Young's modulus $E \approx 1\,\text{TPa}$, and use Eq. (1) to obtain

$$\frac{d^2C(0)}{dx^2} = \frac{\pi\varepsilon_0 L}{d_0^2}\frac{\left(2 + Ln\left(4d_0/r\right)\right)}{Ln^3\left(4d_0/r\right)}. \tag{4}$$

Eq. (3) is in the form of the driven Mathieu equation, and we therefore expect the amplitude-frequency response of the nanotube to exhibit amplification governed by the relative strength of the modulated term $\Delta k(t)$. The solution for amplitude response of Eq. (3) in the limit $Q >> 1$ is $A = F\frac{Q\omega_0}{k}G(\phi)$[8, 20], where the phase-sensitive gain is given by

$$G(V_0, V_p, \phi) = \left[\frac{\cos^2\phi}{(1+Q\Delta k/2k)^2} + \frac{\sin^2\phi}{(1-Q\Delta k/2k)^2}\right]^{1/2} \tag{5}$$

The gain of our amplifier is therefore specified by V_p, V_0, and ϕ. It is a well-known fact that tuning the phase to zero in a parametric amplifier leads to de-amplification, while setting $\phi = \pi/2$ yields maximum amplification. Figure 2 shows $G(\phi)$ for $\phi = \pi/2$ as a function of V_p for device geometries similar that shown in Figure 1 (a) and to those from our previous experiments[14, 15, 16] (e.g. nanotubes of length L = 1 μm, radius r = 5 nm, and Q = 500). For fixed values of V_0, the gain is rather modest ($G \sim 10$) for low V_p but quickly increases to values approaching infinity for finite V_p. This divergence marks the threshold for instability and parametric oscillations[8].

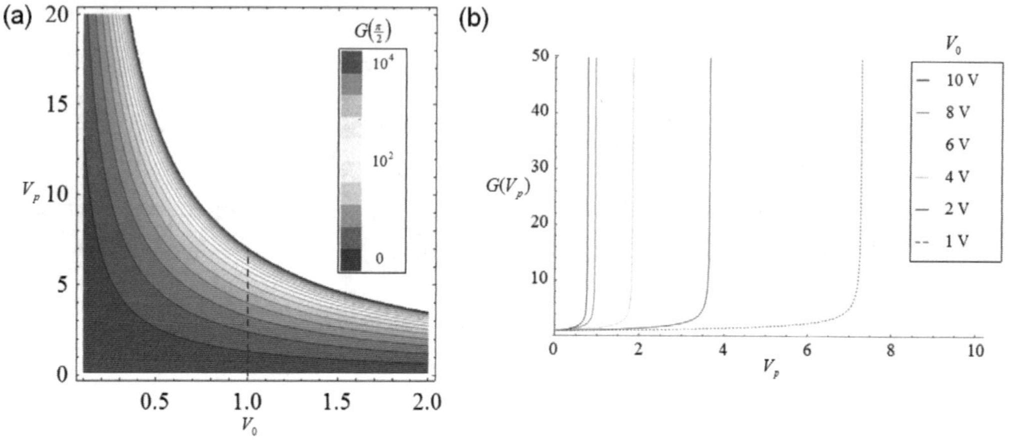

Figure 2: Amplifier Gain: (a) Gain of amplifier as a function of V_0 and V_p. The upper white portion of the graph is a region of resonator instability. (b) Shows gain as a function of V_p for several fixed values of V_0. The dashed line at $V_0 = 1$ V in (a) corresponds to the rightmost dashed line in this plot and aids in visualizing the rapid change in gain near a critical value (about 7.25 V for the dashed curve) of V_p. In all the curves, the gain is near unity for lower values of V_p and then quickly diverges to infinity. In all plots, L = d_0 = 1 μm, r = 5 nm, Q = 500, and $\phi = \pi/2$.

The carbon nanotube NEMS device discussed so far will amplify or detect the Fourier components of *any* signal that are both centered within the bandwidth ($\Delta\omega_0 = \omega_0/Q$) of the nanotube's vibrational resonance frequency and capable of exciting the nanotube's vibrational modes into resonance. In this way, our device functions as a tunable band pass filter while simultaneously serving as an amplifier and a detector (Figure 3).

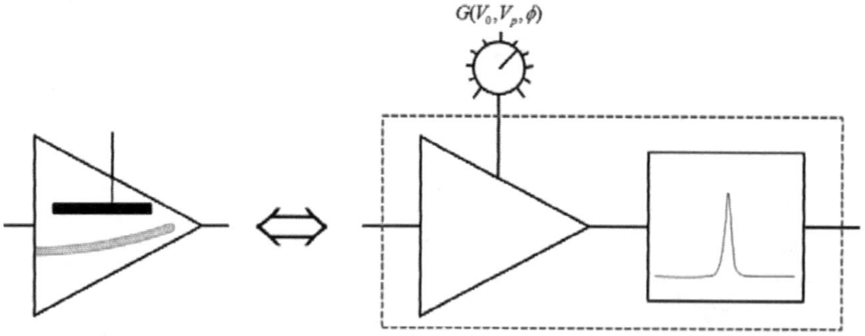

Figure 3: Schematic and Circuit Symbol of Carbon Nanotube-based NEMS Parametric Amplifier. A proposed circuit (left) symbol for our

nanoelectromechanical carbon nanotube paramp which serves as a band-pass-filtered, variable-gain, phase-sensitive amplifier (right).

The bandwidth of the filtering process will be on the order of 100 kHz for a 100 MHz nanotube resonator. While operating in amplification or detection mode, the gain/sensitivity is controlled by V_p, V_0, and ϕ. The resonance frequency, through its dependence on the spring constant k, can be tuned by adjusting V_0. Note that this capacitive frequency tunability is in addition to the electrostatic tensioning of the nanotube with V_{NT}[13, 14]. Chip-based devices, as shown in Figure 1 (a), can be readily fabricated and engineered to operate in the 100 kHz – 1 GHz frequency band.

AC current and voltage signals in the appropriate band will be amplified. For instance, applying an AC voltage to the pump probe or the nanotube electrode at ω_0 will drive the nanotube into resonance (with an amplified response do to Δk) resulting in an AC current through the source-drain circuit path. Similarly, a current driven through the nanotube at ω_0 will modulate the static charge on the tube and, because of V_0, excite oscillations which are in turn amplified by the parametric pump.

Detection of RF electromagnetic radiation by a suspended, field-emitting carbon nanotube has been reported in previous work[14]. Signal transduction occurs because the RF wave drives the nanotube into oscillation resulting, once again, in a current signal at the output of the device. Our parametrically modulated system adds improvements to the nanotube radio because we can enhance the sensitivity to incoming RF photons by several orders of magnitude, possibly by as much as a factor of 10^5-10^8. Furthermore, a high-frequency device operated in the limit of high gain, in order to induce parametric self-oscillations[16], could conceivably be used to generate detectable RF dipole radiation.

3. Conclusion

We have described a novel electromechanical nanotube device that acts as a phase-sensitive, variable-gain filtering amplifier of electronic signals and, at the same time, can operate as a variable-sensitivity detector/transducer of RF electromagnetic waves. The amplifier can exhibit infinite gain at pumping voltages much less than 10 Volts. Additionally, the amplifier's overhead power consumption will be on the order of 10-1000 nW, making it an exceptionally low-power amplifier. Ultimate gain limitations of the described nanotube amplifier will be governed by nonlinear effects and the maximum achievable amplitude of the nanotube resonator. Also, the operational range for incoming signals will be limited by the current capacity of the nanotube, which is typically ~ 10 – 100 μA. We expect this highly-tunable carbon nanotube paramp will find many useful applications in electronics, as a filter and an amplifier, in wireless communications, as a low-power RF receiver and transmitter, and in fundamental science as a versatile NEMS device for studies in condensed matter and as a highly-sensitive RF photon detector in optics.

4. Acknowledgments

We thank W. Regan for discussion and comments on the manuscript. This work was supported in part by the Director, Office of Energy Research, Office of Basic Energy Sciences, Materials Sciences and Engineering Division, of the U.S. Department of Energy under contract no. DE-AC02-05CH11231, which provided for nanotube growth and SEM characterization, and by the National Science Foundation within the Center of Integrated Nanomechanical Systems under grant EEC-0832819, which provided for the design, prototype fabrication, and modeling of the parametric amplifier. B.A. acknowledges support from the UC Berkeley Anselmo John Macchi Fellowship Fund in the Physical Sciences.

References

[1] Faraday 1831 *M Phil. Trans. Roy. Soc.* **121** 299-318
[2] Strutt J and Rayleigh L 1883 *Phil. Mag.* **16** 50-53
[3] Caves C 1982 *Phys. Rev. D* **26** 1817

[4] Hatridge M, Vijay R, Slichter D, Clarke J and Siddiqi I 2011 *Phys. Rev. B* **83** 134501
[5] Mortenson K 1959 *J. Appl. Phys.* **30** 1542
[6] Yurke B, Corruccini L, Kaminsky P, Rupp L, Smith A, Silver A, Simon R and Whittaker E 1989 *Phys. Rev. A* **39** 2519
[7] Giordamaine J and Miller 1965 R *Phys. Rev. Lett.* **14** 973
[8] Rugar D and Grütter P 1991 *Phys. Rev. Lett.* **67** 699–702
[9] Collins P and Zettl A 1996 *Appl. Phys. Lett.* **69** 1969
[10] Collins P and Zettl A 1997 *Phys. Rev. B* **55** 9391
[11] de Heer W, Chatelain A and Ugarte D 1995 *Science* **270** 1179-1180
[12] Fowler R and Nordheim L 1928 *Proc. R. Soc. London, Ser. A* **119** 173-181
[13] Poncharal P, Wang Z, Ugarte D and de Heer W 1999 *Science* **283** 1513-1516
[14] Jensen K, Weldon J, Garcia H and Zettl A 2007 *Nano Letters* **7**(11) 3508-3511
[15] Jensen K, Kim K and Zettl A 2008 *Nature Nanotech.* **3**(9) 533-537
[16] Weldon J, Alemán B, Sussman A, Gannett W and Zettl A 2010 *Nano Lett.* **10** 1728–1733
[17] Hayt W and Buck J 2006 *Engineering Electromagnetics* (Boston, MA: McGraw-Hill)
[18] We use the Electrostatics and MEMS modules in COMSOL Multiphysics 4.0a. The nanotube is placed in a box with 5 charge-free walls and 1 conducting, grounded wall. A bias is applied to the nanotube, it is allowed to deflect, and then the capacitance of the bent tube is calculated by charge integration.
[19] Senturia S 2001 *Microsystem Design* (New York, NY: Springer)
[20] Cleland A 2003 *Foundations of Nanomechanics: from Solid-State Theory to Device Applications* (New York, NY: Springer)

Suppression of Aharonov-Casher spin interference in an InGaAs ring array

Junsaku Nitta[1], Jun Takagi[1], Fumiya Nagasawa[1], and Makoto Kohda[1,2]

[1] *Graduate School of Engineering, Tohoku University, Sendai 980-8579, Japan*
[2] *PRESTO-Japan Science and Technology Agency, Saitama 331-0012, Japan*

E-mail address: nitta@material.tohoku.ac.jp

Abstract. An Aharonov-Casher (AC) spin interference is studied in a gate-fitted InGaAs ring array. The AC interference period is consistent with gate voltage dependence of Rashba spin-orbit interaction parameter in a single subband occupied region. However, the AC interference amplitude and oscillations are suppressed above a particular gate voltage where the second subband is occupied. This suppression is attributed to the decoherence induced by inter-subband scattering.

1. Introduction

An electron acquires a phase around magnetic flux due to the vector potential leading to Aharonov-Bohm (AB) effect [1] in an interference loop. From the view point of inherent symmetries between magnetic field and electric field in the Maxwell equations, Aharonov and Casher have predicted that a magnetic moment acquires a phase around charge flux line [2]. It should be noted that the original Aharonov-Casher (AC) effect has been proposed for charge neutral particles since the electric field modifies the charged particles in the same sense as the original AB effect was predicted in the situation where the magnetic flux should not exist in an electron path. It is pointed out that the AC phase shift can be derived from spin-orbit interaction (SOI) [3]. The major difference between the AB and AC effects is that the AC effect is not observable if the electric field is not in the electron paths. Cimmino *et al.* [4] managed to perform the AC interference experiment in a neutron (having spin 1/2 but no charge) beam loop using a voltage of 45 kV to create the electric field. However, the modified AC phase shift was only 2.2 mrad since the SOI is not strong in vacuum.

Mathur and Stone have theoretically shown that the effects of SOI in disordered conductors are manifestation of the AC effect in the same sense as the effects of weak magnetic fields are manifestation of AB effect [5]. They have proposed the electronic AC effect in a mesoscopic interference loop made of GaAs 2DEG with the Dresselhaus SOI [6]. It is emphasized that a thousand-fold improvement in its experiment can be expected in the electronic AC effect since the SOI in semiconductors is much enhanced compared with that in vacuum. We have proposed an AC spin interference in a mesoscopic ring [7] based on the gate controlled Rashba SOI [8], [9]. However, the spin interference is not easily observable since the gate electric field tunes not only SOI but also carrier density, namely electron wave length which affects the interference. In this paper, we have measured time reversal symmetric Al'tshuler-Aronov-Spivak (AAS) oscillations [10] in an array of mesoscopic loops in order to distinguish spin interference from other spurious effects. The AAS oscillation amplitude which does not depend on the electron wavelength but depends on spin precession shows an oscillatory behaviour as a function of the gate voltage. The observed AC spin interference is consistent with the gate voltage dependence of the Rashba SOI. This result shows that the spin precession can be controlled by the gate voltage. However, the AC spin interference amplitude and oscillatory behaviour are suppressed in the second subband occupied gate voltage region. The spin decoherence induced by inter-subband scattering results in this suppression.

2. Experiment and Discussion

The resistance of a mesoscopic ring is affected by different quantum interference effects. The well known AB effect results in a resistance oscillation with a magnetic flux period of h/e. The AB effect is sample specific and very sensitive to the Fermi wave length, therefore, the interference pattern is rapidly changed by the gate voltage. In order to detect the AC effect we used another quantum interference phenomenon, the AAS effect. The AAS effect is an AB effect of time reversal symmetric paths, where the two wave function parts go all around back to the origin on identical paths, but in opposite directions. In this situation, any phase which is due to path geometry will be identical and will not affect the interference. This also means that it is independent of the Fermi energy (and consequently the carrier density N_s). However, the AAS effect is sensitive to the spin phase when the SOI plays a role. If there is magnetic flux inside the paths the resistance will oscillate with the period of h/2e. When the flux is increased the resistance oscillates with the period h/2e, but the AAS oscillation amplitude decays after a several periods because of averaging between different paths in the ring, with different areas. If there is SOI in the ring, the electron spin will start processing around the effective magnetic field and change the interference at the entry point. Note that the effective magnetic field due to the SOI is much stronger than the external magnetic field to pick up AAS oscillations. The precession axes for the two parts of the wave function are opposite and therefore the relative precession angle is twice the angle of each part. If the relative precession angle is π the spins of the two parts are opposite and can not interfere, and the AAS oscillations disappear. If the relative angle is 2π the two parts will have the same spin but opposite signs because of the 1/2 spin quantum laws (a 4π rotation is required to return to the original wave function), effectively changing the phase of the AAS oscillations by π, which we interpret as a negative amplitude. By using arrays rather than single rings we get a stronger spin signal and we average out some of the universal conductance fluctuations (UCF) and sample specific AB oscillations .

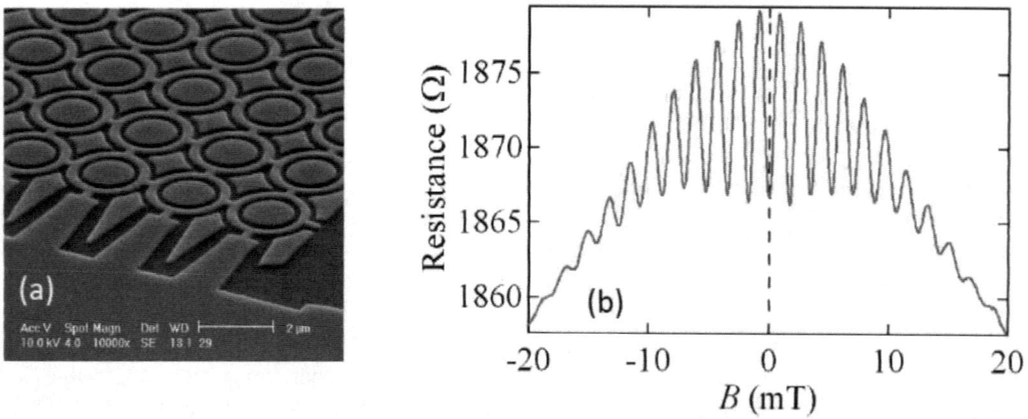

Figure 1. **(a)** SEM photograph of fabricated InGaAs ring array. The whole area of the sample is covered with gate electrode to control the Rashba SOI parameter. (b) AAS oscillation measured at V_g= -2.4 V. AB oscillation is suppressed by ensemble averaging due to 40x40 rings.

The sample was epitaxially grown on (100) InP substrate by metal organic chemical vapour deposition. The quantum well (QW) channel consists of 2.5 nm $In_{0.53}Ga_{0.47}As$ / 10 nm $In_{0.7}Ga_{0.3}As$ for QW / 2.5 nm $In_{0.53}Ga_{0.47}As$. The epitaxial wafer was processed by a reactive ion etching (RIE) into a 40×40 ring array as shown in Fig. 1 (a). The ring array is placed close to the Hall bar to measure the carrier density and the strength of the Rashba SOI. For a gate insulator, a 100 nm Al_2O_3 was deposited on the whole sample area by atomic-layer deposition (ALD). All the measurements were preformed with a ^{3}He cryostat refrigerator at $T = 0.27$ K. We measured magnetoresitance oscillations in the ring array by changing gate bias voltages, V_g. The external magnetic field was applied perpendicular to the QW plane. Figure 1 (b) shows an example of magnetoresistance oscillation measured at V_g = -2.4 V. By increasing the magnetic field, the amplitude of the oscillation is decreasing. This is because the time reversal symmetry is broken by the magnetic field, and it is an essential feature of the AAS interference. The period of AAS oscillation is given by quantized magnetic flux of h/2e, and the experimentally obtained periodic magnetic field is about $\Delta B = 1.7$ mT, which is consistent with the ring radius of 0.62 μm. The sample specific AB oscillation which has h/e period is almost completely suppressed because of the ensemble averaging effect by the 40x40 ring array.

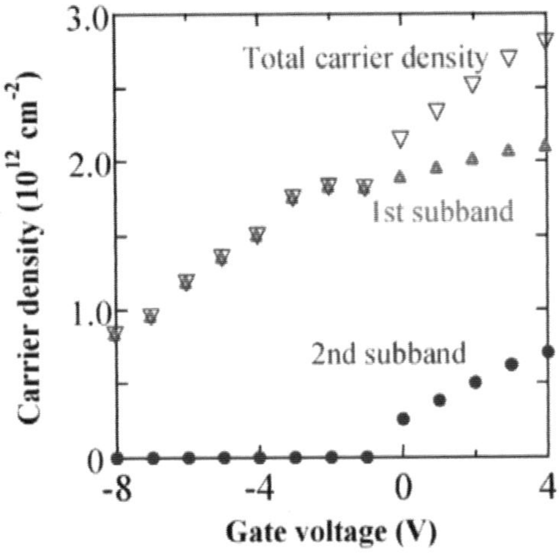

Figure 2. Gate voltage dependence of the carrier density. The carrier density is obtained from the SdH oscillations. The total carrier density is linearly increasing with the gate voltage. The second subband is occupied above V_g= -1 V.

Sheet carrier density, N_s, as a function of gate voltage V_g was determined from Shubnikov de Haas (SdH) oscillations in the Hall bar. The results are shown in Fig. 2. The total carrier density is linearly increasing with the gate voltage. It should be noted that the second subband is occupied above the gate voltage of V_g = -1 V. The effective mass m^*= 0.05 was derived from the temperature dependence of SdH oscillation amplitude. The Rashba SOI parameter α was obtained from the weak antilocalization analysis of the magnetoresitance measurement in the weak magnetic field region. The Rashba SOI parameter is almost linearly decreased such as

$$\alpha(10^{-12}\ eVm) = 7.81 - 3.32 \times N_s(10^{12}\ cm^{-2})$$

Figure 3 shows color-scale plots of the gate voltage dependence of AAS oscillations. It is clearly found that the AAS oscillation phase is periodically changed as increasing the gate voltage. This AAS oscillation phase change can be attributed to the AC spin interference effect where the spin phase is modified by the gate electric field. However, we can see almost no gate voltage dependence above V_g = -1 V, which corresponds to the second subband occupied region.

To investigate it more clearly, the AAS oscillation amplitude at B= 0 mT as a function of the gate voltage (AC spin interference) is plotted as shown in Fig. 4. As is clearly seen, the AC oscillation is suppressed above V_g = -1 V although a clear periodic AC oscillation is observed under a single subband region. The suppression of the AC interference which is reproducible in other two samples will be discussed below. Another interesting feature of this experiment is that the AC oscillation has negative bias of about -3.5 Ω. This negative bias was not observed in our previous AC spin interference experiment with 4x4 and 5x5 ring arrays [11]. It is known that randomization of spin phase in time reversal symmetric interference loops gives rise to positive magnetoresistance near zero magnetic field (resistance dip), and is called the weak antilocalization. The negative bias is explained by the weak antilocalization effect appeared in the large number of rings.

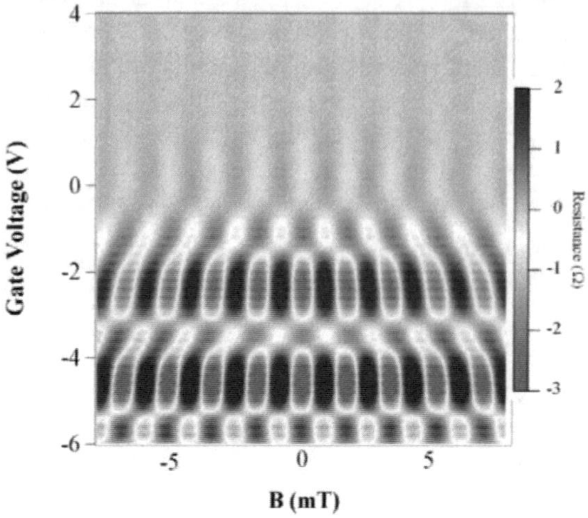

Figure 3. Color plot for the gate voltage dependence of AAS interference. The AAS oscillation phase is periodically changed as increasing the gate voltage. This AAS oscillation phase change can be attributed to the AC spin interference effect.

The modulation of an AC oscillation amplitude in a time reversal interference ring can be expressed as a function of the Rashba SOI parameter α ,

$$\frac{\delta R_{\alpha}}{\delta R_{\alpha=0}} = \cos\left\{ 2\pi \sqrt{1 + \left(\frac{2m^{*}\alpha}{\hbar^{2}} r \right)^{2}} \right\} \quad (1)$$

where δR_{α} and $\delta R_{\alpha=0}$ is the AC amplitude with and without SOI, respectively. A dashed line in Fig. 4 is a calculated result on the basis of the above equation. All parameters used in the calculation are experimentally obtained ones. Here, we assume the negative bias resistance of -3.5 Ω. The experimentally observed AC period is consistent with the theoretical one. However, there is a phase difference between the experiment and the calculation in the AC oscillations. A possible reason is that there exists a finite Dresselhaus SOI which is expected in the III-V compound semiconductors. A phase shift in the AC oscillation is predicted in the presence of the Dresselhaus SOI [12].

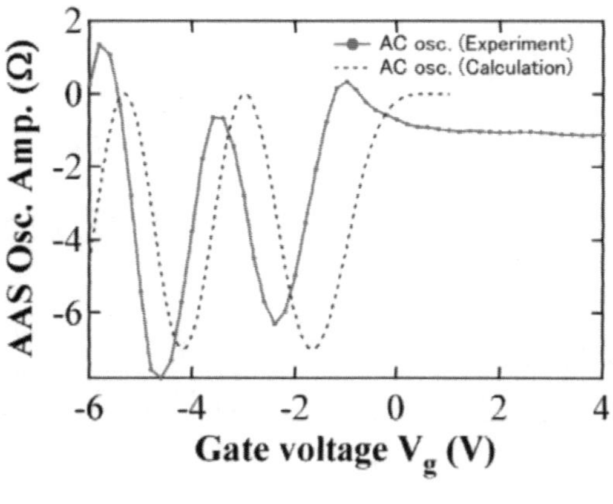

Figure 4. The AC oscillation amplitude as a function of gate voltage. The AC oscillation is suppressed in the second subband occupied region. The dashed line is theoretical calculation based on Eq. (1).

It should be noted that the spin precession angle does not depend on the wave vector. This is an advantage of the spin interference device [7]. The most surprising feature in the present experiment is that the AC oscillation is suppressed in the second subband occupied region. The Rashba parameter α in the second subband is different from that in the first subband [13]. This is because the electric fields which are felt by the first and second subbands are different in a QW due to the difference in electron distributions. It is also expected that the Dresselhaus SOI parameters are different among two subbands because the electron confinement in the QW is crucial for the Dresselhaus SOI. These two subbands provide electron spins with different spin precession angles. We could not observe any oscillatory behaviours in the second subband region. Further theoretical analysis is required for the AC spin interference in the second subband occupied region.

3. Summary

We have investigated AC spin interference in a gate-fitted InGaAs ring array. The AC interference period is consistent with the gate voltage dependence of the Rashba SOI parameter in a single subband occupied region. The AC interference amplitude and oscillations are suppressed in the second subband occupied region. This suppression is attributed to the dephasing induced by inter-subband scattering. Further theoretical analysis is required.

This work was partly supported by Grants-in-Aid from the Japan Society for the Promotion of Science (JSPS) and the Ministry of Education, Culture, Sports, Science and Technology (MEXT). JN acknowledges support from Strategic Japanese-German Joint Research Program.

References

[1] Aharonov Y and Bohm D 1959 *Phys. Rev.* **115** 485
[2] Aharonov Y and Casher A 1984 *Phys. Rev. Lett.* **53** 319
[3] Balatsky A and Al'tshuler B L 1993 *Phys. Rev. Lett.* **70** 1678
[4] Cimmino A, Opat G I, Klein A G, Kaiser H, Werner S A, Arif M, and Clothier R 1989 *Phys. Rev. Lett.* **63** 380
[5] Mathur H, Stone A D 1992 *Phys. Rev. Lett.* **68**, 2964
[6] Dresselhaus G 1955 *Phys. Rev.* **100** 580
[7] Nitta J, Meijer F E, and Takayanagi H 1999 *Appl. Phys. Lett.* **75** 695
[8] Rashba E I 1960 *Sov. Phys. Solid State* **2** 1224
[9] Nitta J, Akazaki T, Takayanagi H, and Enoki T 1997 *Phys. Rev. Lett.* **78** 1335
[10] Al'tshuler B L, Aronov A G, and Spivak B Z 1981 *JETP Lett.* **33** 94
[11] Bergsten T, Kobayashi T, Sekine Y, and Nitta J 2006 *Phys. Rev. Lett.* **97** 196803
[12] Van Veenhuizen M J, Koga T, and Nitta J 2006 *Phys. Rev.* **B73**, 235315
[13] Hu C.-M, Nitta J, Akazaki T, Takayanagi H, Osaka J, Pfeffer P, and Zawadzki W 1999 *Phys. Rev.* **B 60** 7736

Magnetic-Field Induced Quantum Phase Transitions in Triangular-Lattice Antiferromagnets

T Ono[1], H Tanaka[1], Y Shirata[1], A Matsuo[2], K Kindo[2], F Ishikawa[2], O Kolomiyets[2], H Mitamura[2], T Goto[2], H Nakano[3], N A Fortune[4], S T Hannahs[5], Y Yoshida[6] and Y Takano[6]

[1] Department of Physics, Tokyo Institute of Technology, Tokyo 152-8551, Japan
[2] Institute for Solid State Physics, University of Tokyo, Kashiwa, Chiba 277-8581, Japan
[3] Graduate School of Material Science, University of Hyogo, Kamigori, Hyogo 678-1297, Japan
[4] Department of Physics, Smith College, Northampton, Massachusetts 01063, USA
[5] National High Magnetic Field Laboratory, Tallahassee, Florida 32310, USA
[6] Department of Physics, University of Florida, Gainesville, Florida 32611-8440, USA

E-mail: tanaka@lee.phys.titech.ac.jp

Abstract. Cs_2CuBr_4 and $Ba_3NiSb_2O_9$ are magnetically described as quasi-two-dimensional triangular-lattice antiferromagnets with spin-$\frac{1}{2}$ and 1, respectively. We show that both systems exhibit a magnetization plateau at one-third of the saturation magnetization M_s due to the interplay of spin frustration and quantum fluctuation. In Cs_2CuBr_4 that has a spatially anisotropic triangular lattice, successive magnetic-field induced quantum phase transitions including a magnetization plateau $\frac{2}{3}M_s$ were observed. For $Ba_3NiSb_2O_9$, we performed exact diagonalization for rhombic spin clusters with up to 21-sites to analyze the magnetization process. The calculated results are in agreement with experimental observations.

1. Introduction

Triangular-lattice antiferromagnet (TLAF) often exhibits remarkable quantum effect in magnetic field. A typical quantum effect predicted for small spin Heisenberg TLAF is magnetization plateau at one-third of the saturation magnetization M_s [1–3]. For classical spin, the stable state is not uniquely determined, because the number of parameters that determine the spin configuration is more than the number of equations giving the equilibrium conditions. Thus, the ground state is infinitely degenerate. No phase transition accompanied by the anomaly of magnetization arises up to saturation, so the magnetization curve is monotonic, as shown by dashed line in Fig. 1(a). This classical degeneracy can be lifted by the quantum fluctuation that is remarkable for small spin, and a specific spin state is selected as the ground state. Consequently, the *up-up-down* state is stabilized in a finite field range, which leads to a magnetization plateau at $\frac{1}{3}M_s$. Solid line in Fig. 1(a) shows the schematic magnetization curve for the quantum Heisenberg TLAF. The magnetization plateau is a new macroscopic quantum phenomenon.

In contrast to theoretical studies, the experimental studies on the quantum magnetization plateau in TLAF are limited. In the case of spin-$\frac{1}{2}$ and 1, well defined quantum magnetization plateau has not been reported except for Cs_2CuBr_4 and $Ba_3NiSb_2O_9$. In this paper, we show that these two triangular-lattice antiferromagnets exhibit the nonclassical magnetization plateau at $\frac{1}{3}M_s$ [4–7]. In Cs_2CuBr_4, an additional quantum plateau was also observed at $\frac{2}{3}M_s$, as shown below [5, 6]. Furthermore, precise thermodynamic measurements revealed cascade of field-induced quantum phase transitions [8, 9].

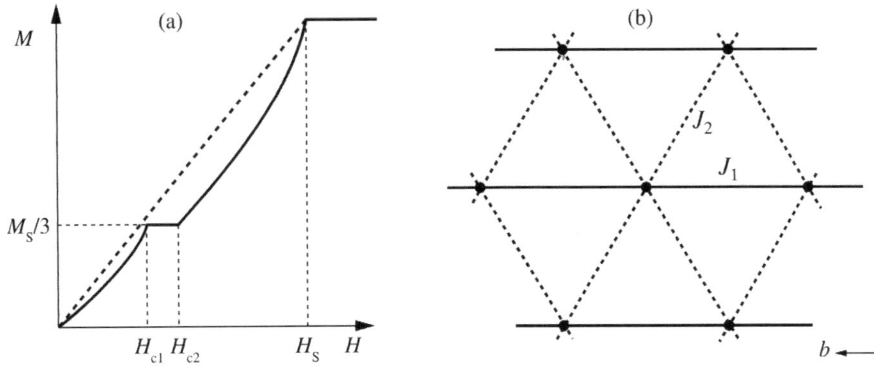

Figure 1. (a) Schematic magnetization curves of triangular-lattice Heisenberg antiferromagnets with classical (dashed line) and quantum spins (solid line). (b) Spatially anisotropic triangular lattice composed of Cu^{2+} ions in Cs_2CuBr_4.

2. Spatially anisotropic S=1/2 triangular-lattice antiferromagnet Cs_2CuBr_4

Cs_2CuBr_4 has an orthorhombic structure with space group $Pnma$ [10, 11]. The structure is composed of $CuBr_4^{2-}$ tetrahedra and Cs^+ ions. Magnetic Cu^{2+} ions that have spin-$\frac{1}{2}$ form a distorted triangular lattice parallel to the bc plane. Figure 1(b) shows the exchange network in the bc plane. There are two kinds of antiferromagnetic nearest-neighbor exchange interaction J_1 and J_2.

Cs_2CuBr_4 undergoes three-dimensional magnetic ordering at $T_N = 1.4$ K [4]. In the ordered phase, spins lie approximately in the bc plane and form an incommensurate helical structure that is characterized by a wave vector $\boldsymbol{Q}_0 = (0, 0.575, 0)$. Within the classical spin model, the value of Q_0 is given by $\cos(\pi Q_0) = -J_2/(2J_1)$. Using $Q_0 = 0.575$, we obtain $J_2/J_1 = 0.467$ for Cs_2CuBr_4. Because of spin-$\frac{1}{2}$, the quantum correction of the helical pitch is significant in the present system, and thus, the value of J_2/J_1 obtained from the classical spin model is different from the real value when $J_2 \neq J_1$. Zheng *et al.* [12] investigated theoretically a spatially anisotropic Heisenberg TLAF with spin-$\frac{1}{2}$, which is equivalent to the model shown in Fig. 1(b). They calculated the ordering vector Q_0 as a function of $J_1/(J_1 + J_2)$. From their result and $Q_0 = 0.575$, we obtain $J_2/J_1 = 0.74$, which is 1.6 times as large as the classical value.

Figure 2(a) shows the magnetization curve and dM/dH vs H measured at $T = 0.4$ K for $H \parallel c$. Notable feature is that the magnetization curve has a plateau at approximately one-third of the saturation magnetization M_s. This is more clearly recognized in dM/dH vs H shown in Fig. 2(a). Derivative susceptibility dM/dH exhibits sharp peaks at the plateau edges. The lower and higher edge fields are $H_{c1} = 13.1$ and $H_{c2} = 14.4$ T, respectively. For $H_{c1} < H < H_{c2}$, dM/dH is small. The $\frac{1}{3}$-magnetization-plateau was also observed for $H \parallel b$, while no plateau was observed for $H \parallel a$. Because the $\frac{1}{3}$-magnetization-plateau in Cs_2CuBr_4 is clearly observed for two different field directions, the plateau cannot be explained in terms of the classical model. Thus, we can conclude that the magnetization plateau arises from the quantum effect. The Dzyaloshinsky-Moriya (DM) interaction with the $\boldsymbol{D}$ vector perpendicular to the triangular lattice should be responsible for the absence of the magnetization plateau for $H \parallel a$.

The magnetic-field dependence of the ordering vector Q_0 was measured through neutron elastic scattering [5, 6]. The value of Q_0 increases with increasing magnetic field, and is locked at $Q_0 \simeq 2/3$ in the plateau region. This result indicates that the *up-up-down* spin structure is realized in the plateau state.

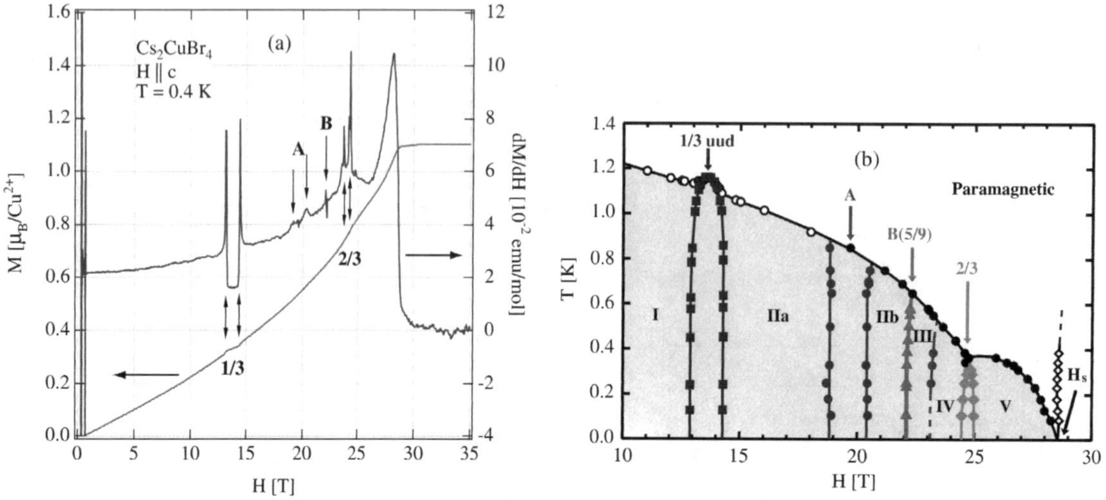

Figure 2. (a) Magnetization curve and dM/dH vs H in Cs_2CuBr_4 measured at $T = 0.4$ K for $H \parallel c$. Arrows denote critical fields. (b) Magnetic phase diagram obtained from the magnetocaloric-effect data measured at various temperatures for $H \parallel c$.

Fujii *et al.* [13, 14] performed ^{133}Cs-NMR experiments on Cs_2CuBr_4 in the magnetic fields that cover the field range of the 1/3 magnetization plateau. For $H \parallel b$, they observed that outside the plateau region, the NMR spectrum has a double-horn type broad shape characteristic of an incommensurate spin structure, while in the plateau region, the spectrum changes into discrete two peaks indicative of a commensurate *up-up-down* spin structure. Their NMR results are consistent with magnetization and neutron scattering measurements.

As shown in Fig. 2(a), dM/dH exhibits additional sharp double peak structure around $H \sim 23$ T, where the magnetization is approximately $\frac{2}{3}M_s$. This is indicative of the second tiny plateau at $\frac{2}{3}M_s$. This additional sharp double peak structure in dM/dH was also observed for $H \parallel b$. The magnetization processes for $H \parallel c$ and b almost coincide when normalized by the g factor. This means that spin-spin interactions are almost isotropic in the triangular-lattice plane.

Miyahara *et al.* [15] investigated the ground state of the spatially anisotropic Heisenberg TLAF with spin-$\frac{1}{2}$ by exact diagonalization. They showed that the $\frac{1}{3}$- and $\frac{2}{3}$-magnetization-plateaus appear for $0.7 \le J_2/J_1 \le 1.3$ and $0.5 \le J_2/J_1 \le 0.8$, respectively. Fortunately, the value of $J_2/J_1 = 0.74$ for Cs_2CuBr_4 satisfies both conditions. This should be the reason why both $\frac{1}{3}$- and $\frac{2}{3}$-magnetization-plateaus were observed in Cs_2CuBr_4.

Fortune *et al.* [9] performed magnetocaloric effect and magnetic-torque measurements and found the cascade of magnetic-field induced quantum phase transitions. Figure 2(b) is the phase diagram for $H \parallel c$ determined by magnetocaloric effect. In addition to the $\frac{1}{3}$- and $\frac{2}{3}$-plateau phases, many quantum phases were observed. The phase transitions to the A phase and the transitions to the very narrow B phase can also be observed as small peaks in dM/dH shown in Fig. 2(a). These successive quantum phase transitions should be attributed to the spatially anisotropic triangular lattice and the Dzyaloshinsky-Moriya interaction with the $\boldsymbol{D}$ vector perpendicular to the triangular lattice [15–19], but the overall explanation is still an open question.

3. S=1 triangular-lattice antiferromagnet $Ba_3NiSb_2O_9$

Figure 3 shows the crystal structure of $Ba_3NiSb_2O_9$. This compound crystalizes in a high symmetric hexagonal structure, $P6_3/mmc$ [20, 21], which is the same as the hexagonal $BaTiO_3$ structure. The structure is composed of single NiO_6 octahedra and face-connected Sb_2O_9 double octahedra, which are linked sharing corners. Magnetic Ni^{2+} ions that have spin-1 form triangular lattice parallel to the c plane.

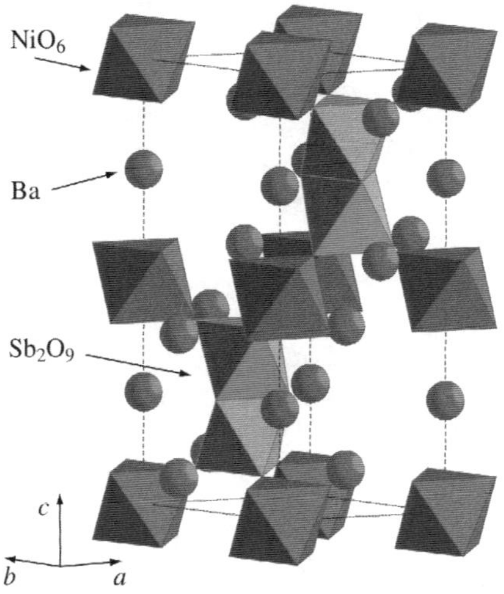

Figure 3. Crystal structure of $Ba_3NiSb_2O_9$. Dotted lines denote the chemical unit cell.

The magnetic properties of $Ba_3NiSb_2O_9$ were investigated by Doi *et al.* [22]. They reported that $Ba_3NiSb_2O_9$ undergoes antiferromagnetic ordering at $T_N = 13.5$ K. In the present study, we first measured specific heat to investigate the nature of the magnetic ordering. We found that $Ba_3NiSb_2O_9$ undergoes two magnetic phase transitions at $T_{N1} = 13.5$ K and $T_{N2} = 13.0$ K. In Heisenberg-like TLAF, the successive phase transition arises when the magnetic anisotropy is of easy-axis type, while a single transition occurs for the easy-plane anisotropy [23]. The successive phase transitions with a very narrow intermediate phase in $Ba_3NiSb_2O_9$ indicates that the magnetic anisotropy is of the easy-axis type and is much smaller than the exchange interaction. The origin of the anisotropy should be single ion anisotropy of the form $D(S^z)^2$.

Figure 4(a) shows the magnetization curve of $Ba_3NiSb_2O_9$ powder and dM/dH vs magnetic field H measured at 1.3 K. A well defined magnetization plateau is observed at $M \simeq 0.8\,\mu_B/Ni^{2+}$. Because the g factor of $Ba_3NiSb_2O_9$ powder is $g \simeq 2.3$ [24], the magnetization at the plateau corresponds to one-third of the saturation magnetization M_s. The magnetization anomalies at the lower and the higher edge fields are smeared because of powder sample.

In the case of classical Heisenberg TLAF with the easy-axis anisotropy along the c axis, the $\frac{1}{3}$-plateau appears for $H \parallel c$, while not for $H \perp c$ [25]. Thus, for powder sample of such classical system, dM/dH at the plateau range should be about two-third of dM/dH for slope region below the lower edge field. In the present system, the minimum of dM/dH at the plateau range is smaller than half of dM/dH for $H < 30$ T. This implies that the $\frac{1}{3}$-plateau appears, irrespective of external field direction. In classical Heisenberg TLAF, thermal fluctuation stabilizes a collinear *up-up-down* state at finite temperatures, so that plateau-like magnetization anomaly occurs in magnetization curve [26]. The magnetic field range of the thermally stabilized collinear state decreases with decreasing temperature and becomes zero at $T = 0$. The $\frac{1}{3}$-plateau in $Ba_3NiSb_2O_9$ has a magnetic field range of about 10 T even at 1.3 K ($\ll T_{N2}$). Thus, the $\frac{1}{3}$-plateau observed in $Ba_3NiSb_2O_9$ arises from the quantum effect. However, the magnetization at the plateau range

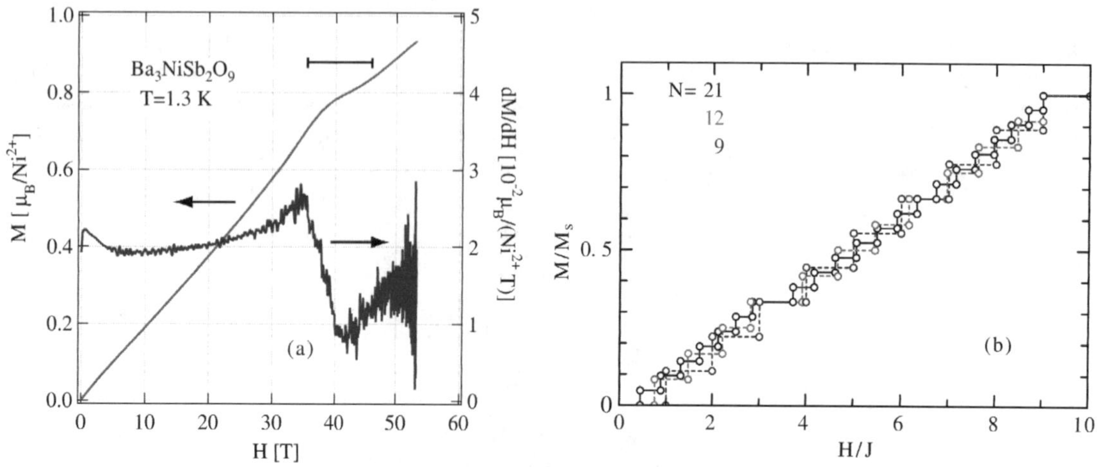

Figure 4. (a) Magnetization curve and dM/dH vs H of $Ba_3NiSb_2O_9$ measured at 1.3 K. A horizontal bar denotes the field range of plateau calculated by exact diagonalization. (b) Magnetization curves for spin-1 Heisenberg TLAF calculated by exact diagonalizations for 9-, 12-, and 21-site rhombic clusters.

is not completely flat. This should be ascribed to the increase of the sample temperature due to the magnetocaloric effect, as observed in Cs_2CuBr_4.

Figure 4(b) shows magnetization curves for spin-1 Heisenberg TLAF calculated by exact diagonalization for 9-, 12- and 21-site rhombic clusters. The calculated magnetization curve shows a plateau at $\frac{1}{3}M_s$, although the magnetic field range of the plateau is fairly small as compared with the spin-$\frac{1}{2}$ case [2, 3]. The lower and higher edge fields tend to converge to $H_{c1} = 2.85J$ and $H_{c2} = 3.72J$, respectively, with increasing cluster size. A horizontal bar in Fig. 4(a) denotes the calculated field range of the $\frac{1}{3}$-plateau. Here we chose the center field of the $\frac{1}{3}$-plateau as $(H_{c1} + H_{c2})/2 = 41$ T. The experimental field range of the plateau is roughly estimated as 70% of the calculated field range. The field range of the $\frac{1}{3}$-plateau observed in $Ba_3NiSb_2O_9$ is consistent with that calculated for spin-1 Heisenberg TLAF.

Läuchli *et al.* [27] also calculated the magnetization curve for spin-1 Heisenberg TLAF using exact diagonalization up to 27-site cluster and showed a small magnetization plateau at $\frac{1}{3}M_s$. However, their calculations for $N \geq 21$ are limited to a certain magnetization range and do not cover all the magnetization value. For spin-1 case, the field range of the $\frac{1}{3}$-magnetization-plateau normalized by the saturation field H_s is calculated as $(H_{c2} - H_{c1})/H_s = 0.097$, while for spin-$\frac{1}{2}$ case, the normalized plateau range is 0.171 [3, 29]. The $\frac{1}{3}$-magnetization-plateau for spin-1 case is considerably suppressed as compared with the spin-$\frac{1}{2}$ case.

For TLAF with $S \geq 1$, the magnetization plateau at $\frac{1}{3}M_s$ can be stabilized with the help of the biquadratic exchange interaction [27, 28], which mainly arises from the spin-lattice coupling as discussed by Penc *et al.* [28]. The biquadratic exchange interaction may be responsible for the magnetization plateau observed in $RbFe(MoO_4)_2$ [30], because the normalized plateau range is as large as 0.10 despite spin-$\frac{5}{2}$. In $Ba_3NiSb_2O_9$, the contribution of the biquadratic exchange interaction should be negligible, because the field range of the magnetization plateau can be explained quantitatively within the the bilinear exchange interaction.

4. Conclusion

We have presented experimental evidence of magnetic-field induced quantum phase transitions in Cs_2CuBr_4, which is described as a spatially anisotropic spin-$\frac{1}{2}$ TLAF. Cs_2CuBr_4 exhibits quantum magnetization plateaus at $\frac{1}{3}M_s$ and $\frac{2}{3}M_s$ for magnetic field parallel to the bc plane. In addition to these magnetization-plateau phases, many quantum phases induced by magnetic field were observed as shown in Fig. 2(b). We have also presented the results of high-field magnetization measurements on $Ba_3NiSb_2O_9$ that is described as a spin-1 Heisenberg TLAF. We observed a well defined nonclassical magnetization plateau at $\frac{1}{3}M_s$. We performed exact diagonalization for rhombic spin clusters with up to 21-sites to analyze the magnetization process. The calculated results are in agreement with experimental observations.

Acknowledgments

This work was supported by a Grant-in-Aid for Scientific Research (A) from Japan Society for the Promotion of Science (JSPS), and the Global COE Program "Nanoscience and Quantum Physics" at TIT funded by the Ministry of Education, Culture, Sports, Science and Technology of Japan.

5. References

[1] Chubukov A V and Golosov D I 1991 *J. Phys.: Condens. Matter* **3** 69.
[2] Honecker A 1999 *J. Phys.: Condens. Matter* **11** 4697.
[3] Farnell D J J, Zinke R, Schulenburg J and Richter J 2009 *J. Phys.: Condens. Matter* **21** 406002.
[4] Ono T, Tanaka H, Aruga Katori H, Ishikawa F, Mitamura H and Goto T 2003 *Phys. Rev. B* **67** 104431.
[5] Ono T, Tanaka H, Kolomiyets O, Mitamura H, Goto T, Nakajima K, Oosawa A, Koike Y, Kakurai K, Klenke J, Smeibidle P and Meißner M 2004 *J. Phys.: Condens. Matter* **16** S773.
[6] Ono T, Tanaka H, Nakagomi T, Kolomiyets O, Mitamura H, Ishikawa F, Goto T, Nakajima K, A. Oosawa, Koike Y, Kakurai K, Klenke J, Smeibidle P, Meißner M and Aruga Katori H 2005 *J. Phys. Soc. Jpn.* **74** Suppl. 135.
[7] Y. Shirata, H. Tanaka, T. Ono, A. Matsuo, K. Kindo and Nakano H unpublished data.
[8] Tsujii H, Rotundu C R, Ono T, Tanaka H, Andraka B, Ingersent K and Takano Y 2007 *Phys. Rev. B* **76** 060406(R).
[9] Fortune N A, Hannahs S T, Yoshida Y, Sherline T E, Ono T, Tanaka H and Takano Y 2009 *Phys. Rev. Lett.* **102** 257201.
[10] Morosin B and Lingafelter E C 1960 *Acta Crystallogr.* **13** 807.
[11] Li T I and Stucky G D 1973 *Inorg. Chem.* **12** 441.
[12] Zheng W, McKenzie R H and Singh R R P 1999 *Phys. Rev. B* **59** 14367.
[13] Fujii Y, Nakamura T, Kikuchi H, Chiba M, Goto T, Matsubara S, Kodama K and Takigawa M 2004 *Physica B* **346-347** 45.
[14] Fujii Y, Hashimoto H, Yasuda Y, Kikuchi H, Chiba M, Matsubara S and Takigawa M 2007 *J. Phys.: Condens. Matter* **19** 145237.
[15] Miyahara S, Ogino K and Furukawa N 2006 *Physica B* **378-380** 587, S. Miyahara private communication.
[16] Alicea J and Fisher M P A 2007 *Phys. Rev. B* **75** 144411.
[17] Alicea J, Chubukov A V and Starykh O A 2009 *Phys. Rev. Lett.* **102** 137201.
[18] Tay T and Motrunich O I 2010 *Phys. Rev. B* **81** 165116.
[19] Starykh O A, Katsura H and Balents L 2010 *Phys. Rev. B* **82** 014421.
[20] A. J. Jacobson and A. J. Calvert 1978 *J. Inorg. Nucl. Chem.* **40** 447.
[21] Treiber U and Kemmler-Sack S 1982 *Z. Anorg. Allg. Chem.* **487** 161.
[22] Doi Y, Hinatsu Y and Ohoyama K 2004 *J. Phys.: Condens. Matter* **16** 8923.
[23] Matsubara F 1982 *J. Phys. Soc. Jpn.* **51** 2424.
[24] Köhl P and Reinen D 1977 *Z. Anorg. Allg. Chem.* **433** 81.
[25] Miyashita S 1986 *J. Phys. Soc. Jpn.* **55** 3605.
[26] Kawamura H and Miyashita S 1985 *J. Phys. Soc. Jpn.* **54** 4530.
[27] Läuchli A, Mila F and Penc K 2006 *Phys. Rev. Lett.* **97** 087205.
[28] Penc K, Shannon N and Shiba H 2004 *Phys. Rev. Lett.* **93** 197203.
[29] Sakai T and Nakano H arXiv: 1102.3486.
[30] Svistov L E, Smirnov A I, Prozorova L A, Petrenko O A, Demianets L N and Shapiro A Ya 2003 *Phys. Rev. B* **67** 094434.

Breakdown of exchange approximation for cross-polarized excitons in carbon nanotubes

Seiji Uryu[1] **and Tsuneya Ando**[2]
[1]Department of Materials Science and Engineering, Iwate University
4–3–5 Ueda, Morioka, Iwate 020-8551, Japan
[2]Department of Physics, Tokyo Institute of Technology
2–12–1 Ookayama, Meguro-ku, Tokyo 152-8551, Japan

E-mail: uryu@iwate-u.ac.jp

Abstract. Cross-polarized excitons in carbon nanotubes where polarization is perpendicular to the tube axis are calculated in a Tamm-Dancoff-type approximation and the results are compared with those obtained in the self-consistent-field method. For the lowest exciton, the oscillator strength for the optical absorption peak is considerably reduced and the energy is overestimated.

1. Introduction

Carbon nanotubes are quasi-one-dimensional systems of rolled up graphenes. For optical absorption, there are two possible polarizations, parallel and perpendicular to the tube axis. For perpendicular (cross) polarization, an excited electron-hole pair is separated into opposite sides of the circular cross-section and a depolarization field is induced. It was shown that this depolarization effect considerably reduces band-to-band optical absorption [1, 2], but distinct peaks can appear in absorption spectra due to strong exciton effects even in its presence [3–7]. The depolarization effect should be considered in the self-consistent-field method where an optically induced current and electric field which electrons feel are determined self-consistently [1, 2, 5]. It is sometimes treated within a Tamm-Dancoff-type approximation where the effect is included in the form of a so-called electron-hole exchange interaction for excitonic Bethe-Salpeter-type equation [3, 4, 8–10]. In this paper we show that the latter approximation does not appropriately describe this depolarization effect in carbon nanotubes.

2. Model and method

In carbon nanotubes, electrons near the Fermi energy exist around two valleys, K and K' points at the corners of the first Brillouin zone. In an effective-mass approximation [11, 12], states can be specified by a set of indices $\alpha = (\pm, n, k, v)$ where $+$ and $-$ denote the conduction and valence band, respectively, n is the band index indicating the wave vector in the circumference direction, k is the wave vector in the axis direction, and $v = K$ and K'.

An exciton wave function at valley v with momentum $2\pi\hbar l/L$ in the circumference direction, where L is the circumference length, is written as [5, 13, 14]

$$|u, l, v\rangle = \sum_{n,k} \psi^l_{n,v}(k) c^\dagger_{+,n+l,k,v} c_{-,n,k,v} |g\rangle, \tag{1}$$

where $c_\alpha^\dagger$ and c_α are the creation and annihilation operators, respectively, and $|g\rangle$ is the ground state. The coefficients $\psi^l_{n,v}(k)$ and energy eigenvalues for the corresponding states, ε_u, are obtained by solving an equation of motion for an electron and hole under attractive interaction, i.e.,

$$\varepsilon_u \psi^l_{n,v}(k) = \Delta\varepsilon_{n,v}(k)\psi^l_{n,v}(k) - \sum_{m,q} V_{(+,n+l,k,v;+,m+l,k+q,v)(-,m,k+q,v;-,n,k,v)}\psi^l_{m,v}(k+q), \tag{2}$$

where $\Delta\varepsilon_{n,v}(k)$ is the one-particle transition energy between states with $(+,n+l,k,v)$ and $(-,n,k,v)$ and $V_{(\alpha;\alpha')(\beta;\beta')}$ is a matrix element of the screened Coulomb interaction between a two-particle state with α and β and that with α' and β'. For perpendicular polarization, $l=\pm1$, and the dynamical conductivity without the depolarization effect, $\sigma^l_{xx}(\omega)$, is obtained from solutions of this equation, for example, with the use of the Kubo formula.

For perpendicular polarization, the depolarization effect must be considered. The effect is included in the self-consistent-field method [1, 2, 5] and the dynamical conductivity $\tilde{\sigma}_{xx}(\omega)$ characterizing the optical absorption is given by

$$\tilde{\sigma}_{xx}(\omega) = \frac{1}{2}\sum_{l=\pm1}\tilde{\sigma}^l_{xx}(\omega), \quad \tilde{\sigma}^l_{xx}(\omega) = \frac{\sigma^l_{xx}(\omega)}{\varepsilon^l_{xx}(\omega)}, \quad \varepsilon^l_{xx}(\omega) = 1 + \frac{4\pi^2 i}{\kappa L\omega}\sigma^l_{xx}(\omega), \tag{3}$$

where κ is an effective dielectric constant describing screening by electrons in σ bands, core states, and the π bands away from the K and K' points.

The depolarization effect may approximately be included based on the Bethe-Salpeter equation for the exciton in terms of an exchange term under the condition that only a single electron-hole pair can be excited [3,4,8], corresponding to a Tamm-Dancoff-type approximation [9,10]. In this scheme, the equation is given by (2) with the following term on the right-hand side:

$$2\sum_{v'}\sum_{m,q} V^0_{(+,n+l,k,v;-,n,k,v)(-,m,k+q,v';+,m+l,k+q,v')}\psi^l_{m,v}(k+q), \tag{4}$$

where $V^0_{(\alpha;\alpha')(\beta;\beta')}$ is a matrix element of the unscreened Coulomb interaction and factor two is due to the electron spin. The dynamical conductivity with the depolarization effect is directly given from the eigenstates and eigenenergies of Eq. (2) with term (4) by using the Kubo formula. It should be noted that term (4) vanishes for the parallel polarization because of the zero center-of-mass momentum of excitons and the charge neutrality.

The Tamm-Dancoff-type approximation is certainly valid when the spectrum is dominated by a single resonance peak well separated from others. In nanotubes in the cross-polarized geometry, however, various resonances such as excitons and interband continuum coexist and strongly interact with each other, and the approximate method can give results significantly different from exact results. In the following, we show that the approximation is actually very poor in describing the absorption intensity in comparison with the absorption energies, using the lowest order effective-mass approximation [11–14].

3. Numerical results

In the effective-mass scheme, the energy is scaled by a typical kinetic energy $2\pi\gamma/L$ where γ is a band parameter describing the slope of the linear dispersion of graphene. Thus, a typical Coulomb interaction is given by $(e^2/\kappa L)(2\pi\gamma/L)^{-1} \approx 0.35/\kappa$. Because κ is considered to be of the order of unity ($\kappa = 2.4$ for graphite, for example), the typical strength of the Coulomb interaction is of the order of 0.1∼0.2. We should introduce a cutoff energy which corresponds to the half of the π-band width $3\gamma_0$, where γ_0 is the resonance integral between nearest neighbor

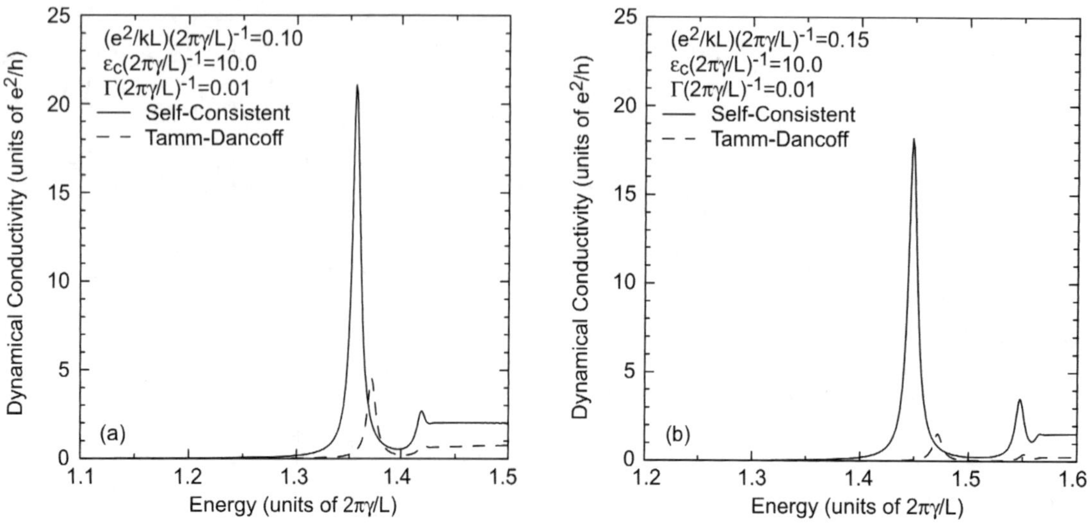

Figure 1. Dynamical conductivity for $(e^2/\kappa L)(2\pi\gamma/L)^{-1} = 0.1$ in (a) and 0.15 in (b). Phenomenological energy broadening is introduced as $\Gamma(2\pi\gamma/L)^{-1} = 0.01$.

sites and related to the band parameter through $\gamma = \sqrt{3}a\gamma_0/2$ with $a = 2.46$ Å being the lattice constant. Therefore, $\varepsilon_c(2\pi\gamma/L)^{-1} \approx (\sqrt{3}/\pi)(L/a) = \sqrt{3}d/a$, with d being the diameter of the nanotube. In the following, $\varepsilon_c(2\pi\gamma/L)^{-1} = 10$ is used, corresponding to a diameter ~ 1.4 nm.

Figure 1 shows the calculated dynamical conductivity near the lowest band edge when the Coulomb interaction parameter is (a) 0.1 and (b) 0.15. Solid and dashed lines denote results in the self-consistent-field method and the Tamm-Dancoff-type approximation, respectively. For the solid lines, sharp peaks appear due to the lowest exciton. For the dashed lines, the conductivity is considerably suppressed in comparison with the solid lines and the peak positions are shifted to the higher energy side.

In Fig. 2 the lowest exciton energy is shown as a function of the Coulomb interaction. Solid and dashed lines are results for the self-consistent-field method and the Tamm-Dancoff-type approximation, respectively, and the dotted line is the band gap. The exciton energy obtained in the Tamm-Dancoff-type approximation is higher than that obtained in the self-consistent-field method. The difference may not look so important in the absolute value, but is significant in the scale of the exciton binding energy, i.e., the difference between the band gap and the exciton level. In Fig. 3 the oscillator strength for the exciton peak is shown. The dashed line obtained in the Tamm-Dancoff-type approximation becomes vanishingly small at $(e^2/\kappa L)(2\pi\gamma/L)^{-1} \approx 0.25$, showing that the approximation becomes invalid in the important parameter region of $0.1 \lesssim (e^2/\kappa L)(2\pi\gamma/L)^{-1} \lesssim 0.2$.

Equation (3) shows that the exciton energy is determined by a zero of $\varepsilon_{xx}^{\pm 1}(\omega)$, where $\sigma_{xx}^{\pm 1}(\omega)$ becomes a pure imaginary. The imaginary part of $\sigma_{xx}^{\pm 1}(\omega)$ vanishes at a point lying between adjacent two discrete energy levels corresponding to poles of $\sigma_{xx}^{\pm 1}(\omega)$. In the Tamm-Dancoff-type approximation, $\sigma_{xx}^{\pm 1}(\omega)$ in $\varepsilon_{xx}^{\pm 1}(\omega)$ is replaced by an approximate expression because all excitations are assumed to have nearly same energy. Thus, $\sigma_{xx}^{\pm 1}(\omega)$ in the numerator and in the denominator of $\tilde{\sigma}_{xx}(\omega)$ effectively become different from each other, leading to vanishingly small imaginary part of $\sigma_{xx}^{\pm 1}(\omega)$ in the numerator at an approximate zero of the denominator. This is the origin of the breakdown of the approximation scheme, particularly, for the oscillator strength, although such severe problems do not manifest themselves in the exciton energy.

In conclusion, cross-polarized excitons have been calculated in the Tamm-Dancoff-type

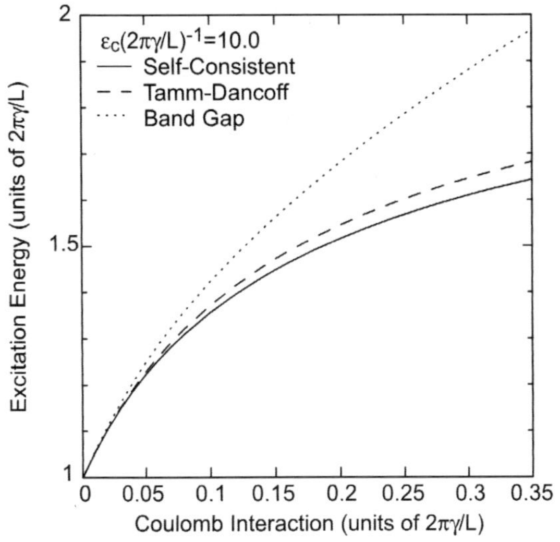

Figure 2. Coulomb interaction dependence of the excitation energy.

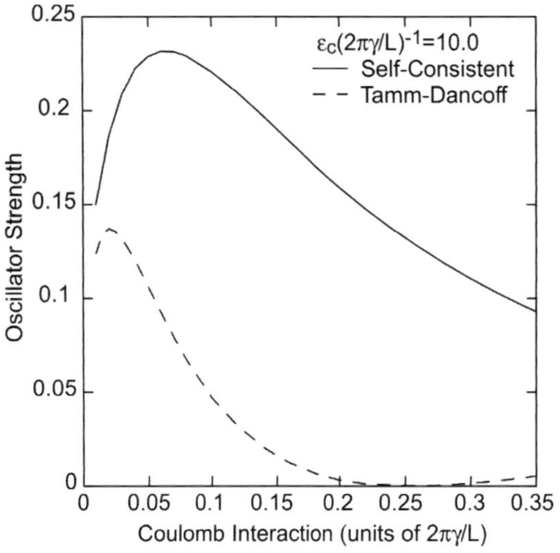

Figure 3. Coulomb interaction dependence of the oscillator strength.

approximation and the results have been compared with those in the exact self-consistent field method in semiconducting carbon nanotubes. For the lowest exciton, the oscillator strength for the optical absorption peak can be considerably reduced and sometimes vanishingly small, and the excitation energies are overestimated.

Acknowledgments

This work was supported in part by a Global Center of Excellence Program at Tokyo Tech "Nanoscience and Quantum Physics," Grants-in-Aid for Scientific Research, and Grant-in-Aid for Scientific Research on Priority Area "Carbon Nanotube Nanoelectronics" from the Ministry of Education, Culture, Sports, Science and Technology, Japan. The numerical calculations were partly performed at the Supercomputer Center, Institute for Solid State Physics, University of Tokyo and at Iwate University Super Computing and Information Sciences Center.

References

[1] Ajiki H and Ando T 1994 *Physica* B **201**, 349
[2] Ajiki H and Ando T 1995 *Jpn. J. Appl. Phys. Suppl.* **34-1**, 107
[3] Chang E, Bussi G, Ruini A, and Molinari E 2004 *Phys. Rev. Lett.* **92**, 196401
[4] Zhao H and Mazumdar S 2004 *Phys. Rev. Lett.* **93**, 157402
[5] Uryu S and Ando T 2006 *Phys. Rev. B* **74**, 155411
[6] Miyauchi Y, Oba M, and Maruyama S 2006, *Phys. Rev.* B **74**, 205440
[7] Lefebvre J and Finnie P 2007 *Phys. Rev. Lett.* **98**, 167406
[8] Wang Z-D, Zhao H-B, and Mazumdar S 2007 *Phys. Rev.* B **76**, 115431
[9] Rohlfing M and Louie S G 1998 *Phys. Rev. Lett.* **80**, 3320
[10] Rohlfing M and Louie S G 2000 *Phys. Rev.* B **62**, 4927
[11] Ajiki H and Ando T 1993 *J. Phys. Soc. Jpn.* **62**, 1255
[12] Ando T 2005 *J. Phys. Soc. Jpn.* **74**, 777
[13] Ando T 1997 *J. Phys. Soc. Jpn.* **66**, 1066
[14] Ando T 2004 *J. Phys. Soc. Jpn.* **73**, 3351

Journal of Physics: Conference Series **302** (2011) 012005 doi:10.1088/1742-6596/302/1/012005

Aharonov-Bohm effect on impurity-bound excitons in semiconducting carbon nanotubes

Yuh Tomio, Hidekatsu Suzuura

Division of Applied Physics, Graduate School of Engineering, Hokkaido University, Sapporo 060-8628, Japan

E-mail: tomio@eng.hokudai.ac.jp

Abstract. We study the Aharonov-Bohm effect on exciton absorption spectra in semiconducting carbon nanotubes with an impurity in an effective-mass theory. In the presence of an impurity potential, there are two impurity-bound exciton states, one is dark and the other is bright. The dark impurity-bound exciton becomes brightened with increasing a magnetic flux through the cross section. The energy level of the bright impurity-bound exciton shows a *blueshift* in the magnetic flux dependence for a long-range impurity; in contrast, it shows a *redshift* for a short-range impurity, due to an enhancement of the coupling between intra- and inter-valley excitons. This suggests that the magnetic flux dependence can distinguish the range of the impurity potential and the strength of the inter-valley coupling of excitons.

1. Introduction

Optical properties of carbon nanotubes (CNs) have attracted much attention because of the strong enhancement of exciton effects arising from the Coulomb interaction in the one-dimensional systems [1, 2]. The valley degeneracy of the electronic state, which causes the fine structures of excitons, also gives a rich variety of phenomena in CNs. In particular, when a magnetic flux is applied through the cross section of CN, the shift of the band edges is opposite between K and K$'$ valleys [3], leading to the Aharonov-Bohm (AB) splitting of absorption peaks [4, 5].

The exciton fine structures are generated by the short-range part of the Coulomb interaction that induces the mixing between several excitons consisting of intra- or inter-valley electron-hole pairs [6, 7, 8, 9]. In possible exciton states, only the bonding state of two singlet intra-valley excitons is bright, called the bright exciton, and the rest are dark, with no magnetic flux.

Impurities give rise to the mixing of excitons as well and largely modifies those energy levels. Moreover, impurity-bound exciton states appear for a strong impurity potential [10]. Generally, it is expected that impurities cause the mixing between the bright exciton and the other dark excitons because of crystal symmetry breaking [11], which means that some dark excitons can be brightened without a magnetic field. Actually, recent photoluminescence studies have reported that lattice defects [12] or local electric fields [13] can brighten dark exciton states. However, the peak assignments are still controversial.

The purpose of this paper is to study the AB effect on exciton absorption in semiconducting CNs with an impurity in an effective-mass theory. In particular, we discuss the difference of the AB effect on impurity-bound excitons between the cases of long- and short-range impurities.

2. Model

Electronic states near the K and K$'$ points are described by the $\boldsymbol{k}\cdot\boldsymbol{p}$ equation with a band parameter γ [1]. By using these electronic states in the absence of impurity, the exciton state of CN is expressed by

$$|\mathrm{K}_1\mathrm{K}_2\rangle_Q^u = \sum_{n,k} \psi_{u,n}^{\mathrm{K}_1\mathrm{K}_2}(k,Q) c_{n+,k+Q/2}^{\mathrm{K}_1\dagger} c_{n-,k-Q/2}^{\mathrm{K}_2} |g\rangle, \tag{1}$$

where $\mathrm{K}_1,\mathrm{K}_2 = \mathrm{K}$ or K', $c_{n+,k}^{\mathrm{K}_1\dagger}$ ($c_{n-,k}^{\mathrm{K}_2}$) is the electron creation (annihilation) operator in the conduction (valence) band specified by the subband index n of the K_1 (K_2) valley with the wavenumber k in the axis direction, and $|g\rangle$ is the ground state wave function. Each exciton state with the center-of-mass wavenumber Q is labeled by the index $u = 1s, 2p, 2s, \cdots$ in order of increasing energy. Solving the Bethe-Salpeter equation,

$$\begin{aligned} E_u^{\mathrm{K}_1\mathrm{K}_2}(Q)\psi_{u,n}^{\mathrm{K}_1\mathrm{K}_2}(k,Q) &= \left(\varepsilon_{n,k+Q/2}^{\mathrm{K}_1} + \varepsilon_{n,k-Q/2}^{\mathrm{K}_2} + \Delta\varepsilon_{n,k,Q}^{\mathrm{K}_1\mathrm{K}_2}\right)\psi_{u,n}^{\mathrm{K}_1\mathrm{K}_2}(k,Q) \\ &\quad - \sum_{m,q} V_{(n+,k+Q/2;m+,k+Q/2+q)(m-,k-Q/2+q;n-,k-Q/2)}^{\mathrm{K}_1\mathrm{K}_2} \psi_{u,m}^{\mathrm{K}_1\mathrm{K}_2}(k+q,Q), \end{aligned} \tag{2}$$

we can obtain the exciton energy dispersion $E_u^{\mathrm{K}_1\mathrm{K}_2}(Q)$ and the corresponding wave function $\psi_{u,n}^{\mathrm{K}_1\mathrm{K}_2}(k,Q)$. One-particle electron energy for the K point $\varepsilon_{n,k}^{\mathrm{K}} = \gamma\sqrt{\kappa_{\nu\varphi}(n)^2 + k^2}$ with $\kappa_{\nu\varphi}(n) = (2\pi/L)(n + \varphi - \nu/3)$ involves a AB magnetic flux ϕ passing through the cross section, where L is the circumference length of CN, $\varphi = \phi/\phi_0$ with the magnetic flux quantum $\phi_0 = ch/e$, and ν is an integer ($\nu = 0, \pm 1$) and $\nu = \pm 1$ for semiconducting CNs. The equation for the K$'$ point is obtained by replacing k by $-k$ and ν by $-\nu$. Within the conventional screened Hartree-Fock approximation [1], the self-energy and the Coulomb matrix elements are considered in $\Delta\varepsilon_{n,k,Q}^{\mathrm{K}_1\mathrm{K}_2}$ and $V_{(n+,k;m+,k+q)(m-,k'+q;n-,k')}^{\mathrm{K}_1\mathrm{K}_2}$, respectively.

We consider the following two bonding and two anti-bonding exciton states within the singlet exciton manifold:

$$|b1\rangle_Q^u = \frac{1}{\sqrt{2}}\left(|\mathrm{KK}\rangle_Q^u + |\mathrm{K'K'}\rangle_Q^u\right), \quad |a1\rangle_Q^u = \frac{1}{\sqrt{2}}\left(|\mathrm{KK}\rangle_Q^u - |\mathrm{K'K'}\rangle_Q^u\right), \tag{3}$$

$$|b2\rangle_Q^u = \frac{1}{\sqrt{2}}\left(\xi|\mathrm{KK'}\rangle_Q^u + \zeta|\mathrm{K'K}\rangle_Q^u\right), \quad |a2\rangle_Q^u = \frac{1}{\sqrt{2}}\left(\xi|\mathrm{KK'}\rangle_Q^u - \zeta|\mathrm{K'K}\rangle_Q^u\right), \tag{4}$$

where ξ and ζ are arbitrary coefficients and in this paper we set $\xi = \zeta^* = e^{-i(\eta + 2\pi/3)}$ with the chiral angle η. The two excitons $|\mathrm{KK}\rangle_0^{1s,2s,\cdots}$ and $|\mathrm{K'K'}\rangle_0^{1s,2s,\cdots}$ consisting of intra-valley electron-hole pairs can be coupled to light polarized parallel to the tube axis because of momentum conservation and its parity. Thus, only the bonding state $|b1\rangle_0^{1s,2s,\cdots}$ is bright, and all others are dark, in the absence of a magnetic flux and an impurity. In the presence of a magnetic flux, $|a1\rangle_0^{1s,2s,\cdots}$ becomes bright, but the inter-valley excitons $|b2\rangle_0^u$ and $|a2\rangle_0^u$ remain dark [5].

Now we shall introduce a single impurity at a carbon B site, described by the Gaussian potential [14] with the strength u_0 and the range d,

$$V_{\mathrm{imp}}(\boldsymbol{r}) = \frac{u_0}{\pi d^2}\exp\left(-r^2/d^2\right). \tag{5}$$

Using the basis of exciton states (3) and (4), we evaluate the matrix elements of the impurity potential and derive the effective exciton Hamiltonian. Performing exact diagonalization of the Hamiltonian, we obtain the eigenstates, some of which with the lower energy can be found as the impurity-bound exciton states.

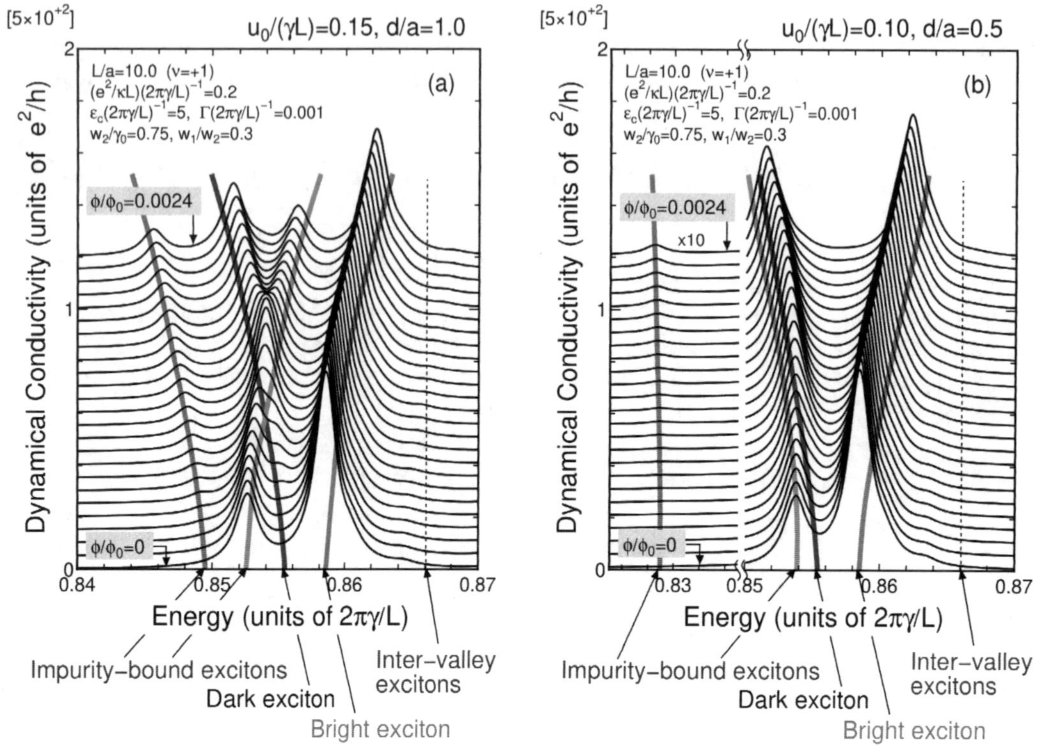

Figure 1. Calculated absorption spectra in the presence of an impurity for (a) $u_0/(\gamma L) = 0.15$ and $d/a = 1.0$. (b) $u_0/(\gamma L) = 0.10$ and $d/a = 0.5$. The flux ϕ is varied with interval $\Delta\phi/\phi_0 = 0.0001$ up to $\phi/\phi_0 = 0.0024$. The peak positions are shown by the thick lines. The energy broadening $\Gamma(2\pi\gamma/L)^{-1} = 0.001$ is used and the energy unit is $2\pi\gamma/L \sim 1.5$ eV for $L/a = 10$.

The matrix elements imply a fact that the anti-bonding states are still decoupled with the bonding states under no AB flux: $\langle b1|V_{\rm imp}|a1\rangle = \langle b2|V_{\rm imp}|a2\rangle = 0$ and $\langle b1|V_{\rm imp}|a2\rangle = \langle b2|V_{\rm imp}|a1\rangle = 0$ for any u and Q. This immediately leads to the result that an impurity does not make any anti-bonding dark states optically active [10]. Further, inter-valley transitions are possible only for $d < a$ where a is the lattice constant and $\langle b1|V_{\rm imp}|b2\rangle \neq \langle a1|V_{\rm imp}|a2\rangle$, while intra-valley transitions give $\langle b1|V_{\rm imp}|b1\rangle = \langle a1|V_{\rm imp}|a1\rangle \simeq \langle b2|V_{\rm imp}|b2\rangle = \langle a2|V_{\rm imp}|a2\rangle$.

3. Numerical results

In our calculation, the short-range part of the Coulomb interaction is introduced by two parameters w_1 and w_2 proposed by Ando [5], giving the energy splitting between the excitons (3) and (4). The parameters are fixed to $w_2/\gamma_0 = 0.75$ and $w_1/w_2 = 0.3$ where γ_0 is the nearest neighbor hopping integral related to the band parameter through $\gamma = \sqrt{3}a\gamma_0/2$. The strength of the long-range part of the Coulomb interaction $(e^2/\kappa L)(2\pi\gamma/L)^{-1} = 0.2$ is used, corresponding to $\kappa = 2$ for $\gamma_0 = 2.7$ eV, where κ is the dielectric constant.

Figures 1 shows some examples of absorption spectra in the presence of an impurity for (a) $u_0/(\gamma L) = 0.15$ and $d/a = 1.0$ (long-range impurity) and (b) $u_0/(\gamma L) = 0.10$ and $d/a = 0.5$ (short-range impurity). With no AB flux, double peaks already appear in the both cases of $d/a = 1.0$ and 0.5. This is because, in addition to the bright exciton state $|b1\rangle_0^{1s}$, there exists a bright impurity-bound exciton state [15]. With the increase of an AB flux, a dark impurity-bound exciton with the lowest energy becomes brightened as well as the dark exciton $|a1\rangle_0^{1s}$, resulting in several peak structures below the bright exciton peak.

What should be emphasized here is that the AB flux dependence of the bright impurity-bound exciton is qualitatively different between the long- and short-range impurities. For the long-range impurity, the bright impurity-bound exciton peak follows the bright exciton peak and is *blueshifted*. By contrast, for the short-range impurity, it follows the dark exciton peak and is *redshifted*. Such a difference can be understood as follows. For the long-range impurity, the bright impurity-bound exciton consists of a linear combination of mainly $|b1\rangle_0^{1s}$ and $|b1\rangle_Q^{2p}$, whose relative energy is unchanged because these two states have the same dependence on the AB flux. On the other hand, for the short-range impurity, it consists of a linear combination of mainly $|b1\rangle_0^{1s}$ and $|b2\rangle_Q^{2p}$, whose relative energy decreases with increasing an AB flux because the inter-valley exciton levels are independent of the AB flux. As a result, the inter-valley coupling between $|b1\rangle_0^{1s}$ and $|b2\rangle_Q^{2p}$ becomes larger and larger as the AB splitting between the bright and dark excitons increases, pushing down the energy level of the bright impurity-bound exciton.

The peaks due to localized excitons were recently observed below the bright exciton peak with no magnetic flux [16]. Our results may be informative and useful for the detailed assignments.

4. Conclusion

We have studied the AB effect on exciton absorption in semiconducting CNs with an impurity. In the presence of an AB flux, the bright impurity-bound exciton shows different behavior depending on the range of the impurity potential: while its peak behaves like a side-peak of the *bright exciton* peak for the long-range potential ($d \gtrsim a$), it behaves like that of the *dark exciton* peak for the short-range potential ($d < a$). This suggests that the AB flux dependence can distinguish and estimate the range of the impurity potential and the strength of the inter-valley coupling of excitons.

Acknowledgment

This work was supported in part by a Grant-in-Aid for Scientific Research on Innovative Area "Optical Science of Dynamically Correlated Electrons" (20104009) and a Grant-in-Aid for Scientific Research (C) (20540307) from Ministry of Education, Culture, Sports, Science and Technology, Japan.

References

[1] Ando T 2005 *J. Phys. Soc. Japan* **74** 777, 1997 *ibid.* **66** 1066, 2004 *ibid.* **73** 3351.
[2] Ichida M, Mizuno S, Tani Y, Saito Y and Nakamura A 1999 *J. Phys. Soc. Japan* **68** 3131.
[3] Ajiki H and Ando T 1993 *J. Phys. Soc. Japan* **62** 1255.
[4] Zaric S, Ostojic G N, Kono J, Shaver J, Moore V C, Strano M S, Hauge R H, Smalley R E and Wei X 2004 *Science* **304** 1129.
[5] Ando T 2006 *J. Phys. Soc. Japan* **75** 024707.
[6] Spataru C D, Ismail-Beigi S, Capaz R B and Louie S G 2005 *Phys. Rev. Lett.* **95** 247402.
[7] Jiang J, Saito R, Samsonidze Ge G, Jorio A, Chou S G, Dresselhaus G and Dresselhaus M S 2007 *Phys. Rev. B* **75** 035407.
[8] Srivastava A, Htoon H, Klimov V I and Kono J 2004 *Phys. Rev. Lett.* **101** 087402.
[9] Matsunaga R, Matsuda K and Kanemitsu Y 2008 *Phys. Rev. Lett.* **101** 147404.
[10] Tomio Y and Suzuura H 2010 *Phys. E* **42** 783.
[11] Perebeinos V, Tersoff J and Avouris P 2005 *Nano Lett.* **5** 2495.
[12] Harutyunyan H, Gokus T, Green A A, Hersam M C, Allegrini M and Hartschuh A 2009 *Nano Lett.* **9** 2010.
[13] Matsuda K, Inoue T, Murakami Y, Maruyama S and Kanemitsu Y 2008 *Phys. Rev. B* **77** 193405.
[14] Ando T and Nakanishi T 1998 *J. Phys. Soc. Japan* **67** 1704,
[15] Note that the absorption intensity of the impurity-bound exciton is proportional to the impurity density and thus should be zero for a single impurity in a CN with infinite length. Because of the numerical discretization of k space, our present results correspond to those for an impurity per length of $100L$.
[16] Comfort E S, Jones D A, Malapanis A, Robinson Z R, Fishman M T and Lee J U 2011 *Phys. Rev. B* **83** 081401.

Structure and stability of hydrogen atom adsorbed on nitrogen-doped carbon nanotubes

Yoshitaka Fujimoto and Susumu Saito

Department of Physics, Tokyo Institute of Technology, Oh-okayama, Meguro, Tokyo 152-8551, Japan

E-mail: fujimoto@stat.phys.titech.ac.jp

Abstract. The adsorption process of hydrogen atom on nitrogen-doped carbon nanotube (CNT) and its effects on the electronic properties are investigated using the first-principles density-functional methods. As possible hydrogen adsorption sites, three different positions are considered to discuss adsorption energies of the hydrogen atom on N-doped (10,0) CNTs. It is found that the favorable hydrogen adsorption site on N-doped (10,0) CNT is not on top of the nitrogen atom but on top of carbon atoms next to the nitrogen atom. Interestingly, it is found that the impurity state induced by doping with nitrogen atom is shifted from conduction-band minimum to valence-band maximum by the adsorption of the hydrogen atom.

1. Introduction

Carbon nanotube (CNT) has been widely investigated since it possesses unique structural and electronic properties such as quasi-one-dimensional form and high electron mobility [1, 2]. The chemical functionalization of CNTs by doping and adsorption of impurity atoms and molecules is of great importance for a realization of nanotube-based nanoelectronics devices. It is well-known that the doping is a practical way to modify electronic structures of CNTs [3]. Nitrogen is a good dopant for carbon-based materials, and nitrogen-doped CNTs have been synthesized experimentally and studied theoretically [4, 5].

The adsorption of hydrogen atoms and molecules on pristine CNTs has been also studied for developments of future nanoelectronics applications such as hydrogen storages and gas sensors [6, 7]. The N-doped CNT adsorbed by the hydrogen atom is expected to exhibit the novel electronic property since impurity defect often serves as a reactive site and the adsorption of the impurity on nanotubes might lead to dramatic changes in their electronic properties [8]. However, the knowledge of hydrogen adsorption effects on N-doped CNT is still limited so far. In order to tailor the desired electronic properties of N-doped CNTs, it is of high importance to understand interactions of hydrogen atom with N-doped CNT and the differences between the electronic structure of N-doped CNTs and that with the adsorbed H atom.

In this work, we report adsorption effects of hydrogen atom upon nitrogen-doped (10,0) CNT using the first-principles total-energy calculations in the framework of the density-functional theory. The adsorption energies of hydrogen atom at several different sites in N-doped CNT are calculated, and it is found that the adsorption of H atom on C atoms next to N atom is energetically favored rather than that on the N atom. The effects of hydrogen-atom adsorption on electronic structures of N-doped CNT are also discussed.

Published under licence by IOP Publishing Ltd

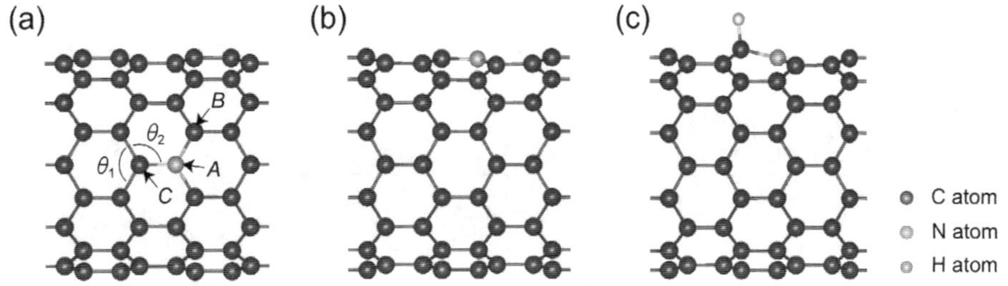

Figure 1. Optimized atomic structures of N-doped (10,0) carbon nanotubes: (a) top view with three possible hydrogen adsorption sites labeled as *A*, *B*, and *C*, and (b) its side view, and (c) N-doped (10,0) nanotube with hydrogen atom adsorbed on the *C* site. θ_1 and θ_2 denote bond angles around carbon atom labeled as *C*.

2. Computational Methods

The model system considered here consists of two unit cells for a zigzag (10,0) CNT, and therefore the N-doped (10,0) CNT contains 79 C atoms and one N atom. In Fig. 1(a), the atomic structure of N-doped (10,0) CNT is illustrated, together with three different hydrogen adsorption sites labeled as *A*, *B*, and *C*. Figures 1(b) and (c) show side views of the optimized geometries of N-doped (10,0) CNT without and with H atom, respectively.

Our calculations are carried out using the first-principles total-energy methods within the density-functional theory [9]. The interactions between the ions and the valence electrons are described by the norm-conserving Troullier-Martins pseudopotentials [10], and exchange-correlation effects are treated using the local density approximation (LDA) parameterized by Perdew and Zunger [11, 12]. The wave functions are expanded in a plane-wave basis and the kinetic-energy cutoff is taken to be 50 Ry [13]. The atomic positions are optimized until the Hellmann-Feynman force on each atom becomes less than 0.05 eV/Å. One-dimensional Brillouin-zone integration is performed with four k-point sampling.

3. Results and Discussion

3.1. Energetics and geometry

To discuss the stability of hydrogen atom adsorbed on the N-doped (10,0) CNT, we calculate the adsorption energies $E_{\rm ad}$ defined as

$$E_{\rm ad} = E_{\rm tot} - E_{\rm N-CNT} - E_{\rm H}, \qquad (1)$$

where $E_{\rm tot}$ and $E_{\rm N-CNT}$ are total energies of N-doped (10,0) CNT with and without hydrogen atom, respectively, and $E_{\rm H}$ is the total energy per atom of the H_2 molecule.

In Table 1, we list the adsorption energy of hydrogen atom at three sites in N-doped CNT. The adsorption of H atom on the N atom (site *A*) is energetically unfavorable since the positive adsorption energy corresponds to endothermic adsorption ($E_{\rm ad} > 0$), whereas each H atom adsorbed on two different C atoms next to the N atom (sites *B* and *C*) is expected to be both stable due to their exothermic behavior ($E_{\rm ad} < 0$). The adsorption energy at site *C* is somewhat lower than that at site *B*, which should be the curvature effect of N-doped CNT. We thus find that the adsorption of H atom upon N-doped CNT is energetically favored when H atom is adsorbed on the C atom rather than on the N atom.

We next study the atomic geometries of N-doped (10,0) CNT without and with H atoms, as shown in Figs. 1(b) and (c), respectively. The bond angles θ_1 and θ_2 of N-doped CNT without H atom are 114.2° and 119.2°, respectively, and the N atom resides in the sp^2-bonding configuration

Table 1. Calculated adsorption energies (E_{ad}) for N-doped (10,0) CNT with hydrogen atom at three different sites. Three sites A, B, and C are shown in Fig. 1.

Position	A	B	C
Adsorption energy E_{ad} (eV)	0.96	-0.28	-0.29

[Fig. 1(b)]. After H atom is adsorbed on N-doped CNT, the C–H bond with the length of 1.11 Å is formed, and the bond angles θ_1 and θ_2 change into 106.4° and 112.3°, respectively. It can be seen that the C atom bonded with H atom protrudes from the original *surface* of the nanotube and the C–H pair forms a covalent bond [see Fig. 1(c)]. By the adsorption of H atom, the configuration of the C atom in N-doped CNT transforms from sp^2 into sp^3-hybridized configurations.

3.2. Electronic structure
We have studied the effect of hydrogen-atom adsorption on the electronic structures of N-doped (10,0) CNT. Figures 2 (a), (b), and (c) show the calculated energy-band structures of pristine (10,0) CNT, N-doped (10,0) CNT, and N-doped (10,0) CNT with H atom, respectively. The pristine (10,0) CNT exhibits semiconducting properties with a band gap of ~ 0.72 eV, as shown in Fig. 2(a) [5]. When carbon atom in pristine (10,0) CNT is substituted by nitrogen atom, the impurity state associated with the N atom appears below conduction-band minimum and would act as a donor state [Fig. 2(b)]. This is because N atom possesses extra one electron compared with C atom. When H atom is adsorbed on the C site in N-doped (10,0) CNT, the originally partially-filled impurity state is completely filled with electrons and the donor state induced by nitrogen doping disappears [Fig. 2(c)]. Thus, the adsorption of H atom causes dramatic changes in the electronic structure of N-doped CNT.

4. Summary
The adsorption effects of hydrogen atom on geometry, energetics, and electronic structure of nitrogen-doped (10,0) carbon nanotube have been investigated using the first-principles density-functional study. Adsorption energies of hydrogen atom at three different adsorption sites are

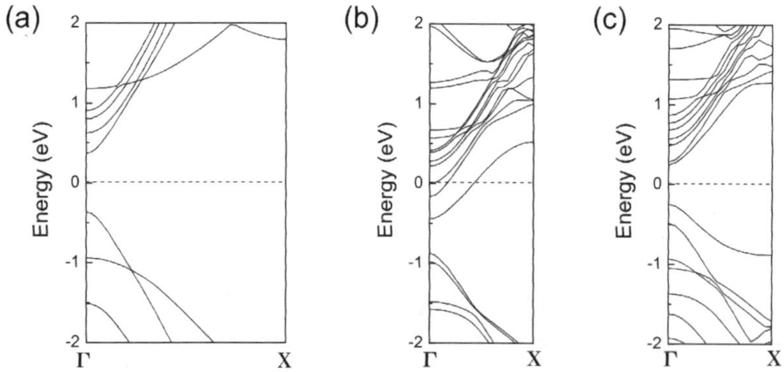

Figure 2. Energy bands of (a) pristine (10,0) CNT, (b) N-doped (10,0) CNT, and (c) N-doped (10,0) CNT with hydrogen atom. The H atom is adsorbed at C site of N-doped (10,0) CNT in Fig. 1(a). The pristine (10,0) CNT considered here consists of 40 carbon atoms. The Fermi level is set to be zero.

calculated and the adsorption energy is found to be much lower when H atom is adsorbed on top of C atoms next to the N atom than when it is on top of the N atom. The energy bands of N-doped (10,0) CNT and that with H atom are also obtained. By adsorption of H atom upon N-doped CNT, the originally half-filled impurity level is to be filled with the electrons, leading to the Fermi-level shift into valence-band maximum.

Acknowledgments

This work was supported by Global Center of Excellence Program by MEXT Japan through the Nanoscience and Quantum Physics Project of the Tokyo Institute of Technology, and a Grant-in-Aid for Scientific Research on Priority Areas of the MEXT Japan, "Carbon Nanotube Nano-Electronics" project, and Elements Science and Technology Project of MEXT Japan. Computations were done at Institute for Solid State Physics, the University of Tokyo, on TSUBAME Grid Cluster at Tokyo Institute of Technology, and at Research Center for Computational Science, National Institute of Natural Sciences.

References

[1] Iijima S 1991 *Nature* **354** 56.
[2] Dürkop T, Getty S A, Cobas E and Fuhrer M S 2004 *Nano Lett.* **4** 35.
[3] Koretsune T and Saito S 2008 *Phys. Rev.* B **77** 165417.
[4] Fang W C 2008 *Nanotechnology* **19** 165705.
[5] Fujimoto Y and Saito S 2011 *Physica* E **43** 677.
[6] Schlapbach L and Züttel A 2001 *Nature* **414** 353.
[7] Akai Y and Saito S 2003 *Jpn. J. Appl. Phys.* **42** 640.
[8] Zhou Z, Gao X, Yan J and Song D 2006 Carbon **44** 939.
[9] Hohenberg P and Kohn W 1964 *Phys. Rev.* **136** B864.
[10] Troullier N and Martins J L 1991 *Phys. Rev.* B **43** 1993.
[11] Kohn W and Sham L J 1965 *Phys. Rev.* **140** A1133.
[12] Ceperley D M and Alder B J 1980 *Phys. Rev. Lett.* **45** 566; Perdew J P and Zunger A 1981 *Phys. Rev.* B **23** 5048.
[13] Computations have been performed using Tokyo Ab-initio Program Package (TAPP) which has been developed by a consortium initiated at the University of Tokyo: Yamauchi J, Tsukada M, Watanabe S and Sugino O 1996 *Phys. Rev.* B **54** 5586; Kageshima H and Shiraishi K 1997 *Phys. Rev.* B **56** 14985; Sugino O and Oshiyama A 1992 *Phys. Rev. Lett.* **68** 1858.

Twisting Effects on Carbon Nanotubes: A First-Principles Study with Helical Symmetry Operations

Koichiro Kato, Takashi Koretsune and Susumu Saito

Department of Physics, Tokyo Institute of Technology,
2-12-1 Oh-okayama, Meguro-Ku, Tokyo 152-8550, Japan

E-mail: kato@stat.phys.titech.ac.jp

Abstract. We report the energetics and the electronic properties of twisted carbon nanotubes (CNTs). We use a real-space density functional theory with helical-symmetry operation, and apply it for several CNTs with the diameters of around 0.8 nm including the experimentally abundant (6,5) nanotube. By using this computational code, one can now obtain the total energies with enough accuracies to optimize the CNT geometries including quasi-continuous twisting levels for any type of nanotube in principle. As a result, it is found that chiral nanotubes possess twisted geometries at their ground states. The electronic structures of CNTs depend sensitively on twisting levels in this diameter region, and the twisting effects on their fundamental gap values can be judged by the value of $\mathrm{mod}(n - m, 3)$, where n and m are the chiral indices.

1. Introduction

Carbon nanotube has been considered to be one of the most important materials in nanoscience and nanotechnology [1]. The reason of this attention mainly comes from their unique electronic properties. It is well known that the electronic transport properties of CNT sensitively depend on the diameter and chirality from the reports by using the tight-binding method [2]. Therefore, it is very important to know the accurate electronic properties of CNTs for any practical electronic application. It is often thought CNT is a perfect cylinder made by rolling up a two-dimensional graphite sheet. In reality, however, there are undeniable possibilities for them to possess structural deformations such as bent, twisted or collapsed region. As for the chiral nanotubes, their ground state geometry may possess intrinsic twisting because of their inherent helical structures. According to the studies by using the density functional theory (DFT), it has been revealed that the electronic structures of CNTs depend sensitively on their geometrical parameters such as bond lengths and bond angles [3 - 5]. Hence, electronic properties of deformed nanotubes should be of high interest and importance. The uniaxial strains of CNTs have been often discussed by using DFT. However, there are no systematic DFT study which has considered twisting effects on CNTs. This is because the most DFT computational codes utilize the translational symmetry together with the plane-waves basis set. The uniaxial strains on CNTs can easily be dealt with by this popular code. On the other hand, applying the twist breaks the original translational symmetry of CNTs and the periodicity of the translational symmetry

Journal of Physics: Conference Series **302** (2011) 012007	doi:10.1088/1742-6596/302/1/012007

changes significantly depending on the direction and the magnitude of twisting. Hence, it is quite difficult to discuss the twisting effects on CNTs by using the popular DFT code.

2. Computational Details and System Studied

Mintmire *et al.* reported that all CNTs can be classified by the two kinds of symmetry operations [6]. One is the helical-symmetry operation and the other is the rotational-symmetry operation. By using these symmetry operations not the translational-symmetry but the helical-symmetry, unit cell of any CNT includes only two carbon atoms. In order to implement these symmetry operations to the computational code, we adopt not the plane waves but the real space grids as basis set. Arbitral boundary conditions can be applied by the use of the real space grids, and all quantities are directly calculated on the grid points. Therefore, this DFT code is classified as so-called "real-space" method. In the DFT code, the twist is applied by inclining the translational vector in the expanded plane. The twist angle (θ_T) is defined as the angle between the original translational vector and the declined translational vector. The number of carbon atoms per unit cell does not change by applying the twist unlike the translational unit-cell method. Therefore, we perform the geometrical optimization including the twisting degree of freedom and discuss the electronic properties of all the optimized CNTs in the same computational cost. The present real-space DFT computational code use the local density approximation [7] parametrized by Perdew and Zunger [8, 9]. The norm-conserving pseudopotential proposed by Troullier and Martins is implemented with the separable form [10, 11].

The system studied is shown in figure 1. We choose the several CNTs having the diameters of about 0.8 nm, and focus on the experimentally abundant nanotubes, *i.e.* (6,5) and (7,5), in the sample produced by using the chemical vapor deposition (CVD) method with the Co-Mo catalyst [12] or the alcohol catalytic CVD method [13].

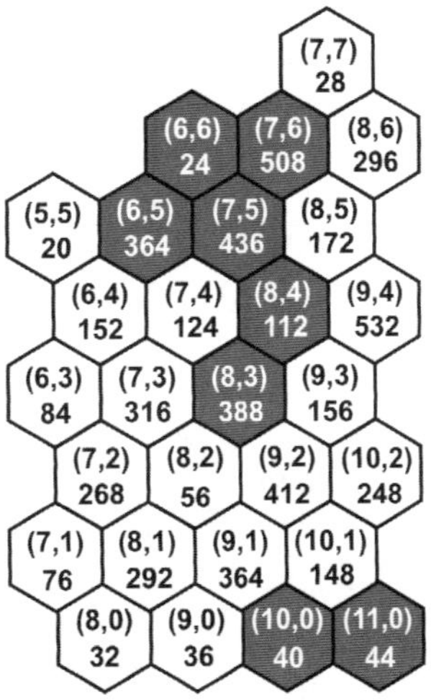

Figure 1. (n, m) indices of the systems studied (color online). In each hexagon are written not only the chiral indices but also the number of atoms in the translational unit cell. We discuss the twisting effects of CNTs corresponding to the red-colored hexagons.

3. Results and Discussion

From the structural optimization, it is found that the total energy of the slightly twisted geometry is generally lower than that of the non-twisted geometry in chiral nanotubes. This

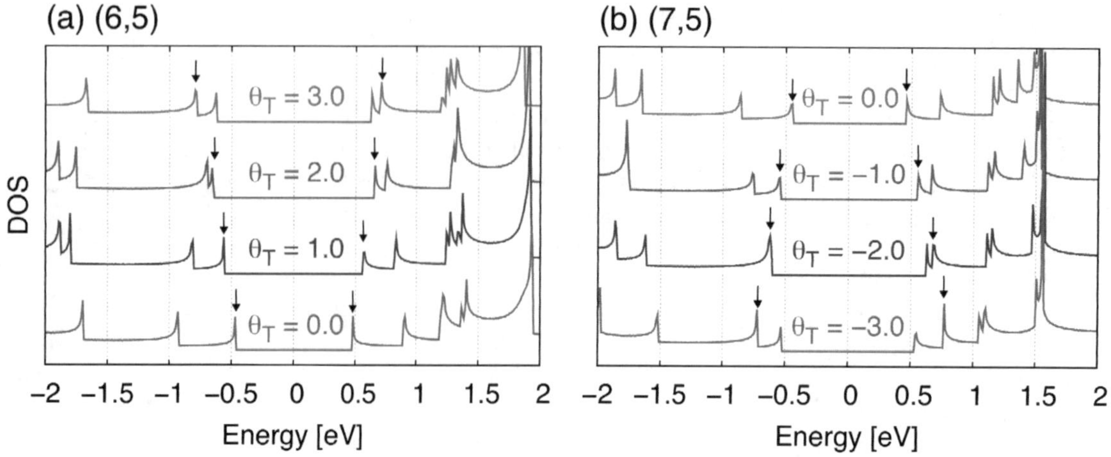

Figure 2. Twist-angle dependences of the density of states of (6,5) and (7,5) nanotubes. The black allows show the van Hove singularities corresponding to the E_{11} under $\theta_T = 0$.

result suggests that chiral nanotubes should possess the intrinsic twisting even in their ground state geometry. Among the studied nanotubes, the (6,5) nanotubes is found to have the largest intrinsic-twisting angle. The estimated intrinsic twisting of the (6,5) nanotube should correspond to the half revolution for each 1 μm.

As for the electronic structures, figure 2 represents the twist-angle dependences of the density of states of (6,5) and (7,5) nanotubes. Interestingly, (6,5) and (7,5) show the opposite twist angle dependences. In addition, E_{11} and E_{22} values, the peak-to-peak energy separation between the valence bands and the corresponding conduction bands, also show the inverse twist-angle dependences in each nanotube. When the positive twist is applied on (6,5) nanotube, E_{11} increases while E_{22} decreases. On the other hand, when the same twist is applied on (7,5) nanotube, E_{11} decreases while E_{22} increases. This effect can be understood by the so-called "zone-folding" method which gives the approximate electronic structure of CNT from that of graphene. In that method, the cutting lines which reflect the circumferential boundary conditions play important roles. The two-dimensional first Brillouin zone (FBZ) of graphene is mapped into the one-dimensional FBZ along the cutting lines. The E_{11} and E_{22} values are associated with the nearest and second nearest cutting lines to the "K" point in the FBZ of graphene. These lines lie across the "K" point. At the "K" point, the valence band top and the conduction band bottom touch each other linearly. It is known that the constant-energy surface around the "K" point is similar to the concentric circle in graphene. Therefore, E_{11} and E_{22} values are nearly proportional to the distance between the "K" point and the cutting lines. When the twist is applied, all cutting lines move to the same direction depending on the direction and the magnitude of the twist in a nanotube. Considering this effect, E_{11} and E_{22} values should be inversely affected by the applied twist. When the distance between the "K" point and the nearest cutting line becomes shorter by the twist, the distance between the "K" point and the second nearest cutting line becomes longer. This property can be generalized for the "moderate-gap" semiconducting CNTs by using the value of $mod(n-m,3)$, where n and m are the chiral indices. E_{11} and E_{22} values of (7,6), (6,5) and (8,4) show the same twist-angle dependences. On the other hand, their twist-angle dependences of (7,5) and (8,3) show the opposite behavior comparing with the above group. It is also found that the magnitude of change of E_{11} and E_{22} values depends on the chiral angle. Armchair and near armchair nanotubes show the large twist-angle dependences of about 0.2 eV/degree, while E_{11} and E_{22} values of zigzag nanotubes scarcely depend on the twist angles. They show almost the constant values in the calculated range of the twist angles. It is interesting to note that, when a sizable twist is applied, the original E_{22} line can become closer to the "K" point than the original E_{11} line and E_{11} can be larger than E_{22}. Actually, figure 2 shows this situation. In the (6,5) nanotube, the difference between E_{11} and E_{22} values becomes smaller with increasing the positive twist. When the applied positive twist exceeds over two degree, the original E_{11} should be called E_{22} and the original E_{22} should be called E_{11}. Also the same exchange is observed in the twisted (7,5) nanotube toward the negative θ_T direction.

4. Summary

We have investigated the twisting effects on the several CNTs using the real-space DFT computational code which can deal with the geometrical optimization including the twisting degree of freedom. Because we use the helical-symmetry and the rotational-symmetry operations, the computational costs for all CNTs are the same with all twisting conditions in principle. As a result, it is found that the intrinsic twisting should occur in the chiral CNTs, and that the twisting strongly affects the electronic structures of CNTs, inducing the significant variations of not only E_{11} but also E_{22} values. Interestingly, we predict that one should observe the exchange of E_{11} and E_{22} states in the (6,5) and (7,5) with relatively small twisting angles. These twisting effects on the electronic structures should be explained by "zone-folding method". These results also suggest that the optical properties of CNTs should be changed considerably just by applying the twist.

Acknowledgements

The original version of the real-space DFT computational code was written by Prof. Yabana and Dr. Iwata. Numerical calculations were performed on TSUBAME Grid Cluster at Global Scientific Information. This work was supported by the Global Center of Excellence Program by MEXT, Japan through the "Nanoscience and Quantum Physics" Project of the Tokyo Institute of Technology. KK also acknowledges JSPS Research Fellowship for Young Scientists.

References

[1] Iijima S 1991 *Nature* **354** 56
[2] Hamada N, Swada S and Oshiyama A 1992 *Phys. Rev. Lett.* **68** 1579
[3] Blase X, Benedics L X, Shirey E L and Louie S G 1994 *Phys. Rev. Lett.* **72** 1878
[4] Kanamitsu K and Saito S 2002 *J. Phys. Soc. Jpn* **71** 483
[5] Kato K and Saito S 2011 *Physica E* **43** 669
[6] White C T, Robertson D H and Mintmire J W 1993 *Phys. Rev.* B **47** 5485
[7] Kohn W and Sham L J 1965 *Phys. Rev.* **140** A1133
[8] Cerperley D M and Alder B J 1980 *Phys. Rev. Lett.* **45** 566
[9] Perdew J P and Zunger A 1981 *Phys. Rev.* B **23** 5048
[10] Troullier N and Martins J L 1991 *Phys. Rev.* B **43** 1993
[11] Kleinman L and Bylander D M 1982 *Phys. Rev. Lett.* **48** 1425
[12] Bachilo S M, Balzano L, Herrera J E, Pompeo F, Resasco D E and Weisman R B 2003 *J. Am. Chem. Soc.* **125** 11186
[13] Miyauchi Y, Chiashi S, Murakami Y, Hayashida Y and Maruyama S 2004 *Chem. Phys. Lett.* **387** 198

NMR study on Ru-based quantum spin systems of $[Ru(acac)_3]$ and $[\{Ru(acac)_2\}_2(\mu\text{-}OEt)_2]$

S. Nakajima, M. Masuko , A. Oosawa, T. Goto, T. Hashimoto, A. Endo, and T. Hayashita

Faculty of Science and Technology, Sophia University
7-1 Kioicho Chiyodaku Tokyo 102-8554 Japan

E-mail: gotoo-t@sophia.ac.jp

Abstract. Tris(β-diketonato)ruthenium(III) complexes of mononuclear $[Ru(acac)_3]$ and binuclear $[\{Ru(acac)_2\}_2(\mu\text{-}OEt)_2]$ are spin-1/2 quantum spin magnets. In both system, Ru and surrounding O_6 atoms consist of octahedral geometry, and in the binuclear complex, two (acetylacetonato) ruthenium complexes are bridged by two oxygen atoms of ethoxyl groups, making up a dimer with Ru-Ru bond. The electronic state of ruthenium is easily controlled by changing the methyl groups on the acetylacetone to the different substituents such as electronegative trifluoromethyl group. We expect to introduce a new degree of freedom, a charge fluctuation into the dimer spin system. In this study, we investigate the spin state of monomer and dimer quantum spin systems by means of ^{13}C,^{19}F,^{101}Ru-NMR at low temperatures. In the former $[Ru(acac)_3]$, we observed Curie-like temperature dependence in the Knight shift down to 1.5 K, indicating that the system was simply in the paramagnetic state. The dimer system $[\{Ru(acac)_2\}_2(\mu\text{-}OEt)_2]$ on the other hand showed the NMR signal with very long T_1 at zero shift position at low temperatures, indicating that the system was the spin singlet state. In order to investigate the effect of substitution, we also report on results of *mer*-$[Ru(ehfa)_3]$ and $[Ru(fhea)_3]$, for which methyl groups are substituted by trifluoromethyl groups.

1. Introduction

In the studies of quantum spin dimer systems reported so far, most interest is concentrated on destructing the singlet ground state by applying an external magnetic field[1], a high pressure[2-3], or substituting non-magnetic impurity[4-7]. These perturbations, by inducing triplet sites or creating unpaired spins, lead the system to the magnetic state, which undergoes a phase transition at low temperatures with an assist of inter-dimer interactions. In particular, a field-induced long range magnetic order is known as the magnon Bose-Einstein condensation (BEC) [8].

We introduce into the dimer system a new degree of freedom, a valence fluctuation. By tuning valence of each atom to be fractional value, we expect the wave function of the dimer to be a superposition of the two states in which an extra electron is located at an either atom, and to bear a possibility of an exotic ground state. As a candidate of valence fluctuating dimer, we have chosen as a kernel atom Ru, which is stable in various valence state 2, 3 and 4. In order to realize a valence fluctuating dimer compound in the solid state form which is stable in atmosphere of air, we have adopted a method developed by one of our the present authors A. E.[9] to substitute methyl groups on the acetylacetone with more electronegative one such as trifluoromethyl group. This method allows one to adjust the potential surrounding Ru ions by changing the number of substituent group in six

Table 1. List of samples with the number of Ru atoms, nominal Ru-valence and number of CF_3 substituted. Results of NMR and the effective valence deduced from them are also shown.

Compound		nominal valence	CF_3	shift	width	effective valence
$[Ru^{III}(acac)_3]$	monomer	III	0	Curie	Curie	III
$[\{Ru^{III}(acac)_2\}_2(\mu\text{-}OEt)_2]$	dimer	III	0	dia	dia	III
mer-$[Ru^{III}(ehfa)_3]$	monomer	III	3	Curie	Curie	III
$[Ru^{III}(fhfa)_3]$	monomer	III	6	Curie (small)	dia	III–Δ
$K^+[Ru^{II}(fhfa)_3]$	monomer	II	6	dia	dia	II

methyl groups in a dimer so as to obtain a fractional valence state without making a solid solution between hetero-valent ions.

In this preliminary report we demonstrate by NMR technique the variation of magnetism in synthesized five compounds of tris(β-diketonato)ruthenium(III) complexes of monomer systems $[Ru(acac)_3]$ and those with different number of CF_3 substituted, and a dimer system $[\{Ru(acac)_2\}_2(\mu\text{-}OEt)_2]$.

2. Experimental

The list of samples investigated is shown in Table 1 and their molecular structure in Fig. 1, where (a) and (b) are those of a monomer and a dimer without CF_3 substitution respectively, (c), the partially-substituted, (d), the fully-substituted and (e), the fully-substituted and with the counter cation of potassium. The nominal valence of Ru is expected to be +3 except for the last one which has +2. The details of the sample synthesis is shown in ref. 10. As for $[\{Ru(acac)_2\}_2(\mu\text{-}OEt)_2]$, there are the two enantiomers with and without a mirror symmetry. We have investigated both types of this sample, meso and racemi.

All the samples were powder with amount 0.2 g for NMR measurements; no enrich treatment was done for ^{13}C-NMR. NMR spectra and longitudinal spin relaxation rate T_1^{-1} were measured at temperature region between 1.6-30 K and under magnetic fields around 1 T for ^{19}F-NMR, 5 T for ^{13}C-NMR and up to 12 T for $^{99/101}$Ru-NMR. NMR spectra have been obtained by plotting the spin-echo amplitude against the either external field or the resonance frequency. The temperature dependence of the resonance line-width and shift was determined from obtained spectra. T_1^{-1} was determined by tracing the evolution of spin-echo amplitude with repetition rates τ and fitting it with the function $1-e^{-\tau/T_1}$. These three quantities the resonance line shift, the line width and T_1^{-1} correspond to the uniform susceptibility, its distribution and fluctuation respectively, and allow one to investigate the Ru-4*d* spin state.

3. Results and Discussion

Figure 2 shows typical NMR spectra of the four monomer systems. First, we demonstrate by Ru-

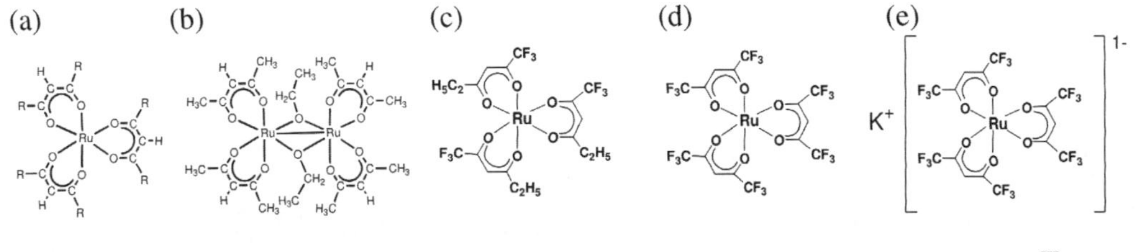

Figure 1. Schematic molecular structure of each sample: (a) $[Ru^{III}(acac)_3]$, (b) $[\{Ru^{III}(acac)_2\}_2(\mu\text{-}OEt)_2]$, (c) *mer*-$[Ru^{III}(ehfa)_3]$, (d)$[Ru^{III}(fhfa)_3]$ and (e) $K[Ru^{II}(fhfa)_3]$.

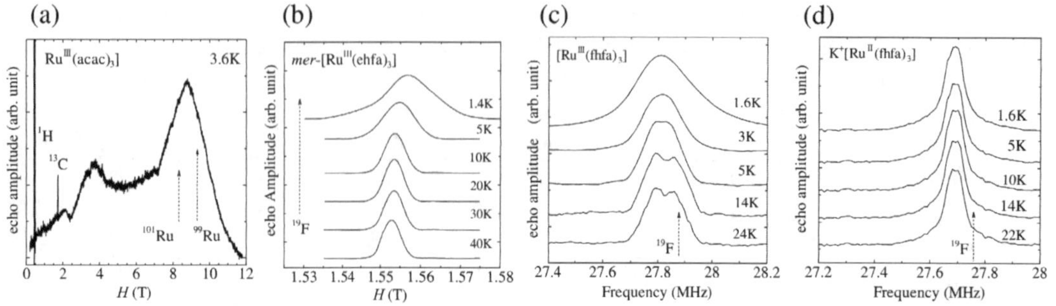

Figure 2. Typical Ru- and F-NMR spectra for (a) [Ru^{III}(acac)$_3$], (b) *mer*-[Ru^{III}(ehfa)$_3$], (c) [Ru^{III}(fhfa)$_3$] measured under constant field 0.693 T, and (d) K[Ru^{II}(fhfa)$_3$] under 0.695 T. The zero-shift position is shown by dashed arrows.

NMR the existence of the paramagnetism in the monomer sample [Ru^{III}(acac)$_3$], where a spectrum with a huge width contributed both from the eqq-interaction and hyperfine field was observed. The center position of the spectrum was shifted from the zero shift position to lower side by approximately 2 T, indicating that 4*d* moments in trivalent Ru are paramagnetic and appreciably polarized in high field. This is consistent with the fact that ^{13}C-NMR Knight shift exhibited the Curie-Weiss-like temperature dependence as shown in Fig. 3(a).

Next, we investigated the effect of trifluoromethyl group substitution. Figure 1(c)-(e) show the samples containing CF_3. The partially substituted system *mer*-[Ru^{III}(ehfa)$_3$] with the nominal Ru-valence of +3 shows a significant broadening and increase in the shift at low temperatures shown in Fig. 2(a) and (b) respectively, indicating that the spin state is nearly the same as in [Ru^{III}(acac)$_3$]. Note here that the negative sign of the shift is a simple consequence of the negative hyperfine coupling between Ru-4*d* spin and ^{13}C/^{19}F-nuclei. The nuclear spin-lattice relaxation of these two was also very fast, that is, below 5 ms at 4 K. However, rigorously speaking, we cannot directly compare the resonance line shift in [Ru^{III}(acac)$_3$] and *mer*-[Ru^{III}(ehfa)$_3$], because of the difference in the hyperfine coupling constant of C and F sites. Therefore, we cannot deny the possibility that the slight change in the spin state between the two.

The compound of K[Ru^{II}(fhfa)$_3$] that has the nominal valence +2 shows a diamagnetic behavior as expected. Both the shift and the line width do not change with temperature; they stay constant at very small value down to 1.8 K. T_1 is very long and exceed 50 sec at 1.8 K, as shown in Fig. 3 (c) and (d). These results indicate that Ru in K[Ru^{II}(fhfa)$_3$] is divalent and with *S*=0 spin state.

As for [Ru^{III}(fhfa)$_3$], which is in the intermediate state between the above two group shows a

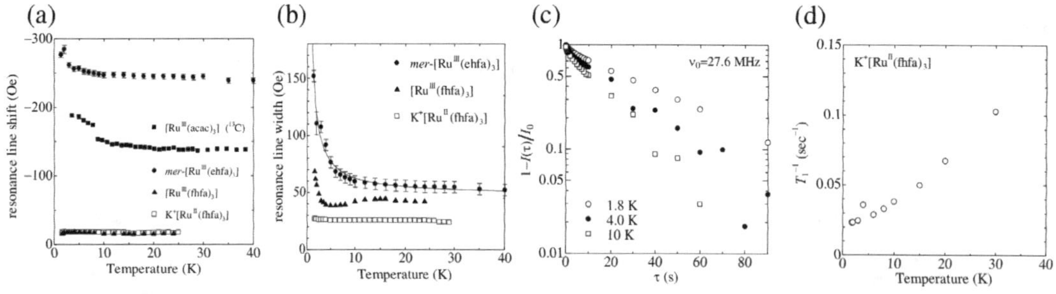

Figure 3. Temperature dependence of the resonance line shift (a) and the line width (b), where solid curve shows the Curie temperature dependence. Typical relaxation curves of the longitudinal spin relaxation (c) and the temperature dependence of T_1^{-1} (d) in K[Ru^{II}(fhfa)$_3$].

peculiar behavior. This compound is fully substituted by CF_3. However, Ru is oxidized by hydrogen peroxide, and expected to be in trivalent state with $S=1/2$. Actually, T_1 is very short below 10 ms, and the temperature dependence of the line width shows Curie-like behavior. However, the shift stays constant at low temperatures, and its magnitude is just identical to mer-K[Ru^{II}(fhfa)$_3$]. These results indicate that the valence state of Ru is effectively reduced to be III–Δ, and hence that the paramagnetism in this system is weakened. The result of these substitution effects is summarized in Table 2.

Finally, as for the dimer compound of [{Ru^{III} (acac)$_2$}$_2$(μ-OEt)$_2$], T_1 is very long, irrespective of the sample is meso and raceme, that is, around 10 s at 4 K, and the resonance line is very sharp; the signal contains FID beat. These results indicate that the system is in non magnetic spinless state owing to the singlet coupling between the two spins in a dimer.

In summary, we have synthesized monomer and dimer compounds of Ru-complex, and investigated by NMR the substitution effect of trifluoromethyl group to find that the effective valence of Ru and hence its spin state is tunable by the substitution.

Acknowledgments
This research was supported by a Grant-in-Aid for Scientific Research on Innovative Areas "New Frontiers of Materials Science Opened by Molecular Degree of Freedom" (Grant No. 21110518) from the Ministry of Education, Science, Sports and Culture.

References
[1] Tanaka H, Oosawa A, Kato T, Uekusa H, Ohashi Y, Kakurai K and Hoser A, 2001 *J. Phys. Soc. Jpn.* **70** 939
[2] Oosawa A, Fujisawa A, Osakabe T, Kakurai K and Tanaka H, 2003 *J. Phys. Soc. Jpn.* **72** 1026
[3] Oosawa A, Kakurai K, Osakabe T, Nakamura M, Takeda M and Tanaka H, 2004 *J. Phys. Soc. Jpn.* **73** 1446
[4] Oosawa A, Ono T, and Tanaka H, 2002 *Phys. Rev.* **B 66**, 020405(R)
[5] Suzuki T, Yamada F, Watanabe I, Matsuzaki T, Goto T, Oosawa A and Tanaka H, 2009 *J. Phys. Soc. Jpn.* **78** 074705
[6] Goto T, Suzuki T, Kanada K, Saito T, Oosawa A, Watanabe I, and Manaka H, 2008 *Phys. Rev.* **B 78** 054422
[7] Saito T, Oosawa A, Goto T, Suzuki T, and Watanabe I, 2006 *Phys. Rev.* **B74** 134423
[8] Nikuni T, Oshikawa M, Oosawa A, and Tanaka H, 2000 *Phys. Rev. Lett.* **84** 5868.
[9] Endo A, Hoshino Y, Hirakata K, Takeuchi Y, Shimizu K, Furushima Y, Ikeuchi H and Satô G.P. 1989 *Bull. Chem. Soc. Jpn.* **62** 709
[10] Hashimoto T, Kawamoto Y, Ishitobi Y, Sasaki T, Fukuda Y, Yamagishi A, Sato H, Shimizu K, and Hayashita T, 2007 *Chem. Lett.* **36**, 1174

^{63,65}Cu-NMR study on Mg-doped quantum spin system $TlCu_{1-x}Mg_xCl_3$

S. Atarashi[1], T. Goto[1], M. Yamada[2], F. Yamada[2], A. Oosawa[2], H. Tanaka[2] and T. Suzuki[3,4]

[1]Department of Physics, Sophia University, Kioi-cho, Chiyoda-ku, Tokyo 102-8554, Japan
[2]Department of Physics, Tokyo Institute of Technology, Oh-okayama, Meguro-ku, Tokyo 152-8551, Japan
[3]Nishina Center, RIKEN, Hirosawa, Wako-shi, Saitama 351-0198, Japan
[4]International Christian University, Osawa, Mitaka-shi, Tokyo 181-8585, Japan

Email:s-ataras@sophia.ac.jp

Abstract. NMR measurements were carried out on the impurity doped quantum spin system $TlCu_{1-x}Mg_xCl_3$ with nominal concentration x = 0.03 and 0.01, in which the magnetic phase transition is expected at 3.6 K and 2.2 K, respectively. For the sample x = 0.03, the peaks of NMR spectra split clearly below T = 3.8 K. However, for the sample x = 0.01, no peak splitting was observed down to T = 1.52 K. This result supports the idea of coherent spin fluctuation in the impurity-induced ordered state as suggested by Suzuki *et al.* [2009 *J. Phys. Soc. Jpn.* **78** 074705.].

1. Introduction

The singlet ground state with an excitation gap has been found in many quantum spin systems such as the Haldane chain, spin-Peierls system, spin ladder, exchange alternating chain, and coupled spin dimer system. These spin gap systems do not have a magnetic ordering down to zero temperature, and possess gapped ground state that arises from the dimerization of spins localized in magnetic ions. Therefore, when a small number of nonmagnetic ions are substituted for magnetic ions, singlet spin dimers are partially broken, so that unpaired spins are produced near the nonmagnetic ions. Unpaired spins can interact through effective exchange interactions that are mediated by singlet dimers [1-5], and hence may show a three-dimensional Néel ordering. Such impurity-induced magnetic ordering is observed in many spin gap systems [6-8].

Recently, impurity-introduced $TlCu_{1-x}Mg_xCl_3$ single crystals have been grown, and new intriguing phenomena have been reported. Oosawa *et al.* and Fujisawa *et al.* reported, by magnetization and neutron scattering experiments, a coexistence of the spin gap and the impurity-induced magnetic order [9-11]. When the magnetic field reaches the critical field H_C, the gap is collapsed and magnons are induced. The system can undergo field-induced magnetic ordering when the concentration of these field-induced magnons reaches the critical value in still higher field. It is not clear whether field- and impurity-induced ordered states have the same spin state. Furthermore, even in the ordered state, no muon spin rotation was observed at temperatures well below the Néel temperature T_N for the sample with an extremely small impurity concentration $x = 0.0047$ [12]. The purpose of our research is to investigate microscopically the difference of spin states between field- and impurity-induced magnetic orders by NMR.

2. Experimental

The single crystals of $TlCu_{1-x}Mg_xCl_3$ used in this study were grown from a melt by the Bridgman method. The details of crystal growth are given in ref. 13. The $TlCu_{1-x}Mg_xCl_3$ crystal is easily cleaved parallel to the planes (0, 1, 0) and (1, 0, $\underline{2}$). The single crystals were cleaved in a glove box filled with dry nitrogen gas to reduce the amount of hydrate phase on the sample surfaces. We measured the temperature dependence of ^{63,65}Cu-NMR spectra for single crystals of $TlCu_{1-x}Mg_xCl_3$ with nominal concentration $x = 0.03$ and 0.01 in the temperature region down to 1.52 K. We also measured the specific heat with $x = 0.03$ by the apparatus of PPMS, Quantum Design, Inc.

3. Results and Discussion

Fig. 1(a) shows the temperature dependence of NMR spectra corresponding to the satellite transition ($I_z = 3/2 \leftrightarrow 1/2$) for the sample with nominal concentration $x = 0.03$. The applied field is much smaller than the critical field, approximately $H_c = 6$ T at $T = 2$ K [13], where the spin gap is collapsed. At high-temperature region above $T > 4$ K, where the system is paramagnetic, sharp six peaks were observed. These two pairs of satellite peaks, #1, #3 and #4, #6, are considered to arise, because the local magnetic environment of the Cu sites nearest-neighbouring or second nearest-neighbouring to the Mg atoms is slightly modified due to the substitution. As lowering temperature, every peak showed a gradual broadening, and below $T < 3.8$ K, these peaks split drastically as indicated on, for example, peak #5 by arrows in Fig. 1(a). Fig. 2 shows the results of specific heat measurements on the present sample. A small peak was observed at $T \cong 3.5$ K and 3.6 K for $H = 0$ T and 5 T, respectively. Therefore, under the field of NMR measurements, T_N was expected to be near 3.6 K which was consistent with the above mentioned temperature $T = 3.8$ K of the beginning of the peak #5 splitting. We plotted the temperature dependence of the split width of peak #5 in Fig. 3. In the impurity-induced magnetic order, the split width continuously increased without a jump below T_N, indicating that it is of the second-order phase transition, while in the field-induced magnetic order [14], the split width shows a finite jump, indicating the involvement of spin-lattice coupling. The hyperfine field in the impurity-induced magnetic order is estimated from the splitting width to be $\Delta\nu/2\gamma \cong 0.07$ T, at the lowest temperature. From a comparison of split widths in field- and impurity-induced magnetic orders [14], the hyperfine field in the impurity-induced magnetic order is a twentieth of that in the field-induced magnetic order. This is consistent with the neutron experiment, which reports that the impurity-induced magnetic moment is much smaller than the full moment of Cu^{2+} [10].

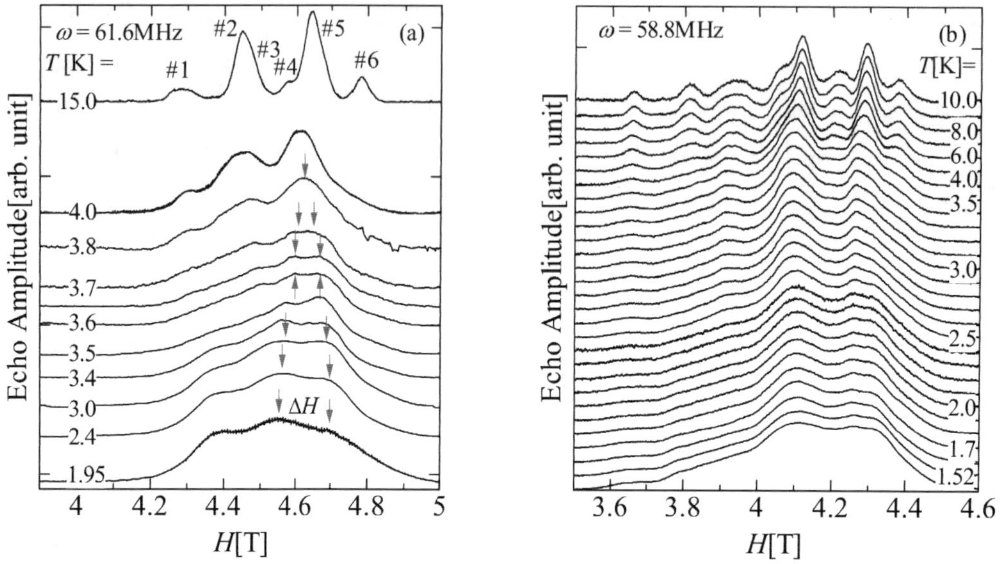

Figure 1. The temperature dependence of NMR spectra. (a) For the sample with $x = 0.03$. (b) For the sample with $x = 0.01$.

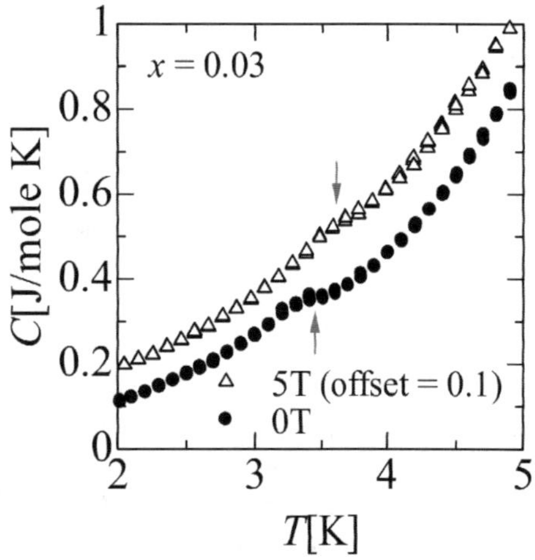

Figure 2. The specific heat measurements for the sample with $x = 0.03$. The Néel temperature is shown by arrows.

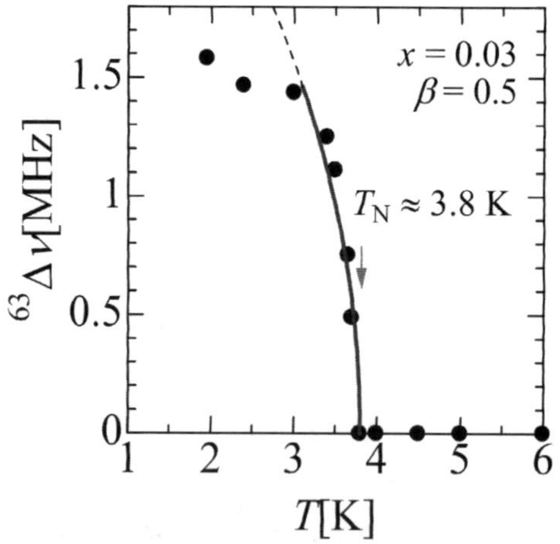

Figure 3. The temperature dependence of the split width of peak #5 under a field 4.6 T. The curve shown is a guide to eyes.

Fig. 1(b) shows the temperature dependence of NMR spectra for the sample $x = 0.01$. Similar to the case of $x = 0.03$, sharp eight peaks were observed at high-temperature region and showed a gradual broadening as lowering temperature. However, peak splitting was not observed down to $T = 1.52$ K, which is lower than the reported $T_N \cong 2.2$ K [12]. Here we insist on that the observed gradual broadening of the peaks starts at around $T = 6$ K, which is much higher than T_N, and hence irrelevant of the magnetic order. This means that the impurity-induced magnetic order is invisible through a probe of NMR for the sample $x = 0.01$, indicating that the ordered moments field in this system might fluctuate dynamically and coherently as suggested in ref. 12. If the fluctuation is faster than the hyperfine field, a splitting should be invisible due to the motional narrowing, although their spatial coherence allows well the Bragg scattering of neutron experiments [10], which have an extremely short time scale.

4. Conclusion

We have presented the results of ^{63,65}Cu-NMR measurements on Mg-doped quantum spin system $TlCu_{1-x}Mg_xCl_3$ with nominal concentration $x = 0.03$ and 0.01. For the sample $x = 0.03$, the peaks of NMR spectra split clearly below $T_N \cong 3.8$ K, demonstrating the existence of the impurity-induced magnetic order. However, for the sample $x = 0.01$, no peak splitting was observed suggesting that the static field in this system coherently fluctuates even in the ordered state [12].

5. Acknowledgement

This work was partly supported by a Grant-in-Aid for Scientific Research on Innovative Areas "New Frontier of Materials Science Opened by Molecular Degrees of Freedom" (No. 20110003) of The Ministry of Education, Culture, Sports, Science, and Technology, Japan.

References

[1] M. Sigrist *et al.* 1996 *J. Phys. Soc. Jpn.* **65** 2385.
[2] M. Imada *et al.* 1997 *J. Phys. Soc. Jpn.* **66** 568.
[3] C. Yasuda *et al.* 2001 *Phys. Rev. B* **64** 092405, and references therein.
[4] H.-J. Mikeska *et al.* 2004 *Phys. Rev. Lett.* **93** 217204.
[5] H. Fukuyama *et al.* 1996 *J. Phys. Soc. Jpn.* **65** 2377.
[6] M. Hase *et al.* 1993 *Phys. Rev. Lett.* **71** 4059.
[7] T. Masuda *et al.* 1998 *Phys. Rev. Lett.* **80** 4566.
[8] M. Azuma *et al.* 1997 *Phys. Rev. B* **55** R8658.
[9] A. Oosawa *et al.* 2002 *Phys. Rev. B* **66** 020405(R).
[10] A. Oosawa *et al.* 2003 *Phys. Rev. B* **67** 184424.
[11] H. Fujiwara *et al.* 2005 *Prog. Theor. Phys. Suppl.* **159** 392.
[12] T. Suzuki *et al.* 2009 *J. Phys. Soc. Jpn.* **78** 074705.
[13] M. Fujisawa *et al.* 2006 *J. Phys. Soc. Jpn.* **75** 033702.
[14] O. Vyaselev *et al.* 2004 *Phys. Rev. Lett.* **92** 207202.

^{63}Cu-NMR study on High-T_c Superconductor $La_{2-x}Ba_xCuO_4$

Y. Shindo[1], T. Goto[1], K. Nakamura[1], T. Takao[1], T. Adachi[2], Y. Koike[2]
[1]Sophia University, 7-1, Kioi-cho, Chiyoda-ku, Tokyo, Japan
[2]Tohoku University, 2-1-1, Katahira, Aoba-ku, Sendai, Miyagi, Japan

E-mail: shindo-y@sophia.ac.jp

Abstract. In order to investigate the local structure in the CuO_2 plane of the La-based high-T_c cuprate $La_{2-x}Ba_xCuO_4$, we have carried out measurements of ^{63}Cu-NMR on single crystals with x=0.08 and 0.10 at low temperatures down to 4 K. In both samples, the local structure deduced from observed NMR spectra agreed with the averaged structure. In high-temperature region around 300K, detailed investigation of NMR line width has shown the existence of the local structure different from the averaged structure; that is, CuO_6 octahedra tilt from the c-axis toward the random direction by approximately 1.4 deg.

1. Introduction

Physical properties of metals are dominated by its crystal structure through two different ways. One is the modification of the band structure by the Bragg scattering of electronic wave in the structure with translational symmetry. The other is the pinning effect, that is, the crystal structure with the specific spatial pattern stabilizes the electronic state with the spatial pattern identical to the crystal structure, which is well known, for example, as the intrinsic pinning effect of superconducting magnetic flux in high T_C cuprates [1]. In La-based high-T_C cuprates, there is another intriguing pinning effect by the crystal structure, the stabilization of the stripe order. The stripe order in high-T_C cuprates is a one dimensional segregation of holes and spins, and strongly suppresses the superconductivity[2]. So far, the stripe order in La-based high-T_C cuprates has been intensively studied by many experimental techniques [2,3,4] and the existence of the strong stabilization effect on the stripe order by the structure of so-called low-temperature tetragonal phase (LTT) has been confirmed [5,6]. Furthermore, it has been reported that the local structure of LTT, which does not have a spatial coherence and hence cannot be detected by scattering techniques, appear in some conditions of temperatures or dopant concentration [7-11]. This incoherent local structure also shows a slight pinning effect on the stripe order if weaker than the LTT structure with the spatial coherence.[8]

Though the important role of the LTT phase both coherent and incoherent in the stabilization of the stripe order is thus recognized, one is still far from the full understanding of the appearance mechanism of the LTT phase itself. Actually, the LTT phase appears in either case when hetero ions with radius larger than La^{3+} such as Ba^{2+} or smaller such as Nd^{3+} are substituted, and its appearance mechanism cannot be simply explained in terms of the epitaxial distortion between the CuO_2 plane and the block layer consisting of the layered structure of La-based high-T_C cuprates[12,13]. As for the local structure of the LTT phase, its existence itself is still in controversy[7-11,14]. The purpose of this paper is devoted to the microscopic detection of the local structure of LTT phase in $La_{2-x}Ba_xCuO_4$ (LBCO) and to find out the appearance condition of the phase.

Before describing the NMR-detection of local structures, we briefly summarize the electronic phase diagram of La-based high-T_C cuprates. First, the parent material La_2CuO_4 is a typical Mott insulator and becomes conductive when provided holes by substitution of La^{3+} with Ba^{2+} or Sr^{2+}. These divalent atoms are much larger than La^{3+} ion and are expected to bring a structural instability into the system [5,12,15]. Particularly, LBCO shows a successive structural phase transition as lowering temperatures; the structure changes from the high-temperature tetragonal phase (HTT, space group *I4/mmm*) to the low-temperature orthorhombic phase (LTO, *Cmca*) at T_{d1} and then, within a limited region of $0.09<x<0.25$, successively to the LTT ($P4_2/ncm$) at T_{d2} [5,16,17].

These three crystal structures are characterized by bucking patterns of the CuO_2 plane. In HTT phase, the plane is flat and CuO_6 octahedra stand upright. In low-temperature phases, the CuO_2 plane buckles along the line of $(110)_T$ in LTO phase, and along $(100)_T$ in LTT phase, respectively in the tetragonal notation. Especially, in the latter, the buckling direction coincides with that of the stripe order and stabilizes it as stated above[2,15,18]. The principle of detecting the LTT structure by Cu-NMR is based on this buckling structure in CuO_2 planes. Since the principal axis of the electric field gradient (EFG) tensor at the Cu site is approximately perpendicular to the basal plane of CuO_6 octahedra, one can obtain the buckling angle directly from the NMR shift, which delicately depends on the angle between the applied field and the principal axis of EFG. NMR is a local probe, and hence is capable of detecting the LTT structure both with and without the spatial coherence. We have previously applied this technique onto $La_{2-x}Sr_xCuO_4$ and related systems to reveal successfully an existence of the incoherent LTT phase at low temperatures [8]. In this article, we report results of NMR measurements on single crystals of LBCO at various temperatures.

2. Experimental

Single crystals of LBCO $x=0.08$, $T_C=24$ K and $x=0.10$, $T_C=28$ K were prepared by the conventional floating zone method [19]. The superconducting critical temperature T_c is defined as the mid point of the resistivity drop. In the sample of x=0.08, the spatially-coherent LTT phase is absent [17]. In x=0.10, the thermodynamic phase transition temperature to LTT phase $T_{d2}\approx 39.2$ K was determined from the small but finite step in the temperature dependence of the magnetic susceptibility [20].

NMR spectra were obtained by plotting the spin echo amplitude against the applied magnetic field. The NMR resonance peak position H_{res} of the central transition $I_z = -1/2 \leftrightarrow +1/2$ is expressed by the angle θ formed by the principal axis of the electric field gradient tensor (EFG) and the magnetic field direction as

$$H_{res} = \frac{\nu}{\gamma(1+K)} - \frac{3\nu_Q^2}{16\nu\gamma(1+K)}\left(1-9\cos^2\theta\right)\left(1-\cos^2\theta\right) \qquad (1)$$

where $K\approx 0$ is the Knight shift, and ν, γ, and ν_Q are the injected radio frequency, the gyromagnetic ratio, and the quadrupole frequency of ^{63}Cu nucleus, respectively. ν_Q was determined to be 32.2 and 32.8 MHz for x=0.08 and 0.10 respectively from the θ-dependence of H_{res}[8,9]. As shown schematically in Fig. 1 (a), the neighboring two Cu nuclei become inequivalent, and spectrum splits into multiple peaks when the buckling

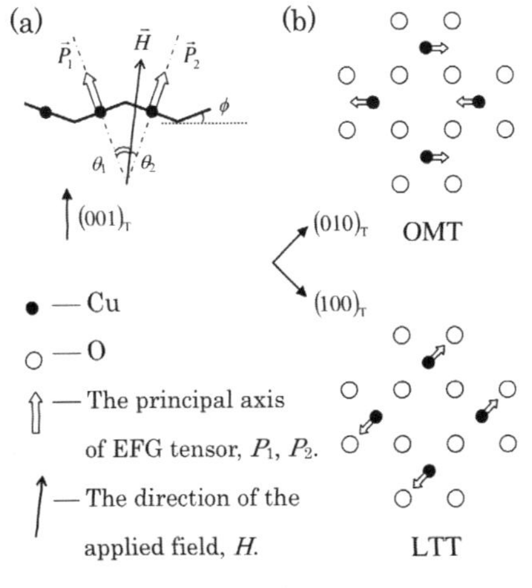

Figure 1. (a) The principle of the NMR detection of the buckling of the CuO_2 plane. (b) Schematic drawings of CuO_2 plane showing the direction of principal axes of EFG tensor from top view.

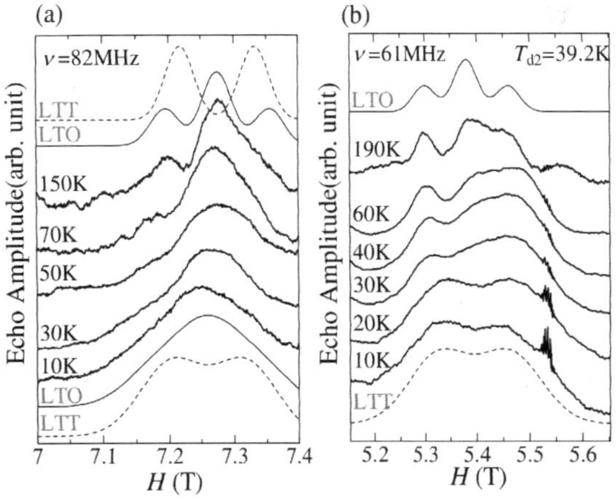

Figure 2. The temperature dependence of the profile of spectra in LBCO (a) x=0.08 and (b) x=0.10. Dashed and solid curves show the calculated spectra by assuming the local structure to be LTO or LTT phase with ϕ of 3.3 and 4 deg respectively. The convoluted Gaussian width was adjusted so that they reproduce the observed spectra.

angle of the CuO_2 plane ϕ is finite and the applied field is tilted from the *c*-axis. The splitting profile depends delicately on the spatial pattern of the buckling shown in Fig. 1 (b). When the field is tilted toward $(110)_T$ direction, the spectrum is expected to split into three peaks in LTO, and two in LTT. This allows one to discriminate the LTO from the LTT irrespective of their spatial coherence. Note that the discrimination still works well even if there exists a twinned structure in the LTO phase[8].

3. Results and discussion

Figure 2 (a) shows ^{63}Cu-NMR spectra in LBCO (x=0.08) measured at various temperatures. We also show for comparison the calculated spectra, which are obtained from the convolution of resonance line positions by Gaussian width. The resonance line positions were calculated by eq. (1) with an assumption of the CuO_6 tilting angle ϕ to be 3.3 deg and 4 deg for LTO and LTT phases, respectively. The width was determined so that the calculation reproduces observed spectra. In the entire temperature region of our measurements, the observed NMR spectra composing a large peak at the center and small peaks at both sides match well with the simulated spectra based on the LTO structure. We insist that the calculated spectrum for the LTT phase consists of two peaks with the same amplitude and never coincides with the observation. The averaged crystal structure of this sample retains to be LTO down to 4.2K [8], so that the present NMR result indicates that there does not exists an incoherent local structure with a buckling pattern different from LTO. In LBCO (x=0.10) with the finite $T_{d2} \approx$ 39.2K, on the contrary, observed NMR spectra changed their profile distinctly from the three-peaked shape (LTO) to the two-peaked shape (LTT) at around its T_{d2} as shown Fig. 2 (b). From these two observations, we can conclude that the local structure is always in accordance with the averaged structure in LBCO. This makes a clear contrast to LSCO, where the existence of the local structure of LTT is reported in samples possessing the averaged structure of LTO [8,9]. In the present case of LBCO, a great disparity between the ionic radii of Ba^{+2} and La^{3+} may stabilize the spatially-coherent LTT structure to make the phase boundary between LTT and LTO distinct.

Next, in order to investigate the local structure in the HTT phase where the averaged structure is the completely flat CuO_2 plane, we investigated the field-direction dependence of the center peak signal of ^{63}Cu of LBCO (x=0.10). The spectral profile at each field direction at room temperature is shown in Fig. 3. Note that both the shift and the width are significantly dependent on the field direction. In order to show the change in width clearly, we compare the two typical spectra at different angles in the inset of Fig. 4. Now, while the θ-dependence of the shift is simply described by eq. (1), the latter dependence is explained in terms of the local tilting of the principal axis of EFG tensor as follows. Generally the width of the eqq-perturbed NMR resonance line, ΔH_{res}, is contributed both by the distribution in the tilting angle $\Delta\theta$ and in the quadrupole frequency $\Delta\nu_Q$ as,

$$\Delta H_{res} = \left|\frac{\partial H_{res}}{\partial \theta}\right| \Delta\theta + \left|\frac{\partial H_{res}}{\partial \nu_Q}\right| \Delta\nu_Q + \text{const.} \qquad (2)$$

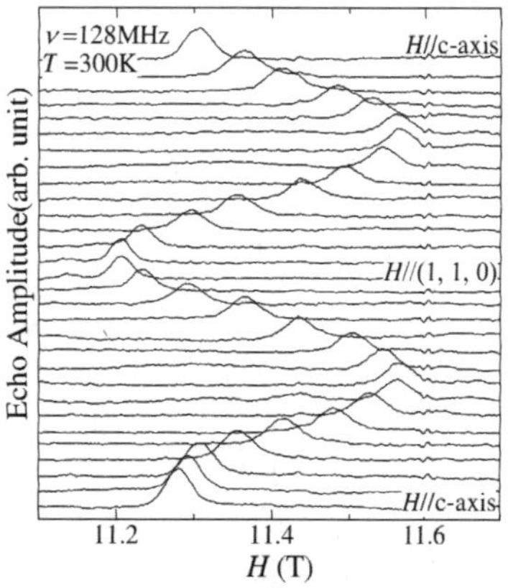

Figure 3. The profile of ^{63}Cu-NMR spectra corresponding to the central transition line at various field direction in the HTT phase of the sample x=0.10.

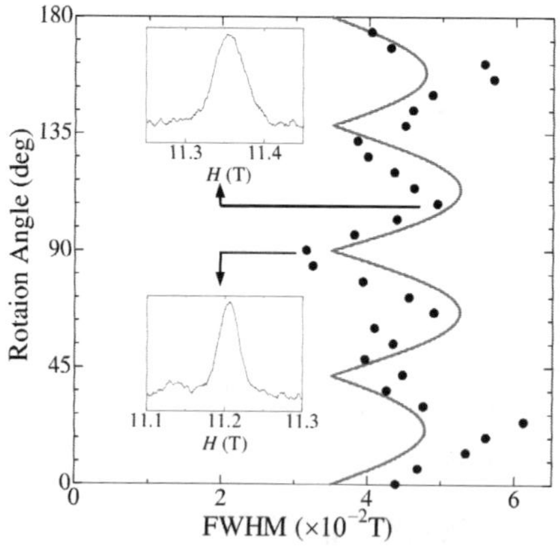

Figure 4. The field-direction dependence of FWHM. The grey curve shows the calculation with $\Delta\theta$=1.4(2) deg and $\Delta\nu_Q$ =0.0(1) MHz. The insets show typical spectra at different angles.

When there is a distribution in either θ or ν_Q, it must contribute to the distribution of ΔH_{res}. The θ-dependence of the two terms in eq. (2), that is, $\partial H_{res}/\partial\theta$ and $\partial H_{res}/\partial\nu_Q$ is quite different, which allows us to determine $\Delta\theta$ and $\Delta\nu_Q$ independently. In Fig. 4, the observed dependence of the width on θ is shown. We fitted eq. (2) to the observed dependence of ΔH_{res} on the field direction to obtain $\Delta\theta$ and $\Delta\nu_Q$ to be 1.4(2) deg and 0.0(2) MHz, respectively.

This result shows that CuO_6 octahedra tilt from the *c*-axis toward the random direction by approximately 1.4 deg even in the HTT phase, where the averaged structure is the completely flat CuO_2 plane. The existence of the local structure in the HTT phase is so far argued by the neutron scattering [7]. However, due to the extremely fast time scale of the neutron experiment one was not able to determine whether the local structure was static or dynamically fluctuating. The result of the present work, which is based on the quasi static probe of NMR spectra, assures that this local structure of the random tilting of octahedra in the HTT phase is static one. This type of static and random local structure has been argued to be one of the triggers for the appearance of LTT phase[13]. We expect that a further investigation on the Ba-concentration dependence of the tilting angle in HTT phase will reveal the critical condition of the appearance of LTT phase.

4. Summary

By NMR measurements on single crystals of LBCO, we have shown that the local structure of LBCO is always in accordance with the averaged structure in LTO and LTT phases, making a clear contrast with the LSCO, where the discrepancy of the two is reported. In HTT phase at room temperature, where the averaged structure of CuO_2 plane is flat, CuO_6 octahedra are found to tilt from the *c*-axis toward the random direction by approximately 1.4 deg.

5. Acknowledgments

This work was partly supported by a Grant-in-Aid for Scientific Research on Innovative Areas "New Frontier of Materials Science Opened by Molecular Degrees of Freedom" (No. 20110003) of The Ministry of Education, Culture, Sports, Science, and Technology, Japan.

References

[1] Tachiki M and Takahashi S, 1989 *Solid State Comm.* **70** 291
[2] Tranquada J M, Sternlieb B J, Axe J D, Nakamura Y, Uchida S 1995 *Nature* **375** 561.
[3] Goto T, Kazama S, Miyagawa K, and Fukase T, 1994 *J. Phys. Soc. Jpn.* **63** 3494.
[4] Nachumi B , Fudamoto Y, Keren A, Kojima K M, Larkin M, Luke G M, Merrin J, Tchernyshyov O, Uemura Y J, Ichikawa N, Goto M, Takagi H, Uchida S, Crawford M K and McCarron E M, MacLaughlin D E, Heffner R H,*et al.* 1998 *Phys. Rev.* B**58** 8760.
[5] Axe J D, Moudden A H, Hohlwein D, Cox D E, Mohanty K M, Moodenbaugh A R, and Xu Y, 1989 *Phys. Rev. Lett.* **62** 2751.
[6] Büchner B, Breuer M, and Freimuth A 1994 *Phys. Rev. Lett.* **73** 1841.
[7] Billinge S J L, Kwei G H and Takagi H 1994 *Phys. Rev Lett.* **72** 2282.
[8] Goto T *et al.* 2008 *J. Phys. Conf. Ser.* **108** 012007.
[9] Tonishi J *et al.* 2006 *J. Phys.: Conf. Ser.* **51** 275
[10] Haskel D *et al.* 1996 Phys. Rev. Lett. **76** 439.
[11] Saini N L, Oyanagi H, Scagnoli V, Ito T, Oka K and Bianconi A, *2003* Europhys. Lett. **63** 125
[12] Suzuki T, Sera M, Hanaguri T, Fukase T, 1994 *Phys. Rev. B***49** 12392.
[13] Maki H, Sera M, Hiroi M, and Kobayashi N, 1996 *Phys Rev **B*** **53** 11324.
[14] Braden M, Meven M, Reichardt W, Pintschovius L, Fernandez-Diaz M T, Heger G, Nakamura F, and Fujita T, 2001 *Phys. Rev.* B**63** 140510(R)
[15] Kimura H, Noda Y, Goka H, Fujita M, Yamada K and Shirane G, 2005 *J. Phys. Soc. Jpn.* **74**, 445.
[16] Moodenbaugh A R, Xu Y, Suenaga M, Folkerts T J and Shelton R N, 1988 *Phys. Rev. B* **38** 4596.
[17] Fukase T, Hanaguri T, Nomoto T, Goto T, Koike Y, Shinohara T, Sato T, Tanaka I and Kojima H, 1992 *JJAP Series* **7**: Mechanisms of Superconductivity 213.
[18] Fujita M, Goka H, Yamada K, Tranquada J M, Regnault L P, 2004 *Phys. Rev.* B **70** 104517.
[19] Adachi T, Noji T, and Koike Y 2001 *Phys. Rev. B* **64**, 144524 .
[20] Adachi T, Kitajima N, Manabe T, Koike Y, Kudo K, Sasaki T and Kobayashi N, 2005 *Phys. Rev. B* **71** 104516 .

Antiferromagnetic resonance modes for the $S = 1/2$ kagome antiferromagnet $Cs_2Cu_3SnF_{12}$

S Sharmin[1], I Umegaki[1], H Tanaka[1], T Ono[1], G. Tanaka[2], H Nojiri[2], M Fujisawa[3], N Matsumi[4], M Tomoo[4], S Okubo[5], H Ohta[5] and T Sakurai[6]

[1]Department of Physics, Tokyo Institute of Technology, Ookayama 2-12-1, Meguro, Tokyo 152-8551, Japan
[2]Institute for Materials Research, Tohoku University, Sendai 980-8577, Japan
[3]Development of Frontier Research and Technology, Meguro-ku, Tokyo 152-8551, Japan
[4]Graduate School of Science, Kobe University, Kobe 657-8501, Japan
[5] Molecular Photoscience Research Center, Kobe University, Kobe 657-8501, Japan
[6]Center for Supports to Research and Education Activities, Kobe University, Kobe 657-8501, Japan

E-mail: sharmin@lee.phys.titech.ac.jp

Abstract. We have investigated the antiferromagnetic resonance modes of the kagome antiferromagnet $Cs_2Cu_3SnF_{12}$ both theoretically and experimentally. High-field electron spin resonance experiments on single crystals were conducted in the ordered phase at different frequencies and temperatures with the external magnetic field parallel to the c axis. Considering three sublattices, we calculated the resonance modes for the motions of the sublattice magnetizations within the framework of the mean field approximation. It was found that the frequency-field diagram is in good agreement with the experimental results.

1. Introduction

A two-dimensional (2-D) spin frustrated system is an interesting subject in condensed matter physics, because of the competition between the frustration of spins and the anisotropy due to the low dimensionality. The basic building block of most geometrically frustrated systems is a set of three spins on a triangle interacting antiferromagnetically. An example of such a system is the kagome lattice which may be seen as a diluted triangular lattice with larger geometrical frustration and lower coordination number than the triangular lattice [1].

Calculating the antiferromagnetic resonance modes (AFMR) provides a microscopic view of the magnetic structure, leading to measurements of magnetic exchange, magnetic anisotropy, spin-flop and critical fields, and spin canting. Also, there is no inversion center at the middle point of two neighboring magnetic ions in the kagome lattice which differs from the situation in the triangular lattice. Thus, in general, the Dzyaloshinskii-Moriya (DM) interaction is allowed in the kagome lattice [3]. AFMR modes should also be able to provide information on the DM interaction. In this work, we present a calculation of the resonance modes of a stoichiometrically pure $S = 1/2$ ideal kagome lattice antiferromagnet $Cs_2Cu_3SnF_{12}$ with full occupancy of spin-1/2 at the Cu sites in the easy-plane anisotropy case and compare the theoretical results with experimental data.

2. Kagome antiferromagnet $Cs_2Cu_3SnF_{12}$

$Cs_2Cu_3SnF_{12}$ is a relatively new antiferromagnetic (AF) kagome structure which possesses quite a number of properties that are not well understood. The spins are 120° apart from each other as expected in the so-called $q = 0$ or $\sqrt{3} \times \sqrt{3}$ spin structures proposed as the possible ground states of a kagome antiferromagnet. The magnetic Cu^{2+} - ions are surrounded by F^--ions octahedrally and CuF_6 octahedra are linked in the ab plane with sharing corners. Since CuF_6 octahedra are elongated along the principal axes approximately parallel to the c axes, the hole orbitals $d(x^2 - y^2)$ of Cu^{2+}-ions spread within the kagome layer. The value of the bond angle α for $Cs_2Cu_3SnF_{12}$ is 138.9° at room temperature [4]. Furthermore, the kagome layers are well separated by non-magnetic Cs^+, Sn^{4+} and F^+ layers. We observe the presence of antiferromagnetic long-range ordering at around 20K and a structural transition $T_t \sim 185$K The g factors determined self-consistently in the temperature range higher than the structural phase-transition temperature are $g_{||} = 2.48$ and $g_\perp = 2.10$ where $g_{||}$ and $g_\perp$ are the g factors for the magnetic field parallel and perpendicular to the c axis. Exchange constant J for $T > T_t$ is 240 K. The spins are aligned parallel to an easy c axis and a weak ferromagnetic moment along the ab-plane has been established in the magnetization measurements [5].

There exists a mirror plane that passes the middle points of neighboring Cu^{2+} ions and is perpendicular to the line connecting these two ions. Therefore the D vector should be parallel to the mirror plane. Because there are twofold screw axes along the [1,0,0], [0,1,0] and [1,1,0] directions, the D vectors change their directions alternately along these directions [5] [see Fig. 1].

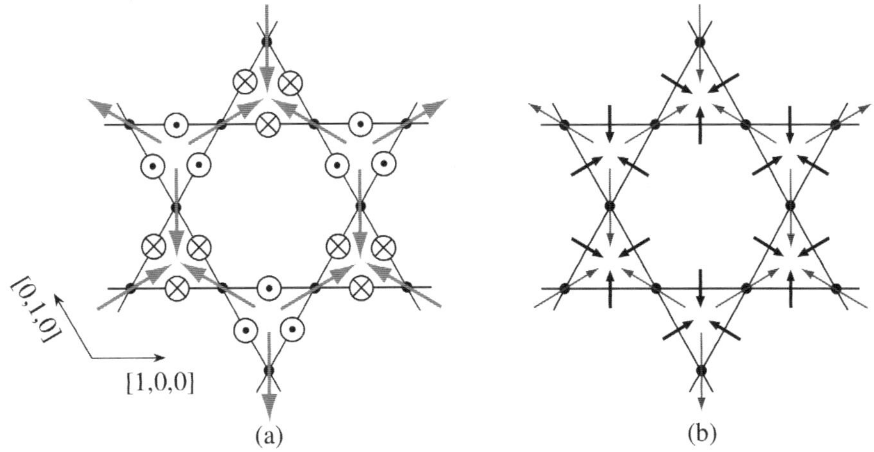

Figure 1. The configuration of D vectors and $q = 0$ spin structure with plus chirality and staggered field directions. Arrangement of the D vectors for $T>T_t$ in $Cs_2Cu_3SnF_{12}$ systems, (a) the c axis component $D^{||}$ and (b) the c plane component $D^{\perp}$. Symbols $\odot$ and $\otimes$ in (a) and short thick arrows in (b) denote the local positive directions of parallel and perpendicular components $D^{||}$ and $D^{\perp}$, respectively. Large arrows in (a) denote the $q = 0$ spin structure stabilized when $D^{||}>0$. Long thin arrows in (b) denote the local direction of staggered field $\boldsymbol{h}_i$

.

3. Antiferromagnetic resonance modes

We derive the resonance modes for $H||c$ within the framework of the mean field approximation. In $Cs_2Cu_3SnF_{12}$, we may consider only three sublattices, which lie in the kagome plane with 120° structure. We assume the $q = 0$ structure with plus chirality as shown in Fig. 1(a).

Journal of Physics: Conference Series **302** (2011) 012011 doi:10.1088/1742-6596/302/1/012011

We consider the Hamiltonian under an external magnetic field H as follows:

$$\mathcal{H} = \sum_{\langle i,j \rangle} J\boldsymbol{S}_i \cdot \boldsymbol{S}_j + \sum_{\langle i,j \rangle} \boldsymbol{D}_{ij} \cdot [\boldsymbol{S}_i \times \boldsymbol{S}_j] - g\mu_{\mathrm{B}} H \sum_i S_i^z - \sum_i g\mu_{\mathrm{B}} \boldsymbol{h}_i \cdot \boldsymbol{S}_i \tag{1}$$

$D^{\perp}$ and $D^{||}$ are respectively the real values of the parallel and perpendicular components of the Dzyloshinskii-Moriya (DM) interaction, J is the exchange constant, $\boldsymbol{h}_i$ the staggered field induced by inclination of the principal axes of the g tensors associated with the low local symmetry of the Cu^{2+} magnetic centers. In this paper, we do not consider exchange interactions between kagome layers.

For a simple three sublattice antiferromagnet, three resonance frequencies should be observed. The resonance conditions of these three observable modes are given by:

$$\frac{\omega_0}{\gamma} = \sqrt{3(d^{\perp})^2 M_0^2 + HM_0 \left\{ \frac{h}{H}(3B + \sqrt{3}d^{||}) + \sqrt{3}d^{\perp} \right\}}\,, \tag{2}$$

$$\frac{\omega_{\pm}}{\gamma} = \sqrt{\frac{3}{2}(B + \sqrt{3}d^{||})M_0 \quad \sqrt{3}d^{||}M_0 + h \quad + \frac{\sqrt{3}}{4}d^{\perp}M_0 \quad \sqrt{3}d^{\perp}M_0 + H \quad + \frac{3(B+\sqrt{3}d^{||})}{4(3B+\sqrt{3}d^{||})}H^2} \pm \frac{3B - \sqrt{3}d^{||}}{2(3B+\sqrt{3}d^{||})}H\,, \tag{3}$$

where M_0 is the magnitude of the sublattice magnetizations, given by $M_0 = (N/3)g\mu_{\mathrm{B}}\langle S \rangle$, where $\langle S \rangle$ is the mean value of spin $\boldsymbol{S}$, h is the magnitude of the staggered field and

$$B = \frac{3}{N}\frac{2J}{(g\mu_{\mathrm{B}})^2}\,, \qquad d^{\perp,||} = \frac{3}{N}\frac{2D^{\perp,||}}{(g\mu_{\mathrm{B}})^2}\,. \tag{4}$$

The ω_0 mode corresponds to the global rotation of spins around the external field H. Thus, its frequency becomes zero in the case of $D^{\perp} = 0$ and $h = 0$. When the staggered field $\boldsymbol{h}_i$ is parallel to the sublattice spin $\boldsymbol{S}_i$, i.e., $h/H > 0$, the frequency of the ω_0 mode increases with increasing H. On the other hand, when $\boldsymbol{h}_i$ and $\boldsymbol{S}_i$ are antiparallel, i.e., $h/H < 0$, the frequency of the ω_0 mode decreases and becomes zero at

$$H_{\mathrm{c}} = -\frac{3(d^{\perp})^2 M_0}{c_{\mathrm{s}}(3B + \sqrt{3}d^{||}) + \sqrt{3}d^{\perp}}\,, \tag{5}$$

where c_{s} is given by $c_{\mathrm{s}} = h/H$. As shown below, $c_{\mathrm{s}} < 0$ in $Cs_2Cu_3SnF_{12}$.

4. Experimental results

High-field ESR experiments were performed on single crystals of $Cs_2Cu_3SnF_{12}$ at different frequencies. Only one antiferromagnetic resonance mode ω_0 was observable below 210 GHz, as shown in Fig. 2. This is to be expected as the frequencies $\omega_{\pm}$ calculated from Equation (3) should be in the terahertz region.

In Figure 2 the field dependence of the frequencies for $H||c$ measured at T = 4.2 and 2 K are shown. The resonance mode can be assigned as the ω_0 mode. The frequency decreases with increasing magnetic field, which indicates that $\boldsymbol{h}_i$ and $\boldsymbol{S}_i$ are antiparallel, $c_s < 0$. The solid lines follow Equation (2) with BM_0 = 178 T. At both temperatures the experimental values fit well to the theoretical equation. The fit gives the values of other interaction parameters: $d^{\perp}M_0$ = 4.93 T and $c_s = -0.057$ for T = 4.2 K, and $d^{\perp}M_0$ = 4.63 T and $c_s = -0.058$ for T = 2.0 K. The values of $d^{\perp}M_0$ evaluated at T = 2.0 and 4.2 K are different. This seems not intrinsic but experimental error. For difference in critical field H_{c}, we have no reasonable explanation at present.

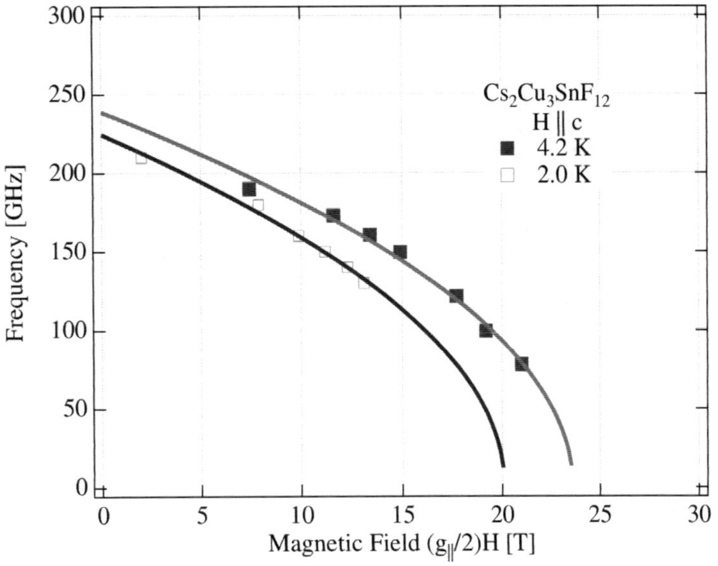

Figure 2. Frequency-field diagram for $H \parallel c$ obtained at $T = 2.0$ and 4.2 K. The solid lines show the fit of Eq. (2) to the experimental data with interaction parameters shown in the text.

5. Conclusion

High-field ESR experiments have been conducted on single crystals of the kagome antiferromagnet $Cs_2Cu_3SnF_{12}$ at different frequencies and at different temperatures. The experimental results were compared with the theoretical data derived from the model proposed using the mean field approximation to calculate the resonant modes in kagome antiferromagnets consisting of three sublattices with easy-plane anisotropy. It was found that the experimental and theoretical data are in good agreement. Values of the magnitudes of the Dzyaloshinskii-Moriya interaction and the staggered field were evaluated.

Acknowledgments

This work was supported by the Global COE Program "Nanoscience and Quantum Physics" at TIT funded by the Ministry of Education, Culture, Sports, Science and Technology of Japan.

6. References

[1] Jiang H C, Weng Z Y and Sheng D N 2008 *Phys. Rev. Lett.* **101** 117203
[2] Kimel A V, Kirilyuk A, Tsvetkov A, Pisarev R V and Rasing Th 2004 *Nature* **429** 850
[3] Moriya T 1960 *Phys. Rev.* **120** 91
[4] Ono T, Morita K, Yano M, Tanaka H, Fujii K, Uekusa H, Narumi Y and Kindo K 2009 J. Phys.: Conf. Ser. **145** 012005
[5] Ono T, Morita K, Yano M, Tanaka H, Fujii K, Uekusa H, Narumi Y and Kindo K 2009 *Phys. Rev.* B **79** 174407

^{19}F-NMR Study on Antiferromagnetic Heisenberg Chain $KCuGaF_6$

Y. Mori[1], T. Goto[1], I. Umegaki[2], H. Tanaka[2]

[1]Department of Physics, Sophia University, Kioi-cho, Chiyoda-ku, Tokyo 102-8554, Japan
[2]Department of Physics, Tokyo Institute of Technology, Oh-okayama, Meguro-ku, Tokyo 152-8551, Japan

E-mail: yasuhi-m@sophia.ac.jp

Abstract. S = 1/2 one-dimensional antiferromagnet $KCuGaF_6$, which has the alternation of both the g-tensor and Dyzaloshinsky-Moriya (DM) vector along the chain, is theoretically expected to show the field-induced spingap. We have measured the longitudinal nuclear spin relaxation rate T_1^{-1} of ^{19}F-NMR and found that it shows the thermal activation type temperature dependence, indicating the existing of the spingap. The field-dependence of the gap is found to be proportional to $H^{2/3}$. These results are consistent with the theory of Affleck and Oshikawa.

1. Introduction

The magnetic ground state of the S = 1/2 antiferromagnetic Heisenberg chain (AFHC) is in general gapless [1]. However, the field-induced gap has recently been observed in some S = 1/2 AFHC systems such as Cu $(C_6H_5COO_2)_2{\cdot}3H_2O$ [2], PM·Cu $(NO_3)_2{\cdot}$ $(H_2O)_2$ (PM = pyrimidine) [3]. The gap $\Delta(H)$ was found to be

$$\Delta(H) = AH^{\alpha}, \tag{1}$$

where the coefficient A depends on the field direction, and α is approximately equal to 2/3. Affleck and Oshikawa investigated theoretically S = 1/2 AFHC, in which both g tensor and Dzyaloshinsky-Moriya (DM) vector alternate along the chain, described as the following model [4]

$$\mathcal{H} = \sum_i [J\vec{S}_i \cdot \vec{S}_{i+1} - g\mu_B HS_i^z + (-1)^i g\mu_B hS_i^x], \tag{2}$$

where $h = c_s H$ is the effective staggered field induced by applied magnetic field, and J is the exchange interaction. The constant c_s depending on the field direction is determined experimentally from the magnetic susceptibility as [6]

$$\chi \cong \chi_u + c_s^2 \chi_s, \tag{3}$$

where χ_u is the uniform susceptibility of the intrinsic part [5] and χ_s is the staggered susceptibility induced by the magnetic field. The former per a spin is written as

$$\chi_u = \frac{g^2 \mu_B^2}{k_B T} F\left(\frac{J}{k_B T}\right), \tag{4}$$

and $F(x)$ is an empirical function proposed by Feyerherm *et al.* [3]

$$F(x) = \frac{1 + 0.08516x + 0.2335x^2}{1 + 0.73382x + 0.1369x^2 + 0.53568x^3}. \tag{5}$$

For $T > \Delta$, χ_s per a spin is given [6] by

$$\chi_s = 0.2779 \cdot \frac{g^2 \mu_B^2}{k_B T} \sqrt{\ln\left(\frac{J}{k_B T}\right)}. \tag{6}$$

So far, Umegaki and Morisaki *et al.* have intensively studied the $S = 1/2$ AFHC system $KCuGaF_6$, which is described by the model (2), to find that the specific heat shows the thermal-activation-type temperature dependence and that the magnetic susceptibility obeys the Curie law below 20 K. These results suggest that the gap and staggered magnetization are induced by the external magnetic field [7] , indicating that their results are consistent with the theory of Oshikawa and Affleck [4], [6], and also with [8], [9] lately developed.

In this paper, we show microscopically by utilizing the NMR technique that $KCuGaF_6$ exhibits both the field-induced gap and the field induced staggered field.

2. Experimental

$KCuGaF_6$ has a monoclinic crystal structure with the space group $P2_1/c$, in which Cu^{2+} and Ga^{3+} ions form a pyrochlore lattice [10]. The single crystal of $KCuGaF_6$ with the size of about 500 mm^3 grown by I. U. was used for present NMR measurements. The temperature dependence of the longitudinal nuclear spin relaxation rate of ^{19}F-NMR under various fields was measured by the saturation-recovery method with a pulse train. The temperature and the field regions of measurements were around 1.8 – 20 K and 2 – 5 T, respectively. The temperature dependence of the uniform magnetization is measured under the field approximately the same as NMR measurements.

3. Results and Discussion

Figure 1 shows typical profiles of ^{19}F-NMR spectra at various temperatures. We observed six NMR resonance lines that correspond to the crystallographically-inequivalent six fluorine sites. The hyperfine coupling constant for each site was obtained by comparing the NMR shift and the uniform magnetization as shown in Fig. 2. The result is listed in Table 1. The variety of both the magnitude and the sign of A comes from the geometrical location of each F site around Cu.

We then analyzed the temperature dependence of the resonance line shift, using the following equation [4]:

$$\nu / {}^{19}\gamma - H = AH(\chi_u + c_s^2 \chi_s), \tag{7}$$

where ν and $^{19}\gamma$ are the resonance frequency and the gyromagnetic ratio of ^{19}F. The temperature dependence of the Knight shift obeys the Curie-like behavior as shown in Fig. 3. With parameters reported by I. U., as $J/k_B = 103$ K, $g = 2.12$, and $c_s = 0.18$ [11], we can reproduce the observed temperature dependence only by modifying the value of A to 40 % larger than the one determined above. We also note that it deviates from the Curie law and tends to saturate below 4 K. This is simply due to the magnetization saturation in high field, and hence does not affect the existence of the Curie term.

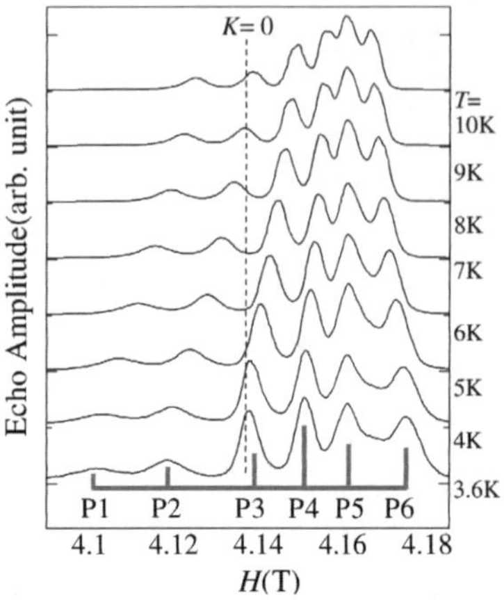

Figure 1. The temperature dependence of ^{19}F-NMR spectra for *c // H*.

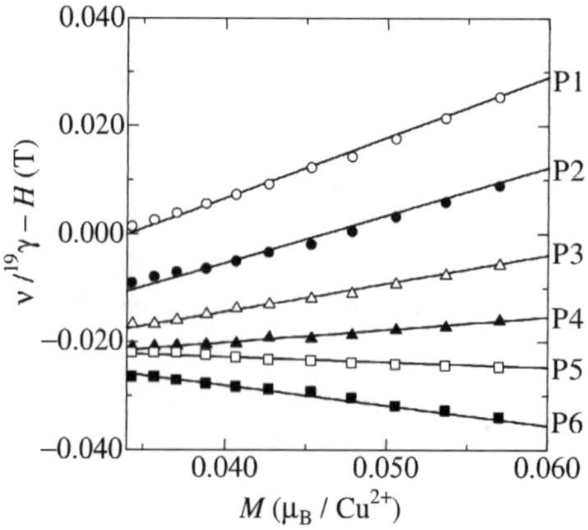

Figure 2. The plot of the resonance line shift vs. the macroscopic magnetic magnetization with the temperature as the implicit parameter. A typical temperature region is between 4 – 16 K.

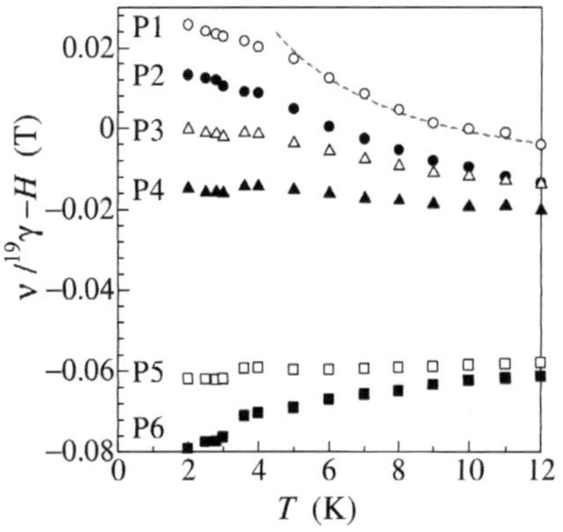

Figure 3. The temperature dependence of the resonance line shift of each peak in ^{19}F-NMR spectrum. The dashed curve shows the Curie-Weiss temperature dependence of eq. (7).

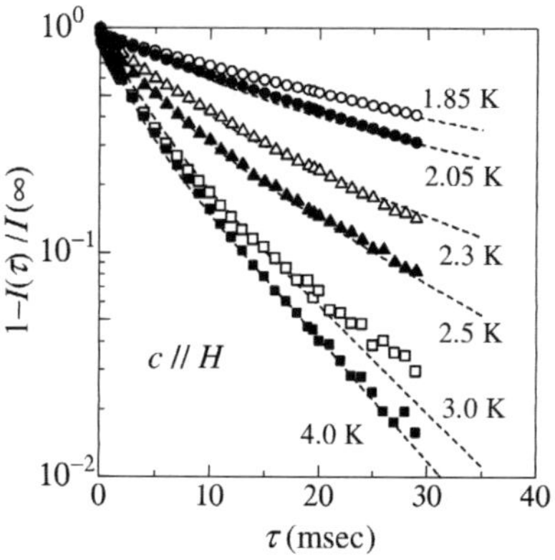

Figure 4. Typical longitudinal nuclear spin relaxation curves for *c // H*. Dashed curves are the fitting function.

Table 1. Hyper fine coupling constants *A* for each site.

Peak	P1	P2	P3	P4	P5	P6
A (T/μ_B)	1.12	0.88	0.53	0.23	−0.098	−0.37

Next, T_1 was measured at the peak of P3, that is, at H = 4.91 T. Figure 4 shows the typical relaxation curves; they are described with the two components short and long, T_{1S} and T_{1L} respectively. The fraction of these two components was found to be temperature independent and approximately unity. Figure 5 shows the temperature dependence of log T_1^{-1}'s for the two components. The values of gap Δ were estimated from gradients of log T_1^{-1} to be approximately 8.7 K, for both components measured at 4.91 T. We also investigated the field dependence of the gap in the region 2 and 5 T as shown in Fig. 6. The gap increased with increasing field *H*, and obeyed a power law $H^{2/3}$, which agrees with the ESR measurement reported by I. U. [11].

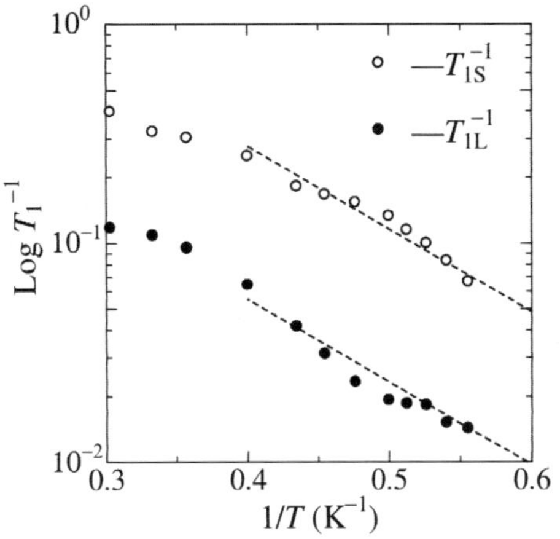

Figure 5. Temperature dependence of T_1^{-1} measured at $H = 4.91$ T. The dashed lines show the thermal activation type function, from which the gap was estimated to be $\Delta =$ 8.7 K.

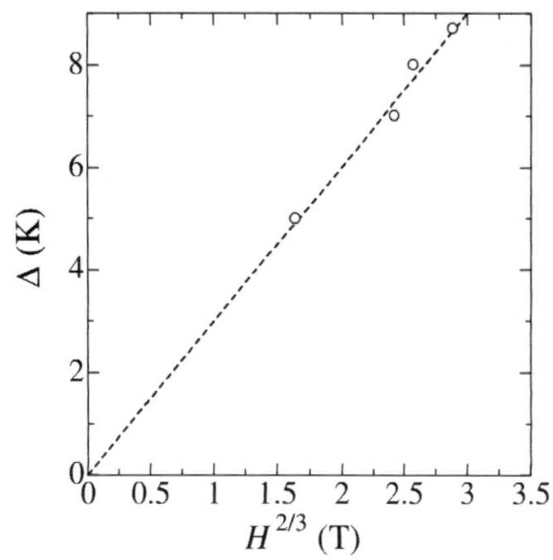

Figure 6. Field dependence of the gap measured in the region 2 and 5 T for *c // H*. The dashed line denotes the function AH^{α} with $\alpha = 2/3$ and $A = 3$ ($K/T^{2/3}$).

4. Conclusion

By utilizing ^{19}F-NMR technique as a microscopic probe, the existence of the field-induced gap in the $S = 1/2$ AFHC $KCuGaF_6$ was confirmed. The field dependence of the gap was found to be proportional to $H^{2/3}$, which is consistent with the theory of Affleck and Oshikawa.

Acknowledgments

This work was partly supported by a Grant-in-Aid for Scientific Research on Innovative Areas "New Frontier of Materials Science Opened by Molecular Degrees of Freedom" (No. 20110003) of The Ministry of Education, Culture, Sports, Science, and Technology, Japan.

References

[1] des Cloizeaux J and Pearson J J 1962 *Phys. Rev.* **128** 2131.
[2] Dender D C, Hammar P R, Reich Daniel H, Briholm C, and Aepplil G 1997 *Phys. Rev. Lett.* **79** 1750.
[3] Feyerherm R, Abens S, Günther D, Ishida T, Meißner M, Meschke M, Nogami T, and Steiner M 2000 *J. Phys. Condens. Matter* **12** 8495.
[4] Oshikawa M and Affleck I 1997 *Phys. Rev. Lett.* **79** 2883.
[5] Eggert S, Affleck I, and Takahashi M 1994 *Phys. Rev. Lett.* **73** 332.
[6] Affleck I and Oshikawa M 1999 *Phys. Rev. B* **60** 1038.
[7] Morisaki R, Ono T, Tanaka H, and Uekusa H 2006 *J. Phys. Conf. Ser.* **51** 17.
[8] Hikihara T, and Furusaki A 2004 *Phys. Rev. B* **69** 064427.
[9] Takayoshi S, and Sato M 2010 *Phys. Rev. B* **82** 214420.
[10] Dahlke P, Pebler J, and Babel D 2005 *Z. Anorg. Allg. Chem.* **631** 115.
[11] Umegaki I, Ono T, Uekusa H, and Nojiri H 2009 *Phys. Rev. B* **79** 184401.

Exotic Quantum Critical Phenomena of the Spin Nanotubes

Tôru Sakai[A,B,C], Masahiro Sato[D], Kouichi Okunishi[E], Kiyomi Okamoto[F] and Chigak Itoi[G]

[A]Japan Atomic Energy Agency, SPring-8, Sayo, Hyogo 679-5148, Japan
[B]Graduate School of Material Science, University of Hyogo, Kamigori, Hyogo 678-1297, Japan
[C]TRIP JST, Saitama 332-0012, Japan
[D]Department of Physics and Mathematics, Aoyama Gakuin University, Kanagawa 229-8558, Japan
[E]Department of Physics, Niigata University, Niigata 950-2181, Japan
[F]Department of Physics, Tokyo Institute of Technology, Tokyo 152-8551, Japan
[G]Department of Physics, Nihon University, Tokyo 101-8308, Japan

E-mail: sakai@spring8.or.jp

Abstract. Recently some quantum spin systems on tube lattices, so called spin nanotubes, have been synthesized. They are expected to be interesting low-dimensional systems like the carbon nanotubes. As the first step of theoretical study on the spin nanotube, we investigate the S=1/2 three-leg spin tube, which is the simplest one, using the numerical exact diagonalization (ED), combined with the recently developed level spectroscopy analysis. Particularly, we introduce the lattice distortion from the regular triangle unit to the isosceles one. We revealed that the S=1/2 isosceles triangle spin tube exhibits the quantum phase transition between the spin gapped and gapless spin liquid phases.

1. Introduction

The quantum spin nanotube [1] are one of interesting objects in the nanoscience and the nanotechnology. Among them the three-leg spin tubes have been extensively studied both theoretically and experimentally in recent years, because they have a strong spin frustration at each unit cell. The density matrix renormalization group (DMRG) analysis [2] indicated a large spin gap in the strong rung coupling limit, in contrast to the three-leg spin ladder which is gapless. After the synthesization of the $S = 1/2$ three-leg spin tube $[(CuCl_2tachH)_3Cl]Cl_2$ [3], several theoretical works on the $S = 1/2$ three-leg spin tube have been published [4–10]. In our previous study [4, 9, 10], the asymmetric rung interaction was introduced and the quantum phase transition from the gapped to the gapless phases with respect to the lattice distortion was investigated. Although the phase boundary was clearly determined by the DMRG method for strong rung coupling region, it is still ambiguous for small one because the DMRG calculation is not converged well. In this paper, using the level spectroscopy analysis which is one of the most precise method to the phase boundary of the Berezinskii-Kosterlitz-Thouless (BKT) transition, we present a phase diagram including the week rung coupling region.

2. Model

We consider the $S = \frac{1}{2}$ asymmetric three-leg spin tube, shown in Fig. 1, described by the Hamiltonian

$$\hat{H} = J_1 \sum_{i=1}^{3}\sum_{j=1}^{L} \vec{S}_{i,j} \cdot \vec{S}_{i,j+1} + J_{\rm r} \sum_{i=1}^{2}\sum_{j=1}^{L} \vec{S}_{i,j} \cdot \vec{S}_{i+1,j} + J_{\rm r}' \sum_{j=1}^{L} \vec{S}_{3,j} \cdot \vec{S}_{1,j}, \qquad (1)$$

where $\vec{S}_{i,j}$ is the spin-$\frac{1}{2}$ operator and L is the length of the tube in the leg direction. The exchange interaction constant J_1 is for the neighboring spin pairs along the legs, while $J_{\rm r}$ and J_r' are the rung interaction constants. All the exchange interactions are supposed to be antiferromagnetic (namely, positive). The ratio $\alpha = J_r'/J_{\rm r}$ stands for the degree of the asymmetry of the rung interactions. We will vary α and J_1 to investigate the quantum phase transitions. Throughout this paper, we fix $J_{\rm r}$ to one.

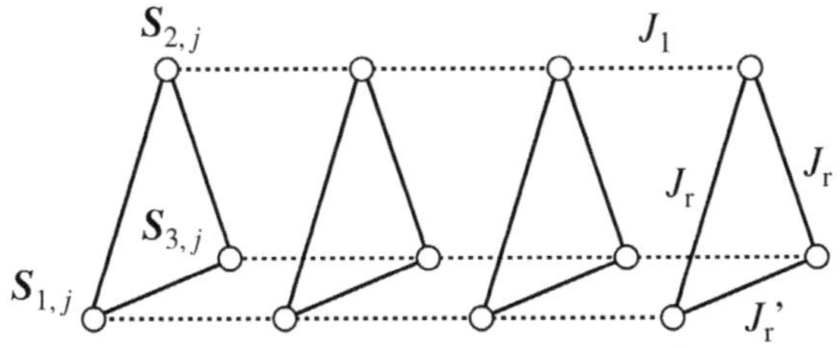

Figure 1. $S = 1/2$ three-leg quantum spin tube.

The present model includes three typical models as limiting cases; (a) $\alpha = 0$: the three-leg spin ladder, (b) $\alpha = 1$: the symmetric spin tube, and (c) $\alpha \to \infty$: the single chain plus rung dimers. Since the system is gapless in the cases (a) and (c), while gapful in the case (b), at least two quantum phase transitions should occur with increasing α from 0 to infinity. As we already mentioned, the one-site translational symmetry along the leg ($\vec{S}_{i,j} \to \vec{S}_{i,j+1}$) is spontaneously broken in the symmetric spin tube at least in the strong-rung-coupling regime. [2, 8] On the other hand, the weak J_r limit is also gapless, because the system consists of three independent chains. This gapless phase is denoted as the phase I. We also denote the gapless phase around (a), the gapful one around (b) and the gapless one around (c) as the phases II, III, and IV, respectively.

3. Level spectroscopy

The level-spectroscopy method [11–14] is a very powerful tool to determine the critical point of the BKT transition in one-dimensional quantum systems. For the SU(2)-symmetric cases including the present spin tube, its strategy becomes easier [11]. According to this method, the critical point can be determined as an intersection between the singlet and the triplet excitation gaps, where their logarithmic finite-size corrections vanish.

The phase boundaries II-III and III-IV estimated by this method are shown in Fig. 2. Here we have applied the results of the numerical diagonalization up to $L = 10$ for $J_1 = 0.1$. The numerical data are well converged to those of the thermodynamic limit by use of the $1/L^2$ extrapolation. The $1/L^2$ extrapolation is justified by considering the most important finite-size correction to the excitation gaps Δ next to the logarithmic term [11].

On the other hand, when we consider the transition between two phases III and I, we search for the level cross value of $1/J_1$ with fixed α. In Fig. 3, the size-dependent level cross points are plotted versus $1/L^2$ for α =0.98 and 1.02.

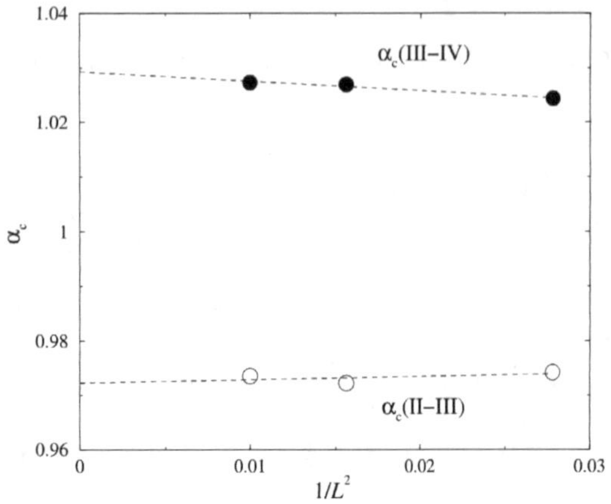

Figure 2. Level cross values of α for L =6, 8 and 10 with fixed J_1 to 0.1.

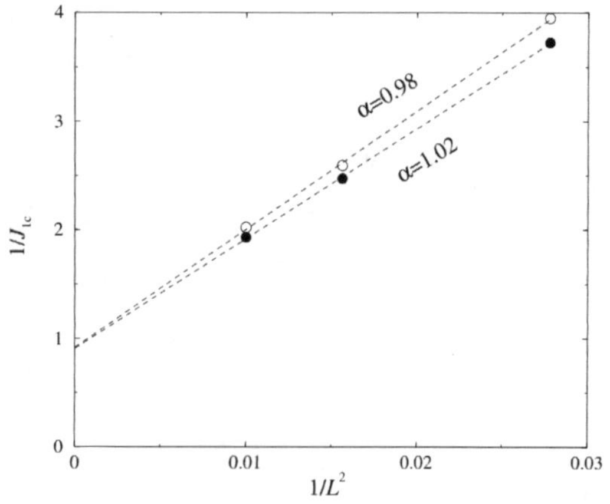

Figure 3. Level cross values of $1/J_1$ for L =6, 8 and 10 with fixed α to 0.98 and 1.02. They correspond to the boundary between the I and III phases.

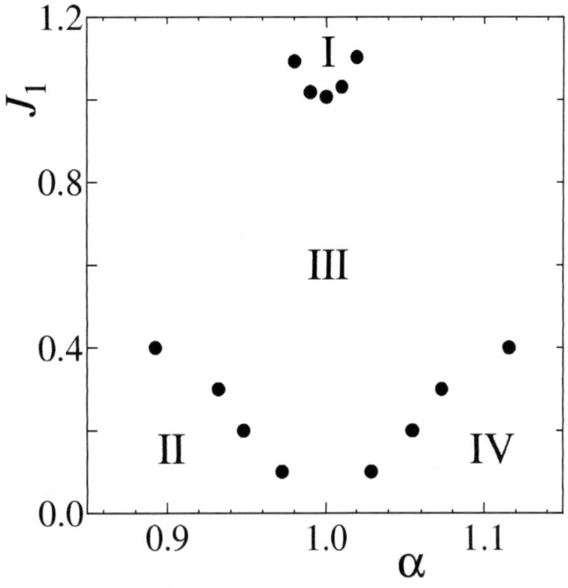

Figure 4. Phase diagram on the J_1-α plane by the level spectroscopy method.

4. Phase diagram

All the estimated phase boundaries in the thermodynamic limit are plotted in Fig. 4. The phase diagram justifies that the gapless phase (I) lies in a finite region.

The phase I in the present study is larger than the phenomenological renormalization up to L = 10. It is consistent with the fact that the phenomenological renormalization always underestimates gapless phases for the boundaries among the phases II, III and IV, as shown in our previous work [9]. The present level spectroscopy study qualitatively justifies the phase diagram obtained by the effective Hubbard Hamiltonian approach in the work [9].

The present system also exhibits the 1/3 magnetization plateau for sufficiently small J_1. This plateau was disucussed in detail in [1] and the J_1-α phase diagram determined by numerical methods was presented in our recent paper [15].

5. Summary

The $S = 1/2$ three-leg spin tube with the asymmetric interaction is investigated using the level spectroscopy method combined with the numerical diagonalization of finite-size clusters up to $L = 10$. The present result indicates that the gapless phase I lies in a finite area continued to the infinite J_1 limit. It is consistent with our previous phenomenological renormalization study and effective theory [9].

Acknowledgments

We wish to thank Prof. I. Affleck for fruitful discussions. This work was partly supported by Grants-in-Aid (No. 20340096 and No. 22014012) for Scientific Research and Priority Areas "Novel States of Matter Induced by Frustration" from the Ministry of Education, Culture, Sports, Science and Technology of Japan. Non-hybrid thread-parallel calculations in numerical diagonalizations were based on the TITPACK ver.2 coded by H. Nishimori. A part of the computations was performed using facilities of the Information Technology Center, Nagoya University, Department of Simulation Science, National Institute for Fusion Science, and the Supercomputer Center, Institute for Solid State Physics, University of Tokyo.

References

[1] See, for example, Sakai T, Sato M, Okunishi K, Okamoto K and Itoi C 2010 *J. Phys.: Condens. Matter* **22** 403201
[2] Kawano K and Takahashi M 1997 *J. Phys. Soc. Jpn.* **66** 4001
[3] Schnack J, Nojiri H, Kögerler P, Cooper G J T and Cronin L 2004 *Phys. Rev. B* **70** 174420
[4] Sakai T, Matsumoto M, Okunishi K, Okamoto K and Sato M 2005 *Physica E* **29** 633
[5] Fouet J -B, Läuchli A, Pilgram S, Noack R M and Mila F 2006 *Phys. Rev. B* **73** 014409
[6] Okunishi K, Yoshikawa S, Sakai T and Miyashita S 2005 *Prog. Theor. Phys. Suppl.* **159** 297
[7] Sato M and Sakai T 2007 *Phys. Rev. B* **75** 014411
[8] Nishimoto S and Arikawa M 2008 *Phys. Rev. B* **78** 054421
[9] Sakai T, Sato M, Okunishi K, Otsuka Y, Okamoto K and Itoi C 2008 *Phys. Rev. B* **78** 184415
[10] Sakai T, Okunishi K, Okamoto K, Itoi C and Sato M 2010 *J. Low. Tenp. Phys.* **159** 55
[11] Okamoto K and Nomura K 1992 *Phys. Lett.* **169** 433.
[12] Nomura K and Okamoto K 1994 *J. Phys. A* **27** 5773
[13] Okamoto K 2002 *Prog. Theor. Phys, Suppl.* No. **145** 113
[14] Nomura K and Kitazawa A 2002 in *Proc. French-Japanese Symp. on Quantum Properties of Low-Dimensional Antiferromagnets* ed Ajiro Y and Bouhcer J -P (Kyushuu University Press); cond-mat/020172
[15] Okamoto K, Sato M, Okunishi K, Sakai T and Itoi C 2011 *Physica E* **43** 769

Ground-state phase diagram of S = 2 quantum spin chain with the XXZ and on-site anisotropies

Kiyomi Okamoto[1], Takashi Tonegawa[2], Hiroki Nakano[3], Tôru Sakai[3,4,5], Kiyohide Nomura[6] and Makoto Kaburagi[2]

[1]Department of Physics, Tokyo Institute of Technology, Meguro-ku, Tokyo 152-8551, Japan
[2]Professor Emeritus, Kobe University, Kobe 657-8501, Japan
[3]Graduate School of Material Science, University of Hyogo, Hyogo 678-1297, Japan
[4]Japan Atomic Energy Agency, SPring-8, Hyogo 679-5148, Japan
[5]Transformative Research-Project on Iron Pnictide, JST, Saitama 332-0012, Japan
[6]Department of Physics, Kyushu University, Fukuoka 812-8581, Japan

E-mail: kokamoto@phys.titech.ac.jp

Abstract. By use of mainly numerical methods, we investigate the ground-state phase diagram of an $S = 2$ quantum spin chain with the XXZ and on-site anisotropies described by $\mathcal{H} = \sum_j (S_j^x S_{j+1}^x + S_j^y S_{j+1}^y + \Delta S_j^z S_{j+1}^z) + D \sum_j (S_j^z)^2$, where Δ denotes the XXZ anisotropy parameter of the nearest-neighbor interactions and D the on-site anisotropy parameter. In the phase diagram with $\Delta > 0$ and $D > 0$ there appear the XY phase, the Haldane/large-D (LD) phase, the intermediate-D (ID) phase, and the Néel phase. Among them the ID phase was proposed by Oshikawa in 1992 and has been considered to be absent since the DMRG studies in the latter half of 1990's. And also, the Haldane state and the LD state belong to the same phase, which should be named Haldane/LD phase.

1. Introduction

In this paper, using mainly numerical diagonalization method, we [1,2] investigate the ground-state phase diagram of the $S = 2$ quantum spin chain described by the Hamiltonian

$$\mathcal{H} = \sum_j (S_j^x S_{j+1}^x + S_j^y S_{j+1}^y + \Delta S_j^z S_{j+1}^z) + D \sum_j (S_j^z)^2 , \qquad S = 2 \tag{1}$$

where Δ and D are, respectively, the XXZ anisotropy parameter of the nearest-neighbor interactions and the on-site anisotropy parameter. We restrict ourselves to the $\Delta > 0$ and $D > 0$ case in this paper.

Schulz [3] first discussed the ground-state phase diagram of this model and concluded that it was essentially the same as that of $S = 1$ case. Namely, it consists of four phases, the Néel phase, the XY phase, the Haldane phase and the large-D (LD) phase. The Haldane state and the LD state are sketched in Fig.1(a) and (c), respectively. On the other hand, for $S \geq 2$ integer quantum spin cases, Oshikawa [4] predicted the existence of $S-1$ intermediate-D (ID) phases in 1992. The valence bond picture of the $S = 2$ ID state is shown in Fig.1(b). After that, however, by use of the density-matrix renormalization-group (DMRG) calculation, Schollwöck et al. [5,6] and Aschauer and Schollwöck [7] concluded the absence of the ID phase. By use of the level spectroscopy (LS) analysis of numerical results of exact-diagonalization calculations, Nomura

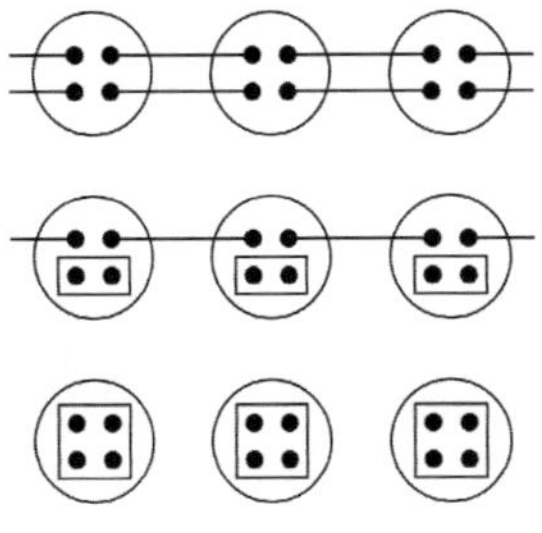

Figure 1. Valence bond pictures of (a) the Haldane state, (b) the ID state and (c) the LD state. Big circles denote $S = 2$ spins and dots $S = 1/2$ spins. Two $S = 1/2$ spins connected by solid lines are in the $(S_{\rm tot}, S^z_{\rm tot}) = (0, 0)$ state. Two $S = 1/2$ spins in rectangles are in the $(S_{\rm tot}, S^z_{\rm tot}) = (1, 0)$ state and similarly four $S = 1/2$ spins in squares are in the $(S_{\rm tot}, S^z_{\rm tot}) = (2, 0)$ state.

and Kitazawa [8] investigated the $\Delta = 1$ case. According to their results, with the increase of D from zero, the transition from the Haldane state to the XY state occurs at $D_{\rm c1} = 0.043$, and that from the XY state to the LD state at $D_{\rm c2} = 2.39$. Then the ID phase does not exist at least in the $\Delta = 1$ case. Thus, since these works, it has been considered for a long time that the ID phase does not exist in the phase diagram of the present $S = 2$ model.

As is well known, the DMRG is a very powerful numerical method for spin chains, especially when the magnitude of the spin gap itself is targeted. However, we have difficulties in discussing the phase transition point because it is determined by the extrapolated values of the finite-size being equal to zero. Namely, for instance, if the extrapolated value of the gap is 0.0001 times of the typical energy unit, we cannot decide whether the system is gapped or gapless. Thus, it is somewhat hard to obtain accurate results if the magnitude of the gap is very small (i.e. the correlation length is very long). This zero-or-finite problem is highly serious for the Berezinskii-Kosterlitz-Thouless transition [9–12] which occurs at the boundaries between the gapless XY phase and the gapped phases (Haldane, ID and LD). This difficult problem is not special to the DMRG method but may be common to almost all numerical methods. On the other hand, the LS method is conspicuous on this point because the phase transition point can determined from the crossing point of two related excitations, which is free from the zero-or-finite problem.

The LS method [8, 13] is firmly based on the effective Hamiltonian, renormalization group method and conformal field theory. We have given a physical and intuitive explanation how to distinguish the Haldane, ID and LD states in the LS method in [2].

2. Ground state phase diagram of an $S = 2$ chain

In the following, $E_0(N, M; {\rm pbc})$ and $E_1(N, M; {\rm pbc})$ are, respectively, the lowest and second-lowest energy eigenvalues of the Hamiltonian (1) under the periodic boundary condition (PBC) within the subspace characterized by N (number of spins) and $M \equiv \sum_j S^z_j$. We also denote by $E_0(N, M, P; {\rm tbc})$ the lowest energy eigenvalue of the Hamiltonian (1) under the twisted boundary condition (TBC), within the subspace characterized by N, M and P. In the twisted boundary condition the coupling between the N-th spin and the 1st spin is $-S^x_N S^x_1 - S^y_N S^y_1 + S^z_N S^z_1$. The quantity $P(= 1$ or $-1)$ is the eigenvalue of the space inversion operator $1 \Leftrightarrow N,\ 2 \Leftrightarrow N-1, \cdots$.

Following the LS method [8, 13], we should compare three energies, $E_0(N, 0, +1; {\rm tbc})$, $E_0(N, 0, -1; {\rm tbc})$ and $E_0(N, 0; {\rm pbc})$, Namely, the ground state is one of the Haldane/LD, ID and XY states, according as the first, second or third energy is the lowest among them. We note that the Haldane state and the LD state belong to the same phase named Haldane-LD phase, as discussed later. Examples of the size dependence of the critical values of Δ are summarized in Table 1, from which we see that the finite-size effects are not so serious in our LS analyses. The $N \to \infty$ extrapolations were done by using the data with $N = 8, 10, 12$, assuming that their N-dependences are quadratic functions of N^{-2}. Finally, the phase transition between the Néel and Haldane/LD states is of the 2D Ising type since the Z_2 symmetry is broken in the Néel state while it is not broken in the Haldane-LD state. In this case the phenomenological renormalization group (PRG) method [14] is a useful one to determine the phase boundary line.

We have numerically solved the PRG equation,

$$(N-2)\{E_1(N-2,0;\mathrm{pbc})-E_0(N-2,0;\mathrm{pbc})\}=N\{E_1(N,0;\mathrm{pbc})-E_0(N,0;\mathrm{pbc})\} \quad (2)$$

to obtain the solution $\Delta_c^{(\mathrm{H,N})}(N)$ for a given value of D.

Our final phase diagram is shown in Figs. 2 and 3. The remarkable features of the phase diagram are: (1) there exists the ID phase which was predicted by Oshikawa in 1992 and has been believed to be absent for a long time; (2) the Haldane state and the LD state belong to the same phase. The latter feature is consistent with the discussion by Pollmann et al. [15].

N	$\Delta_c^{(\mathrm{H,ID})}(N)$	$\Delta_c^{(\mathrm{ID,LD})}(N)$	$\Delta_c^{(XY,\mathrm{LD})}(N)$
6	2.17687	2.14971	2.08931
8	2.22262	2.16106	2.12246
10	2.23529	2.16527	2.13607
12	2.23793	2.16639	2.14265
∞	2.241 ± 0.001	2.167 ± 0.001	2.156 ± 0.001

Table 1. Examples of critical values of Δ in the case of $D = 1.82$ obtained by the LS method

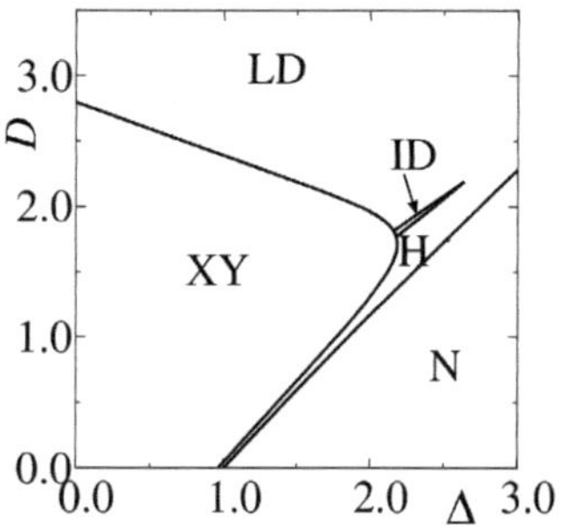

Figure 2. Phase diagram of an $S = 2$ chain. H and N are the abbreviations of Haldane and Néel, respectively.

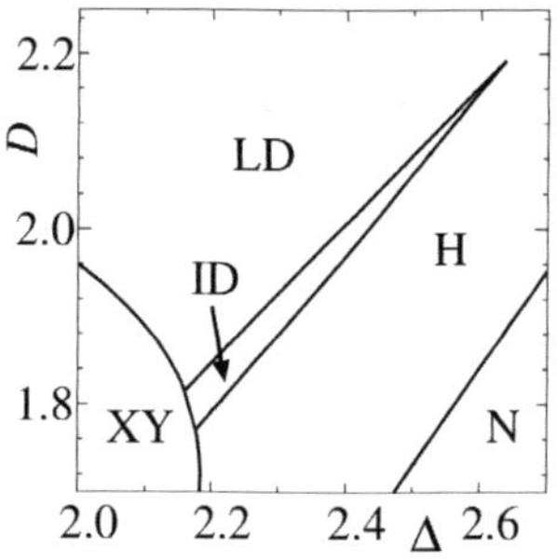

Figure 3. Enlarged phase diagram near the ID phase.

3. Discussion

Here we discuss why the ID phase was not found by previous DMRG studies. As is already stated, it is very difficult to find the phase with very small gap by usual method other than the LS method. This difficulty was mentioned also by Ashauer and Schollwöck [7] (see Fig.7 of their paper). The apparent boundaries between the $XY1$ phase and the gapped phases determined by the DMRG shift into the gapped region, because the system with $\xi > N$ behaves as if it is gapless, where ξ is the correlation length. In fact, for the case of $\Delta = 1$, the XY-LD transition point was estimated as $D_{c2} \sim 3.0$ by the DMRG [6], which is rather larger than Nomura-Kitazawa value $D_{c2} = 2.39$ [8] obtained by the LS method. As can be seen from the phase diagram, the ID phase is very narrow, and thus the whole region of the ID phase was completely masked by the "apparent XY region" in the DMRG calculation. We have tried to detect the edge spin which is expected in the ID phase with the open boundary chain, but in vain. We have also calculated the central charge c from the finite-size correction of the ground-state energy under the PBC [16–18] up to $N = 12$ system

$$E_0(N,0;\mathrm{pbc})/N \simeq \lim_{N\to\infty} E_0(N,0;\mathrm{pbc})/N - \pi v_s c/6N^2 \quad (3)$$

where v_s is the spin wave velocity. The central charge c is defined only for the gapless region, and it becomes zero in the gapped region if the above formula is formally applied. Along the line $D = 1.90$, where the ground-state changes as $XY1 \Rightarrow$ LD $\Rightarrow$ ID $\Rightarrow$ Haldane $\Rightarrow$ Néel with the increase of Δ from zero, we found the values of "apparent c" are $c \simeq 1$ for $0 < \Delta < 2.7$ and $c \simeq 0$ for $\Delta > 2.7$. This result shows that our system behaves as if it is in the gapless XY state when $\Delta < 2.7$ along the $D = 1.90$ line. Thus we have successfully explained why the ID phase was not detected by the DMRG in 1990's. The central charge c can be obtained also from the finite-size correction of the ground-state energy under the boundary condition with arbitrary twist angle [19], which will be discussed in future publications.

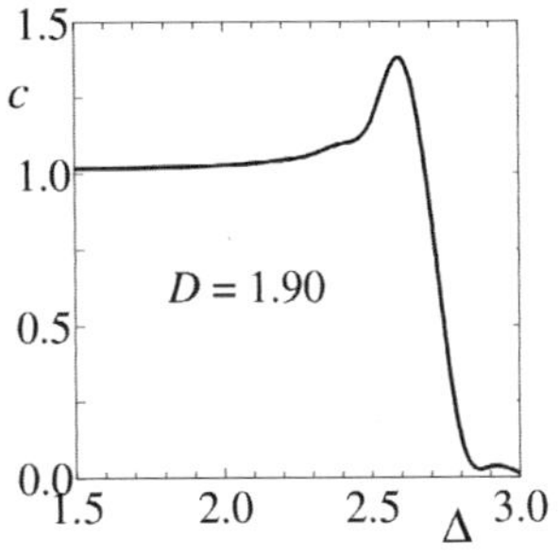

Figure 4. Behavior of the central charge c as a function of Δ along the $D = 1.90$ line.

Acknowledgments
We would like to express our appreciation to Professor Masaki Oshikawa and Dr. Frank Pollmann for their invaluable discussions and comments. We also thank the Supercomputer Center, Institute for Solid State Physics, University of Tokyo, the Computer Room, Yukawa Institute for Theoretical Physics, Kyoto University and the Cyberscience Center, Tohoku University for computational facilities. The present work has been supported in part by a Grant-in-Aid for Scientific Research (B) (No. 20340096), and a Grant-in-Aid for Scientific Researches on Priority Areas "Novel states of matter induced by frustration" from the Ministry of Education, Culture, Sports, Science and Technology.

References
[1] Tonegawa T, Okamoto K, Nakano H, Sakai T, Nomura K and Kaburagi M 2011 *J. Phys. Soc. Jpn* **80** 043001
[2] Okamoto K, Tonegawa T, Nakano H, Sakai T, Nomura K and Kaburagi M 2011 arXiv:1101.2799
[3] Schulz H J 1986 *Phys. Rev. B* **34** 6372
[4] Oshikawa M 1992 *J. Phys.: Cond. Matter* **4** 7469
[5] Schollwöck U and Jolicœur Th 1995 *Europhys. Lett.* **30** 493
[6] Schollwöck U, Golinelli O and Jolicœur Th 1996 *Phys. Rev. B* **54** 4038
[7] Aschauer H and Schollwöck U 1998 *Phys. Rev. B* **58** 359
[8] Nomura K and Kitazawa A 1998 *J. Phys. A: Math. Gen.* **31** 7341
[9] Berezikskii Z L 1972 *Zh. Eksp. Teor. Fiz.* **61** 1144; 1972 *Sov. Phys.-JETP* **34** 610.
[10] Kosteralitz J M and Thouless D J 1973 *J. Phys. C* **6** 1181
[11] Kosterlitz J M 1974, *J. Phys. C* **7** 1046
[12] Kogut J B 1979 *Rev. Mod. Phys* **51** 659
[13] Kitazawa A 1997 *J. Phys. A: Math. Gen.* **30** L285
[14] Nightingale M P 1976 *Physica A* **83** 561
[15] Pollmann F, Berg E, Turner A M and Oshikawa M 2009 arXiv:0909.4059
[16] Cardy J L 1984 *J. Phys. A: Math. Gen* **17** L385
[17] Blöte H W, Cardy J L and Nichtingale M P 1986 *Phys. Rev. Lett.* **56** 742
[18] Affleck I 1986 *Phys. Rev. Lett.* **56** 746
[19] Destri C and De Vega H J 1989 *Phys. Lett. B* **223** 365

Charged impurity scattering in graphenes: Effects of environmental screening, band gap, and AA stacking

Tsuneya Ando

Department of Physics, Tokyo Institute of Technology
2–12–1 Ookayama, Meguro-ku, Tokyo 152-8551, Japan

E-mail: `ando@phys.titech.ac.jp`

Abstract. The conductivity is calculated for dominant charged-impurity scattering in graphenes within a self-consistent Born approximation. In monolayer graphene, environmental dielectric screening is shown to change their effective potential from short to long range. In bilayer graphene, effects of band-gap opening due to asymmetry between two layers caused by perpendicular electric field and effects of AA stacking are considered.

1. Introduction

Recently, atomically thin graphenes consisting of a few layers of monolayer graphite sheet were experimentally fabricated using mechanical exfoliation [1, 2] and epitaxial growth [3, 4]. Several reviews have already been published on their intriguing electronic properties [5, 6, 7, 8]. Transport properties of graphenes were often studied in a self-consistent Born approximation assuming model short-range scatterers because of simplicity [9, 10, 11, 12, 13, 14, 15]. Calculations were extended to the case of long-range scatterers in monolayer graphene [16] and in bilayer graphene [17]. The purpose of this paper is to generalize this scheme to study effects of charged-impurity scattering in the presence of environmental dielectric material for monolayer graphene, effects of asymmetry induced band-gap in bilayer graphene, and properties of bilayer graphene with AA stacking.

2. Environmental effect in monolayer graphene

In graphene, a unit cell contains two carbon atoms, which are denoted by A and B. Two bands having approximately linear dispersion cross the Fermi level (chosen at $\varepsilon=0$) at K and K' points of the first Brillouin zone. In an effective-mass or $\mathbf{k}\cdot\mathbf{p}$ scheme, the electron motion near the K point is described by

$$\gamma(\vec{\sigma}\cdot\hat{\mathbf{k}})\mathbf{F}(\mathbf{r}) = \varepsilon\mathbf{F}(\mathbf{r}), \tag{1}$$

where $\vec{\sigma} = (\sigma_x, \sigma_y)$ are the Pauli spin matrices, $\hat{\mathbf{k}} = -i\vec{\nabla}$, and $\mathbf{F}(\mathbf{r})$ is a two-component wave function consisting of F_A and F_B describing slowly-varying parts of the amplitude at A and B sites, respectively. This is exactly the same as Weyl's equation for a neutrino, except that the velocity, given by $v = \gamma/\hbar$, is much smaller than light velocity c ($v \approx c/300$). For the K' point, we should replace $\vec{\sigma}$ with its complex conjugate $\vec{\sigma}^*$. This equation of motion is quite useful for the description of characteristic features of electronic states in graphene.

Journal of Physics: Conference Series **302** (2011) 012015   doi:10.1088/1742-6596/302/1/012015

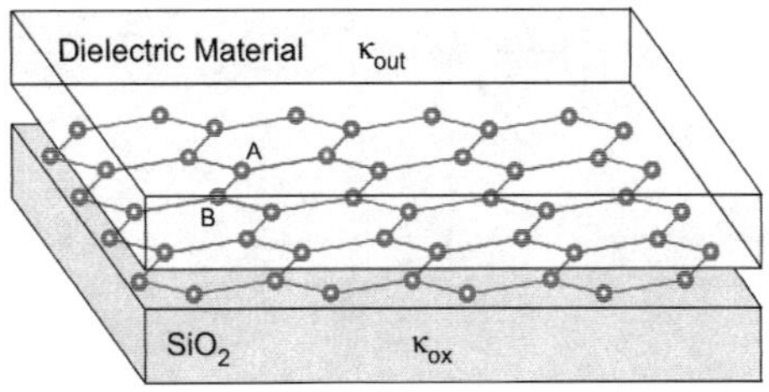

Figure 1. Schematic illustration of graphene on SiO_2 substrate covered by dielectric material on top.

The energy dispersion is given by $\varepsilon_{\pm}(\mathbf{k}) = \pm\gamma|\mathbf{k}|$ and the density of states becomes $D(\varepsilon) = g_v g_s|\varepsilon|/(2\pi\gamma^2)$, where $g_s = 2$ is the spin degeneracy and $g_v = 2$ is the valley degeneracy corresponding to the K and K' points. The density of states vanishes at $\varepsilon = 0$ and therefore graphene is often called a zero-gap semiconductor. As will later become clear, however, this is quite inappropriate.

We consider graphene on SiO_2 covered by dielectric material with dielectric constant $\kappa_{\rm out}$ on the top side as schematically illustrated in Fig. 1. When screening effect is included within a Thomas-Fermi approximation, the potential of charged impurities in the vicinity of graphene is

$$v_{\rm i}(\mathbf{q}) = \frac{2\pi e^2}{\kappa(q+q_{\rm s})}, \tag{2}$$

where $q_{\rm s}$ is the Thomas-Fermi screening constant given by $q_{\rm s} = (2\pi e^2/\kappa)D(\varepsilon_{\rm F})$ and κ is the effective dielectric constant given by

$$\kappa = \frac{1}{2}(\kappa_{\rm out} + \kappa_{\rm ox}), \tag{3}$$

with $\kappa_{\rm ox} = 3.9$ for SiO_2. Then, the calculation of the density of states and the conductivity is performed by a straightforward extension of the previous work for $\kappa_{\rm out} = 1$ [16].

Figure 2 shows some examples of (a) calculated density of states and (b) conductivity for $1 \leq \kappa_{\rm out} \leq 50$. Here, $\varepsilon_{\rm c}$ is the cutoff energy of the order of $\gamma_0 \sim 3$ eV, where $-\gamma_0$ is the hopping integral in the simplest tight-binding model, and $n_{\rm c} = \varepsilon_{\rm c}^2/(4\pi\gamma^2)$ is the concentration roughly corresponding to an impurity per several unit cells. With the increase of $\kappa_{\rm out}$, the density of states in the vicinity of the Dirac point exhibits clear decrease, indicating the reduction of disorder. The conductivity does not increase so much, presumably due to the fact that the decrease of potential by dielectric screening is partly canceled by the reduction in the screening by electrons in graphene.

Figure 2 (c) shows the minimum conductivity at the Dirac point as a function of $\kappa_{\rm out}$. It increases gradually with $\kappa_{\rm out}$. The reason is that the effective potential tends to become of long range due to less screening by electrons in graphene. In fact, previous calculations for scatterers with Gaussian potential show that the minimum conductivity increases with the range except in the clean limit where the scattering strength is sufficiently small [16]. In fact, states at the Dirac point have higher k components because of strong forward scattering caused by long-range scatterers. These higher k states are weakly scattered in backward direction and therefore tend to have large contribution to the conductivity. For short-range scatterers, higher k components do not contribute much to the conductivity and the minimum conductivity remains small.

Figure 2 (c) also shows ratio of transport relaxation time $\tau_{\rm tr}$ determining the conductivity and self-energy relaxation time τ corresponding to the inverse of the total scattering probability,

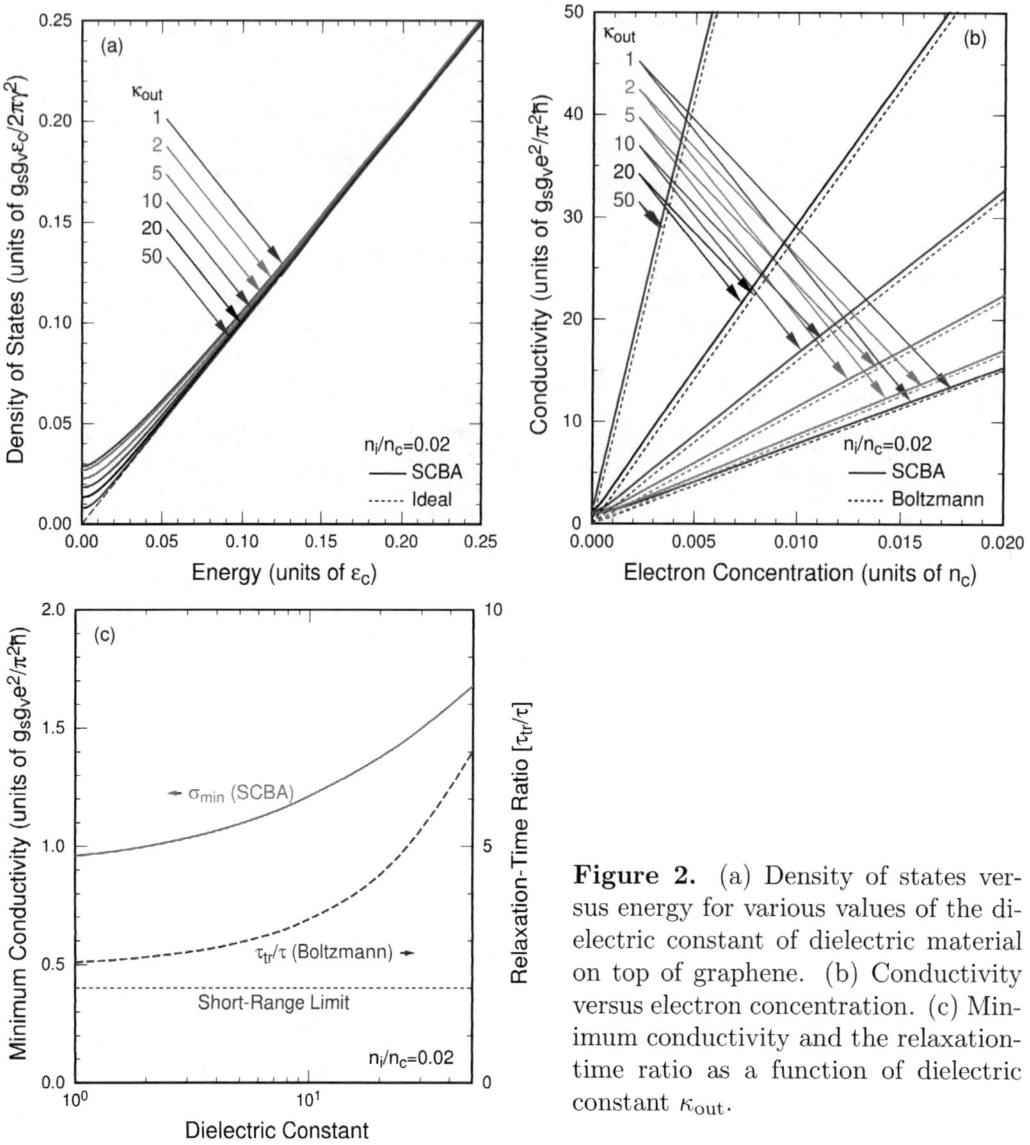

Figure 2. (a) Density of states versus energy for various values of the dielectric constant of dielectric material on top of graphene. (b) Conductivity versus electron concentration. (c) Minimum conductivity and the relaxation-time ratio as a function of dielectric constant κ_{out}.

calculated using the Boltzmann transport equation [18]. For short-range scatterers we have $\tau_{tr}/\tau = 2$ because of the absence of backscattering [9]. For $\kappa_{out} = 1$ without environmental dielectric material, we have $2 < \tau_{tr}/\tau < 3$ not so much different from that for short-range scatterers. This shows that charged impurities should be regarded as short-range scatterers for $\kappa_{out} = 1$ in contrast to the claim otherwise [19]. For sufficiently large κ_{ox}, for which τ_{tr}/τ is much larger than two, charged impurities tend to have the nature of long-range scatterers. Experiments under different dielectric environments were reported [20, 21, 22], but results vary among them and therefore seem to remain inconclusive yet.

3. Band-gap opening in bilayer graphene

We consider a bilayer graphene which is arranged in the AB (Bernal) stacking as shown in Fig. 3. The upper layer is denoted as 1 and the lower layer denoted as 2. In each layer, the unit cell contains two carbon atoms denoted by A_1 and B_1 in layer 1 and A_2 and B_2 in layer 2. For the

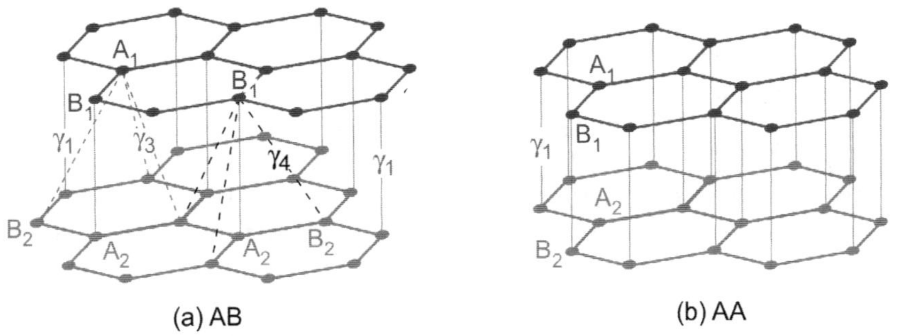

Figure 3. Lattice structure of (a) AB stacked and (b) AA stacked graphenes. The nearest-neighbor interlayer hopping integral denoted by γ_1 determines essential features of the band structure. The next-nearest-neighbor hopping integrals in AB graphene denoted by γ_3 and γ_4 are neglected because of simplicity.

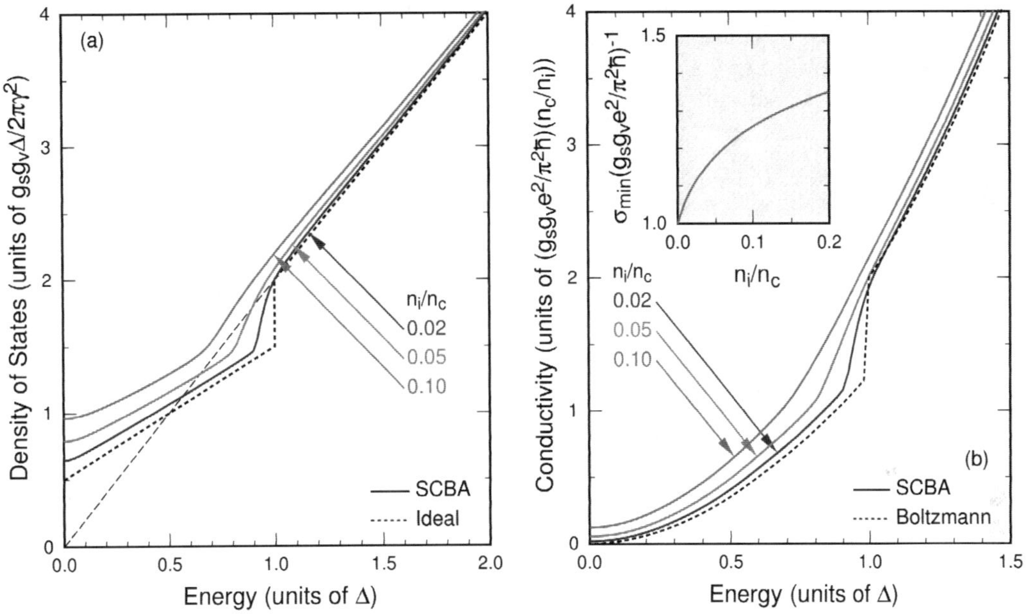

Figure 4. Calculated (a) density of states and (b) conductivity versus the Fermi energy for charged scatterers screened in the Thomas-Fermi approximation. The inset in (b) shows the minimum conductivity versus the impurity concentration.

inter-layer coupling, we include only the coupling between vertically neighboring atoms. Then, electronic states are described by the **k**·**p** Hamiltonian [23, 12]:

$$\mathcal{H} = \begin{array}{c} \begin{array}{cccc} \mathrm{A}_1 & \mathrm{B}_1 & \mathrm{A}_2 & \mathrm{B}_2 \end{array} \\ \begin{pmatrix} +eFd/2 & \gamma\hat{k}_- & 0 & 0 \\ \gamma\hat{k}_+ & +eFd/2 & \Delta & 0 \\ 0 & \Delta & -eFd/2 & \gamma\hat{k}_- \\ 0 & 0 & \gamma\hat{k}_+ & -eFd/2 \end{pmatrix} \end{array}, \tag{4}$$

where Δ ($= \gamma_1 \approx 0.4$ eV) represents the inter-layer coupling between sites B_1 and A_2 and eFd represents the potential difference between layers 1 and 2 (F is the effective electric field and

$d = 0.334$ nm is the interlayer distance). This potential difference should be determined by a self-consistent calculation [24, 25, 26]. Very roughly, F becomes reduced by about factor two from field externally applied [26].

Figure 4 shows some examples of calculated (a) density of states and (b) conductivity for a symmetric bilayer with $\Delta = 0$. The density of states in the low energy region becomes larger with the concentration of scatterers. Further, the excited conduction band has a long tail in the density of states, its amount being roughly proportional to the concentration. The minimum conductivity at zero energy becomes slightly larger than the universal value for short-range scatterers [12] and increases with the impurity concentration. However, it approximately remains universal because the dependence on $n_{\rm i}$ is weak as shown in the inset of Fig. 4 (b). When the energy exceeds the bottom of the excited conduction band, the Boltzmann conductivity exhibits a discrete jump. This jump is also present in the conductivity in the self-consistent Born approximation, although it is considerably smoothed out. This increase when the excited band is occupied comes from the sudden increase in the screening effect.

Figure 5 shows some examples for an asymmetric bilayer with nonzero gap. The density of states does not easily vanish and has low energy tail. This is partly due to the self-consistency between the screening constant and the density of states. In fact, smaller density of states causes less screening and causes effectively large scattering strength giving rise to long tail in the density of states. In spite of the considerable density of states, the conductivity within the gap region is much smaller than that outside, in particular for large gap $eFd/\Delta = 1$.

4. AA stacked bilayer graphene

The structure of an AA stacked bilayer graphene is shown in Fig. 3 (b). When we consider the strongest nearest-neighbor interlayer interaction, the Hamiltonian becomes

$$\hat{\mathcal{H}}_0 = \begin{array}{c} \begin{array}{cccc} \mathrm{A}_1 & \mathrm{B}_1 & \mathrm{A}_2 & \mathrm{B}_2 \end{array} \\ \begin{pmatrix} +eFd/2 & \gamma\hat{k}_- & \Delta & 0 \\ \gamma\hat{k}_+ & +eFd/2 & 0 & \Delta \\ \Delta & 0 & -eFd/2 & \gamma\hat{k}_- \\ 0 & \Delta & \gamma\hat{k}_+ & -eFd/2 \end{pmatrix} \end{array}. \tag{5}$$

Let us define

$$\Delta = \tilde{\Delta}\sin\psi, \quad \frac{eFd}{2} = \tilde{\Delta}\cos\psi, \quad \tilde{\Delta} = \sqrt{\Delta^2 + \Big(\frac{eFd}{2}\Big)^2}, \tag{6}$$

and

$$V = \begin{pmatrix} \cos(\psi/2) & 0 & -\sin(\psi/2) & 0 \\ 0 & \cos(\psi/2) & 0 & -\sin(\psi/2) \\ \sin(\psi/2) & 0 & +\cos(\psi/2) & 0 \\ 0 & \sin(\psi/2) & 0 & +\cos(\psi/2) \end{pmatrix}. \tag{7}$$

Then, we have

$$V^{-1}\hat{\mathcal{H}}_0 V = \begin{pmatrix} \tilde{\Delta} & \gamma\hat{k}_- & 0 & 0 \\ \gamma\hat{k}_+ & \tilde{\Delta} & 0 & 0 \\ 0 & 0 & -\tilde{\Delta} & \gamma\hat{k}_- \\ 0 & 0 & \gamma\hat{k}_+ & -\tilde{\Delta} \end{pmatrix}. \tag{8}$$

This immediately shows that the band structure consists of two monolayer-like bands relatively shifted in energy by amount $\pm\tilde{\Delta}$. Further, scatterers with potential same in the top and bottom layers cannot cause scattering between the two bands, i.e., the conductivity is the sum of those of two monolayer bands. When the scattering potential is different between the top and bottom layers, interband scattering appears and can modify the conductivity. Possible

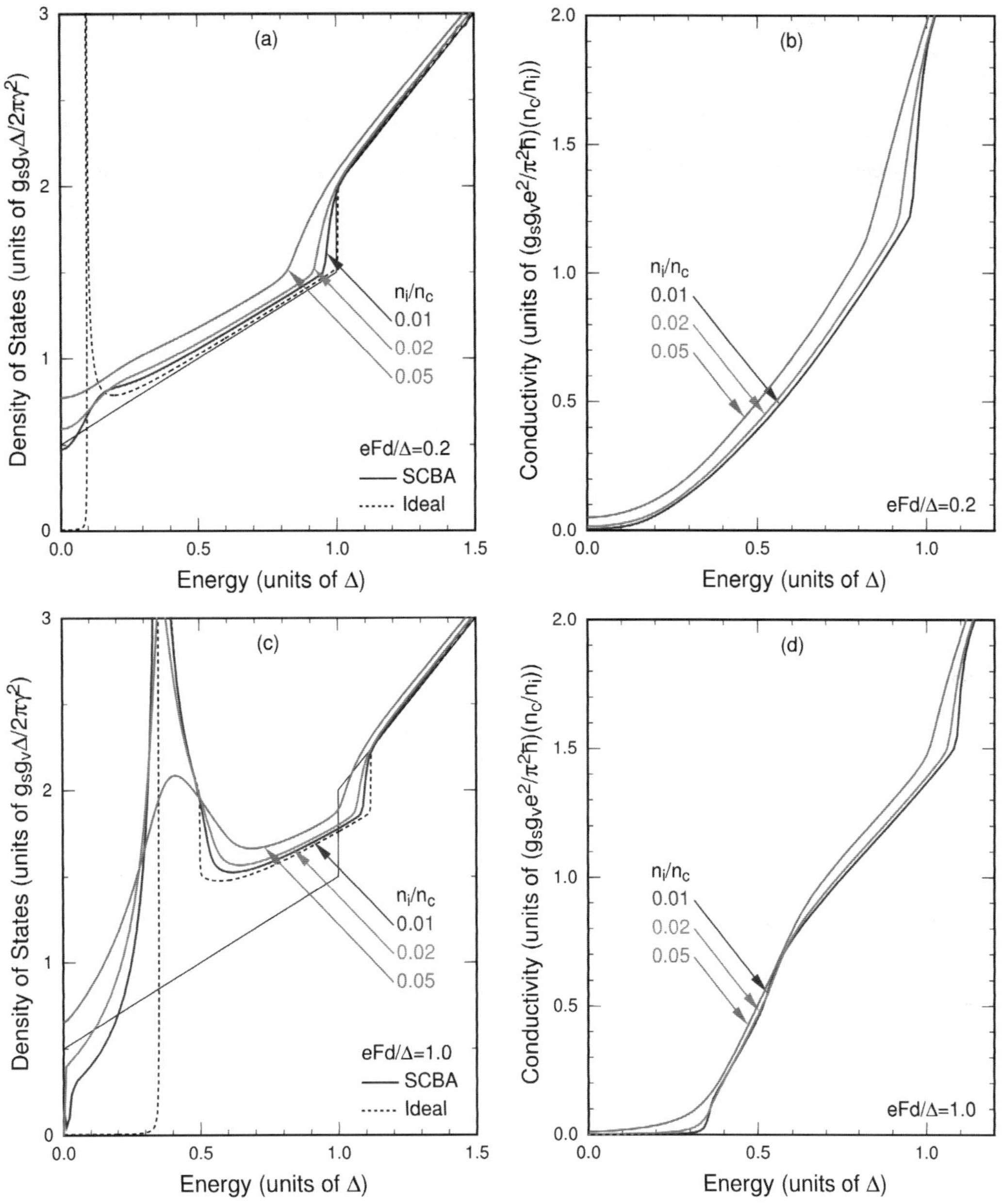

Figure 5. Calculated density of states (a) and (c), and conductivity (b) and (d) of asymmetric bilayer graphene. (a) and (b) $eFd/\Delta = 0.2$. (c) and (d) $eFd/\Delta = 1$.

asymmetry between the top and bottom layers due to perpendicular electric field does not affect this conclusion and contribute only to the change of Δ into $\tilde{\Delta}$.

Figure 6 shows some examples of calculated (a) density of states and (b) conductivity as a function of energy for symmetric bilayer $eFd = 0$. Results for asymmetric bilayer can be obtained by replacing Δ with $\tilde{\Delta}$ given in Eq. (6). The density of states in the clean limit is independent of energy in the region given by $-\Delta < \varepsilon < +\Delta$ and starts to increase linearly outside this region, as shown by a dotted line in Fig. 6 (a). In the presence of charged impurities,

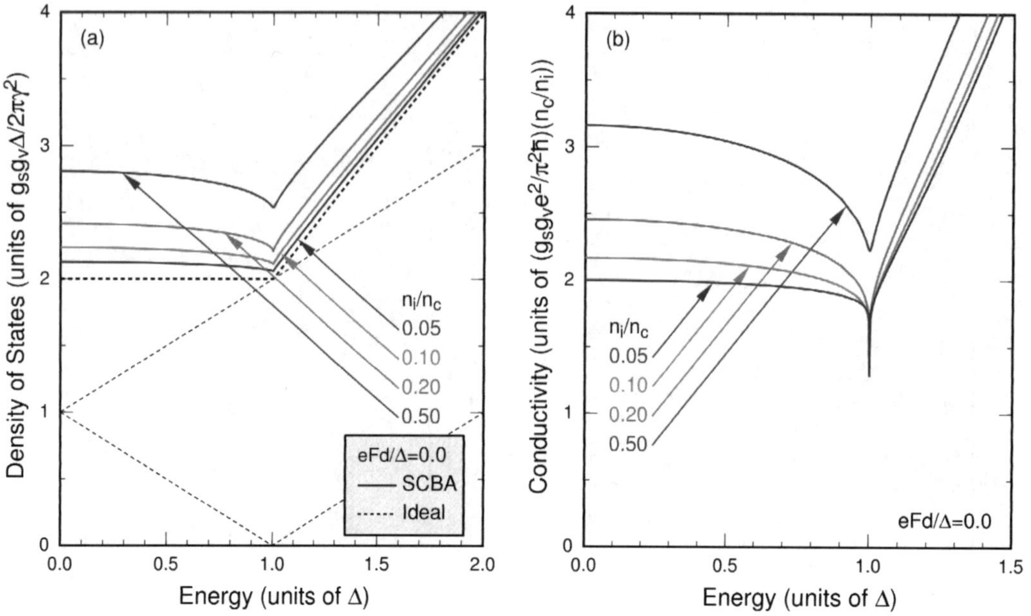

Figure 6. Calculated density of states (a) and conductivity (b) of AA-stacked bilayer graphene for charged impurities with screening effect being included within the Thomas-Fermi approximation.

the density of states is enhanced symmetrically around $\pm\Delta$ corresponding to the Dirac points of the two bands given by Eq. (8). As a result, the total density of states is most strongly enhanced near zero energy.

For small impurity concentration $n_{\rm i}/n_{\rm c} = 0.05$, the conductivity exhibits a singularly sharp drop at $\pm\Delta$. This is a manifestation of the zero-mode anomaly appearing in the conductivity for short-range scatterers with strength independent of energy or electron concentration predicted previously [9]. In fact, the screening effect is essentially independent of the energy in the vicinity of $\pm\Delta$ as shown in Fig. 6 (a), making charged impurities effectively "short-range scatterers" with energy independent strength. Therefore, the zero-mode anomalies characteristic of massless Dirac electrons can be experimentally observed in an AA stacked bilayer graphene even if scattering is dominated by charged impurities.

5. Summary

We have studied the conductivity in the presence of dominant charged impurities in graphenes, using the self-consistent Born approximation. In monolayer graphene, effects of environmental dielectric screening have been considered, showing that charged impurities should be regarded as short-range scatterers except in the case of very large dielectric constant. In bilayer graphene, effects of band-gap opening due to asymmetry between two layers caused by perpendicular electric field are considered, showing that the conductivity remains small even in the presence of appreciable density of states within the gap. The Hamiltonian in AA-stacked bilayer graphene can be decomposed into those of two monolayer graphenes with relative displacement of the Dirac points by the amount determined by the nearest-neighbor inter-layer interaction and asymmetry between two layers. Because of the near independence of the screening effect, a singularity at the Dirac point predicted in monolayer graphene with short-range scatterers with energy-independent strength appears also for charged impurities.

Acknowledgments
This work was supported in part by Grant-in-Aid for Scientific Research on Priority Area "Carbon Nanotube Nanoelectronics," by Grant-in-Aid for Scientific Research, and by Global Center of Excellence Program at Tokyo Tech "Nanoscience and Quantum Physics" from Ministry of Education, Culture, Sports, Science and Technology Japan. The author thanks Mr. Masaki Noro and Dr. Mikito Koshino for collaboration on some of the works presented here.

References

[1] Novoselov K S, Geim A K, Morozov S V, Jiang D, Zhang Y, Dubonos S V, Grigorieva I V, and Firsov A A 2004 *Science* **306**, 666
[2] Novoselov K S, McCann E, Morozov S V, Falko V I, Katsnelson M I, Zeitler U, Jiang D, Schedin F, and Geim A K 2006 *Nature Phys.* **2**, 177
[3] Berger C, Song Z, Li T, Li X, Ogbazghi A Y, Feng R, Dai Z, Marchenkov A N, Conrad E H, First P N, and de Heer W A 2004 *J. Phys. Chem. B* **108**, 19912
[4] Ohta T, Bostwick A, Seyller T, Horn K, and Rotenberg E 2006 *Science* **313**, 951
[5] Ando T 2007 *Physica E* **40**, 213
[6] Ando T 2009 *NPG Asia Mater.* **1**, 17
[7] Castro Neto A H, Guinea F, Peres N M, Novoselov K S, and Geim A K 2009 *Rev. Mod. Phys.* **81**, 109
[8] Abergel D S L, Apalkov V, Berashevich J, Ziegler K, and Chakraborty T 2010 *Adv. Phys.* **59**, 261
[9] Shon N H and Ando T 1998 *J. Phys. Soc. Jpn.* **67**, 2421
[10] Zheng Y and Ando T 2002 *Phys. Rev. B* **65**, 245420
[11] Ando T, Zheng Y, and Suzuura H 2002 *J. Phys. Soc. Jpn.* **71**, 1318
[12] Koshino M and Ando T 2006 *Phys. Rev. B* **73**, 245403
[13] Koshino M and Ando T 2007 *Phys. Rev. B* **75**, 235333
[14] Koshino M 2009 *New J. Phys.* **11**, 095010
[15] Fukuzawa T, Koshino M, and Ando T 2009 *J. Phys. Soc. Jpn.* **78**, 094714
[16] Noro N, Koshino M, and Ando T 2010 *J. Phys. Soc. Jpn.* **79**, 094713
[17] Ando T 2011 *J. Phys. Soc. Jpn.* **80**, 014707
[18] Ando T 2006 *J. Phys. Soc. Jpn.* **75**, 074716
[19] Monteverde M, Ojeda-Aristizabal C, Weil R, Bennaceur K, Ferrier M, Gueron S, Glattli C, Bouchiat H, Fuchs J N, and Maslov D L 2010 *Phys. Rev. Lett.* **104**, 126801
[20] Jang C, Adam S, Chen J-H, Williams E D, Das Sarma S, and Fuhrer M S 2008 *Phys. Rev. Lett.* **101**, 146805
[21] Ponomarenko L A, Yang R, Mohiuddin T M, Katsnelson M I, Novoselov K S, Morozov S V, Zhukov A A, Schedin F, Hill E W, and Geim A K 2009 *Phys. Rev. Lett.* **102**, 206603
[22] Chen F, Xia J, Ferry D K, and Tao N 2009 *Nano Lett.* **9**, 2574
[23] McCann E and Falko V I 2006 *Phys. Rev. Lett.* **96**, 086805
[24] McCann E 2006 *Phys. Rev. B* **74**, 161403
[25] Min H, Sahu B, Banerjee S K, and MacDonald A H 2007 *Phys. Rev. B* **75**, 155115
[26] Ando T and Koshino M 2009 *J. Phys. Soc. Jpn.* **78**, 104716

Journal of Physics: Conference Series **302** (2011) 012016 doi:10.1088/1742-6596/302/1/012016

One-dimensional structural irregularities in graphene: chiral edges and grain boundaries

Oleg V. Yazyev[1,2], Rodrigo B. Capaz[1,3] and Steven G. Louie[1,2]

[1] Department of Physics, University of California, Berkeley, California 94720, USA
[2] Materials Sciences Division, Lawrence Berkeley National Laboratory, Berkeley, California 94720, USA
[3] Instituto de Física, Universidade Federal do Rio de Janeiro, Caixa Postal 68528, Rio de Janeiro, RJ 21941-972, Brazil

Abstract. Many properties of graphene strongly depend on the presence of disorder and confinement effects. Using computational approaches based on model Hamiltonians and first-principles formalisms, we investigate the structure and electronic properties of chiral edges and grain boundaries in graphene. We show that magnetic edge states are inherent to low-symmetry graphene edges and exhibit a number of systematic structure-property relations. Certain grain boundary structures are shown to dramatically affect electronic transport in polycrystalline graphene by reflecting charge carriers in considerable energy ranges.

1. Introduction

Graphene has attracted considerable interest due to its extraordinary physical properties. One striking feature of this two-dimensional material is a sensitive dependence of its properties on the presence of disorder and confinement [1]. For instance, one-dimensional nanostructures made of graphene, the so-called graphene nanoribbons, exhibit dramatic dependence of their electronic structure on the crystallographic orientation of the edges [2,3]. While graphene nanoribbons with the edges cut along the armchair high-symmetry direction are semiconductors with band gap depending on the width and terminating structure, nanoribbons having zigzag high-symmetry edges would have flat band of zero-energy localized electronic edge states in an independent-particle picture neglecting many-electron effects. Electron-electron interactions are expected to remove the divergent density of state and the electronic instability associated with these zero-energy states resulting in a novel type of magnetic ordering [3-5]. The edge-state magnetism in graphene has been a matter of intense investigations in the last several years. A recent experimental study of graphene nanoribbons with ultrasmooth low-symmetry (chiral) edges revealed both the presence of edge states and a clear indication of magnetic ordering [6].

It has been realized recently that large-scale samples of graphene tend to be polycrystalline. In the case of graphene produced by means of micromechanical cleavage of graphite, the size of single-crystalline samples of graphene is limited by the size of graphite crystallites. Similarly, substrate imperfections and multiple nucleation centers limit the size of crystalline domains in graphene produced by epitaxial growth techniques. Indeed, recent experimental studies have revealed the polycrystalline nature of epitaxial graphene samples showing the presence of one-dimensional grain-boundary defects [7,8]. Issues of interest include: What determines the structure of defects in polycrystalline graphene? And how do these defects affect the charge transport properties of graphene?

In this article, we briefly describe our recent theoretical studies directed towards understanding the two above-mentioned types of one-dimensional structural irregularities in graphene – chiral edges and grain boundaries. We show that although these defects can affect the properties of graphene in undesirable ways, they can also be used for purposeful engineering graphene nanostructures and materials with novel functionalities.

2. Chiral graphene edges and nanoribbons

We investigate the electronic structure of graphene nanoribbons using a mean-field approximation to the Hubbard model Hamiltonian

$$H = -t\sum_{\langle i,j\rangle,\sigma}[c^{\dagger}_{i\sigma}c_{j\sigma} + \text{h.c.}] + U\sum_{i}(n_{i\uparrow}\langle n_{i\downarrow}\rangle + \langle n_{i\uparrow}\rangle n_{i\downarrow} - \langle n_{i\uparrow}\rangle\langle n_{i\downarrow}\rangle), \tag{1}$$

in which the strength of electron-electron interactions responsible for the onset of magnetic ordering is controlled by the magnitude of the on-site Coulomb repulsion U relative to the hopping integral t. In graphene, the magnitude of U is expected to be comparable to or less than that of t. Moreover, we point out that this effective parameter U should also incorporate environmental factors such as the enhancement of screening due to the proximity of dielectric or metallic substrate [6].

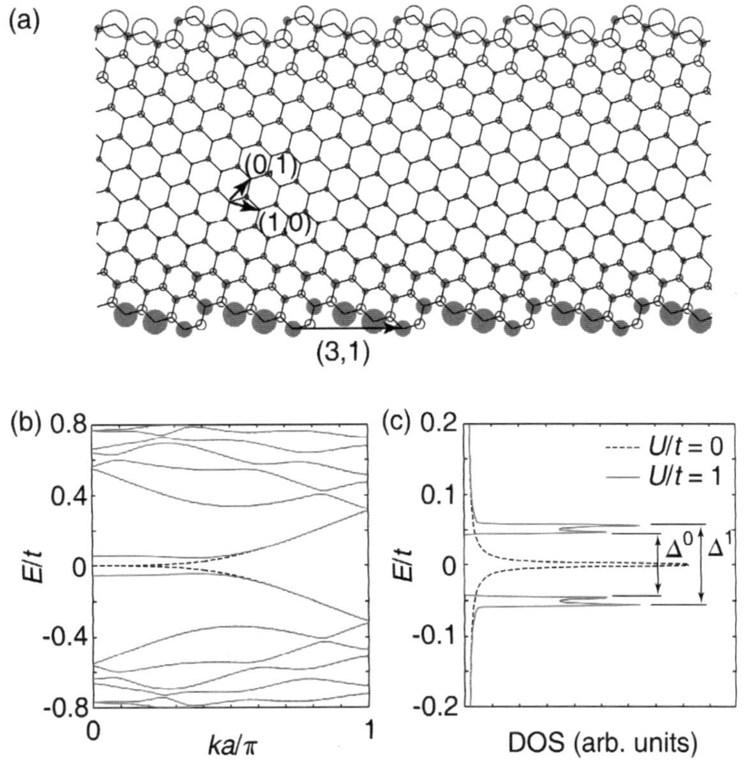

Figure 1. (a) Chiral graphene nanoribbon with edge defined by (3,1) vector (θ = 13.9°). Circle areas correspond to the local magnetic moments from the mean-field Hubbard model solution obtained at U/t = 1 (closed circles – spin-up; open circles – spin-down). (b) Band structure and (c) density-of-states plot for the nanoribbon shown in panel (a) obtained at U/t = 0 (dashed line) and U/t = 1 (solid line).

Figure 1(a) shows a representative example of chiral graphene nanoribbon with the edge oriented along the translational vector (3,1) of the graphene lattice. This edge orientation characterized by chirality angle θ = 13.9° is situated roughly in the middle between the high-symmetry zigzag (θ = 0°) and armchair (θ = 30°) directions. In the absence of electron-electron interactions (U/t = 0) the band structure features a zero-energy flat band of edge states (Fig. 1(b)). At U/t = 1, the ground state solution for this system is a semiconducting magnetic state. The density-of-states plot shown in Figure 1(c) has one peak at E = 0 in the absence of electron-electron interactions which splits into four peaks characterized by the energy splittings Δ^0 and Δ^1 once the Hubbard term is introduced into the model Hamiltonian in Eq. (1). This scenario is qualitatively very similar to the case of zigzag graphene nanoribbons. However, the magnetic moment per edge unit length M is about twice smaller for this chiral graphene nanoribbon than that in the zigzag case. The energy splittings, in particular Δ^1, are reduced in comparison with the corresponding splittings in the case of zigzag graphene nanoribbon of equivalent width.

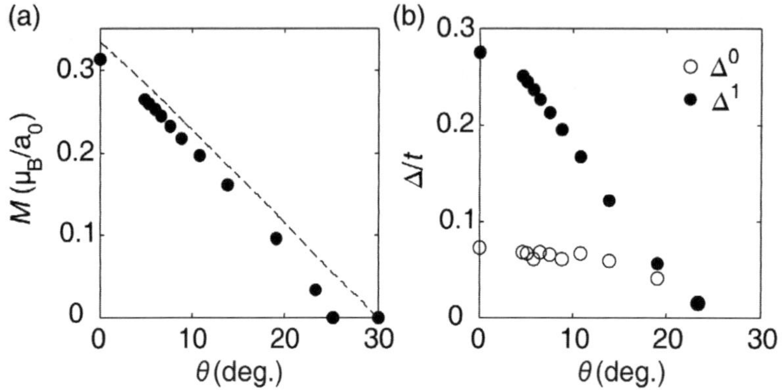

Figure 2. Dependence of (a) magnetic moment per edge unit length M and (b) energy splittings Δ^0 and Δ^1 on chirality angle θ for 5-nm-wide nanoribbons obtained from the mean-field Hubbard model (U/t = 1) calculations. The dotted line shows magnetic moments in the limit of infinite width (cf. Eq. (2)).

Figure 2 shows the dependence of magnetic moment per edge length M and energy splittings Δ^0 and Δ^1 on chirality angle θ. Remarkably, the semiconducting magnetic ground-state solutions span almost the whole range of chiralities with exception of a small region close to the armchair direction (θ = 30°). The values of M show monotonic dependence on θ which reflects a change in the density of edge states as a function of θ in the limit of infinite width is [9]:

$$\rho(\theta) = \frac{2}{3a_0}\cos\left(\theta + \frac{\pi}{3}\right). \tag{2}$$

The energy splitting Δ^1 follows a very similar trend (Fig. 2(b)) as M, while the band gap Δ^0 shows only weak variations in a broad range of chiralities. Interestingly, the dependence on the nanoribbons width is completely opposite for the two splittings (not shown), owing to the fact that Δ^0 arrives from inter-edge interaction whereas Δ^1 arrives from intra-edge interaction. The magnetic moments per edge length M are also mostly independent on width. These revealed structure-property relations thus can serve as unambiguous signatures of edge-state magnetism in graphene [10].

3. Grain boundaries and electronic transport in polycrystalline graphene

The properties of polycrystalline materials are determined by the presence of grain boundaries – one-dimensional defects separating single-crystalline domain of different orientations. The structure of these one-dimensional structural irregularities is dictated by the misorientation angle of the two crystallites. In this sense, misorientation angle is a topological invariant which introduces non-local disorder into polycrystalline materials while the local atomic structure of grain boundary defects is determined by thermodynamic arguments at fixed misorientation angle. We found that in graphene for any possible misorientation angle, it is possible to construct a grain boundary structure in which all carbon atoms maintain their three-fold coordination [11]. Such structures have inherently low formation energies. The basic building blocks of grain boundaries in graphene are the dislocations arranged in one-dimensional arrays. The shortest Burgers vector dislocation in graphene represents a pair of edge-sharing five- and seven-membered rings in the honeycomb lattice of graphene. Two examples of large-angle grain boundaries are shown in Figure 3.

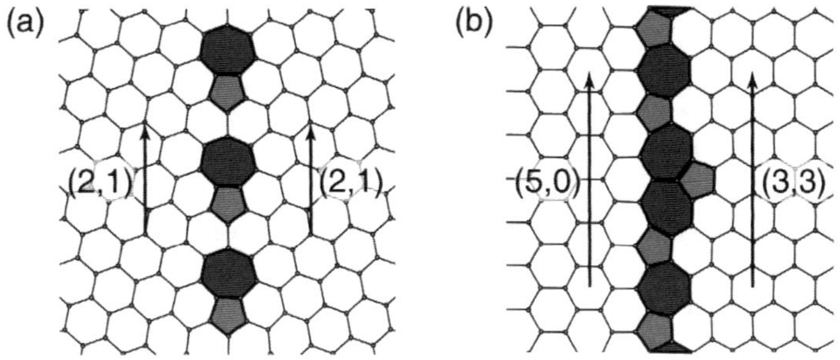

Figure 3. Two examples of grain boundaries in graphene: (a) a low energy symmetric grain boundary characterized by misorientation angle 21.8° and (b) an asymmetric 30° grain boundary that has a significant transport gap.

The grain boundary defects inevitably affect the transport properties of graphene. Remarkably, the periodic grain boundary structures are predicted to show two very different transport behaviors [12]. One way to describe such periodic structures is to refer to the corresponding matching vectors which are translational vectors on the two-dimensional graphene lattice in the corresponding single-crystalline domains, (n_L, m_L) and (n_R, m_R). If $n_L - m_L = 3p$ and $n_R - m_R = 3q$, or $n_L - m_L \neq 3p$ and $n_R - m_R \neq 3q$ $(p,q \in \mathbb{Z})$, the band structure of graphene dictates that, for any charge carrier with allowed values of momentum and energy in one domain, there exists a conduction channel in the other domain. That is, no transmission channel is blocked by the specific translation symmetry of the grain boundary defect. An example of such structure is shown in Figure 3(a): $(n_L, m_L) = (n_R, m_R) = (2, 1)$. However, selection by symmetry takes place when $n_L - m_L = 3p$ and $n_R - m_R \neq 3q$, or $n_L - m_L \neq 3p$ and $n_R - m_R = 3q$. In this case, transmission is blocked in the range of energies $E \in (-E_g/2, E_g/2)$ with transport gap

$$E_g = \hbar v_F \frac{2\pi}{3d} \approx \frac{1.38}{d\,[\mathrm{nm}]}\,[\mathrm{eV}], \tag{3}$$

which is dependent only on the periodicity of grain boundary d. An example of such structure is shown in Figure 3(b). The matching vectors defining the structure of this defect are (5,0) and (3,3). Our first-principles non-equilibrium Green's function calculations for this case predict a large transport gap of 1.1 eV in good agreement with Eq. (3).

The possibility of engineering grain boundaries in graphene characterized by large transport gaps may find important practical applications [12]. Owing to the absence of a band gap and the Klein tunneling, electrostatic barriers in graphene are ineffective. This is the major obstacle on the way of introducing graphene into digital electronics. Controlling electronic transport by means of engineered grain boundaries relies on the momentum mismatch rather than on tailoring potential barrier profiles. The proposed approach thus may help circumventing the above-mentioned difficulties thus paving a way towards practical graphene electronics.

4. Conclusions

We have shown that one-dimensional structural irregularities in graphene, chiral edges and grain boundaries, exhibit a range of novel structure-property relations. Similarly to zigzag graphene nanoribbons, nanoribbons with chiral edges also host spin-polarized edge states. Moreover, their electronic and magnetic properties can be tuned by changing the edge chirality and the width. Grain boundaries in polycrystalline graphene are shown to affect electronic transport in a dramatic manner. Certain grain boundary structures perfectly reflect charge carriers in large energy ranges. We anticipate that these novel structures can be used for purposeful engineering graphene nanostructures and materials with novel functionalities.

Acknowledgments

We wish to thank C. Tao and M. F. Crommie for discussions. This work was supported by NSF Grant No. DMR10-1006184 (numerical simulations of GNRs) and by the Director, Office of Science, Office of Basic Energy Sciences, Divisionof Materials Sciences and Engineering Division, US Department of Energy under Contract No. DE-AC02-05CH11231 (software development of electron correlation effects). O. V. Y. was partially supported by the Swiss National Science Foundation (grant No. PBELP2-123086). R.B.C. acknowledges financial support from Brazilian agencies CNPq, CAPES, FAPERJ and INCT – Nanomateriais de Carbono and the ONR MURI program.

References

[1] Castro Neto A H, Guinea F, Peres N M R., Novoselov K S and Geim A K 2009 *Rev. Mod. Phys.* **81** 109
[2] Nakada K, Fujita M, Dresselhaus G and Dresselhaus M S 1996 *Phys. Rev. B* **54** 17954
[3] Son Y-W, Cohen M L and Louie S G 2006 *Phys. Rev. Lett.* **97** 216803
[4] Fujita M, Wakabayashi K, Nakada K and Kusakabe K 1996 *J. Phys. Soc. Jpn.* **65** 1920
[5] Son Y-W, Cohen M L and Louie S G 2006 *Nature* **444** 347
[6] Tao C, Jiao L, Yazyev O V, Chen Y-C, Feng J, Zhang X, Capaz R B, Tour J M, Zettl A, Louie S G, Dai H and Crommie M F 2011 Nature Phys. (Published online: 8 May 2011 | DOI: 10.1038/NPHYS1991)
[7] Huang P Y, Ruiz-Vargas C S, van der Zande A M, Whitney W S, Levendorf M P, Kevek J W, Garg S, Alden J S, Hustedt C J, Zhu Y, Park J, McEuen P L and Muller D A 2011 *Nature* **469** 389
[8] Kim K, Lee Z, Regan W, Kisielowski C, Crommie M F and Zettl A 2011 *ACS Nano* **5** 2142
[9] Akhmerov A R and Beenakker C W J 2008 *Phys. Rev. B* **77** 085423
[10] Yazyev O V, Capaz R B and Louie S G *Preprint* arXiv:1102.4886
[11] Yazyev O V and Louie S G 2010 *Phys. Rev. B* **81** 195420
[12] Yazyev O V and Louie S G 2010 *Nature Mater.* **9** 806

Journal of Physics: Conference Series **302** (2011) 012017 doi:10.1088/1742-6596/302/1/012017

Dirac Electrons in Molecular Solids

Hidetoshi Fukuyama

Faculty of Science and Research Institute for Science and Technology

Tokyo University of Science, Tokyo 162-8601

Akito Kobayashi

Institute for Advanced Research, Nagoya University, Nagoya 464-8602

Yoshikazu Suzumura

Department of Physics, Nagoya University, Nagoya 464-8602

fukuyama@rs.kagu.tus.ac.jp

Abstract. Electrons in solids are characterized by the energy bands, which indicate that electrons are considered to be "elementary particles" with specific effective masses and g-factors reflecting features of each solid. There are cases where these particles obey dispersion relationship similar to those of Dirac electrons. Examples include graphite and bismuth both of which are known for many years, together with graphene, a single layer of graphite, recently addressed intensively after its realization. Another recent example is a molecular solid, α-ET_2I_3, which is described by an equation similar to Weyl equation with massless Dirac cones but the coordinate axis is tilted because of the location of cones at off-symmetry points. Orbital susceptibility of such Dirac electrons in graphite and bismuth has been known to have striking features not present in ordinary band electrons but resulting from the inter-band matrix elements of magnetic field. Results of theoretical studies on not only orbital susceptibility but also Hall effect of such Dirac electrons in molecular solids with tilting are introduced in this paper.

1. Introduction

Energy bands of electrons in solids have diversity reflecting particular features of solids and disclose specific energy-momentum relationship, which can be viewed as particular "elementary particles" at least locally in momentum space with effective masses and g-factors generally with spatial anisotropy. There are cases where these elementary particles obey equations similar to Dirac electrons. In this context graphite[1] and bismuth (Bi)[2, 3] have been known for many years; 2x2 Weyl equation for the former and 4x4 Dirac equation for the latter. Main physical interests in these systems had been the orbital susceptibility in weak magnetic field especially in bismuth (Bi) [4]. The difficulty with Landau-Peierls formula based on the pioneering work by Peierls [5], which takes only intra-band contribution of the effects of magnetic field on Bloch electrons, had long been noticed in view of experimental findings and roles of inter-band effects have been recognized [6]. Accordingly since the first derivation of orbital susceptibility by Wilson [7] there have been many efforts to derive exact formula of orbital susceptibility of Bloch electrons [8]. Theoretical understanding of experimentally observed anomalous orbital susceptibility has been achieved for graphite[9] and Bi[10] by taking these inter-band contributions of vector potential based on k-p Hamiltonian (Luttinger–Kohn representation) [11] which is suited to explore such intricate effects in a precise and transparent way. A simple and exact formula [12] for the orbital susceptibility for Bloch electrons has been derived based on this k-p Hamiltonian.

Besides orbital susceptibility, which is thermodynamic, there is general interest in such inter-band contribution on transport properties, typically weak field Hall conductivity and then Hall effect [13, 14]. These have been explicitly studied for the recent example of Dirac electrons in graphene [15, 16], a single layer of graphite. Based on the proper scheme [17] in extracting correctly (in a gauge-invariant way) the contributions in the linear order of vector potential in the Kubo formula Hall conductivity has been evaluated [18] together with conductivity and orbital susceptibility at absolute zero as a function of chemical potential under the assumption of energy independent electron damping (lifetime). The results indicate that Hall conductivity vanishes and change signs at the crossing of Dirac cone. This feature at the crossing energy is very natural in view of the symmetry between "electrons" and "holes". This fact is, however, completely opposite to the conventional understanding based on the band electrons, since the effective carrier density would be vanishing if the Fermi energy is located at the crossing.

2. Experimental puzzles in molecular solids, α-ET_2I_3

There had been various puzzles in the experimental results in α-ET_2I_3 [19], which is one of polytypes of charge–transfer molecular solids consisting of ET molecules, ET_2X, with X being anions. At ambient pressure the resistivity of α-ET_2I_3 shows a sharp increase at around 150K as the temperature is lowered [19]. This metal-insulator transition is suppressed by external pressure and vanishes completely at around 20 GPa leading to almost temperature independent resistivity below room temperature. The measurement of Hall effect in this pressure region of temperature-independent resistivity discloses that the effective carrier number decreases very strongly (several orders of magnitude between room temperature and few degree K) as temperature is lowered, whereas the mobility increases accordingly. Even more, there are cases where Hall constant changes sign at low temperature very sharply [20], which is sample-dependent.

3. Electronic states of molecular solids, α-ET_2I_3: tilted Weyl particles

It is now established that electronic states of molecular solids are, in general, well described by the tight-binding approximation based on molecular orbitals [21]. Molecular solids of charge transfer type with particular ratio of 2:1 between cations and anions, which lead to the quarter-filling of holes, have a great diversity in electronic properties. Microscopic and systematic theoretical understanding of the diversity is now possible by the extended Hubbard model which is based on the tight-binding approximation in terms of molecular orbitals for the band structure together with Coulomb interactions U on the same molecules and V between them [22,23]. The characteristic electronic states of quarter-filled band are due to coordinated effects of dimerization and frustration (from square to triangular) of lattice structures resulting in two typical insulating states, dimer Mott and charge ordering [23]. Typical example of the former is the κ type which shows competition between antiferromagnetic Mott insulator and superconductivity [24] and the realization of spin liquid [25] for the first time, while α-ET_2I_3 is the example of the latter[26].

For α-ET_2I_3, using the transfer energy experimentally obtained under uniaxial strain [27], the gapless behaviour in the density of states has been disclosed [28]. Further, detailed analysis of such band structure within the tight-binding approximation [29] and later by first-principles calculations [30, 31] have led to the remarkable finding of the possible accidental degeneracy of bands (Dirac cones). This finding has led to the breakthrough of microscopic understanding of mysteries of this system.

Since Dirac cones in this case are located at two points (with inversion symmetry) in the Brillouin zone away from symmetry points in contrast to graphenes, the band energy is described by the following equation similar to Weyl equation but with diagonal component linear in wave vector , k, measured from the crossing point leading to "tilting " of cones[32].

$$H = \sum_{\rho=0,x,y,z} \mathbf{k}\cdot\mathbf{v}_\rho \sigma_\rho \qquad (1)$$

Here $\hbar=h/2\pi$ taken as unit with h being the Plank constant and σ_ρ are 2x2 matrices with σ_0 being unit matrix and the Pauli matrix for ρ=x, y, z, respectively. It turns out that velocity $\mathbf{v}_\rho$ are given by $\mathbf{v}_0=(v_0,0)$, $\mathbf{v}_x=(v_x,0)$, $\mathbf{v}_y=0$, $\mathbf{v}_z=(0,v_z)$ and that v_x is close to v_z and then $v_x=v_z$ is assumed in the explicit calculations. The existence of finite v_0 represents tilting, which is characterized by the dimensionless parameter, $\alpha=v_0/v_c$. Here v_c is defined by $v_c= v_x=v_z$ [32]. It turns out that α ~0.8 in the case of α -ET_2I_3.

4. Effects of tilting on orbital susceptibility and Hall conductivity

The effects of tilting on conductivity and Hall conductivity have been studied [33] in terms of proper formulas applied to the Hamiltonian, eq. (1). The results are shown in Fig.1 and 2 at absolute zero as a function of chemical potential under the assumption of energy independent electron damping (lifetime). Here $X=\mu/\Gamma_0$, with μ and Γ_0 being the chemical potential and damping, respectively. In the case of conductivity, there are two different types of contributions, i.e. $\sigma_{xx}= \sigma_{xx}^a + \sigma_{xx}^b$, where both σ_{xx}^a and σ_{xx}^b have contributions from only at the Fermi energy (on-shell contributions) but have

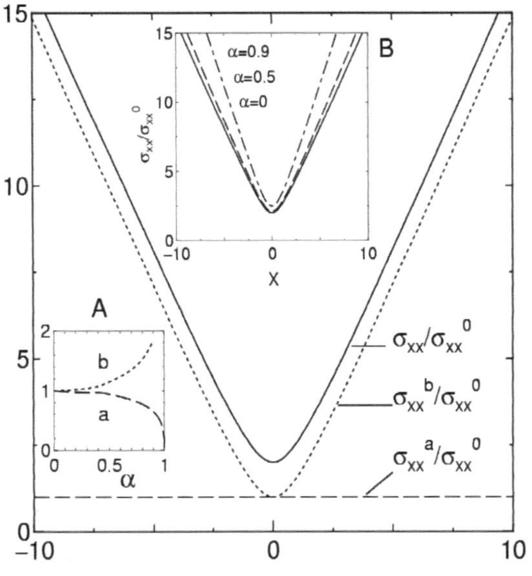

Fig. 1. The conductivity $\sigma_{xx}/\sigma_{xx}^0$ (the solid line), $\sigma_{xx}^a/\sigma_{xx}^0$ (the dashed line) and $\sigma_{xx}^b/\sigma_{xx}^0$ (the dotted line) for α=0, where $\sigma_{xx}^0=e^2/2\pi^2$. The inset A shows α-dependences of enhancement factors, $\sigma_{xx}^a(\alpha)$ / $\sigma_{xx}^a(\alpha=0)$ (the dashed line) and $\sigma_{xx}^b(\alpha)$ / $\sigma_{xx}^b(\alpha=0)$ (the dotted line). The inset B shows X-dependences of total σ_{xx} for several degrees of tilting, α=0 (the solid line), 0.5 (the dashed line) , and 0.9 (the dot-dashed line). [33]

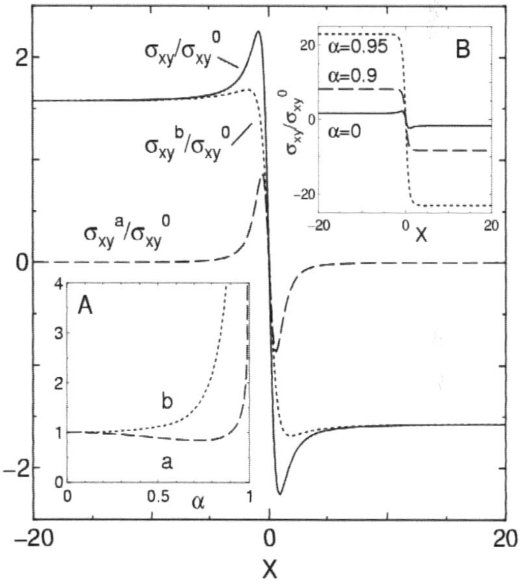

Fig. 2. The Hall conductivity $\sigma_{xy}/\sigma_{xy}^0$ (the solid line), $\sigma_{xy}^a/\sigma_{xy}^0$ (the dashed line) and $\sigma_{xy}^b/\sigma_{xy}^0$ (the dotted line) in the absence of tilting, where $\sigma_{xy}^0=e^3v_c^2H/4\pi^2c\Gamma_0^2$. The inset A shows α–dependences of enhancement factors, $\sigma_{xy}^a(\alpha)$ / $\sigma_{xy}^a(\alpha=0)$ (the dashed line) and $\sigma_{xy}^b(\alpha)$ / $\sigma_{xy}^b(\alpha=0)$ (the dotted line). The inset B shows X-dependences of total σ_{xy} for several degrees of tilting. [33]

different dependences on α. (There is a finite anisotropy in the conductivity but it is small as is expected from the weak α dependences of σ_{xx} , especially near the crossing, X=0, as seen in the inset of Fig.1.) Similarly the Hall conductivity has two contributions, $\sigma_{xy}= \sigma_{xy}^{a} + \sigma_{xy}^{b}$. It is to be noted, however, that there is a qualitative difference from conductivity; σ_{xy}^{a} contains all contributions below the Fermi energy whereas σ_{xy}^{b} has only on-shell contributions. In Fig.3 is shown the result for orbital susceptibility, in which case tilting affects only the over-all prefactor. In all these singular behaviour toward α =1 is due to the fact that the velocity in one direction is vanishing in this limit.

It is seen that finite tilting affects both orbital susceptibility and Hall conductivity quantitatively but qualitative features are unchanged except close to α=1. It is to be noted that the present assumption of energy-independent damping of electrons leads to increase of conductivity as a function of energy. This result will lead to the increase of conductivity as a function of temperature in the case where the Fermi energy is located at the crossing energy at the absolute zero. This is not consistent with the experimental results. This will imply that the damping is increasing as a function of temperature (energy) almost linearly. If this is the case, the Hall coefficient will be strongly suppressed as the temperature is increased as seen in the experiment, although it is a future problem to explore these dependences on energy at absolute zero and on temperature at fixed carrier density.

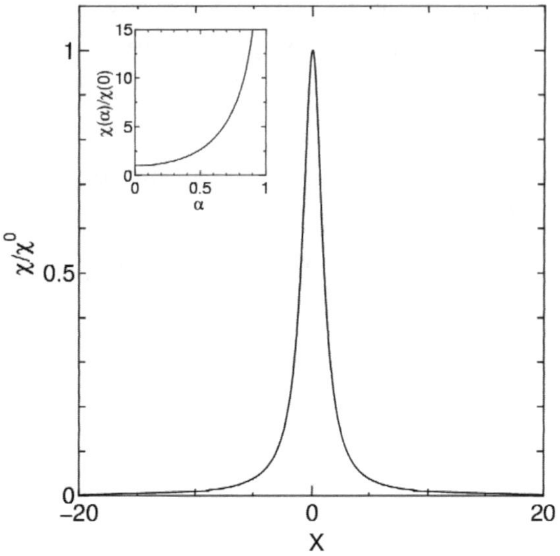

Fig. 3. The orbital susceptibility χ/χ^0 as the function of X in the absence of tilting, where $\chi^0 = - e^2 v_c^2 / 3 \pi^2 c^2 \Gamma_0$. The inset shows α–dependence of enhancement factor $\chi(\alpha) / \chi(\alpha=0)$. [33]

Regarding the sharp change of sign in Hall coefficient at low temperatures, it has been indicated [33] that the possible small amount of deficiency of anions (I_3^-) of the order of only ppm together with asymmetry of energy dependences of density of states relative to the crossing energy can be the cause.

5. A possible Berezinski-Kosterlitz-Thouless phase transition under strong magnetic field

So far the orbital susceptibility and Hall conductivity have been studied in weak magnetic field. Under strong magnetic field of the order of 10 Tesla applied perpendicularly, resistivity measurement indicates interesting temperature dependences[19], which shows the existence of two characteristic temperatures, T_0 and T_1, for its sharp change. It has been proposed [34] that these characteristic temperatures will correspond to the crossover temperature of the formation of valley-order similar to the XY ferromagnetic order and the Berezinski-Kosterlitz-Thouless (BKT) transition at lower temperature as will be briefly explained below.

Magnetic field applied perpendicularly to the plane leads to the quantized Landau levels. Note that there exists a particular Landau level at the crossing energy (N=0) independent of the strength of magnetic field. The finite tilting does not affect this qualitatively [35] and the energy level separation between N=0 and N=1 is of the order of 50K at H=10T.

The spin Zeeman splitting is much smaller than the Landau level splitting since the g-factor is g~2. Because there are two Dirac cones in the Brillouin zone, these states (valleys) are doubly degenerate (characterized by the pseudo-spin) if two valleys are independent. Once mutual Coulomb interactions are taken into account, electrons in different valleys interact and get scattered among them. In order to treat such many-body effects in the presence of Landau quantization, ordinary Landau wave functions, which are plane wave and then infinitely spread in one direction, is not convenient and then orthonormal basis sets, which are localized in space similar to "Wannier functions", introduced some time ago are employed.

It turns out that finite tilting plays crucial roles by introducing processes of mixing different valleys. If the chemical potential is fixed to the crossing energy and this mixing is strong enough compared with the spin Zeeman splitting, some kind of valley order state described as ferromagnetism of pseudo-spins is possible at $T=T_0$ in the mean-filed approximation. The order parameters in this case are complex and their phases represent degrees of freedom leading to vortices which can have BKT phase transition at lower temperature $T=T_1$. Theoretical analysis concludes $T_1/T_0 \sim 1/4$. If T_0 and T_1 are associated with the two step changes of resistivity in the experiment [19] this ratio between T_0 and T_1 is close to experimental results. A phase transition similar to BKT had been indicated in resistivity measurement of graphene [36] at much lower temperatures. Since tilting is absent in graphene, some mechanism other than the present tilting-driven inter-valley mixing by Coulomb interactions, *e. g.* an electron-lattice interaction [37], will be operating in graphene.

Acknowledgment

One of authors (HF) dedicates hearty gratitude to late Professor Ryogo Kubo for his guidance in very early stage based on the deep insight into inter-band effects of magnetic field on Bloch electrons. Authors are grateful to Mark Goerbig for collaborations for the subject in Sec.5.

References

[1] P. R. Wallace, Phys. Rev. 71, 622 (1947)
[2] M. H. Cohen and E. I. Blount, Phil. Mag. 5, 115 (1960)
[3] P. A. Wolff, J. Phys. Chem. Solids 25, 1057 (1964).
[4] D. Schoenberg and M. Z. Uddin, Proc. Roy. Soc. A156, 687,701 (1936).
[5] R. Peierls, Z. Phys. 80, 763 (1933).
[6] E. N. Adams, Phys. Rev. 89, 633 (1953).
[7] A. H. Wilson, Proc. Camb. Phil. Soc. 49, 292 (1953).
[8] For example, G. H. Wannier and U. N. Upadyaya, Phys. Rev. 136, A803 (1964); R. Kubo. Proc. Intern. Conf. Magnetism, Notthingam 1964 (The Institute of Physics and Physical Society) p.35, and references therein.
[9] J. W. McClure, Phys. Rev. 104, 666 (1956).
[10] H. Fukuyama and R. Kubo, J. Phys. Soc. Jpn. 28, 570 (1970).
[11] J. M. Luttinger and W. Kohn, Phys. Rev. 97, 869 (1955).
[12] H. Fukuyama, Prog. Theor. Phys. 45, 704 (1971).
[13] R. Kubo and H. Fukuyama, Proc. of the "10th International Conference on the Physics of Semiconductors" (Ed. E. P. Keller and J. C. Hensel and F. Stern, Published by the United State Atomic Energy Commission, 1970).
[14] H. Fukuyama, Ann. Phys. (Leipzig) 15, 520 (2006).
[15] K. S. Novoselov, A. K. Geim, S. V. Morozov, D. Jiang, M. I. Katsnelson, I. V. Grigorieva, S. V. Dubonos, and A. A. Firsov: Nature 438, 197 (2005).
[16] T. Ando: J. Phys. Soc. Jpn. 74, 777 (2005).
[17] H. Fukuyama, H. Ebisawa and Y. Wada, Prog. Theor. Phys. 42, 494 (1969).
[18] H. Fukuyama, J. Phys. Soc. Jpn. 76, 043711 (2007).
[19] N. Tajima, S. Sugawara, M. Tamura, Y. Nishio and K. Kajita, J.Phys. Soc. Jpn. 75, 0510101 (2005).
[20] N. Tajima, S. Sugawara, M. Tamura, R. Kato, Y. Nishio and K. Kajita, private communications.
[21] T. Mori, A. Kobayashi, T. Sasaki, H. Kobayashi, G. Saito and H. Inokuchi, Chem. Lett. 13, 957 (1984) and references therein.

[22] H. Kino and H. Fukuyama, J.Phys. Soc. Jpn. 65, 2157 (1996).
[23] H. Seo, C. Hotta and H. Fukuyama, Chem. Rev. 104, 5005 (2004).
[24] K. Kanoda, J. Phys. Soc. Jpn. 75, 051007 (2005).
[25] K. Kanoda and R. Kato, Annu. Rev. of Condens. Matter Phys. 2011. 2:167-88
[26] H. Seo, J. Phys. Soc. Jpn. 69, 805 (2000).
[27] R. Kondo, S. Kagoshima, and J. Harada, Rev. Sci. Instrum. 76, 090302 (2005).
[28] A. Kobayashi, S. Katayama, K. Noguchi, and Y. Suzumura, J. Phys. Soc. Jpn 73, 3135 (2004).
[29] S. Katayama, A. Kobayashi, and Y. Suzumura, J. Phys. Soc. Jpn 75, 054705 (2006).
[30] H. Kino and T. Miyazaki, J. Phys. Soc. Jpn 75, 034704 (2006).
[31] S. Ishibashi, T. Tamura, M. Kohyama, and K. Terasaki, J. Phys. Soc. Jpn 75, 015005 (2006).
[32] A. Kobayashi, S. Katayama, Y. Suzumura, and H. Fukuyama,J. Phys. Soc. Jpn 76, 034711 (2007).
[33] A. Kobayashi, Y. Suzumura, and H. Fukuyama, J. Phys. Soc. Jpn 77, 064718 (2008).
[34] A. Kobayashi, Y. Suzumura, H. Fukuyama, and M. O. Goerbig, J. Phys. Soc. Jpn 78, 11471, (2009).
[35] T. Morinari, T. Himura, and T. Tohyama, J. Phys. Soc. Jpn 78, 023704 (2009).
[36] J. G. Checkelsky, L. Li, and N. P. Ong, Phys. Rev. B 79, 115434 (2009).
[37] K. Nomura, S. Ryu, and D-H Lee : Phys. Rev. Lett. 103 (2009) 216801.

Electronic properties of graphene and boron-nitride based nanostructured materials

Masahiro Sakurai, Yuki Sakai, and Susumu Saito

Department of Physics, Tokyo Institute of Technology

2-12-1 Oh-okayama, Meguro-ku, Tokyo 152-8551, Japan

saito@stat.phys.titech.ac.jp

Abstract. We study the electronic properties of graphene with periodic structural modifications using the local density approximation within the framework of the density functional theory. It is found that the double-layer graphene with nanotube arrays which connect the two sheets seamlessly is a direct-gap semiconductor. Also the single-layer graphene with periodic holes of 24-atomic-site vacancy is found to have a sizable direct fundamental gap as well. These graphene-based materials should be of high importance for nanoelectronics applications in the future. In addition, we study the electronic structure of the superlattice composed of graphene and boron-nitride layers which are stacked alternately. The material is found to be metallic even though each graphene layer is sandwiched by insulating BN layers and is very far from other graphene layers. The material may show interesting electronic properties being different from those of graphene.

1. Introduction

Ever since the macroscopic production of spherical carbon clusters (C_{60} fullerenes) [1], various low-dimensional nanostructured carbon materials have been produced so far: Larger fullerenes such as C_{70}, C_{76}, and C_{84}, carbon nanotubes with different diameters and chiralities [2], and fullerene polymers having one- and two-dimensionally polymerized C_{60} networks [3]. Most recently, graphene, the single-layer graphite sheet, was fabricated experimentally via mechanical cleavage [4].

In the case of fullerenes and nanobutes, low-dimensional structural units were produced first, and their bulk solid phases, e.g., solid C_{60} and nanobute bundles, were produced later. On the other hand, "solid graphene", i.e., graphite, has been well known as the ground-state phase of carbon and has been studied for many decades both theoretically and experimentally before the production of its constituent unit, graphene. Therefore, the basic electronic structure of graphene has been known theoretically for decades as well. However, most of the theoretical studies of graphene before its experimental production were carried out in order to understand the electronic properties of graphite. Only a couple of theoretical groups have studied graphene as realistic materials in order to clarify the electronic properties of graphene itself [5] or to design new graphene-based nanostructured carbon materials [6]. In the latter work of ours, three kinds of the periodic structure modifications of the graphene sheet were proposed (figure 1), and their electronic properties were studied in detail using the generalized and transferable tight-binding method. Importantly, most of the periodically modified graphene systems are found to be semiconductors with sizable fundamental-gap values. This is of high importance from the viewpoint of the nanoelectronics applications of graphene, since the pristine and perfect graphene has metallic transport properties and is not suitable for device materials.

In the present work, we study these periodically modified graphene sheets more systematically in the framework of the density-functional theory, and clarify their energetics and electronic band structures further. In addition, we newly study graphene with regularly spaced holes. Also we report a different kind of graphene-based materials design, i.e. the superlattice of graphene and boron-nitride layers.

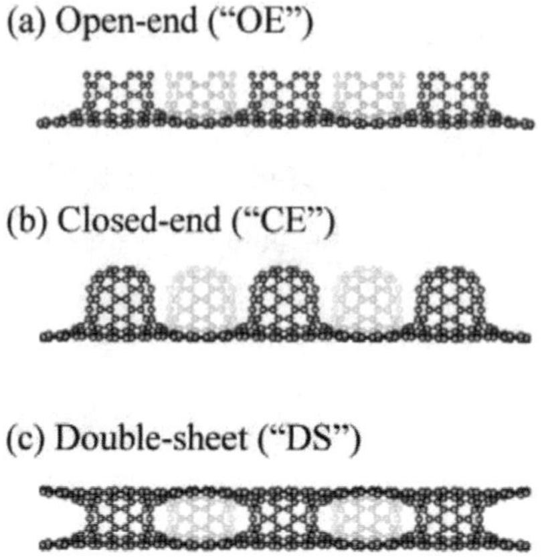

Figure 1 Periodically modified graphene using finite-length carbon nanotubs [6]

2. Periodically modified graphene sheets

In our previous work [6], we proposed three kinds of periodic structural modifications of graphene using finite-length (6,6) carbon nanotubes, which are called open-end (OE), closed-end (CE), and double-sheet (DS) systems according to their geometries (figure 1). It has been clarified using the tight-binding method that most of these OE, CE, and DS nanostructured materials are semiconductors although the structural constituent units, i.e. graphene and (6,6) nanotube both possess metallic electronic structures.

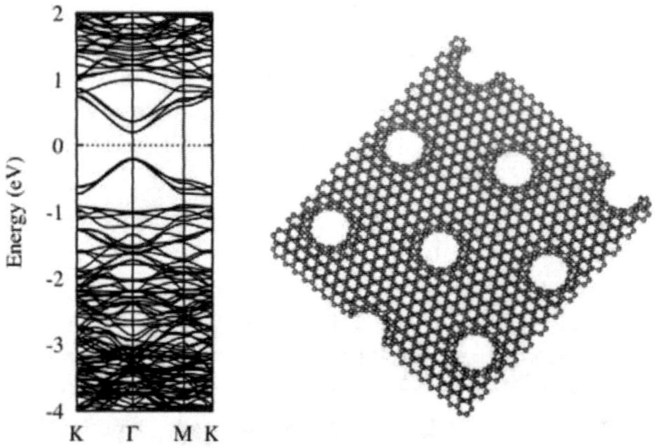

Figure 2 Electronic band structure of DS(5,0) (left panel). Geometry is shown in the right panel.

In figure 2, the electronic band structure of one of optimized DS geometries, DS(5,0) is shown (see Reference 6 for the detailed definition of the "(n,m)" indices following "DS", "OE", or "CE"). Here, the local-density approximation (LDA) in the framework of the density-functional theory (DFT) is used [7]. As can be seen from the figure, the material is predicted to be a direct-gap semiconductor with the fundamental gap of about 0.4 eV in LDA. Considering the usual underestimation of the fundamental-gap value of semiconductors in LDA, one can expect that this DS(5,0) material should have a sizable gap of about 1 eV. Interestingly enough, the systematic LDA study has clarified that the gap value is inversely proportional to the squared lattice constant of not the original graphene but the superstructure with periodic structural modification [8].

3. Graphene with holes

We also study the periodic modification of graphene using holes, i.e. vacancy clusters using the LDA (figure 3). As for the σ dangling-bond states associated with edge atoms at holes, hydrogen atoms are introduced to terminate them. On the other hand, the π states may possess dangling-bond characters in some cases and may reside around the Fermi level as in the case of one-dimensional graphene nanoribbons [9]. The so-called zigzag edge is found to induce nearly flat-band states just at the Fermi level, giving rise to spin polarization with ferromagnetic coupling along the edge direction while antiferromagnetic coupling between the opposite-side edges.

In this graphene system with holes, it should be note that, due to the presence of the curvature along the circle-shape edge at each hole, it is not straightforward to judge each edge atom to be a zigzag-edge atom or an armchair-edge atom. In the case of the holes corresponding to the 24-site vacancies, all the edge atoms are equivalent as is shown in figure 3. This graphene sheet with the hexagonal-cell arrangements of "V24" holes is found to be a semiconductor having the direct gap at Γ point (figure 4), being similar to the DS structure shown in figure 2. Although all the C atoms in the DS structure possess the sp^2 configuration with three neighbor C atoms and there is no explicit edge atom, this similar gap structure should indicate that opening of the fundamental gap in the two systems might have similar origins. This is an interesting issue to be addressed in the future. In addition, experimental works to introduce periodic geometry modifications into graphene were reported recently [10,11]. The comparative study with experimental reports should be another future important issue in this field.

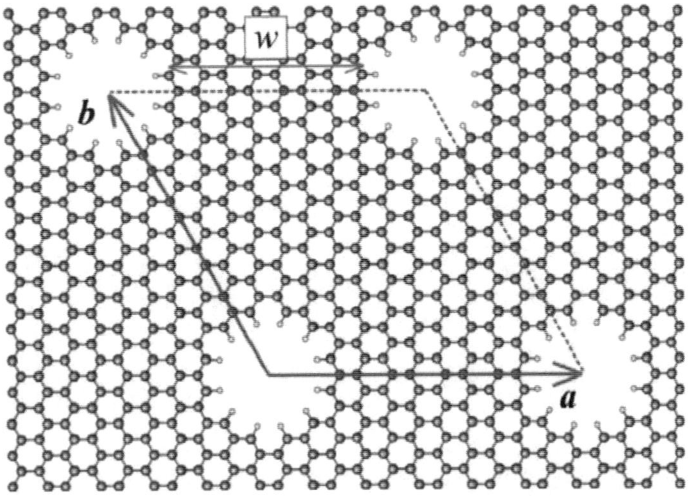

Figure 3 Graphene with holes of "V24" with a=25.3 Å. The width between two holes w is 15.4 Å.

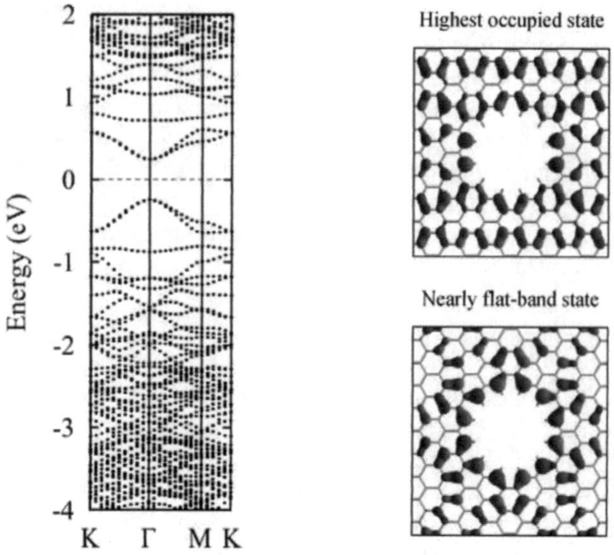

Figure 4 Electronic energy band of graphene with periodic holes of V24. The spatial distributions of the highest occupied state (at Γ point) and the nearly flat-band state inside the valence band (also at Γ point) are also shown.

4. Superlattice of graphene and BN layers

Recently, based on the DFT energetics, we have designed the superlattice of graphene and boron-nitride layers which are stacked alternately along the direction perpendicular to the layers [12]. It is known that, in the case of graphite, the usual stacking sequence is called "*AB* stacking" although the "*ABC* stacking" graphite (rhombohedral graphite) is also known to be present. In the case of hexagonal BN solid, on the other hand, the stacking sequence can be called *aa*' stacking where below and above each B (N) atom of each BN layer there are N (B) atoms on neighboring layers. Here, capital letters indicate graphene layer configurations while small letters indicate those of BN layers. As for *a*' layer, boron atoms are replaced by nitrogen atoms compared with the *a* layer, and vice versa. Interestingly, the most stable stacking sequence of the alternating graphene-BN superlattice is found to possess *Ab* stacking (figure 5) where below and above each B atom there always are C atoms on neighboring layers. On the other hand, there are center of carbon hexagons below and above each N atom. Therefore, the B-C interaction is considered to be more attractive than other pairs.

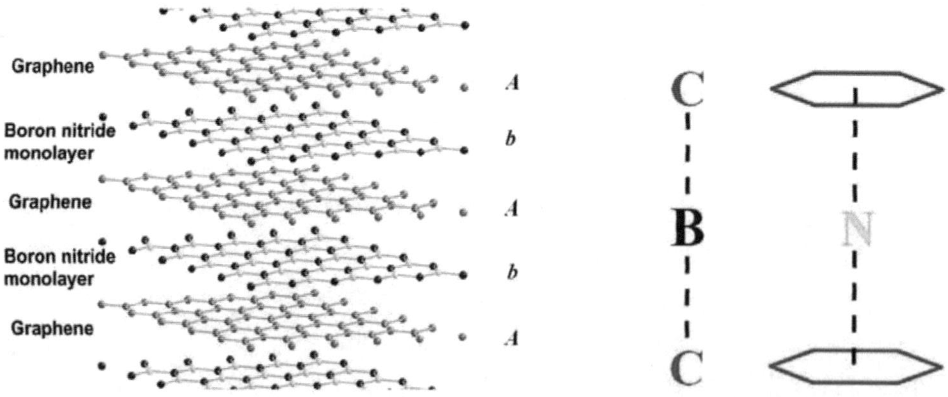

Figure 5 Schematic picture of the *Ab*-stacking superlattice of alternating graphene and BN layers

This *Ab*-stacking superlattice is predicted to be metallic, although each graphene sheet is sandwiched by insulating BN layers and the graphene-graphene distance is rather long. Moreover, in this superlattice two C atoms in the unit cell are not geometrically equivalent with each other any more due to the effect of the neighboring BN layers, and one may expect the semiconducting electronic properties. These rather unexpected properties might also be due to a rather strong interlayer B-C interaction which might mediate the graphene-graphene interaction in the *Ab*-stacking superlattice. Also it is predicted that *AbAc*'-stacking superlattice should be as stable as the *Ab*-stacking superlattice because in this superlattice the geometry below and above each B (N) atom is the same as those in *Ab*-stacking superlattice.

In the present work, we study the energetics and the electronic structure of the *AbAc*'-stacking superlattice in the DFT. The total energy per atom of this superlattice is actually confirmed to be the same as that of *Ab*-stacking superlattice. Its band structure is shown in figure 6. Interestingly, this newly identified stable superlattice is also found to be metallic, supporting the concept of the strong B-C interlayer interaction proposed previously.

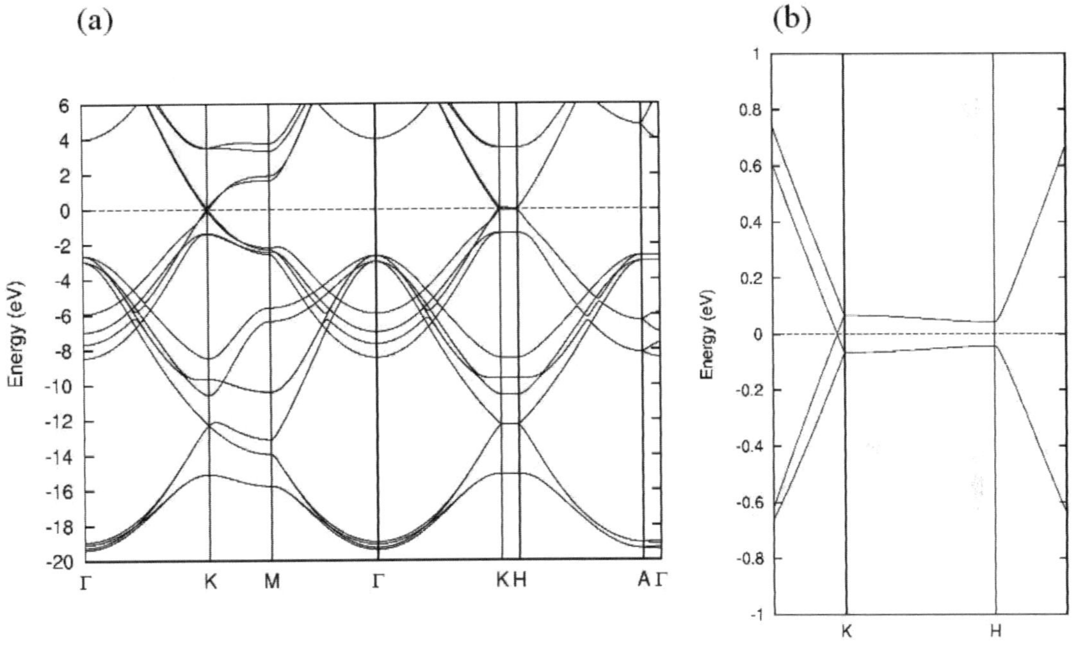

Figure 6 (a) Electronic band structure of the *AbAc*'-stacking superlattice of graphene and BN layers. Expanded fiture around the KH line is shown in (b).

5. Concluding remarks

Graphene is now the material of the highest interest in nanoscience and nanotechnology due partly to its large potential as a nanodevice material. In order to use graphene to the electronic device in the future, however, in most cases one has to modify its electronic structure from metallic to semiconducting in some way. In this respect, periodic geometry modification of graphene addressed in ths issure is of high importance as the systems are found to possess the direct band gap. Also the superlattice of graphene and BN layers studied in this work is to be of high importance in the near future since not only graphene but also monoatmic layer of boron nitride have been produced, and the superlattice of these two layers will be fabricated soon. Since the stable superlattice of graphene and BN layers is found to be metallic, they may show unique and interesting electronic transport properties in their thin-film phase as in the case of graphene.

Acknowledgments
This work was partly supported by the Global Center of Excellence Program by the Ministry of Education, Culture, Sports, Science and Technology of Japan ("MEXT") through the *Nanoscience and Quantum Physics* project of Tokyo Institute of Technology, and also by a Grant-in-Aid for Scientific Research on Priority Area of the MEXT, Japan, *Carbon Nanotube Nano-Electronics*.

References
[1] Krätschmer W, Lamb L D, Fostiropoulous K and Hoffman D R 1990, *Nature* **347** 354
[2] Iijima S 1991, *Nature* **354** 56
[3] Nunez-Regueiro M, Marques L, Hodeau J-L, Bethoux O and Perroux M 1995, *Phys. Rev. Lett.* **74**, 278
[4] Novoselov K S, Geim A K, Morozov S V, Jiang D, Zhang Y, Dubonos S V, Grigorieva I V and Firsov A A 2004, *Science* **306**, 666
[5] Zheng Y and Ando T 2002, *Phys. Rev.* B **65**, 245420
[6] Matsumoto T and Saito S 2002, *J. Phys. Soc. Jpn.* **71**, 2765
[7] http://www.quantum-espresso.org/ [see also Giannozzi P et al. 2009, *J. Phys. Condens. Matter* **21**, 395502]
[8] Sakurai M and Saito S, to be published.
[9] Fujita M, Wakabayashi K, Nakada K and Kusakabe K 1996, *J. Phys. Soc. Jpn* **65**, 1920
[10] Bai J, Zhong X, Jiang S, Huang Y and Duan X 2010, *Nat. Nanotechnol.* **5**, 190
[11] Liang X, Jung Y-S, Ismach A, Wu S, Olynick D L, Cabrini S and Bokor J 2010, *Nano Lett.* **10**, 2454
[12] Sakai Y, Koretsune T and Saito S, *Phys. Rev.* B (in press).

Journal of Physics: Conference Series **302** (2011) 012019 doi:10.1088/1742-6596/302/1/012019

Two-dimensional topological insulators and their edge states

Shuichi Murakami

Department of Physics, Tokyo Institute of Technology, Ookayama, Meguro-ku, Tokyo 152-8551, Japan

PRESTO, Japan Science and Technology Agency (JST), Kawaguchi, Saitama 332-0012, Japan

E-mail: murakami@stat.phys.titech.ac.jp

Abstract. Topological insulators are nonmagnetic insulators in the bulk, but have gapless edge/surface states characterized by Z_2 topological numbers. In this paper we review basic properties of topological insulators. We then focus on the two-dimensional topological insulators, and describe various properties of the edge states.

1. Introduction

Topological insulators (quantum spin Hall systems) are new quantum states theoretically proposed in [1, 2, 3, 4, 5], and then have been experimentally observed by various ways. They are nonmagnetic insulators in the bulk, but have gapless edge/surface states. They can be realized in two dimensions (2D) (Fig. 1(a)) and in three dimensions (3D) (Fig. 1(b)) . In the topological insulators, the time-reversal symmetry is assumed, i.e. nonmagnetic. The edge/surface states consist of pairs of states which have opposite spins and propagate in the opposite directions; they are degenerate by the Kramers theorem, and form pure spin currents as has been discussed in the context of spin Hall effect [6, 7].

Any nonmagnetic insulators are classified into either topological insulators or ordinary insulators by using Z_2 topological numbers, which will be discussed in the next section. In the 2D topological insulators, the Z_2 topological number ν can take two values: $\nu = 1$ for the topological

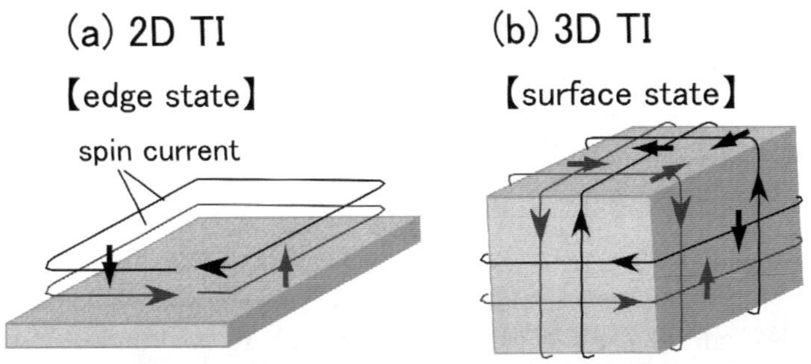

Figure 1. Schematic figures for (a) a 2D topological insulator and (b) a 3D topological insulator.

Published under licence by IOP Publishing Ltd

insulator, and $\nu = 0$ for the ordinary insulator. Nonmagnetic insulators without the spin-orbit coupling are ordinary insulators, and when the spin-orbit coupling becomes stronger, insulators may become topological insulators. Hence, the edge/surface states of topological insulators arise from the spin-orbit coupling. Roughly speaking the mechanism for the emergence of topological insulators from the spin-orbit coupling is the following. The spin-orbit coupling acts as a "spin-dependent magnetic field", and it gives rise to spin-dependent quantum Hall effect [2, 3]. The edge states from this spin-dependent quantum Hall effect consist of counterpropagating states with opposite spins.

These edge/surface states are called "helical". They are degenerate due to the Kramers theorem, and these degenerate pairs of states are called Kramers pairs. The important point for these helical edge/surface states is that any time-reversal-symmetric perturbation, such as nonmagnetic impurities or electron-electron interaction, cannot open a gap [8, 9]. This can be understood in the following way: the Z_2 topological number cannot change continuously when nonmagnetic perturbation is added, and the system remains the topological insulator, even in the presence of perturbation. This robustness of topological insulator against perturbations is called topological protection.

2. Topology and topological insulators

Not all materials with strong spin-orbit coupling are topological insulators; in order to distinguish between topological and ordinary insulators, one has to calculate the Z_2 topological numbers [2, 10]. First we introduce time-reversal invariant momenta (TRIM), as the momenta satisfying $\mathbf{k} \equiv -\mathbf{k}$ (mod$\mathbf{G}$), where $\mathbf{G}$ is a reciprocal lattice vector. In the two-dimensional Brillouin zone, there are four TRIM $\mathbf{k} = \frac{1}{2}(n_1\mathbf{b}_1 + n_2\mathbf{b}_2)$, $(n_1, n_2 = 0, 1)$, where $\mathbf{b}_1$ and $\mathbf{b}_2$ are the reciprocal primitive vectors. Let $\mathbf{k} = \mathbf{\Gamma}_i$ $(i = 1, 2, 3, 4)$ denote the four TRIM. Here we explain how to calculate the Z_2 topological number for inversion-symmetric systems. In this case the Bloch Hamiltonian $H(\mathbf{k})$ satisfies $PH(\mathbf{k})P^{-1} = H(-\mathbf{k})$, where P is the inversion operator (parity). Therefore, at the TRIM $\mathbf{\Gamma}_i$, $PH(\mathbf{\Gamma}_i) = H(\mathbf{\Gamma}_i)P$ is satisfied, and the eigenstates at TRIM $\Psi(\mathbf{\Gamma}_i)$ is an eigenstate of P: $P\Psi(\mathbf{\Gamma}_i) = \xi\Psi(\mathbf{\Gamma}_i)$. Because $P^2 = 1$, the parity eigenvalue ξ is either $\xi = 1$ (symmetric) or $\xi = -1$ (antisymmetric).

Let us pick up one TRIM $\mathbf{k} = \mathbf{\Gamma}_i$, and consider the product of all parity eigenvalues of Kramers pairs below the Fermi energy:

$$\delta_i = \prod_{m=1}^{N} \xi_{2m}(\mathbf{\Gamma}_i), \tag{1}$$

where $\xi_m(\mathbf{\Gamma}_i)$ denotes the parity eigenvalues of the m-th eigenstate from the lowest energy states at $\mathbf{k} = \mathbf{\Gamma}_i$. The $(2m-1)$-th and $(2m)$-th states are Kramers degenerate and share the same parity eigenvalues: $\xi_{2m-1}(\mathbf{\Gamma}_i) = \xi_{2m}(\mathbf{\Gamma}_i)$. As a result the Z_2 topolological number ν in 2D is expressed as a product of these indices δ_i $(i = 1, 2, 3, 4)$ over the four TRIM:

$$(-1)^{\nu} = \prod_{i=1}^{4} \delta_i. \tag{2}$$

If this expression is $+1$, it gives $\nu = 0$ and it is an ordinary insulator; otherwise, when it is -1, it gives $\nu = 1$ and it is a topological insulator.

To illustrate the differences of edge states between ordinary and topological insulators, we consider a semiinfinite plane of a 2D system. The translational symmetry along the boundary of the semiinfinite plane (edge) is assumed, thereby the Bloch wavenumber k along this direction is a good quantum number. Schematics for the edge states of the ordinary insulators and topological insulators are shown in Fig. 2(a) and (b), respectively. We can see the difference

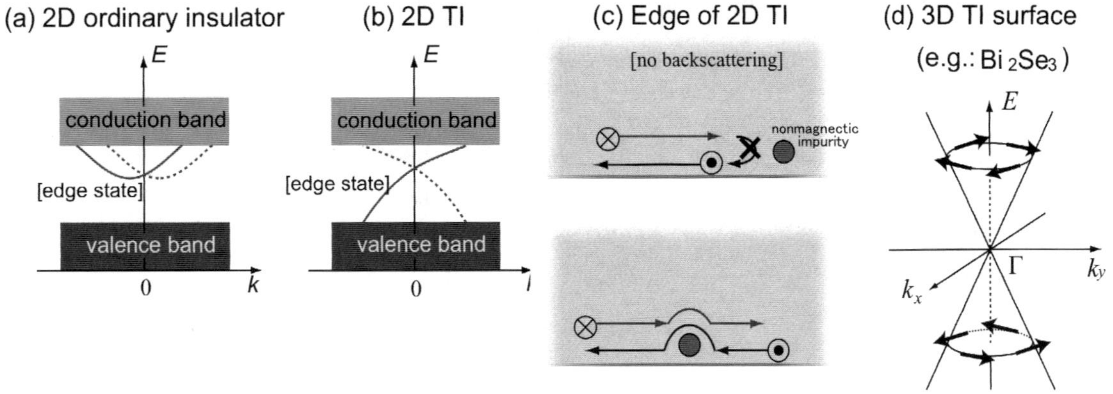

Figure 2. Schematic figure of edge states for (a) 2D ordinary insulator and (b) topological insulator. (c) Backscattering is prohibited for edge states of 2D topological insulators. (d) Dirac cone of the surface states of 3D toplogical insulators.

how the edge states are connected to bulk valence or conduction bands. In ordinary insulators the both sides of the edge-state dispersion are connected to the same bulk bands, whereas in topological insulators the edge-state dispersion connects between the bulk conduction and bulk valence bands. We note that because of the spin-orbit coupling, the edge states are spin-split in Fig. 2(a)(b), which is called the Rashba splitting. From the time-reversal symmetry, the edge states which are symmetric with respect to $k = 0$ are Kramers degenerate, and have opposite spins. It is noted that Figs. 2(a) and (b) cannot be deformed into each other continuously, without closing the bulk gap. This is guaranteed when the time-reversal symmetry is preserved, and it is a manifestation that these two insulators belong to different topological phasees.

Figures 2 (a) and (b) are the simplest cases, and we can consider other types of edge states. Any cases of edge states in nonmagnetic insulators can be classified either into ordinary or topological insulators, by counting the number of Kramers pairs of edge states on the Fermi energy. Even and odd numbers of Kramers pairs correspond then to the ordinary insulators ($\nu = 0$) and topological insulators ($\nu = 1$), respectively [10]. In Fig. 2(a), the number of pairs is either two or zero, depending on the value of the Fermi energy within the gap, and in (b) it is one. Thus (a) belongs to the ordinary insulator and (B) does to the topological insulator, in accordance with the previous explanations. This topological number ν will not change under continuous change of parameters, unless the bulk gap closes. When the bulk gap closes at some point, the topological number may change. One can classify how the bulk gap closes by changing parameters, and can associate this classification with the change of the topological number [11, 12]. As a result, universal phase diagrams between the topological and ordinary insulators have been found [13].

Here we briefly mention the 3D topological insulators. One of the typical form of surface states of the 3D topological insulators is the single Dirac cone. On the (111) surface of Bi_2Se_3 or Bi_2Te_3, the surface Fermi surface is a single Fermi surface encircling the Γ point. The surface-state dispersion is linear in the wavenumber, schematically shown in Fig. 2(d). This is called a Dirac cone. In Bi_2Se_3 and Bi_2Te_3 [14, 15, 16, 17], the surface states observed in experiments form a single Fermi surface around the Γ point, and these surface states form a single Dirac cone (Fig. 2(d)).

3. Transport on Edge states of 2D topological insulators

In the edge-state dispersion of topological insulators in Fig. 2(b), we note that the slope of the edge-state dispersion corresponds to the electron velocity $v = \frac{1}{\hbar}\frac{\partial E}{\partial k}$. Therefore, the two edge states in Fig. 2(b) are propagating in the opposite directions, and have the opposite spins due to the time-reversal symmetry. Thus these edge states carry pure spin current [6, 7]. The edge states have fixed velocity directions for each spin. Therefore, nonmagnetic impurities cannot flip spins of the edge-state electrons and hence they cannot backscatter edge states (see Fig. (2(c)). [8, 9]. As a result the edge states form perfectly conducting channels.

These perfectly conducting channels of edge states have been observed in CdTe/HgTe/CdTe quantum well [18, 19, 20]. The CdTe/HgTe/CdTe quantum well was theoretically proposed to be a 2D topological insulator [21] in 2007. This prediction is based on the special band structure of HgTe; due to the strong spin-orbit coupling, HgTe has an inverted band structure, and the gap is zero at the Γ point. Because the gap is zero, HgTe itself is not a topological insulator. To open a gap, one should break the cubic symmetry because the degeneracy between the valence and the conduction bands at the Γ point is protected by the cubic symmetry. Indeed the gap is opened by making it into a quantum well with CdTe. Detailed calculations show that the system will be a topological insulator when the well thickness d exceeds $d_c = 60\text{Å}$[21]. If the well thickness d is less than d_c it is an ordinary insulator. These predictions have been verified in transport experiments. When the Fermi energy is in the bulk band gap, there are no conducting channels for $d < d_c$ and two conducting channels for $d > d_c$. Hence the conductance is predicted to be $G = 2e^2/h$ for $d > d_c$ and $G = 0$ for $d < d_c$, which has been observed in experiments [18, 19]. These perfectly conducting channels has further been verified by nonlocal transport measurements in multiterminal geometry [20],

These perfectly conducting channels can contribute also to thermoelectric transport. In fact, there is an interesting overlap between topological insulator materials and efficient thermoelectric materials. To clarify the reason for this overlap, the thermoelectric transport has been calculated for various cases of special boundary states, such as edge states of 2D topological insulators [22], surface states of 3D topological insulators [23], and helical states on the dislocations [24] of 3D topological insulators [25]. For the edge states of 2D topological insulators [22] or the dislocation states of the 3D topological insulators [25], the electrons do not undergo elastic backscattering, which is good for thermoelectric transport. Nevertheless, in finite temperatures the inelastic scattering causes decoherence of these 1D helical states, and the otherwise good thermoelectric transport will be reduced. It is therefore predicted that at lower temperature, such helical 1D states become gradually dominant over the transport by the bulk carriers, and thermoelectric figure of merit increases.

4. Penetration depth of edge states

To measure or to control the edge states, we should note that the edge states have a finite penetration depth ℓ to the bulk. It sets the critical system size, below which the edge states from the both sides of the sample will hybridize and open a gap. To see the penetration depth, previous works used the ribbon geometry [28], where the sample has a finite width in one direction and is infinite in the other direction. In this approach, however, the penetration depth is expressed as a solution of transcendental equation, from which it is difficult to see generic behavior of the penetration depth. Instead we employ the semiinfinite plane, i.e. the $y \leq 0$ region in the xy plane. From this we can get a closed simple formula for the penetration depth ℓ [27], and can discuss generic properties of the penetration depth ℓ.

For the calculation we use the following Hamiltonian for the HgTe quantum well.

$$\mathcal{H}(k_x, k_y) = \begin{pmatrix} H(\mathbf{k}) & 0 \\ 0 & H^*(-\mathbf{k}) \end{pmatrix}, \tag{3}$$

where $H(\mathbf{k}) = \epsilon_{\mathbf{k}}\mathbf{1} + d^a(\mathbf{k})\sigma^a$, $\mathbf{1}$ is a 2×2 unit matrix, σ_a the Pauli matrices, $\epsilon_{\mathbf{k}} = C - D\left(k_x^2 + k_y^2\right)$, $d^1 = Ak_x$, $d^2 = Ak_y$, and $d^3 = \mathcal{M}(k) = M - B\left(k_x^2 + k_y^2\right)$. The eigenenergies are given by $C - Dk^2 \pm |\mathbf{d}(\mathbf{k})|$, and the bulk gap at $\mathbf{k} = 0$ is given by $2M$. For calculation on the semiinfinite plane $y \leq 0$, we replace k_y by $-i\partial_y$, and put the boundary condition for the wavefunction to vanish at $y = 0$. Then the allowed eigenstates for the surface states will be the linear combination of the states proportional to $e^{\lambda y}$ (Re$\lambda > 0$). The detailed calculation in [27] show that there are two allowed values for λ, which we call λ_1 and λ_2. The resulting wavefunction is proportional to $e^{\lambda_1 y} - e^{\lambda_2 y}$, where $\lambda_1, \lambda_2 = N \pm \sqrt{N^2 + (k_x - k_x^+)(k_x - k_x^-)}$, and $N = A/(2\sqrt{B^2 - D^2})$. $k_x^\pm$ denote the wavenumbers where the edge-state dispersion is absorbed into the bulk bands. For the parameters of HgTe quantum well, for example, λ_1 and λ_2 are real and positive. Because $\lambda_1 > \lambda_2$, as we go from the edge ($y = 0$) into the bulk, the wavefunction proportional $e^{\lambda_1 y} - e^{\lambda_2 y}$ rapidly increases within the length λ_1^{-1} and then will decrease gradually in the length scale λ_2^{-1}. Therefore the penetration depth ℓ is given by $\ell^{-1} = \lambda_2 = N - \sqrt{N^2 + (k_x - k_x^+)(k_x - k_x^-)}$. Its behavior as a function of wavenumber k is schematically shown in Fig. 3(a). Along the edge-state dispersion, when the wavenumber k approaches the bulk band, the penetration depth ℓ diverges, which is naturally expected. In the middle of the edge-state dispersion, the penetration depth ℓ becomes minimum, and it is shorter when $k_x^+ - k_x^-$ is larger, i.e. when the edge state extends in a larger region in k.

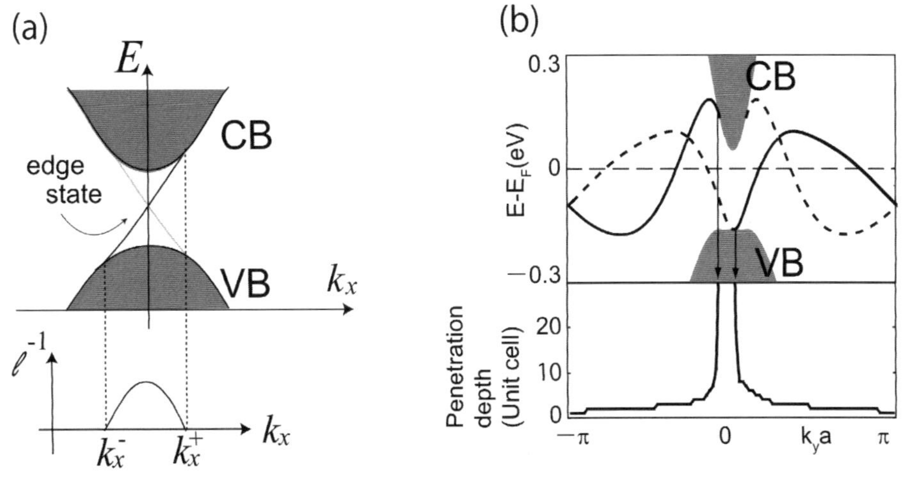

Figure 3. (a) Penetration depth ℓ for the effective model in the semiinfinite plane. CB (VB) denotes the bulk conduction (valence) band. (b) Penetration depth of the edge states on the zigzag edge of the Bi(111).

By extending this discussion, one can expect that when the dispersion extends over a wider range in k, the penetration depth in the middle of the bulk gap becomes larger. One example is the bismuth (111) bilayer film proposed by the author [26, 27]. We show in Fig. 3(b) the band structure for the zigzag edge and the penetration depth of edge states. As we can see, the edge-state dispersion extends over most of the Brillouin zone. Correspondingly, at the middle of the gap, the penetration depth becomes as short as a few lattice constants. From Fig. 3(b), the edge states consists of three Kramers pairs, and two-terminal conductance is predicted to be $G = 3 \cdot \frac{2e^2}{h} = \frac{6e^2}{h}$ [27].

5. Summary

Topological insulators belong to the class of topological phases as quantum Hall systems do. It is surprising that such topological phases can be realized in zero magnetic field, by the spin-orbit coupling only. The close interaction among theory, experiment, and first-principle calculations is strongly supporting the rapid progress in the field of topological insulators in these several years. In the coming years, we can also expect more discoveries and surprises in this field of topological insulators.

Acknowledgements

This research is supported in part by the "Nanoscience and Quantum Physics" project of Tokyo Institute of Technology through the Global Center of Excellence Program, and by Grant-in-Aid for Scientific Research (No. 21000004 and 22540327), both from the Ministry of Education, Culture, Sports, Science and Technology (MEXT).

References

[1] Kane C L and Mele E J 2005 *Phys. Rev. Lett.* **95** 226801
[2] Kane C L and Mele E J 2005 *Phys. Rev. Lett.* **95** 146802
[3] Bernevig B A and Zhang S-C 2006 *Phys. Rev. Lett.* **96** 106802
[4] Hasan M Z and Kane C L 2010 *Rev. Mod. Phys.* **82** 3045
[5] Qi X-L and Zhang S-C *arXiv* 1008.2026
[6] Murakami S, Nagaosa N and Zhang S-C 2003 *Science* **301** 1348
[7] Sinova J, Culcer D, Niu Q, Sinitsyn N A, Jungwirth T and MacDonald A H 2004 *Phys. Rev. Lett.* **92** 126603
[8] Wu C J, Bernevig B A and Zhang S C 2006 *Phys. Rev. Lett.* **96** 106401
[9] Xu C and Moore J E 2006 *Phys. Rev.* B **73** 045322
[10] Fu L and Kane C L 2006 *Phys. Rev.* B **74** 195312
[11] Murakami S, et al. 2007 *Phys. Rev.* B **76** 205304
[12] Murakami S 2007 *New J. Phys.* **9** 356
[13] Murakami S and Kuga S 2008 *Phys. Rev.* B **78** 165313
[14] Hsieh D, et al. 2009 *Phys. Rev. Lett.* **103** 146401
[15] Hsieh D, et al. 2009 *Nature* **460** 1101
[16] Zhang H J, Liu C X, Qi X L, Dai X, Fang Z and Zhang S C 2009 *Nature Phys.* **5** 438
[17] Xia Y, et al. 2009 *Nature Phys.* **5** 398
[18] König M, Wiedmann S, Brune C, Roth A, Buhmann H, Molenkamp L W, Qi X L and Zhang S C 2007 *Science* **318** 766
[19] König M, Buhmann H, Molenkamp L W, Hughes T, Liu C X, Qi X L and Zhang S C 2008 *J. Phys. Soc. Jpn.* **77** 031007
[20] Roth A, Brüne C, Buhmann H, Molenkamp L W, Maciejko J, Qi X L and Zhang S C 2009 *Science* **325** 294
[21] Bernevig B A, Hughes T L and Zhang S-C 2006 *Science* **314** 1757
[22] Takahashi R and Murakami S 2010 *Phys. Rev.* B **81** 161302(R)
[23] Ghaemi P, Mong R S K and Moore J E 2010 *Phys. Rev. Lett.* **105** 166603
[24] Ran Y, Zhang Y and Vishwanath A 2009 *Nature Phys.* **5** 298
[25] Tretiakov O A, Abanov Ar, Murakami S and Sinova J 2010 *Appl. Phys. Lett.* **97** 073108
[26] Murakami S 2006 *Phys. Rev. Lett.* **97** 236805
[27] Wada M, Murakami S, Freimuth F and Bihlmayer G 2011 *Phys. Rev. B* **83** 121310
[28] Zhou B, et al. 2008 *Phys. Rev. Lett.* **101** 246807

Beyond the Tao-Thouless limit of the fractional quantum Hall effect: spin chains and Fermi surface deformation

Masaaki Nakamura[1], Zheng-Yuan Wang[1] and Emil J. Bergholtz[2]

[1]Department of Physics, Tokyo Institute of Technology, Tokyo 152-8551, Japan,
[2]Max-Planck-Institut für Physik komplexer Systeme, Nöthnitzer Straße 38, D-01187 Dresden, Germany

Abstract. We discuss the relationship between the fractional quantum Hall effect in the vicinity of the thin-torus, a.k.a. Tao-Thouless (TT), limit and quantum spin chains. We argue that the energetics of fractional quantum Hall states in Jain sequence at filling fraction $\nu = p/(2p+1)$ (and $\nu = 1 - p/(2p+1)$) in the lowest Landau level is captured by $S = 1$ spin chains with p spins in the unit cell. These spin chains naturally arise at sub-leading order in $e^{-2\pi^2/L_1^2}$ which serves as an expansion parameter away from the TT limit ($L_1 \to 0$). We also corroborate earlier results on the smooth Fermi surface deformation of the gapless state at $\nu = 1/2$, interpolating between a state described by a critical $S = 1/2$ chain and the bulk.

1. Introduction

Ultra cold electrons in two dimensions in a strong perpendicular magnetic field, the quantum Hall (QH) system [1], exhibits a fascinating phase diagram including phases with fractionalized excitations and topological order [2, 3, 4, 5, 6]. Ever since its discovery three decades ago, the QH system has inspired a huge amount of experimental and theoretical effort, not least due to its richness in phenomenology and mathematical structure. New developments include the observation of the fractional quantum Hall effect in graphene [7] and ideas of applications in the context of topological quantum computing [8]. Moreover, it has been realized that the theoretical description of a system of rapidly rotating bosons is formally very similar to that of an electron gas in a magnetic field [9].

A key property of fractional quantum Hall (FQH) states is their topological order [10]. One consequence thereof is that their physical properties are insensitive to smooth deformations of the manifold on which we choose to study them. This fact has been exploited in a series of recent studies of the interacting many-body problem in the limit geometry of a thin torus, referred to as the Tao-Thouless (TT) limit [11, 12, 13, 14] (see also Refs. [15, 16, 17] for precursory studies and Ref. [22] for some recent related approaches).

In this proceeding we study the FQH system beyond the TT limit and argue that some characteristics that are not readily understood in the TT limit states are however manifest in the leading quantum fluctuations away from this limit. While several of the ideas presented here have been published earlier [11, 19, 20, 18], we corroborate these results and also present a number of new results.

Published under licence by IOP Publishing Ltd

We study the structure of excitation spectra of two-dimensional electrons in a magnetic field with Coulomb interactions as a function of the torus circumference, L_1, and contrast the cases of even and odd denominator filling fractions. For $\nu = 1/2$ we interpret these results in terms of a deforming Fermi sea [11]. Then, by extending the results of Refs. [19, 20] for the $\nu = 1/3$ FQH state, we further establish that FQH systems in a Jain sequence[5], which has filling factor $\nu = p/(2p+1)$ in the lowest Landau level can be described by $S = 1$ spin chains with p sites unit cell.

The remainder of this proceeding is organized as follows. In Section 2 we review the lattice describing interacting particles in a Landau level on the torus and present numerical spectra thereof as a function of the circumference, L_1. In Section 3, we briefly motivate and restate known results about the energy spectra in the TT limit and then introduce spin models that provide an intuitive understanding of the numerical data away from this limit. In particular, these considerations shed light on why odd denominator fractions remain gapped while even denominator states tend to be gapless.

2. One-dimensional description and numerical simulations

We consider a model of N interacting electrons on a torus pierced by N_s flux quanta. In the Landau gauge, a complete basis of N_s degenerate single-particle states in the lowest Landau level, labeled by $k = 0, \dots, N_s - 1$, can be chosen as

$$\psi_k(x) = (\pi^{1/2}L_1)^{-\frac{1}{2}} \sum_{n=-\infty}^{\infty} \mathrm{e}^{\mathrm{i}(k_1+nL_2)x_1} \mathrm{e}^{-\frac{1}{2}(x_2+k_1+nL_2)^2}, \tag{1}$$

where L_i are the circumferences of the torus, x_i the corresponding coordinates, and $k_1 = 2\pi k/L_1$ the momentum along the L_1-cycle. We have set the magnetic length $l_\mathrm{B} \equiv \sqrt{\hbar/eB}$ equal to unity.

In this basis, any translation-invariant two-dimensional two-body interaction Hamiltonian assumes the following model on a one-dimensional discrete lattice,

$$\mathcal{H} = \sum_{k>|m|} \hat{V}_{km}, \quad \hat{V}_{km} \equiv V_{km} \sum_i c^\dagger_{i+m} c^\dagger_{i+k} c_{i+m+k} c_i, \tag{2}$$

where the matrix-element V_{km} specifies the amplitude for a process where particles with separation $k+m$ hop m steps to a separation $k-m$ (note that m can be negative). The number of the lattice sites is fixed by the area, $N_s = L_1L_2/2\pi$. At the filling $\nu = p/q$ the Hamiltonian commutes with the center-of-mass magnetic translations [23] T_1 and T_2^q along the cycles, which implies, in particular, that the total momentum K_1 along the L_1-cycle is conserved modulo N_s in this gauge. In this system (2), two conservation numbers are given as

$$T_1 : \mathrm{e}^{\mathrm{i}2\pi K_1} = \exp\left(\mathrm{i}\frac{2\pi}{N_s}\sum_{j=1}^{N_s} j n_j\right), \quad T_2^q : \mathrm{e}^{\mathrm{i}2q\pi K_2/N_s}, \tag{3}$$

where $n_j \equiv c_j^\dagger c_j$ and $K_2 = 0, 1, \cdots, N_s/q$.

We have calculated energy spectra of this system for $\nu = 1/2, 1/3, 2/5$ using exact diagonalization in subspaces labeled by (K_1, K_2). Using these symmetries, the Hilbert space splits into $(N_s/q)^2$ sectors that are all of comparable size. This enable us to treat larger systems. The matrix elements of V_{km} are calculated assuming Coulomb interaction between the electrons. For the half-filled Landau level $\nu = 1/2$, we have calculated the energies up to $N_s = 22$ which is larger than the former analysis [11] (see Fig. 1). Increasing L_1 from the TT limit ($L_1 \to 0$), the first level-crossing in the ground state energy takes place at $L_1 \simeq 5.3$

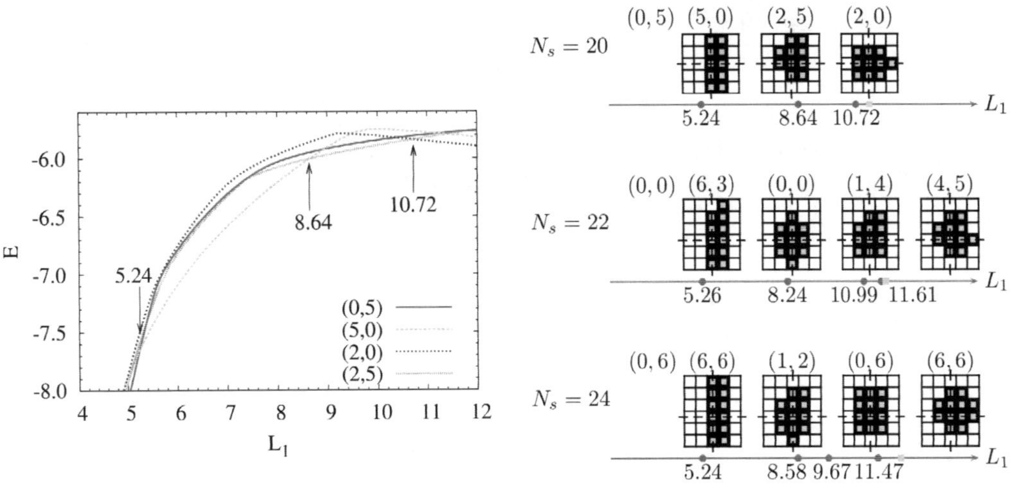

Figure 1. The left panel shows energy spectra of the Hamiltonian (2) with Coulomb potential for $\nu = 1/2$ with $N_s = 20$ as a function of L_1. The corresponding phase diagram with "Fermi sea" representation of the gapless states labeled by quantum numbers (K_1, K_2) [11] is shown in the right panel along with analogous phase diagrams for $N_s = 22$ and $N_s = 24$. The initial phase transition at $L_1 \approx 5.2 - 5.3$ as well as the level crossings between the different Fermi seas are denoted by red circles while green boxes denote the symmetric points $L_1 = L_2$ which occur at $L_1 = 11.21$ ($N_s = 20$), $L_1 = 11.76$ ($N_s = 22$) and $L_1 = 12.28$ ($N_s = 24$).

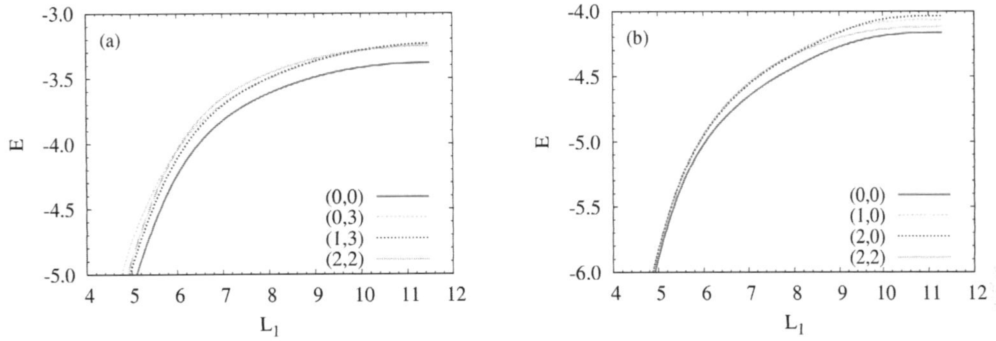

Figure 2. Energy spectra of the Hamiltonian (2) for (a) $\nu = 1/3$ with $N_s = 21$ and (b) $\nu = 2/5$ with $N_s = 20$. The energy gaps never close as functions of the circumference L_1 of the torus.

which corresponds to a phase transition from a charge-density-wave (CDW) to a gapless phase. Further level crossings occur among different gapless states, and finally L_1 arrives at the duality point $L_1 = L_2$ ($L_1 = 11.21$ for $N_s = 20$, $L_1 = 11.76$ for $N_s = 22$ and $L_1 = 12.28$ for $N_s = 24$). The spectrum is invariant under $L_1 \leftrightarrow L_2$ (although quantum numbers change) and hence we only present data for $L_1 \leq L_2$. We have calculated the energies of the $\nu = 1/3$ and $\nu = 2/5$ cases, and confirmed that energy gaps do not close as L_1 changes, and level-crossings occur only in excited states. This is consistent with the fact that these states have energy gaps in the bulk systems as is evident in the thin-torus limit (see below).

3. Beyond the Tao-Thouless limit: Effective spin chains

3.1. Tao-Thouless limit and leading quantum fluctuations

From the shape of the single particle states (1) it follows that the distance between adjacent states, the lattice constant $2\pi/L_1$, depends on L_1 while the extent of the wave functions is constant. For small L_1 the overlap between different single particle states decreases rapidly and the matrix elements V_{km} simplify considerably. As $L_1 \to 0$ one finds that

$$V_{km} \sim \mathrm{e}^{-2\pi^2 m^2/L_1^2} V_{k0}, \tag{4}$$

thus the $m \neq 0$ terms are exponentially suppressed for generic interactions in this limit. The remaining ($m = 0$) problem is exactly solvable: ground states at any $\nu = p/q$ are gapped periodic crystals (with a unit cell of p electrons on q sites) and the fractionally charged excitations appear as domain walls between degenerate ground states. The fractal structure (a.k.a. Devils staircase) of the Abelian Haldane-Halperin hierarchy [3, 4] is manifest for generic convex two-body interactions and provides an explicit estimate for the relative stability of odd denominator states in excellent agreement with experiment [14, 28].

The leading fluctuations are the $m = 1$ terms describing correlated hopping of two particles one site each in opposite directions. In a series of earlier investigations, some of us have shown that these fluctuations ultimately lead to a melting of the gapped state for $\nu = 1/2$ resulting in a gapless state [11], while for $\nu = 1/3$ the excitation structure and gap is essentially unaltered, even for very strong fluctuations [19, 20]. Below, we discuss how the numerical results at finite L_1 can be interpreted in terms of effective spin chains and extend this picture to include the most prominent ($m = 1$) Jain series, $\nu = p/(2p + 1)$ and $\nu = (p + 1)/(2(p + 1) + 1)$. The strategy is as follows. First, we truncate the interaction to only include the leading $m = 1$ term, $\hat{V}_{21}$. This is especially sensible for a pseudo potential interaction where $V_{km} = P(k, m)\mathrm{e}^{-2\pi^2(k^2+m^2)/L_1^2}$ with $P(k, m)$ being a polynomial[1] in k and m. In this case our expansion in $\mathrm{e}^{-2\pi^2/L_1^2}$ is well controlled, and we expect it to capture the physics also for more general interactions. Second, we restrict the local Hilbert space by including only local configurations that correspond to the TT ground state and local applications of $\hat{V}_{21}$. This naturally leads to a spin chain representation of the Hamiltonian and enables us to deform it to connect to well studied spin chain models and obtain physical insights by drawing analogies between the physics of spin chains and FQH states.

3.2. $S = 1/2$ chain for $\nu = 1/2$

The electron system on a torus can be described in terms of $S = 1/2$ variables in the following way [11]: In the vicinity of the TT limit, the relevant physics is captured by a model containing the three dominant terms as

$$\mathcal{H}_t = \sum_i [V_{10} n_i n_{i+1} + V_{20} n_i n_{i+2} + V_{21}(c_{i+1}^\dagger c_i c_{i+2}^\dagger c_{i+3} + \mathrm{H.c.})], \tag{5}$$

where $V_{10} > V_{20} > V_{21}$. In the TT limit, the ground state is the charge-density-wave (CDW) state where electrons are located on every two sites $|\cdots 010101010 \cdots\rangle$, since the matrix elements of the electrostatic term V_{10} is dominant. Away from the TT limit, the competition between V_{10}, V_{20} and V_{21} can be included as an interaction of the local spin states $|10\rangle \to |\uparrow\rangle$ and $|01\rangle \to |\downarrow\rangle$ which means that $c_{2n}^\dagger c_{2n+1} \to S_n^+$, $c_{2n+1}^\dagger c_{2n} \to S_n^-$, $c_{2n}^\dagger c_{2n} \to 1/2 + S_n^z$, $c_{2n+1}^\dagger c_{2n+1} \to 1/2 - S_n^z$, where n is the index of the unit cell. The effective spin Hamiltonian is then given by the $S = 1/2$ XXZ chain,

$$\mathcal{H}_{XXZ} = \sum_n \frac{1}{2}(S_n^+ S_{n+1}^- + S_n^- S_{n+1}^+) + \Delta S_n^z S_{n+1}^z, \tag{6}$$

[1] See Ref. [21] for explicit expressions.

where $\Delta = (2V_{20} - V_{10})/4V_{21}$. It is well known that this model has a phase transition from a ferromagnetic (CDW) phase to a gapless (Tomonaga-Luttinger liquid) phase at $\Delta = -1$. This describes the phase transition from the TT state to the gapless phase for the half-filled Landau level $\nu = 1/2$. The obtained solution corresponds to the (first) elongated Fermi seas in Fig. 1 appearing for $L_1 \gtrsim 5.3$.

3.3. $S = 1$ chain for $\nu = 1/3$

Next we consider spin-mapping for $\nu = 1/3$ state. In the TT limit, the ground state of the system is the three-fold degenerate CDW state $|\cdots \underline{010}\ \underline{010} \cdots\rangle$ where the underlines denote unit cells. To capture the effect of the leading quantum fluctuations, we introduce $S = 1$ spin variables as $|010\rangle \to |0\rangle$, $|100\rangle \to |+\rangle$, $|001\rangle \to |-\rangle$, where the operators are related as

$$c^\dagger_{3n} c_{3n+1} \to \frac{1}{\sqrt{2}} S^z_n S^+_n, \quad c^\dagger_{3n+1} c_{3n+2} \to \frac{1}{\sqrt{2}} S^+_n S^z_n. \tag{7}$$

In this mapping, the three-fold degeneracy of the ground state is hidden. The degenerate states have different center-of-mass quantum numbers and there are no nonvanishing matrix elements between states in different momentum sectors. It follows that there are three distinct sectors which are described by the same effective Hamiltonian,

$$\mathcal{H}_{1/3} = \sum_n -\frac{V_{21}}{2} S^z_n S^+_n S^z_{n+1} S^-_{n+1} + \text{H.c.} = \sum_n \frac{V_{21}}{2} S^+_n S^-_{n+1} [1 - (S^z_n)^2][1 - (S^z_{n+1})^2] + \text{H.c.} \tag{8}$$

This model is given by four spin interactions or the XY model with projectors to the state $|\cdots 0000 \cdots\rangle$. Note that this Hamiltonian does not have the space inversion and spin reversal symmetries: the exchange process $|00\rangle \leftrightarrow |+-\rangle$ exists, but $|00\rangle \leftrightarrow |-+\rangle$ does not. The spin rotational symmetry is also broken. Actually, numerical analysis shows that this Hamiltonian well describes the FQH state around $L_1 \simeq 7l$, by checking the overlap between the truncated Hamiltonian and full Hamiltonian with Coulomb interaction and the Trugmann-Kivelson type potential [24] which provides an exact parent Hamiltonian for the Laughlin state. There is also a term given by a function $\sum_n f(S^z_{n+1} - S^z_n)$ which stems from the electrostatic terms, but this term does not alter any essential features of the present model (8) [31]. In Refs. [19, 20], this model with extra parameters

$$\mathcal{H}(\lambda, D) = \lambda \mathcal{H}_{1/3} + (1 - \lambda)\mathcal{H}_{\text{Hei}}(D), \tag{9}$$

$$\mathcal{H}_{\text{Hei}}(D) = \sum_n \{\boldsymbol{S}_n \cdot \boldsymbol{S}_{n+1} + D(S^z_n)^2\} \tag{10}$$

has been analyzed numerically to study adiabatic continuity from a conventional spin chain ($\lambda = 0$). For $\lambda = 0$, it is known that there is a phase transition at $D = D_c (\simeq 0.968)$ between the Haldane ($D < D_c$) and large-D ($D > D_c$) phases [29]. In Refs. [19, 20] it was shown that the ground state of the model (8) is smoothly connected both to the Haldane phase and the large-D phase without closing the energy gap under the change of parameters. This indicates that the spin model describing the $\nu = 1/3$ FQH state has the nature of these two phases simultaneously, and there are indeed some similarities between the $\nu = 1/3$ FQH effect and the $S = 1$ Haldane-gap state as noticed early on [26]. The reason for the absence of a phase transition is due to the breaking of the discrete symmetries. According to Ref. [30], Haldane and large-D phases are smoothly connected if the dihedral group, time reversal and parity symmetries are broken. Since the present model (9) with $\lambda \neq 0$ breaks these three symmetries, our result is consistent with this general argument. Most properties of a finite chain of this model (9) are similar to those of a usual large-D phase where no edge spins appear, however this is not inconsistent with the absence of a phase transition [30].

$c^\dagger_{5n}c_{5n+1}$	$2^{-1/2}T^z_nT^+_n$	$c^\dagger_{5n+1}c_{5n}$	$2^{-1/2}T^-_nT^z_n$
$c^\dagger_{5n+1}c_{5n+2}$	$-2^{-1/2}T^+_nT^z_n$	$c^\dagger_{5n+2}c_{5n+1}$	$-2^{-3/2}T^z_nT^-_nS^-_nS^+_n$
$c^\dagger_{5n+2}c_{5n+3}$	$2^{-3/2}S^z_nS^+_nT^+_nT^-_n$	$c^\dagger_{5n+3}c_{5n+2}$	$2^{-1/2}S^-_nS^z_n$
$c^\dagger_{5n+3}c_{5n+4}$	$-2^{-1/2}S^+_nS^z_n$	$c^\dagger_{5n+4}c_{5n+3}$	$-2^{-1/2}S^z_nS^-_n$

Table 1. Relationship between fermion and spin-1 operators for the $\nu = 2/5$ state.

3.4. Jain fractions

Let us now consider extentions of the above $S = 1$ mapping to other filling factors. As discussed by Jain [5], $\nu = p/(2mp+1)$ FQH states may be described by the composite fermion picture, where $2m$ quantum flux are attached to non-interacting electrons of the p-th Landau level, and projection onto the lowest Landau level. For definiteness, we will consider the positive $m = 1$ Jain sequence $\nu = p/(2p+1)$, $p > 0$. The results are easily transfered to the negative Jain series by noting that our formulation is particle-hole symmetric, hence the $\nu = p/(2p+1)$ is equivalent to the $\nu = 1-p/(2p+1) = (p+1)/(2(p+1)-1)$ case. This is in contrast to the wave function based approaches and the composite fermion phenomenology, which lack this symmetry. Before giving the general result, let us consider the $\nu = 2/5$ state ($p = 2$) as the simplest example. In the TT limit, the ground state is given by the CDW state with the configuration $|\cdots \underline{01010}\ \underline{01010}\cdots\rangle$. When $\hat{V}_{21}$ and $\hat{V}^\dagger_{21} = \hat{V}_{2,-1}$ are applied to this CDW state several times, one finds that the number of electrons in each unit cell is always conserved, and the configuration $|\cdots 000 \cdots\rangle$ never appears. Therefore, configurations such as $|\cdots \underline{00100}\cdots\rangle$, $|\cdots \underline{00011}\cdots\rangle$, $|\cdots \underline{10001}\cdots\rangle$, $|\cdots \underline{11000}\cdots\rangle$ are absent in the truncated Hilbert space. This means that the $\nu = 2/5$ states can be mapped to two $S = 1$ variables by inserting 0 appropriately (between the two 1's) in each unit cell:

$$|01010\rangle \to |01[00]10\rangle \to |00\rangle, \tag{11}$$

$$|00110\rangle \to |00[10]10\rangle \to |{-0}\rangle, \tag{12}$$

$$|01100\rangle \to |01[01]00\rangle \to |0+\rangle. \tag{13}$$

The relationship between fermion and spin-1 operators (S^α_n, T^α_n) for the $\nu = 2/5$ state is summarized in Table 1. We now obtain an effective spin chain as

$$\mathcal{H}_{2/5} = \frac{V_{21}}{2}\sum_n \left[T^-_nT^z_nS^z_nS^+_n + T^z_nT^-_nS^+_nS^z_n - S^z_nS^-_nT^z_{n+1}T^+_{n+1}\right] + \text{H.c.} \tag{14}$$

Similarly, we can extend the $S = 1$ spin mapping for general p. For $p > 0$ cases, the configuration in the TT limit is given by $|\cdots \underline{0(10)_p}\cdots\rangle$, and we obtain $S = 1$ chains with p-sites unit cell (see also Fig. 3),

$$\mathcal{H}_{\frac{p}{2p+1}} = \frac{V_{21}}{2}\sum_n \left[\sum_{j=1}^{p-1}\left(S^-_{j,n}S^z_{j,n}S^z_{j+1,n}S^+_{j+1,n} + S^z_{j,n}S^-_{j,n}S^+_{j+1,n}S^z_{j+1,n}\right) - S^z_{p,n}S^-_{p,n}S^z_{1,n+1}S^+_{1,n+1}\right] + \text{H.c.} \tag{15}$$

where $S^\alpha_{j,n}$ means $S = 1$ operators of j-th site in n-th unit cell. We have again ignored the contribution from the electrostatic terms, for simplicity. Note that this model includes the $\nu = 1/3$ model in eq. (8), but is not appropriate for the $\nu = 1/2$ state. Since this model also breaks the three discrete symmetries mentioned above, we expect that it has similar low-energy properties as $S = p$ quantum spin chains with Haldane gaps. For $p \leq -2$ cases, effective spin chains can be obtained by replacing p in eq. (15) by $|p| - 1$. The detailed derivation and further analysis of this model will be published elsewhere [31].

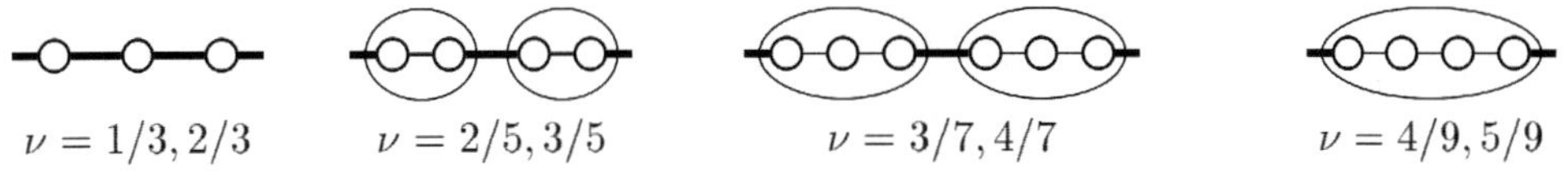

Figure 3. Effective $S = 1$ spin chains for $\nu = p/(2p+1)$ $(p \geq 1)$ FQH states given by p spins in a unit cell. Mapping of negative series can also be possible via particle-hole transformation.

4. Conclusion

We have studied the fractional quantum Hall (FQH) effect based on the one-dimensional description with torus boundary conditions and illustrated how energy spectra of interacting electrons in a magnetic field changes as functions of the circumference of the torus for various filing factors using exact diagonalization. In an effort to understand these results microscopically, we discussed spin chain mappings of the FQH states. By extending earlier results for the $\nu = 1/2$ and the $\nu = 1/3$ cases, we mapped the Jain sequences $\nu = p/(2p+1)$ and $\nu = (p+1)/(2(p+1)-1)$ to $S = 1$ chains with p spins in the unit cell that we conjecture to be gapped for all p. This further strengthens the known analogies between the FQH physics and the quantum spin chains. We have also studied the gapless state at $\nu = 1/2$ numerically. By going to slightly larger system sizes than before (from 10 to 12 particles), we confirmed that the evolution from the thin torus to the bulk is, after an initial first order transition that is accurately describe by an $S = 1/2$ spin chain model, smooth and described by a natural deformation of the quasi-particle Fermi sea.

5. Acknowledgments

We thank A. Karlhede and J. Suorsa for related collaborations, and K. Okamoto and M. Oshikawa for discussions. M. N. acknowledges support from Global Center of Excellence Program "Nanoscience and Quantum Physics" of the Tokyo Institute of Technology by MEXT.

6. References

[1] Klitzing K v, Dorda G and Pepper M 1980 *Phys. Rev. Lett.* **45** 494.; Tsui D C, Stormer H L and Gossard A C 1982 *Phys. Rev. Lett.* **48** 1559.
[2] Laughlin R B 1983 *Phys. Rev. Lett.* **50** 1395.
[3] Haldane F D M 1983 *Phys. Rev. Lett.* **51** 605.
[4] Halperin B I 1984 *Phys. Rev. Lett.* **52** 1583.
[5] Jain J K 1989 *Phys. Rev. Lett.* **63** 199; *Composite fermions*, (Cambridge University Press, 2007).
[6] Moore G and Read N 1991 *Nucl. Phys.* B **360** 362; Read N and Rezayi E H 1999 *Phys. Rev.* B **59** 8084.
[7] Du X, Skachko I, Duerr F, Luican A and Andrei E Y 2009 *Nature* **462** 192; Bolotin K I, Ghahari F, Shulman M D, Stormer H L and Kim P 2009 *Nature* **462** 196.
[8] Nayak C., Nayak, Simon S H, Stern A, Freedman M, Das Sarma S, 2008 *Rev. Mod. Phys.* **80**, 1083.
[9] Wilkin N K, Gunn J M F and Smith R A 1998 *Phys. Rev. Lett.* **80** 2265; Cooper N R and Wilkin N K 1999 *Phys Rev.* B **60** 16279(R); Regnault N and Jolicoeur Th 2003 *Phys. Rev. Lett.* **91** 030402; For a recent review see Viefers S 2008 *J. Phys.: Condens. Matter* **20** 123202.
[10] See e.g. X.-G. Wen, *Quantum Field Theory of Many-body Systems*, (Oxford, 2004).
[11] Bergholtz E J and Karlhede A 2005 *Phys. Rev. Lett.* **94** 026802; 2006 *J. Stat. Mech.* L04001; 2008 *Phys. Rev.* B **77** 155308.
[12] Seidel A, Fu H, Lee D H, Leinaas J M and Moore J 2005 *Phys. Rev. Lett.* **95** 266405.
[13] Bergholtz E J, Kailasvuori J, Wikberg E, Hansson T H and Karlhede A 2006 *Phys. Rev.* B **74** 081308(R); Seidel A and Lee D H 2006 *Phys. Rev. Lett.* **97** 056804; Read N 2006 *Phys. Rev.* B **73** 245334; Ardonne E, Bergholtz E J, Kailasvuori J and Wikberg E 2008 *J. Stat. Mech.* P04016; Ardonne E 2009 *Phys. Rev. Lett.* **102**, 180401; Seidel A 2010 *Phys. Rev. Lett.***105**, 026802.
[14] Bergholtz E J, Hansson T H, Hermanns M and Karlhede A 2007 *Phys. Rev. Lett.* **99** 256803.
[15] Tao R and Thouless D J 1983 *Phys. Rev.* B **28** 1142.

[16] Anderson P W 1983 *Phys. Rev.* B **28** 2264.
[17] Rezayi E H and Haldane F D M 1994 *Phys. Rev.* B **50** 17199.
[18] Wikberg E, Bergholtz E. J. and Karlhede A, J. Stat. Mech. (2009) P07038.
[19] Nakamura M, Bergholtz E J and Suorsa J 2010 *Phys. Rev.* B **81** 165102.
[20] Bergholtz E J, Nakamura M and Suorsa J 2011 *Physica* E **43** 755.
[21] Bergholtz E J and Karlhede A 2009 *J. Stat. Mech.* P04015.
[22] Wen X -G and Wang Z, 2008 *Phys. Rev.* B **78** 155109; Barkeshli M and Wen X -G 2009 *Phys. Rev.* B **79** 195132; Bernevig B A, Gurarie V and Simon S H 2009 *J. Phys. A: Math. Theor.* **42** 245206; Bernevig B A and Haldane F D M 2009 *Phys. Rev. Lett.* **102** 066802; Hermanns M 2010 *Phys. Rev. Lett.* **104**, 056803; Thomale R, Estienne B, Regnault N and Bernevig B A arXiv:1010.4837.
[23] Haldane F D M 1985 *Phys. Rev.* B **31** 2529.
[24] Trugman S A and Kivelson S A 1985 *Phys. Rev.* B **31** 5280.
[25] Halperin B I, Lee P A and Read N 1993 *Phys. Rev.* B **47** 7312; Rezayi E H and Read N 1994 *Phys. Rev. Lett.* **72** 900.
[26] Girvin S M and Arovas D P 1989 *Physica Scripta.* T **27** 156.
[27] Haldane F D M 1983 *Phys. Lett.* **93**A 464; 1983 *Phys. Rev. Lett.* **50** 1153.
[28] Pan W, Störmer H L, Tsui D C, Pfeiffer L N, Baldwin K W and West K W 2003 *Phys. Rev. Lett.* **90** 016801; Pan W, Xia J S, Störmer H L, Tsui D C, Vicente C, Adams E D, Sullivan N S, Pfeiffer L N, Baldwin K W and West K W 2008 *Phys. Rev.* B **77** 075307.
[29] Chen W, Hida K and Sanctuary B C 2000 *J. Phys. Soc. Jpn.* **69** 237; 2003 *Phys. Rev* B **67** 104401; 2008 *J. Phys. Soc. Jpn.* **77** 118001.
[30] Pollmann F, Berg E, Turner A and Oshikawa M, arXiv:0909.4059 (unpublished); Pollmann F, Turner A, Berg E and Oshikawa M, 2010 *Phys. Rev.* B **81** 064439.
[31] Wang Z Y, Nakamura M and Bergholtz E J 2011 *in preparation.*

Role of evanescent wave in valley polarization through junction of mono- and bi-layer graphenes

Takeshi Nakanishi[1], Mikito Koshino[2,3] and Tsuneya Ando[2]

[1]Nanosystem Research Institute, AIST, 1–1–1 Umezono, Tsukuba 305-8568, Japan
[2]Department of Physics, Tokyo Institute of Technology, 2–12–1 Ookayama, Meguro-ku, Tokyo 152-8551, Japan

E-mail: t.nakanishi@aist.go.jp

Abstract. The origin of strong valley polarization of electron wave transmitted through boundary between mono- and bi-layer graphenes can be ascribed to an evanescent wave in the bilayer graphene. The valley polarization is further enhanced across a ribbon-like region formed by partially overlapping of two monolayer graphenes.

Graphene consists of a two-dimensional hexagonal crystal of carbon atoms, in which electron dynamics is governed by the Dirac equation [1]. The electronic states have various intriguing features. In fact, the wave functions are characterized by spinor whose orientation is inextricably linked to the direction of the electron momentum in a different manner between monolayer and bilayer graphenes [2, 3, 4]. The purpose of this paper is to elucidate the origin of valley polarization [8] induced in transmission probability through the boundary between monolayer and bilayer graphenes.

In graphenes, states associated with K and K' points or valleys are degenerate. In a graphene sheet with a finite width, localized edge states are formed, when the boundary is in a certain specific direction under appropriate conditions, and only a single right- and left-going wave can carry current at each of the K and K' points [5]. A way to make valley filtering has been proposed with the explicit use of this fact [6]. Edge states in bilayer graphene were also studied [7].

We consider a straight boundary (with zigzag form) of monolayer and bilayer graphenes arranged in the AB (Bernal) stacking as illustrated in Fig. 1 and we choose the y axis along the boundary. Electronic states are described in an effective-mass scheme. In monolayer graphene, a unit cell contains two carbon atoms denoted by A and B, and for states in the vicinity of the K point, the Schrödinger equation and the corresponding wave function are given by

$$\gamma(\vec{\sigma}\cdot\hat{\mathbf{k}})\mathbf{F}^K(\mathbf{r}) = \varepsilon\mathbf{F}^K(\mathbf{r}), \quad \mathbf{F}^K(\mathbf{r}) = \begin{pmatrix} F_A^K(\mathbf{r}) \\ F_B^K(\mathbf{r}) \end{pmatrix} = \begin{pmatrix} se^{-i\theta} \\ 1 \end{pmatrix} e^{i\mathbf{k}\cdot\mathbf{r}}, \tag{1}$$

where γ is a band parameter, $\hat{\mathbf{k}} = (\hat{k}_x, \hat{k}_y) = -i\vec{\nabla}$ a wave vector operator, $(k_x, k_y) = k(\cos\theta, \sin\theta)$ a wave vector, $s = +1$ and -1 for the conduction and valence band, respectively, σ_x and σ_y the Pauli matrices, and F_A^K and F_B^K slowly-varying envelope functions describing amplitudes at A and B sites, respectively [1].

[3] Present address: Department of Physics, Tohoku University, Sendai 980-8578, Japan

In a bilayer graphene, the bottom layer is denoted as 1 and the top layer as 2. A unit cell contains two carbon atoms denoted by A_1 and B_1 in layer 1, and A_2 and B_2 in layer 2. For the inter-layer coupling, we include coupling γ_1 between vertically neighboring atoms B_1 and A_2. The Schrödinger equation becomes

$$\begin{pmatrix} \gamma(\vec{\sigma}\cdot\hat{\mathbf{k}}) & \frac{1}{2}\gamma_1\sigma_- \\ \frac{1}{2}\gamma_1\sigma_+ & \gamma(\vec{\sigma}\cdot\hat{\mathbf{k}}) \end{pmatrix} \mathbf{F}^K(\mathbf{r}) = \varepsilon \mathbf{F}^K(\mathbf{r}), \tag{2}$$

where $\sigma_\pm = \sigma_x \pm i\sigma_y$ and $\mathbf{F}^K(\mathbf{r})$ is a four component vector consisting of F^K_{A1}, F^K_{B1}, F^K_{A2}, and F^K_{B2}.

In the energy region close to the Dirac point $\gamma k/\gamma_1 \to 0$, wave functions are mainly described by two major components on A1 and B2, and other minor components are small due to the interlayer couplings and can be eliminated. In addition to a traveling mode denoted by $\tilde{\mathbf{F}}^K$, we have evanescent modes $\mathbf{G}^K$ decaying or growing exponentially in the positive x direction in low energy region $|\varepsilon| < \gamma_1$. The major components of the traveling mode and the decaying evanescent mode are given by

$$\begin{pmatrix} \tilde{F}^K_{A1} \\ \tilde{F}^K_{B2} \end{pmatrix} = \begin{pmatrix} se^{-2i\theta} \\ 1 \end{pmatrix} e^{i\mathbf{k}\cdot\mathbf{r}}, \quad \begin{pmatrix} G^K_{A1} \\ G^K_{B2} \end{pmatrix} = \begin{pmatrix} s\gamma(\kappa_x - k_y)/\gamma_1 \\ \gamma(\kappa_x + k_y)/\gamma_1 \end{pmatrix} e^{-\kappa_x x + ik_y y}, \tag{3}$$

with $\kappa_x = \sqrt{|\varepsilon|(\gamma_1 - |\varepsilon|)/\gamma^2 + k_y^2}$. For the traveling mode, the wave function for $k_y < 0$ is complex conjugate of that for $k_y > 0$. For the evanescent mode, however, the absolute value of the amplitude is quite asymmetric between positive and negative k_y. This asymmetry can be seen by ratio G^K_{A1}/G^K_{B2} shown in the upper panel in Fig. 2.

For the K' point the Schrödinger equations are obtained by replacing $\hat{k}_y$ with $-\hat{k}_y$ and therefore the wave functions by replacing k_y with $-k_y$ in both monolayer and bilayer graphenes. Therefore, the asymmetry of $G^{K'}_{A1}/G^{K'}_{B2}$ for the K' point is opposite to that of the K point. This asymmetry is the origin of valley polarization of transmitted wave, as will be shown below.

We can derive the boundary condition for wave functions $\mathbf{F}^K$ and $\mathbf{F}^{K'}$, using their relation to the amplitude of the wave function in a tight-binding model [1]. The results for the boundary shown in Fig. 1 (a) are [8]

$$\text{(i) } F^{\upsilon}_{A1}(0,y) = F^{\upsilon}_{A}(0,y); \quad \text{(ii) } F^{\upsilon}_{B1}(0,y) = F^{\upsilon}_{B}(0,y); \quad \text{(iii) } F^{\upsilon}_{B2}(0,y) = 0, \tag{4}$$

where $\upsilon = K, K'$. These conditions do not cause mixing between the K and K' points, leading to the absence of inter-valley transmission through the boundary.

We consider transmission of electron wave injected from the K valley in the monolayer side at the Fermi level in an oblique direction with wave vector $\mathbf{k}$ in the case that electron concentration is the same in the monolayer and bilayer graphenes. The transmission of the electron wave through the boundary can explicitly be calculated by considering right- and left-going traveling modes in the monolayer graphene and a right-going traveling mode and an evanescent mode decaying in the positive x direction in the bilayer graphene.

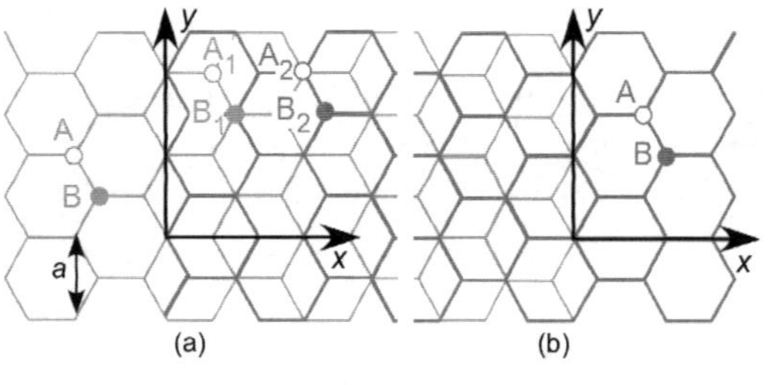

Figure 1. Atomic structure near boundaries between monolayer and bilayer graphene. Red (thick) and green (thin) lines represent the top layer with A_2 and B_2 sites, and bottom layer with A_1 and B_1 sites, respectively.

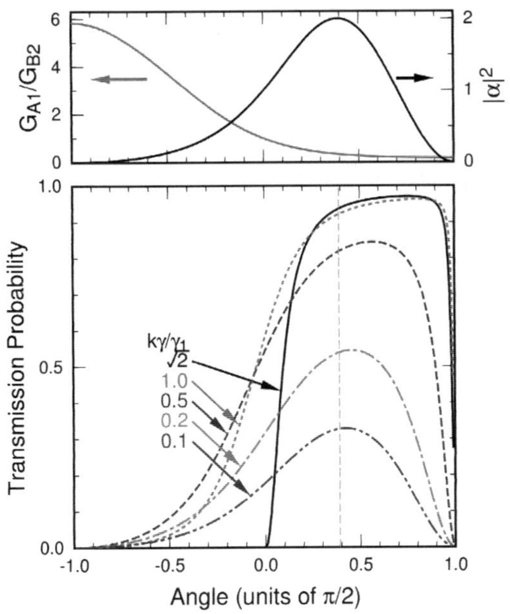

Figure 2. Upper panel: Ratio of G^K_{A1}/G^K_{B2} of the evanescent wave and $|\alpha|^2$ in the limit $k \to 0$. Lower panel: Calculated transmission probability for several charge densities specified by k in the monolayer graphene. The vertical dotted line shows θ_0 in the limit $k \to 0$.

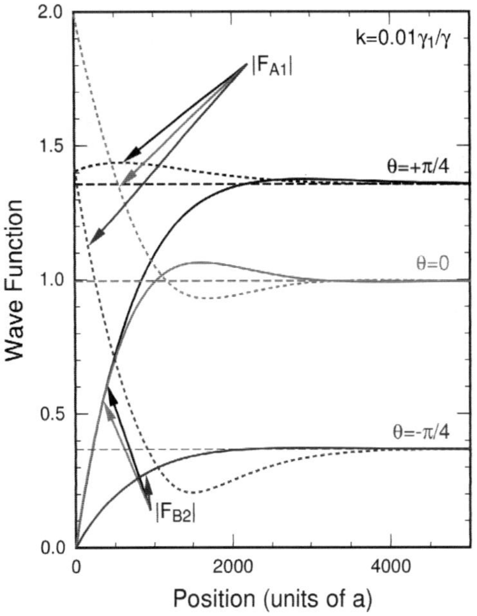

Figure 3. Calculated wave function in bilayer graphene near a boundary for incident angle $\theta = -\pi/4$, 0, and $+\pi/4$. The amplitude is normalized by that of the incident wave.

In the low energy region $\gamma k/\gamma_1 \to 0$, an approximate boundary condition is straightforwardly written down for wave functions in monolayer graphene and major components in bilayer graphene [8]. First, we note that $F^v_B(0,y)$ vanishes, because of the condition (ii) of (4) and F^v_{B1} becomes negligible for $\varepsilon \to 0$. With the use of wavefunctions (1) for incident and reflected waves, we immediately see that the reflection coefficient becomes $r_K = -1$ and therefore $F^v_A(0,y) = 2e^{ik_y y}\cos\theta$ for $k \to 0$. Then, the remaining conditions (i) and (iii) are written as

$$\alpha\tilde{F}^v_{A1}(0,y) + \beta G^v_{A1}(0,y) = F^v_A(0,y), \quad \alpha\tilde{F}^v_{B2}(0,y) + \beta G^v_{B2}(0,y) = 0. \tag{5}$$

Then, we immediately have

$$|\alpha|^2 = \frac{4\cos^2\theta}{1 + 2(G_{A1}/G_{B2})\cos(2\theta) + (G_{A1}/G_{B2})^2}, \tag{6}$$

which is also shown in the upper panel in Fig. 2. The amplitude is suppressed for $\theta < 0$ and enhanced for $\theta > 0$, corresponding to the asymmetry of G^K_{A1}/G^K_{B2}. Apart from this asymmetry, the wave function has considerable amplitude in the bilayer except at $\theta = \pm\pi/2$ in spite of the fact $r_K = -1$.

The transmission probability $T^K(\theta)$ is obtained by multiplying $|\alpha|^2$ by the ratio of the group velocity. Then, $T^K(\theta)$ vanishes for $k \to 0$ and increases in proportion to k, because the velocity is proportional to k in the bilayer side but constant in monolayer side. It takes a maximum at $\theta = s\theta_0$, with $\theta_0 = \sin^{-1}(1/\sqrt{3}) \approx 0.196\pi$. For the K' point, $T_{K'}$ is obtained by replacing θ with $-\theta$. The opposite asymmetry between the K and K' points gives rise to strong valley polarization across the interface of monolayer and bilayer graphenes.

The lower panel in Fig. 2 shows an example of calculated transmission probability as a function of incident angle θ. The electron density is specified by k corresponding to the Fermi energy in the monolayer and the results in the low-density regime $\gamma k/\gamma_1 \leq \sqrt{2}$ are shown. At the bottom of the first excited conduction band, i.e., $k\gamma/\gamma_1 = \sqrt{2}$, the transmission completely vanishes in the region $\theta \leq 0$. This is closely related to the presence of a perfectly reflecting state, which emerges only for the zigzag boundary [8]. For sufficiently small $k\gamma/\gamma_1$, the result agrees with approximate $T^K(\theta)$ obtained above.

Figure 3 shows some examples of the wave function on A1 (dotted lines) and B2 (solid lines) sites in the bilayer graphene as a function of position along incident direction θ. At the boundary chosen as the origin, F_{B2} vanishes and $|F_{A1}| = \sqrt{2}$ and 0 for $\theta = \pm\pi/4$ and 0, respectively, in the low energy region $\gamma k/\gamma_1 \ll 1$. The wave functions consist of traveling and evanescent waves, and the boundary conditions are satisfied by the presence of considerable amplitude of the evanescent mode. In fact, the spatially-varying amplitude in the bilayer graphene mostly consists of the evanescent mode. The constant amplitude at the position away from the boundary corresponds to amplitude $|\alpha|$ of the transmitted wave. It is small for $\theta = -\pi/4$ than that for $\theta = +\pi/4$ as discussed above.

Next, we consider an interface shown in Fig. 1 (b), i.e., opposite to that shown in Fig. 1 (a). The boundary conditions become

$$F^v_{A2}(0,y) = F^v_A(0,y); \quad F^v_{B2}(0,y) = F^v_B(0,y); \quad F^v_{A1}(0,y) = 0, \tag{7}$$

giving $T_K(\theta)$ same as that for interface (a) for the transmission probability from the bilayer into the monolayer with incident angle θ. The situation is the same for boundaries with other three kinds of atomic structure of zigzag or armchair considered previously [8]. This may be derived with the use of symmetry relation between (a) and (b), although not discussed here.

The valley polarization of waves transmitted through a single boundary is increased when waves go through a ribbon-shaped bilayer region formed by partially overlapping monolayer graphenes, i.e., when (a) and (b) in Fig. 1 are connected with each other, In fact, the total transmission probability through the bilayer ribbon becomes $\propto T^K(\theta)^2$, when interference effects are neglected [9]. This is quite in contrast to the case that a ribbon-shaped monolayer graphene is placed on top of a monolayer graphene, where the valley polarization nearly cancels out after transmission through two boundaries [8].

In conclusion, the significant valley polarization of transmitted waves through a boundary between monolayer and bilayer graphenes is ascribed to the evanescent wave in the bilayer graphene possessing strong asymmetry between the K and K' points. Further, the valley polarization can be accumulated in sequence of appropriate types of boundaries.

This work was supported in part by Grant-in-Aid for Scientific Research on Priority Area "Carbon Nanotube Nanoelectronics," by Grant-in-Aid for Scientific Research, and by GCOE Program at Tokyo Tech "Nanoscience and Quantum Physics" from Ministry of Education, Culture, Sports, Science and Technology Japan.

References

[1] T. Ando, J. Phys. Soc. Jpn. **74**, 777 (2005).
[2] T. Ando, T. Nakanishi, and R. Saito, J. Phys. Soc. Jpn. **67**, 2857 (1998).
[3] E. McCann and V. I. Falko, Phys. Rev. Lett. **96**, 086805 (2006).
[4] M. Koshino and T. Ando, Phys. Rev. B **73**, 245403 (2006).
[5] M. Fujita, K. Wakabayashi, K. Nakada. and K. Kusakabe, J. Phys. Soc. Jpn. **65**, 1920 (1996).
[6] A. Rycerz, J. Tworzydo, and C. W. J. Beenakker, Nat. Phys. **3**, 172 (2007).
[7] E. V. Castro, N. M. R. Peres, J. M. B. Lopes dos Santos, A. H. Castro Neto, and F. Guinea, Phys. Rev. Lett. **100**, 026802 (2008).
[8] T. Nakanishi, M. Koshino, and T. Ando, Phys. Rev. B **82**, 125428 (2010).
[9] J. W. Gonzalez, H. Santos, M. Pacheco, L. Chico, and L. Brey, Phys. Rev. B **81**, 195406 (2010).

Spontaneous mass gap generation in monolayer graphene with strong coupling expansion of square/honeycomb lattice gauge theory

Yasufumi Araki
Department of Physics, The University of Tokyo, Tokyo 113-0033, Japan
E-mail: araki@nt.phys.s.u-tokyo.ac.jp

Abstract. Sublattice symmetry of the honeycomb lattice of monolayer graphene, which can be extended to a continuous chiral symmetry of Dirac fermions in the continuum limit, may be spontaneously broken due to effectively strong Coulomb interaction between electrons on the vacuum-suspended graphene. We analyze this mechanism by strong coupling expansion of lattice gauge theory, which is constructed with the original honeycomb lattice structure. Spontaneous sublattice symmetry breaking in the strong coupling limit is shown, and is compared with the result obtained from the square lattice formulation qualitatively and quantitatively. We also introduce the Kekulé distortion of the honeycomb lattice externally, and show a restoration of the sublattice symmetry at finite distortion amplitude.

1. Introduction

Since its first experimental isolation in 2004 [1], graphene, a 2-dimensional electron system easy to design and observe, has become one of the most active research areas [2]. Monolayer graphene behaves as a semimetal at half-filling, where its charge carriers (quasiparticles) can be described as massless Dirac fermions with a small Fermi velocity [3, 4]. Since the Coulomb interaction between the charged quasiparticles is effectively enhanced by such a small Fermi velocity, exciton condensate may be created and the system may become an insulator in vacuum, where the Coulomb interaction is unscreened. This mechanism is similar to the spontaneous breaking of chiral symmetry, which is a continuous extension of the sublattice symmetry of honeycomb lattice, in the strong coupling gauge theory such as quantum chromodynamics (QCD). Making use of this analogy, it has been studied via various approaches [5]. In our previous work, this mechanism has been verified around the strong coupling limit by the techniques of strong coupling expansion of lattice gauge theory, with an effective model regularized on the square lattice [6].

In this work, we study the mechanism of spontaneous breaking of sublattice symmetry by strong coupling expansion, preserving the exact honeycomb lattice structure of graphene. Results from the honeycomb lattice formulation and the square lattice formulation are compared qualitatively and quantitatively, and the validity of the linear dispersion approximation is discussed. We also investigate the behavior of the system in the presence of external lattice distortion (Kekulé distortion), which can be introduced by external materials or addition of atoms on the layer. Restoration of the sublattice symmetry is shown, at finite distortion amplitude.

2. Strong coupling expansion on the honeycomb lattice

In order to construct an effective lattice gauge theory model of graphene with the original lattice structure, we start from the conventional tight-binding Hamiltonian [3]

$$H = h \sum_{\mathbf{r}\in A} \sum_{i=1,2,3} \left[a^\dagger(\mathbf{r}) b(\mathbf{r}+\mathbf{s}_i) + b^\dagger(\mathbf{r}+\mathbf{s}_i) a(\mathbf{r}) \right], \qquad (1)$$

with the nearest-neighbor hopping amplitude h. Here, $a^\dagger$ (a) and $b^\dagger$ (b) denote creation (annihilation) operators on the triangular sublattices A and B respectively. $\mathbf{s}_i$ ($i = 1,2,3$) corresponds to the relative position between two neighboring sites, where the lattice spacing $a_{\rm Hc} \equiv |\mathbf{s}_i| = 1.42$Å. Its energy eigenvalue $E(\mathbf{k}) = \pm h|\Phi(\mathbf{k})|$, with $\Phi(\mathbf{k}) \equiv \sum_{i=1,2,3} e^{i\mathbf{k}\cdot\mathbf{s}_i}$, vanishes at two points $\mathbf{K}_\pm$ in the momentum space, which are called "Dirac points". The dispersion relation around the Dirac points is gapless and linear in the leading order of momentum, with the Fermi velocity $v_F = (3/2)a_{\rm Hc}h$ about 300 times smaller than the speed of light. In Lagrangian formalism, we can construct an Euclidean action from the tight-binding Hamiltonian in Eq.(1), which is spatially discrete and temporally continuous, with the temporal scale transformation $\tau \to \tau'/v_F$, as done in Ref.[6]. By discretizing the temporal kinetic term with lattice spacing $a_{\tau'}(\sim a_{\rm Hc})$ and requiring local U(1) gauge invariance, we obtain a "gauged" honeycomb lattice model [7]

$$\begin{aligned} S_F &= \frac{1}{2}\sum_{\mathbf{r}\in A,\tau'} \left[a^\dagger(\mathbf{r},\tau') U_{\tau'}(\mathbf{r},\tau') a(\mathbf{r},\tau'+a_{\tau'}) - \text{H.c.} \right] + \frac{1}{2}\sum_{\mathbf{r}\in B,\tau'} \left[b^\dagger(\mathbf{r},\tau') U_{\tau'}(\mathbf{r},\tau') b(\mathbf{r},\tau'+a_{\tau'}) - \text{H.c.} \right] \\ &+ \frac{a_{\tau'}h}{v_F} \sum_{\mathbf{r}\in A,\tau'} \sum_{i=1,2,3} \left[a^\dagger(\mathbf{r},\tau') U_i(\mathbf{r},\tau') b(\mathbf{r}+\mathbf{s}_i,\tau') + b^\dagger(\mathbf{r}+\mathbf{s}_i,\tau') U_i^*(\mathbf{r},\tau') a(\mathbf{r},\tau') \right], \end{aligned} \qquad (2)$$

where the U(1) link variables read $U_{\tau'}(\mathbf{r},\tau') = \exp\left[ie\int_{\tau'}^{\tau'+a_{\tau'}} d\tau' A_4\right]$ and $U_i(\mathbf{r},\tau') = \exp\left[ie\int_{\mathbf{r}}^{\mathbf{r}+\mathbf{s}_i} d\mathbf{r}\cdot\mathbf{A}\right]$, with e being the electric charge. As a result of discretization, a pair of fermion doublers are generated in the temporal direction, which we identify here as spin (up/down) degrees of freedom. Since the electromagnetic field propagates much faster than the fermions (quasiparticles), we can neglect the retardation (magnetic) effect of the electric field. Therefore, here we set the spatial link variables U_i to be unity, which is referred to as "instantaneous approximation".

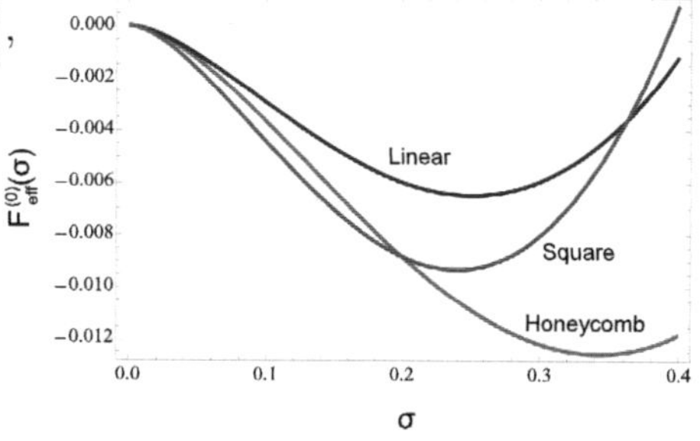

Figure 1. The free energy per one pair of A- and B-sites in the strong coupling limit, $F^{(0)}_{\rm eff}$, as a function of σ. "Honeycomb" shows $F^{(0)}_{\rm eff}(\sigma)$ in Eq.(3) with the exact form of $\Phi(\mathbf{k})$. "Linear" shows $F^{(0)}_{\rm eff}(\sigma)$ in Eq.(3) with the linear approximation $\Phi(\mathbf{K}_\pm + \mathbf{k}) \simeq (3/2)a_{\rm Hc}(\pm k_x + ik_y)$. "Square" shows the effective potential obtained by the strong coupling expansion with the square lattice formulation [6] per four square lattice sites, corresponding to a pair of honeycomb lattice sites.

The kinetic term of the gauge field, S_G, is proportional to the inverse effective coupling strength $\beta = \epsilon_0 v_F/e^2$, where ϵ_0 is the dielectric constant of vacuum. Since the value of β is 0.037 in the vacuum-suspended graphene, where the Coulomb interaction is unscreened, we expect that the strong coupling expansion by β works well. In this work, we only consider the leading order in the strong coupling expansion, so that the gauge term S_G does not need to be taken into account. By integrating out the link variables first in the path integral, we obtain the effective action S_χ only in terms of fermions. Since the U(1) link integration drops the terms in which a link variable is not canceled by its complex conjugate, 4-fermi coupling terms, which are local in the spatial directions and non-local by one lattice spacing in the temporal direction, are generated from the temporal hopping terms.

In order to treat these 4-fermi terms analytically, we introduce an auxiliary field $\sigma = \left\langle a^\dagger a - b^\dagger b \right\rangle$, which denotes the charge density imbalance between two sublattices. Namely, it serves as the order parameter of sublattice (chiral) symmetry breaking, which is referred to as "chiral condensate" in terms of the continuum effective field theory. By the mean-field approximation over the auxiliary field σ, integration by fermions can be successfully performed and we obtain the free energy per one pair of A- and B-sites in the strong coupling limit,

$$F_{\text{eff}}^{(0)}(\sigma) = \frac{1}{2}\sigma^2 - \int_{\mathbf{k}\in\Omega} d\mathbf{k} \ln\left[\left(\frac{\sigma}{2}\right)^2 + \left|\frac{a_{\tau'}h}{v_F}\Phi(\mathbf{k})\right|^2 \right], \tag{3}$$

where $\int_{\mathbf{k}\in\Omega}$ means 2-dimensional momentum integration over the Brillouin zone Ω. The first term, the tree level effect of σ, becomes dominant at large $|\sigma|$, while the second term, coming from fermion one-loop effect, becomes dominant around $\sigma = 0$. In Fig.1, we compare $F_{\text{eff}}^{(0)}(\sigma)$ in Eq.(3) with the exact form of $\Phi(\mathbf{k})$, $F_{\text{eff}}^{(0)}(\sigma)$ with $\Phi(\mathbf{k})$ approximated by a linear dispersion around the Dirac points, and $F_{\text{eff}}^{(0)}(\sigma)$ obtained from the square lattice formulation [6]. It is shown that the sublattice symmetry is spontaneously broken in the strong coupling limit, with the inter-sublattice charge density imbalance $\sigma = 0.3429$. Qualitatively, the sublattice symmetry breaking structure does not depend on the lattice formulation or the approximation of dispersion relation. It is because the most dominant singularity emerges from the excitations in vicinity of the Dirac points, which are kept unchanged in any approximations. Quantitative difference in $F_{\text{eff}}^{(0)}(\sigma)$ comes from the higher order terms in momentum, which becomes dominant far from the Dirac points in the momentum space.

3. External Kekulé distortion

Let us introduce a so-called "Kekulé distortion" pattern to the honeycomb lattice as an external field, which is characterized by the alternating pattern of strong and weak bonds, similar to a benzene molecule [8, 9, 10]. It is described by the addition of the direction-dependent hopping amplitude, $\delta h_i(\mathbf{r}) = (\Delta/3)e^{i\mathbf{G}\cdot\mathbf{r}} + e^{i\mathbf{K}_-\cdot\mathbf{s}_i}e^{-i\mathbf{G}\cdot\mathbf{r}}$, where Δ denotes the amplitude of the Kekulé distortion and $\mathbf{G} \equiv \mathbf{K}_+ - \mathbf{K}_-$. Since this distortion pattern is periodic by three times the unit lattice in the real space, the Brillouin zone Ω is split into three hexagonal regions $\tilde{\Omega}_\pm$ and $\tilde{\Omega}$, surrounding the Dirac points $\mathbf{k} = \mathbf{K}_\pm$ and the original point $\mathbf{k} = 0$ respectively. By introducing the electromagnetic field and following the strong coupling expansion in the previous section, fermion one-loop effect in $F_{\text{eff}}^{(0)}$ [the second term in Eq.(3)] is modified by the Kekulé distortion. Here we neglect the terms with the order $O(\frac{\Delta}{h}|\mathbf{k}|)$ and consider $\Phi(\mathbf{k})$ to be linear around the Dirac points in the region $\tilde{\Omega}_\pm$. If we approximate the hexagonal regions $\tilde{\Omega}$ and $\tilde{\Omega}_\pm$ by the circular regions with the same area respectively [with the radius $\rho = (8\sqrt{3}\pi/27a_{\text{Hc}}^2)^{1/2}$], the momentum integration can be analytically performed, yielding the effective potential

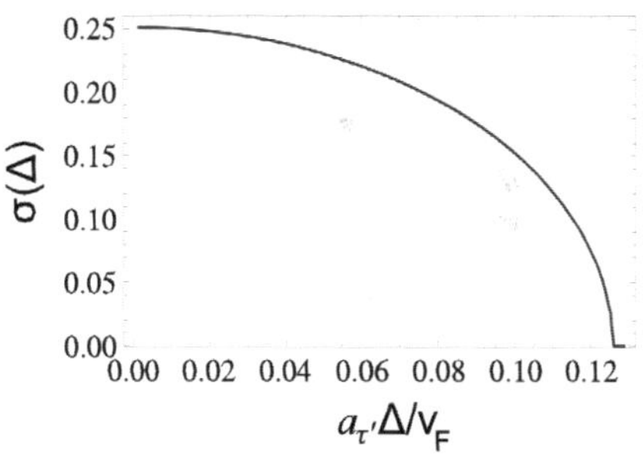

Figure 2. The order parameter for sublattice symmetry breaking σ as a function of the Kekulé distortion amplitude Δ. σ gets reduced for larger value of Δ and vanishes at a critical value Δ_C.

$$F_{\text{eff}}^{(0)}(\sigma) \simeq \frac{\sigma^2}{2} - \frac{4\pi}{\sqrt{3}}\left(\frac{3}{4\pi}\frac{a}{a_{\tau'}}\right)^2 \left\{ \left[\left(\frac{\sigma}{2}\right)^2 + \left(\frac{a_{\tau'}\Delta}{v_F}\right)^2 + a_{\tau'}^2\rho^2 \right] \ln\left[\left(\frac{\sigma}{2}\right)^2 + \left(\frac{a_{\tau'}\Delta}{v_F}\right)^2 + a_{\tau'}^2\rho^2 \right] \right.$$

$$-\left[\frac{\sigma}{2}^2+\frac{a_{T'}\Delta}{v_F}^2\right]\ln\left[\frac{\sigma}{2}^2+\frac{a_{T'}\Delta}{v_F}^2\right]\Bigg\}-\frac{1}{3}\ln\left[\frac{\sigma}{2}^2+\frac{3a_{T'}h}{v_F}^2\right]. \quad (4)$$

Since the second term in Eq.(4) is suppressed as $|\Delta| \to \infty$, fermion one-loop effect becomes less dominant at larger $|\Delta|$. It can be seen that the potential minimum reaches to $\sigma = 0$ as the distortion amplitude $|\Delta|$ increases, and finally vanishes at the finite value $|\Delta_C| = 0.126(v_F/a_{T'})$. As seen from Fig.2, this is the second-order phase transition with the mean-field exponent $z = 1/2$, where $\sigma \propto |\Delta - \Delta_C|^z$ around $\Delta = \Delta_C$.

4. Conclusion

In this work, we have shown the behavior of the sublattice symmetry of monolayer graphene analytically at the strong coupling limit of the Coulomb interaction. We have constructed a lattice Euclidean action including the electromagnetic field from the conventional tight-binding Hamiltonian defined on the honeycomb lattice. With the strong coupling expansion techniques of lattice gauge theory, we have derived the effective potential of the system as a function of the charge density imbalance between two sublattices, and have shown that the sublattice symmetry is spontaneously broken at the strong coupling limit. The structure of spontaneous breaking of the sublattice symmetry is same as that shown from the square lattice formulation [6], but they have a certain quantitative difference due to the higher order corrections in momentum neglected in the continuum effective theory.

We have also investigated the effect of external lattice distortion (Kekulé distortion) on the sublattice symmetry. In presence of external Kekulé distortion, we have shown that the spontaneous breaking of the sublattice symmetry gets suppressed and vanishes at a finite distortion amplitude. Since both the sublattice symmetry breaking and the Kekulé distortion generate a spectral gap, the sublattice symmetry gets restored although the gap remains finite at large distortion amplitude. It may be interesting to compare our results with Ref.[10], which concludes that the electron self-energy is independent of Kekulé distortion at large momentum.

Since the effect of Kekulé distortion is studied with the linear dispersion approximation in this work, quantitative accuracy may not be reliable according to the first part of this work, although qualitative symmetry breaking structure can be checked even with such an approximation. Quantitative investigation should be performed with the exact dispersion relation within the whole Brillouin zone. Moreover, higher order corrections in the strong coupling expansion should also be taken into account with the exact honeycomb lattice structure.

Acknowledgments

The author thanks C. DeTar, T. Hatsuda and S. Sasaki for valuable comments and discussions. This work is supported by Grant-in-Aid for Japan Society for the Promotion of Science (DC1, No.22.8037).

References

[1] K. S. Novoselov et al., Science **306**, 666 (2004).
[2] See, e.g. A. H. Castro Neto et al., Rev. Mod. Phys. **81**, 109 (2009).
[3] P. E. Wallace, Phys. Rev. **71**, 622 (1947).
[4] G. W. Semenoff, Phys. Rev. Lett. **53**, 2449 (1984).
[5] See, e.g. A. H. Castro Neto, Physics **2**, 30 (2009).
[6] Y. Araki and T. Hatsuda, Phys. Rev. B **82**, 121403(R) (2010); Y. Araki, arXiv:1010.0847[cond-mat.str-el] (to be published in Ann. Phys.).
[7] A. Giuliani, V. Mastropietro and M. Porta, Phys. Rev. B **82**, 121418 (2010).
[8] N. A. Viet, H. Ajiki and T. Ando, J. Phys. Soc. Jpn. **63**, 3036 (1994).
[9] C. Chamon, Phys. Rev. B **62**, 2806 (2000).
[10] R. Dillenschneider, Phys. Rev. B **78**, 115417 (2008).

Anomalous magnetoresistance on the topological surface

Takehito Yokoyama[1], Yukio Tanaka[2], and Naoto Nagaosa[3,4]

[1]Department of Physics, Tokyo Institute of Technology, Tokyo, 152-8551, Japan
[2]Department of Applied Physics, Nagoya University, Nagoya, 464-8603, Japan
[3]Department of Applied Physics, University of Tokyo, Tokyo 113-8656, Japan
[4]Cross Correlated Materials Research Group (CMRG), ASI, RIKEN, WAKO 351-0198, Japan

E-mail: `yokoyama@stat.phys.titech.ac.jp`

Abstract. We report theoretical study of charge transport in two-dimensional ferromagnet/ferromagnet junction on a topological insulator. The conductance across the interface shows anomalous dependence on the directions of the magnetizations of the two ferromagnets. This stems from the way how the wavefunctions connect between both sides. It is found that the conductance depends strongly on the in-plane direction of the magnetization. Moreover, in stark contrast to the conventional magnetoresistance effect, the conductance at the parallel configuration can be much smaller than that at the antiparallel configuration.

1. Introduction

Recently, topological insulator has received much attention both theoretically and experimentally [1, 2, 3, 4, 5] because it belongs to a new state of matter in the time-reversal symmetric systems. 3D topological insulators are characterized by two-dimensional metallic states on the surface of the sample. These surface states are protected by the time-reversal symmetry and the topology of the bulk gap, and are robust against the disorder scattering and electron-electron interactions.

On the surface of 3D topological insulator, the electrons obey the 2D Dirac equations. This corresponds to the infinite mass Rashba model[6]. This has been observed by the spin- and angle-resolved photoemission spectroscopy [7, 8]. Therefore, the next step in this field is to unveil unique property of the surface state of the topological insulators, in particular that relevant to magnetism [9, 10, 11, 12, 13, 14, 15, 16]. One remarkable feature of the Dirac fermions is that the Zeeman field acts like vector potential: the Dirac Hamiltonian is transformed as $\mathbf{k} \cdot \sigma \to (\mathbf{k} + \mathbf{m}) \cdot \sigma$ by the Zeeman field $\mathbf{m}$. Therefore, we can expect unusual spin related phenomena by the magnetic field in topological insulator.

In this paper, we study charge transport in 2D topological ferromagnet/ferromagnet junction. The ferromagnet is made of the topological surface with a ferromagnetic insulator on the top. We find anomalous magnetoresistance in this spin-valve: The conductance strongly depends on the in-plane rotation of the magnetization. In stark contrast to the conventional magnetoresistance effect, the conductance may have its minimum at the parallel configuration, while it may take a maximum near antiparallel configuration. This is attributed to the connectivity of the wavefunction across the junction. [13]

2. Formulation

We consider 2D ferromagnet/ferromagnet junctions which is abbreviated as F1/F2 below. We focus on charge transport at the Fermi level inside the bulk gap of the topological insulator, which is described by the 2D Dirac Hamiltonian

$$H = \begin{pmatrix} m_z & k_x + m_x - i(k_y + m_y) \\ k_x + m_x + i(k_y + m_y) & -m_z \end{pmatrix} \tag{1}$$

where m_x, m_y and m_z are exchange field and we set $v_F = \hbar = 1$. The ferromagnetism is induced due to the proximity effect by the ferromagnetic insulators deposited on the top. The interface is parallel to y-axis and located at $x = 0$. We choose the exchange field in the F1 side as $\mathbf{m}_1 = (m_x, m_y, m_z) = m_1(\sin\theta\cos\varphi, \sin\theta\sin\varphi, \cos\theta)$ while in the F2 side, we set $m_x = m_y = 0$ and $m_z = m_2$.

We consider the junction between different ferromagnets. This type of interface should contain a built-in electric field. Thus, we take into account the potential drop V in F2 which represents the difference of the Fermi energies in the two ferromagnets. Then, with the above Hamiltonian, we have the wave function in the F1 side:

$$\psi(x \le 0) = \frac{1}{\sqrt{2E(E - m_z)}} e^{i\tilde{k}_x x} \begin{pmatrix} \tilde{k}_x - i\tilde{k}_y \\ E - m_z \end{pmatrix} + \frac{r}{\sqrt{2E(E - m_z)}} e^{-i\tilde{k}_x x} \begin{pmatrix} -\tilde{k}_x - i\tilde{k}_y \\ E - m_z \end{pmatrix} \tag{2}$$

with $\tilde{k}_x = k_x + m_x$ and $\tilde{k}_y = k_y + m_y$, and also that in the F2 side reads

$$\psi(x \ge 0) = \frac{t}{\sqrt{2E'(E' - m_2)}} e^{ik'_x x} \begin{pmatrix} k'_x - ik_y \\ E' - m_2 \end{pmatrix} \tag{3}$$

with $E' = E - V$, where $E = \sqrt{m_z^2 + (k_x + m_x)^2 + (k_y + m_y)^2} = \pm\sqrt{m_2^2 + k_x'^2 + k_y^2} + V$. Here, $\pm$ sign corresponds to the upper and lower bands. Below, "n" and "p" mean that the Fermi level crosses the upper and the lower bands, respectively. Also, r and t are reflection and transmission coefficients, respectively. It should be noticed that the Fermi surface in the F1 is shifted by $(-m_x, -m_y)$ from the origin. Due to the translational invariance along the y-axis, the momentum k_y is conserved. Hence, the common factor $e^{ik_y y}$ is omitted above.

By matching the wavefunctions at the interface $x = 0$, we obtain the transmission coefficient t. Here, we omit the expression of t because it is rather complicated.

We parametrize $k_x + m_x = k_F \cos\phi$, and $k_y + m_y = k_F \sin\phi$. Then, we have $E = \sqrt{m_z^2 + (k_x + m_x)^2 + (k_y + m_y)^2} = \sqrt{m_z^2 + k_F^2}$.

Finally, we obtain the normalized tunneling conductance as

$$\sigma = \int_{-\frac{\pi}{2}}^{\frac{\pi}{2}} d\phi \mathrm{Re}\left[\psi^\dagger(x \ge 0) v_x \psi(x \ge 0)\right] = \frac{1}{2}\int_{-\frac{\pi}{2}}^{\frac{\pi}{2}} d\phi\, |t|^2\, \mathrm{Re}\left[\frac{k'_x}{E}\right] \tag{4}$$

where $v_x = \frac{\partial H}{\partial k_x} = \sigma_x$.

3. Results

In the following, we will fix m_1 as $m_1 = \sqrt{0.9}E$. In Fig. 1, the normalized tunneling conductance σ in n-n junction is shown for (a) $m_2 = 0$ and (b) $m_2 = \sqrt{0.9}E$. In Fig. 1 (a), the F2 is no longer ferromagnetic. However, the conductance strongly depends on the direction of the magnetization in the F1. At $\theta = 0$ or π, the mismatch of the wavefunctions between the two sides and that of the sizes of Fermi surfaces suppress σ. At $\theta = \pi/2$, on the other hand, the wavefunctions and the sizes of the Fermi surfaces are the same on both sides except the shift of Fermi surface

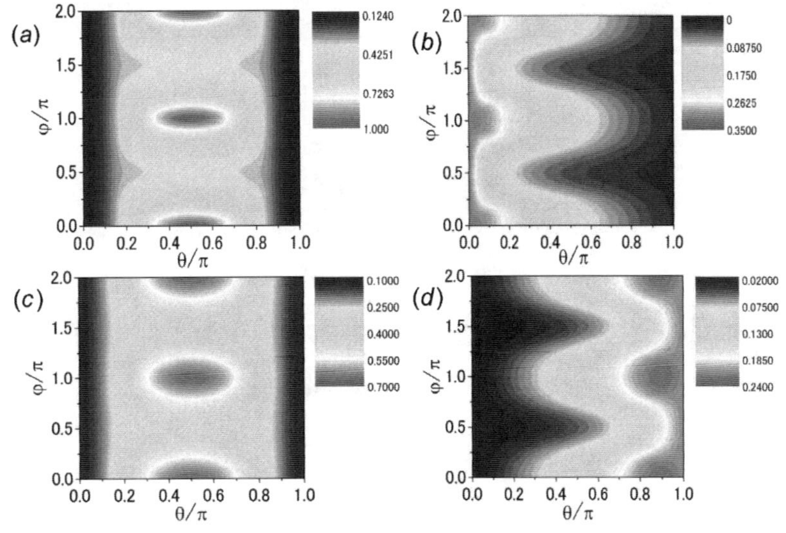

Figure 1. (Color online) tunneling conductance σ for $m_2 = 0$ ((a) and (c)), and $m_2 = \sqrt{0.9}E$ ((b) and(d)). n-n junction at $V = 0$ in (a) and (b). p-n junction at $V = 2E$ in (c) and (d).

in the momentum space due to the in-plane component of the magnetization as shown in the upper panel of Fig. 2. However, this difference of the in-plane momentum between the two sides provides a strong dependence of σ on the in-plane rotation angle φ, which is not seen in the conventional magnetoresistance effect. Since k_y is conserved, the positions of the Fermi surfaces strongly influence the charge transport: if exchange field points to x-axis, there is no evanescent wave. On the other hand, when exchange field is in y-direction, the Fermi surface moves to the k_y direction and hence the overlap region of k_y's between F1 and F2 is reduced. Therefore, the number of the evanescent modes increases and hence the conductance is strongly suppressed. Thus, we can obtain giant magnetoresistance in this system.

In Fig. 1 (b), the conductance is large at the parallel configuration ($\theta = 0$) while it is small for antiparallel configuration ($\theta = \pi$). This θ dependence, similar to the conventional magnetoresistance effect, can be explained by the overlap intergral of the wavefunctions on both sides, as discussed later.

We show tunneling conductance in p-n junction with $V = 2E$ for $m_2 = 0$ in Fig.1 (c) and $m_2 = \sqrt{0.9}E$ in Fig. 1 (d). In Fig. 1 (c), a similar tendency to Figure 1(a) is seen. In Fig. 1 (d), *in contrast to the conventional magnetoresistance effect, the conductance takes minimum at the parallel configuration ($\theta = 0$) while it takes maximum near antiparallel configuration ($\theta = \pi$).*

To understand these results intuitively, we describe the underlying physics in Fig. 2 where the arrows indicate the spin directions in the limiting case of $|m_z| \to \infty$, showing the connection of the wavefunctions on both sides. In the n-n junctions, the tunneling amplitude is determined by the overlap of the same eigenfunctions for parallel configuration ($\theta = 0$), while for antiparallel configuration ($\theta = \pi$) it is given by the overlap of the different eigenfunctions, as shown in Figs. 2 (a) and (b). Thus, the tunneling amplitude takes its maximum at $\theta = 0$, and this explains the θ dependence of the conductance in Fig. 1 (b). In a similar way, in p-n junctions, we find that the tunneling amplitude at $\theta = \pi$ becomes larger than that at $\theta = 0$ as indicated in Figs. 2 (c) and (d). This is the origin of the anomalous θ dependence of the conductance in Fig. 1 (d).

Journal of Physics: Conference Series **302** (2011) 012023 doi:10.1088/1742-6596/302/1/012023

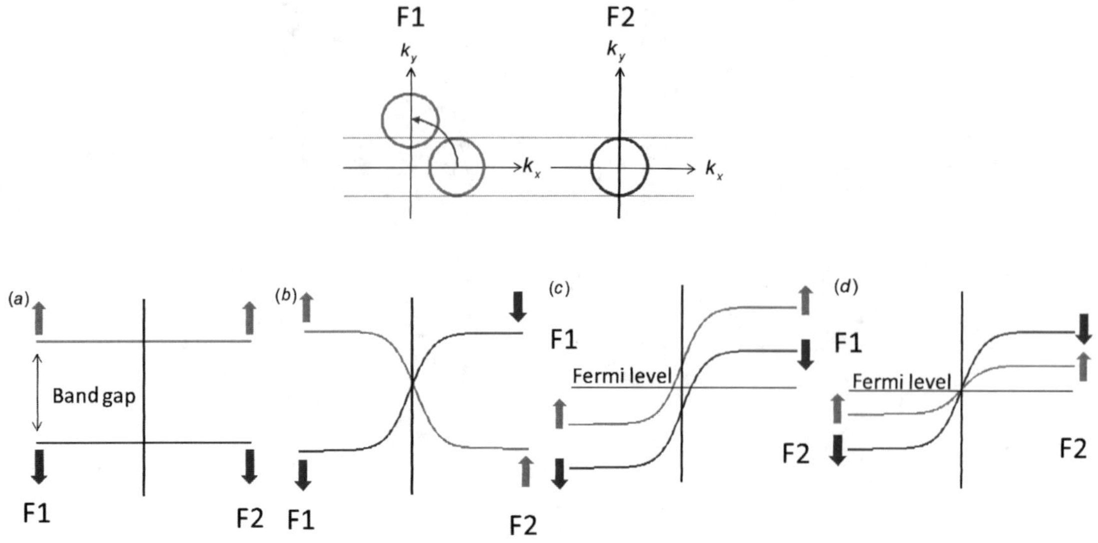

Figure 2. (Color online) (upper) Positions of Fermi surfaces. On the F1, the Fermi surface moves as illustrated, as $\mathbf{m}_1$ rotates around z-axis. (lower) Connectivity of the wavefunction across the n-n junction ((a) and (b)), and p-n junction ((c) and (d)). The magnetizations are parallel ($\theta = 0$) in (a) and (c), while they are antiparallel ($\theta = \pi$) in (b) and (d). The arrows represent the electron's spin.

4. Summary

We investigated charge transport in 2D topological ferromagnet/ferromagnet junction. The ferromagnet is made of the topological surface with a ferromagnetic insulator on the top. We found anomalous magnetoresistance in this system.

References

[1] J. E. Moore and L. Balents, Phys. Rev. B **75**, 121306(R) (2007).
[2] L. Fu, C. L. Kane, and E. J. Mele, Phys. Rev. Lett. **98**, 106803 (2007); L. Fu and C. L. Kane, Phys. Rev. B **76**, 045302 (2007).
[3] J. C. Y. Teo, L. Fu, and C. L. Kane, Phys. Rev. B **78**, 045426 (2008).
[4] X.-L. Qi, T. L. Hughes, and S.-C. Zhang, Phys. Rev. B **78**, 195424 (2008).
[5] D. Hsieh, D. Qian, L. Wray, Y. Xia, Y. S. Hor, R. J. Cava, and M. Z. Hasan, Nature **452**, 970 (2008).
[6] E. I. Rashba, Sov. Phys. Solid State **2**, 1109 (1960); Yu. A. Bychkov and E. I. Rashba, J. Phys. C **17**, 6039 (1984).
[7] D. Hsieh, Y. Xia, L. Wray, D. Qian, A. Pal, J. H. Dil, J. Osterwalder, F. Meier, G. Bihlmayer, C. L. Kane, Y. S. Hor, R. J. Cava, and M. Z. Hasan, Science **323**, 919 (2009).
[8] A. Nishide, A. A. Taskin, Y. Takeichi, T. Okuda, A. Kakizaki, T. Hirahara, K. Nakatsuji, F. Komori, Y. Ando, and I. Matsuda, Phys. Rev. B **81**, 041309 (2010).
[9] Q. Liu, C.-X. Liu, C. Xu, X.-L. Qi, and S.-C. Zhang, Phys. Rev. Lett. **102**, 156603 (2009).
[10] X.-L. Qi, R. Li, J. Zang, S.-C. Zhang, Science **323**, 1184 (2009).
[11] J. Maciejko, C. Liu, Y. Oreg, X.-L. Qi, C. Wu, S.-C. Zhang, Phys. Rev. Lett. **102**, 256803 (2009).
[12] T. Yokoyama, Y. Tanaka, and N. Nagaosa, Phys. Rev. Lett. **102**, 166801 (2009).
[13] T. Yokoyama, Y. Tanaka, and N. Nagaosa, Phys. Rev. B **81**, 121401(R) (2010).
[14] I. Garate and M. Franz, Phys. Rev. Lett. **104**, 146802 (2010).
[15] T. Yokoyama, J. Zang, and N. Nagaosa, Phys. Rev. B **81**, 241410(R) (2010).
[16] T. Yokoyama and S. Murakami, Phys. Rev. B **83**, 161407(R) (2011).

Meissner effect in the layered Kane-Mele model with Hubbard interaction

Jun Goryo and Nobuki Maeda[1]

Institute of Industrial Science, the University of Tokyo, 4-6-1 Komaba, Tokyo, Japan
[1]Department of Physics, Hokkaido University, Sapporo, Japan

E-mail: `jungoryo@iis.u-tokyo.ac.jp`

Abstract. We investigate the magnetic response in the quantum spin Hall phase of the layered Kane-Mele model with Hubbard interaction, and argue a condition to obtain the Meissner effect. The effect of Rashba spin orbit coupling is also discussed.

The Kane-Mele (KM) model is proposed to describe the quantum spin Hall effect (QSHE) of electrons on the two-dimensional honeycomb lattice[1, 2]. Here, we will show that, in a certain parameter region, the Meissner effect is obtained from the effective field theory of the layered KM model with an electronic correlation[3]. We use $\hbar = c = 1$ unit and the Minkovskian metric $g^{\mu\nu} = diag(1, -1, -1)$, where $\mu, \nu = 0, x, y$.

One of the essential ingredients of the KM model[1, 2] is the intrinsic spin-orbit (SO) coupling $\lambda_{\rm SO}$, which gives an effective magnetic field depending on spin and also an excitation gap $\Delta = 3\sqrt{3}\lambda_{\rm SO}$ to the electron[2]. Thus, the term yields quantization of the spin Hall conductivity (SHC)[1, 2],

$$\sigma^s_{xy} = \frac{e}{2\pi d}\frac{\Delta}{|\Delta|}, \tag{1}$$

where d is the interlayer distance. The model can also have the Rashba extrinsic SO coupling λ_R, which breaks the inversion symmetry and is induced by an electric field perpendicular to the honeycomb lattice plane. The term also breaks the conservation of electron spin $s_z/2$. We consider $\lambda_R = 0$ case, first.

We add the on-site Coulomb repulsion U. The Hamiltonian per layer is,

$$\begin{aligned} H &= t\sum_{<ij>} c_i^\dagger c_j + i\lambda_{\rm SO}\sum_{<<ij>>}\nu_{ij}c_i^\dagger s_z c_j \\ &\quad + U\sum_i n_{i\uparrow}n_{i\downarrow}, \end{aligned} \tag{2}$$

where c_i ($c_i^\dagger$) is the annihilation (creation) operator of an electron with spin at the i-th site and t is the nearest neighbor hopping. The second term is the intrinsic SO term consisting of the next nearest neighbor hopping, and $\nu_{ij} = \frac{2}{\sqrt{3}}(\hat{\boldsymbol{d}}_1 \times \hat{\boldsymbol{d}}_2)_z = \pm 1$, where $\hat{\boldsymbol{d}}_1$ and $\hat{\boldsymbol{d}}_2$ are unit vectors along the two bonds where the electron moving from site j to i passes. We assume that U is lower than the critical value U_c, above which the system tends toward the other insulating phase.[3]

Journal of Physics: Conference Series **302** (2011) 012024 doi:10.1088/1742-6596/302/1/012024

Namely, we consider the phase which is continuously connected to the quantum spin Hall state $U = 0$.

Let us discuss how to deal with the electron correlation U. On-site coulomb repulsion can be written by the on-site spin-spin interaction;

$$U n_{i\uparrow} n_{i\downarrow} = \frac{U}{2}(n_{i\uparrow} + n_{i\downarrow}) - \frac{U}{6}(c_i^\dagger \vec{s} c_i)^2. \tag{3}$$

The first term just gives the renormalization for the chemical potential and can be neglected. We introduce the auxiliary field $\vec{\varphi}_i$, which is a three-component vector in the spin space, and use the Stratonovich-Hubbard transformation [4]

$$H \to H_{SH} = H + \Delta H,$$

where,

$$\Delta H = \frac{U}{6}\sum_i (c_i^\dagger \vec{s} c_i - \frac{3}{2U}\vec{\varphi}_i)^2, \tag{4}$$

and therefore,

$$\begin{aligned} H_{SH} &= t\sum_{<ij>} c_i^\dagger c_j + i\lambda_{\mathrm{SO}} \sum_{<<ij>>} \nu_{ij} c_i^\dagger s_z c_j \\ &\quad - \sum_i \vec{\varphi}_i \cdot c_i^\dagger \frac{\vec{s}}{2} c_i + \frac{3}{8U}\sum_i |\vec{\varphi}_i|^2. \end{aligned} \tag{5}$$

The spin-spin interaction is eliminated in appearance, but we have a coupling between $\vec{\varphi}_i$ and the electron spin, and a quadratic term of $\vec{\varphi}_i$ instead.

We consider the continuum limit and take into account the low energy electronic excitations around K and K' points in the Brillouin Zone [1, 2]. We introduce the electromagnetic $U(1)$ gauge field [5] A_μ and $SU(2)$ spin gauge field $\vec{a}_\mu$ via covariant derivative,

$$iD_\mu = i\partial_\mu - eA_\mu + \vec{a}_\mu \cdot \frac{\vec{s}}{2}, \tag{6}$$

where $\vec{a}_0 = \vec{\varphi}$ (the auxiliary field in the continuum limit) and $\vec{\boldsymbol{a}}$ is a constant external field introduced artificially to estimate the spin current. We define a parameter

$$g = \frac{4Ua^2 d}{3} > 0, \tag{7}$$

where a is the lattice constant, and the microscopic Lagrangian density is[6],

$$\begin{aligned} \mathcal{L} &= \Psi^\dagger \left\{ iD_0 - iv(D_x \tau_z \sigma_x + D_y \sigma_y) + \Delta \tau_z \sigma_z s_z \right\} \Psi \\ &\quad + \frac{\epsilon_0 E^2}{2} - \frac{B^2}{2\mu_0} - \frac{1}{2g}|\vec{a}_0|^2, \end{aligned} \tag{8}$$

where $\Psi = \Psi_{\tau\sigma s}$ is the eight-component Fermion field which is labeled by the eigenvalues of the diagonal components of valley spin $\vec{\tau}$, sublattice spin $\vec{\sigma}$ and real spin $\vec{s}/2$. ϵ_0 and μ_0 denote the dielectric constant and magnetic permeability, respectively. Note that, except for the last term, the Lagrangian (8) possesses the $U(1)_{\mathrm{em}} \times U(1)_z$ local gauge symmetry. The $SU(2)$ gauge symmetry is broken down to $U(1)_z$, since the SO term contains s_z.

The calculation which will be shown below is quite similar to one presented in Ref.[7], although the physical meaning of the spin gauge field is different. We integrate out Ψ and obtain the one-loop effective Lagrangian for the gauge fields in the low energy and long wavelength region. The result is[7],

$$\mathcal{L}_{\text{eff}} = -\frac{1}{2g}a_0^{z2} + \mathcal{L}_{\text{ind}}, \tag{9}$$

$$\begin{aligned}\mathcal{L}_{\text{ind}} &= \sigma^s_{xy}\epsilon^{\mu\rho\nu}a^z_\mu\partial_\rho A_\nu + \frac{\epsilon E^2}{2} - \frac{B^2}{2\mu} + \frac{\delta\epsilon}{2}(\nabla a_0^z)^2 \\ &\quad +(\text{terms independent of } a^z_\mu \text{ and } A_\mu),\end{aligned} \tag{10}$$

where $\mathcal{L}_{\text{ind}}$ stands for the induced part of the effective Lagrangian. The first term in Eq. (10) is the BF term[8]. The coefficient is the quantized SHC given in Eq. (1). Note that only a^z_μ couples to the electromagnetic gauge fields. This comes from the fact that the $SU(2)$ symmetry is broken down to $U(1)_z$ symmetry by the SO coupling. Maxwell term is renormalized as[9, 7]

$$\begin{aligned}\epsilon &= \epsilon_0 + \delta\epsilon \\ &= \epsilon_0 + \frac{e^2}{6\pi|\Delta|d},\end{aligned} \tag{11}$$

$$\frac{1}{\mu} = \frac{1}{\mu_0} + \frac{e^2v^2}{6\pi|\Delta|d}. \tag{12}$$

The elastic term for $a_0^z(=\varphi^z)$ is also induced. We can recognize that any potential terms (i.e., zeroth-order terms with respect to the derivative ∂_μ) of A_μ and also a^z_μ in $\mathcal{L}_{\text{ind}}$ are absent because of $U(1)_{\text{em}} \times U(1)_z$ gauge symmetry in the Fermionic part of microscopic Lagrangian (8). Thus, the low energy and long wavelength physics of A_μ and a^z_μ is described definitely by Eq. (9).

The solution of the static equation of motion obtained from the Lagrangian (9) has been investigated [10, 11]. The argument indicates that when the dimensionless parameter satisfies

$$\sigma^{s2}_{xy}\mu g > 1, \tag{13}$$

we obtain the perfect diamagnetism as the energetically favorable solution. Namely, we obtain

$$B(x) = B(0)\frac{\mu}{\mu_0}e^{-\kappa_0 x}, \tag{14}$$

where,

$$\kappa_0^{-1} = \frac{1}{2e\sigma^s_{xy}}\sqrt{\frac{\delta\epsilon}{\mu}\frac{\sigma^{s2}_{xy}\mu g}{\sigma^{s2}_{xy}\mu g - 1}}. \tag{15}$$

shows the penetration depth. We also obtain

$$a_0^z(x) = \frac{B(0)}{\sigma^s_{xy}\mu_0}e^{-\kappa_0 x}. \tag{16}$$

We can see from the BF term in eq. (10) that electric current flows perpendicular to the gradient of $a_0(x)$ [11], and this current screens the applied magnetic field.

Let us take into account the Rashba coupling.[1, 2] We add

$$-\frac{\lambda_R}{2}\Psi^\dagger(\tau_z\sigma_x s_y - \sigma_y s_x)\Psi, \tag{17}$$

to the microscopic Lagrangian (8). Since the Rashba term breaks $U(1)_z$ symmetry, the symmetry-breaking terms are induced additionally in the effective Lagrangian Eq. (9). Up to the Gaussian approximation, the important alteration is the quadratic term of a_0^z;

$$-\frac{1}{2g}a_0^{z2} \rightarrow \frac{1}{2}\ m_a^2 - \frac{1}{g}\ a_0^{z2}, \tag{18}$$

where [7]

$$m_a^2 = \frac{\lambda_R^2}{6\pi|\Delta|d} + \mathcal{O}(\lambda_R^3). \tag{19}$$

Eq. (18) indicates that the parameter g is renormalized as

$$g \rightarrow \frac{g}{1 - gm_a^2}. \tag{20}$$

The other parameters σ_{xy}^s, ϵ and μ in Eq. (9) also receive renormalization of the order $\mathcal{O}(\lambda_R^2)$[7]. The essential point is that Eq. (13), which is the condition for the Meissner effect, is changed as (see eq. (20))

$$\frac{\sigma_{xy}^{s2}\mu g}{1 - gm_a^2} > 1. \tag{21}$$

Compared with the condition (13), Eq. (21) becomes more attainable. gm_a^2 is positive and proportional to λ_R^2, which can be changed by the electric field perpendicular to the two-dimensional plane.

The authors are grateful to N. Hatano, D. S. Hirashima, K.-I. Imura, S. Kurihara, T. Oka, M. Sato and M. Sato for their fruitful discussions and stimulated comments. J.G. is financially supported by Grant-in-Aid for Scientific Research from Japan Society for the Promotion of Science under Grant No. 18540381, and also supported by Core Research for Evolutional Science and Technology (CREST) of Japan Science and Technology Agency.

References

[1] C.L. Kane and E.J. Mele, Phys. Rev. Lett. **95**, (2005) 146802.
[2] C.L. Kane and E.J. Mele, Phys. Rev. Lett. **95**, (2005) 226801.
[3] S. Rachel and K. Le Hur: Phys. Rev. B **82** (2010) 075106.
[4] G. D. Mahan, *Many-Particle Physics, Third edition* (Kluwer Academic, New York, 2000).
[5] We suppose that the electromagnetic field is also two-dimensional.
[6] Hereafter, we use the Lagrange formalism. The reason is that, the BF term, which will be induced later and plays the crucial role in our discussion, does not appear in the Hamilton formalism [8]. The effects which come from the BF term is expressed by the constraint and the formulation becomes complicated in the Hamilton formalism.
[7] J. Goryo, N. Maeda, and K.-I. Imura, arXiv:0905.2296.
[8] M. Blau and G. Thompson, Ann. Phys. **205**, (1991) 130; G. Y. Cho and J. E. Moore: arXiv:1011.3485 (To appear in Ann. of Phys.)
[9] G. W. Semenoff, P. Sodano, and Y.-S. Wu, Phys. Rev. Lett. **62**, (1989) 715.
[10] J. Goryo and N. Maeda, J. Phys. Soc. Japan **79** (2010) 095001.
[11] J. Goryo and N. Maeda, arXiv:01074671 (Accepted for publication in J. Phys. Soc. Jpn)
[12] M. C. Diamantini *et. al.*, Eur. Phys. J. B **53** (2006) 19.

Berry phase and thermal transport coefficients in magnon systems

R Matsumoto[1] **and S Murakami**[1,2]

[1] Department of Physics, Tokyo Institute of Technology, Tokyo 152-8551, Japan
[2] PRESTO, Japan Science and Technology Agency (JST), Kawaguchi, Saitama 332-0012, Japan

E-mail: r-matsumoto@stat.phys.titech.ac.jp

Abstract. We theoretically calculate transverse thermal transport coefficients of the magnon system in the clean limit by the linear response theory in analogy with the electron. We find that there are additional correction terms to what are calculated by the usual Kubo formula. These correction terms arise because external fields change not only the distribution function but also the current operators in thermal transport. We show that the total thermal transport coefficients are expressed by the Berry curvature in momentum space; therefore the thermal Hall effect of magnon is totally due to the property of the magnon band structure.

1. Introduction

Linear response theory is one of the most useful and general formalisms to calculate various coefficients. For example, an electrical conductivity can be calculated as the linear response of the system to an external electric field, and then one obtains the general expression for the electrical conductivity expressed by the correlation functions; this is called the Kubo formula [1]. Such derivation is often called as a "mechanical" approach because Hamiltonian of both the system and the interaction with the external fields are well-defined. However, to consider the thermal transport, one cannot use this formula as such, because the thermal gradient cannot be taken into the Hamiltonian as an external field. To put it another way, the thermal gradient is not a dynamical force which exerts force to particles directly, but a statistical force which affects only through the distribution function of particles. To get rid of this difficulty, Luttinger [2] introduced a gravitational potential whose space- or time-variation produces an energy current (or a heat current), and it can be easily incorporated into the Hamiltonian as a perturbation. Replacement of the thermal gradient with this gravitational potential allows us to use the Kubo formula to calculate the thermal transport coefficients. In this way, Smrčka and Středa [3] obtained thermal coefficients in the electron system with the linear response theory, and Oji and Středa [4] obtained more general expressions by including the interactions of electrons.

In this paper, we calculate transverse thermal transport coefficients in a boson system, e.g., a magnon system, in the same manner as in a fermion system. We derive simple expressions of the thermal coefficients, using the Bloch wave function. In this expression, we show that the thermal coefficients and thermal Hall conductivity are described as a Berry curvature in momentum space; therefore we find that the thermal Hall effect of magnon in the clean limit [5, 6] is purely due to the magnon band structure. This calculation gives a theoretical basis for understanding of dynamics of magnon wavepackets in ferromagnets [7].

2. Electron system

First we review the Smrčka and Středa's work [3, 4] for the electron systems. The relation of both the electric current and the energy current to the fields can be written as

$$\boldsymbol{J} = (L^F)_{11}\left[\boldsymbol{E}+\tfrac{T}{e}\nabla\tfrac{\mu}{T}\right]+(L^F)_{12}\left[T\nabla\tfrac{1}{T}-\tfrac{\nabla\psi}{c^2}\right], \quad (1)$$

$$\boldsymbol{J}_E = (L^F)_{12}\left[\boldsymbol{E}+\tfrac{T}{e}\nabla\tfrac{\mu}{T}\right]+(L^F)_{22}\left[T\nabla\tfrac{1}{T}-\tfrac{\nabla\psi}{c^2}\right], \quad (2)$$

where we label the thermal coefficients with superscript "F", meaning a fermion, $\boldsymbol{E}$ is an electric field, $-e$ $(e>0)$ is the electron charge, μ is the chemical potential, T is the temperature, $(L^F)_{ij}$ is the transport coefficients $(i,j=1,2)$, c is the speed of light and ψ is the gravitational field. In the presence of these fields, the electric and heat current operators, which are defined by the equation of continuity, acquire additional terms. These additional terms also produce correction terms to the thermal transport coefficients. Thus the thermal transport coefficients $(L^F)^{\alpha\beta}_{ij}$, which are expressed as $(L^F)^{\alpha\beta}_{ij}=(S^F)^{\alpha\beta}_{ij}+(M^F)^{\alpha\beta}_{ij}$ $(\alpha,\beta=x,y,z)$, can be obtained as the followings:

$$(S^F)^{\alpha\beta}_{ij} = \frac{i\hbar}{V}\int f(\eta)\mathrm{Tr}\left[j_i^\alpha\frac{dG^+}{d\eta}j_j^\beta\delta(\eta-H)-j_i^\alpha\delta(\eta-H)j_j^\beta\frac{dG^-}{d\eta}\right]d\eta, \quad (3)$$

$$(M^F)^{\alpha\beta}_{11}=0,\quad (M^F)^{\alpha\beta}_{12}=-\frac{e}{2V}\int f(\eta)\mathrm{Tr}[\delta(\eta-H)(r^\alpha v^\beta-r^\beta v^\alpha)]d\eta, \quad (4)$$

$$(M^F)^{\alpha\beta}_{22}=\frac{1}{V}\int \eta f(\eta)\mathrm{Tr}\delta(\eta-H)(r^\alpha v^\beta-r^\beta v^\alpha)d\eta+\frac{i\hbar}{4V}\int f(\eta)\mathrm{Tr}\delta(\eta-H)[v^\alpha,v^\beta]d\eta. \quad (5)$$

Here H is the unperturbed Hamiltonian of the system, V is a volume of the system, $f(\eta)=(e^{(\eta-\mu)/k_BT}+1)^{-1}$ is the Fermi distribution function, $G^\pm$ is the Green's function $G^\pm(\eta)=(\eta-H\pm i\epsilon)^{-1}$, $\boldsymbol{j}_1=-e\boldsymbol{v}$, $\boldsymbol{j}_2=\frac{1}{2}(H\boldsymbol{v}+\boldsymbol{v}H)$, and v is the velocity. $(S^F)^{\alpha\beta}_{ij}$ represent results from the usual Kubo formula since the equation (3) is indeed the Kubo formula itself; $(M^F)^{\alpha\beta}_{ij}$ represent the correction terms [3, 4].

From these results derived by Smrčka and Středa, we derive some useful equations which are applicable for generic systems including magnon systems. First, we find that $(S^F)^{\alpha\beta}_{ij}$ is expressed in terms of Berry phase by taking the trace. For example, we can write $(S^F)^{\alpha\beta}_{12}$ as follows:

$$(S^F)^{\alpha\beta}_{12}=-\frac{e}{\hbar V}\mathrm{Im}\sum_\lambda f(\varepsilon_\lambda)\left\langle\frac{\partial u_\lambda}{\partial k_\alpha}\right|(H+\varepsilon_\lambda)\left|\frac{\partial u_\lambda}{\partial k_\beta}\right\rangle. \quad (6)$$

Here the indices λ, ν denote both the band index n and wave number $\boldsymbol{k}$, and u is the periodic part of the Bloch wave function. Second, at zero temperature, $(L^F)^{\alpha\beta}_{12}$ and $(S^F)^{\alpha\beta}_{12}$ is written as $(L^F)^{\alpha\beta}_{12}=\frac{\mu}{-e}(L^F)^{\alpha\beta}_{11}=-\frac{2e\mu}{\hbar V}\sum_\lambda\Theta(\mu-\varepsilon_\lambda)\mathrm{Im}\left\langle\frac{\partial u_\lambda}{\partial k_\alpha}\middle|\frac{\partial u_\lambda}{\partial k_\beta}\right\rangle$, and $(S^F)^{\alpha\beta}_{12}=-\frac{2e}{\hbar V}\mathrm{Im}\sum_\lambda\Theta(\mu-\varepsilon_\lambda)\left\langle\frac{\partial u_\lambda}{\partial k_\alpha}\right|\frac{H+\varepsilon_\lambda}{2}\left|\frac{\partial u_\lambda}{\partial k_\beta}\right\rangle$. We note that the Fermi distribution function $f(\eta)$ becomes the step function $\Theta(\mu-\eta)$ at zero temperature. Using the relation $(L^F)^{\alpha\beta}_{12}=(S^F)^{\alpha\beta}_{12}+(M^F)^{\alpha\beta}_{12}$ and equation (4) in the zero temperature limit, we obtain the following relation:

$$\begin{aligned}\mathrm{Tr}[\delta(\mu-H)(r^\alpha v^\beta-r^\beta v^\alpha)] &= \frac{d}{d\mu}\int_{-\infty}^{\mu}\mathrm{Tr}[\delta(\eta-H)(r^\alpha v^\beta-r^\beta v^\alpha)]d\eta=\frac{2V}{-e}\frac{d}{d\mu}(M^F)^{\alpha\beta}_{12}\Big|_{T\to0}\\ &=-\frac{2}{\hbar}\frac{d}{d\mu}\sum_\lambda\Theta(\mu-\varepsilon_\lambda)\mathrm{Im}\left\langle\frac{\partial u_\lambda}{\partial k_\alpha}\right|(H+\varepsilon_\lambda-2\mu)\left|\frac{\partial u_\lambda}{\partial k_\beta}\right\rangle.\end{aligned} \quad (7)$$

This equation is useful for the later calculation in the next section. We note that the l.h.s. of equation (7) corresponds to the orbital angular momentum. Hence by following the theories in [8, 9, 10, 11, 12], one can also derive equation (7) directly. It turns out that equation (7) holds whichever the particle is boson or fermion. Therefore we can apply this equation (7) to the magnon system as well.

3. Magnon system

Next we apply these formalisms to the magnon system. Since the magnon has no charge, the electron charge $-e$ is dropped and the electric field $\boldsymbol{E}$ is replaced by a gradient of a confining potential $-\nabla U(\boldsymbol{r})$ [7]. The equation (3)-(5) can be recast to

$$(S^B)^{\alpha\beta}_{ij} = \frac{i\hbar}{V}\int \rho(\eta)\mathrm{Tr} \quad j^\alpha_i \frac{dG^+}{d\eta} j^\beta_j \delta(\eta - H) - j^\alpha_i \delta(\eta - H) j^\beta_j \frac{dG^-}{d\eta} \quad d\eta, \tag{8}$$

$$(M^B)^{\alpha\beta}_{11} = 0, (M^B)^{\alpha\beta}_{12} = \frac{1}{2V}\int \rho(\eta)\mathrm{Tr}[\delta(\eta - H)(r^\alpha v^\beta - r^\beta v^\alpha)]d\eta, \tag{9}$$

$$(M^B)^{\alpha\beta}_{22} = \frac{1}{V}\int \eta\rho(\eta)\mathrm{Tr}\delta(\eta - H)(r^\alpha v^\beta - r^\beta v^\alpha)d\eta + \frac{i\hbar}{4}\int \rho(\eta)\mathrm{Tr}\delta(\eta - H)[v^\alpha, v^\beta]d\eta, \tag{10}$$

where $\boldsymbol{j}_1 = \boldsymbol{v}$, and $\boldsymbol{j}_2 = \frac{1}{2}(H\boldsymbol{v}+\boldsymbol{v}H)$, $\rho(\eta) = (e^{(\eta-\mu)/k_\mathrm{B}T} - 1)^{-1}$ is the Bose distribution function, and we labeled the thermal coefficients in the magnon system with superscript "B", meaning a boson. As equation (6), we can similarly express $(S^B)^{\alpha\beta}_{ij}$ with the Bloch wave function. The calculation of correction terms $(M^B)^{\alpha\beta}_{12}$ and $(M^B)^{\alpha\beta}_{22}$ in terms of Berry phase is technical, because the expectation value of the position operator $r^{\alpha,\beta}$ is not well-defined in the Bloch representation. To calculate these terms, we apply equation (7) to equation (9) and (10). For example, $(M^B)^{\alpha\beta}_{12}$ is calculated as the following:

$$\begin{aligned}(M^B)^{\alpha\beta}_{12} &= \frac{1}{2V}\int_{-\infty}^{\infty} \rho(\eta) \quad -\frac{2}{\hbar}\frac{d}{d\eta}\sum_\lambda \Theta(\eta - \varepsilon_\lambda)\mathrm{Im}\left\langle \frac{\partial u_\lambda}{\partial k_\alpha}\right|(H + \varepsilon_\lambda - 2\eta)\left|\frac{\partial u_\lambda}{\partial k_\beta}\right\rangle\Bigg) d\eta \\ &= \frac{1}{\hbar V}\sum_\lambda \int_{\varepsilon_\lambda}^{\infty} \frac{d\rho(\eta)}{d\eta} \quad \mathrm{Im}\left\langle \frac{\partial u_\lambda}{\partial k_\alpha}\right|(H + \varepsilon_\lambda - 2\mu)\left|\frac{\partial u_\lambda}{\partial k_\beta}\right\rangle - 2(\eta - \mu)\mathrm{Im}\left\langle \frac{\partial u_\lambda}{\partial k_\alpha}\middle|\frac{\partial u_\lambda}{\partial k_\beta}\right\rangle \quad d\eta \\ &= \frac{1}{\hbar V}\sum_\lambda \quad -c_0(\rho(\varepsilon_\lambda))\mathrm{Im}\left\langle \frac{\partial u_\lambda}{\partial k_\alpha}\right|(H + \varepsilon_\lambda - 2\mu)\left|\frac{\partial u_\lambda}{\partial k_\beta}\right\rangle + 2c_1(\rho(\varepsilon_\lambda))k_\mathrm{B}T\mathrm{Im}\left\langle \frac{\partial u_\lambda}{\partial k_\alpha}\middle|\frac{\partial u_\lambda}{\partial k_\beta}\right\rangle \quad ,\end{aligned} \tag{11}$$

where $c_q(\rho) = \int_{\varepsilon_{n\boldsymbol{k}}}^\infty d\varepsilon(\beta\varepsilon)^q \quad -\frac{d\rho}{d\varepsilon} \quad \Big|_{\mu=0} = \int_0^\rho \log(1+t^{-1}) \quad ^q dt$. For example, $c_0(\rho) = \rho$, $c_1(\rho) = (1+\rho)\log(1+\rho) - \rho\log\rho$, $c_2(\rho) = (1+\rho) \quad \log\frac{1+\rho}{\rho} \quad ^2 - (\log\rho)^2 - 2\mathrm{Li}_2(-\rho)$, and $\mathrm{Li}_2(z)$ is the polylogarithm function.

Similarly we obtain the convenient expressions of equation (8)- (10),

$$(S^B)^{\alpha\beta}_{ij} = \frac{2}{\hbar V}\mathrm{Im}\sum_{n,\boldsymbol{k}} \rho_n \left\langle \frac{\partial u_n}{\partial k_\alpha}\right| \quad \frac{H + \varepsilon_{n\boldsymbol{k}}}{2} \quad ^q\left|\frac{\partial u_n}{\partial k_\beta}\right\rangle, \tag{12}$$

$$(M^B)^{\alpha\beta}_{11} = 0, \quad (M^B)^{\alpha\beta}_{12} = -(S^B)^{\alpha\beta}_{12} + \frac{2}{\hbar V}\mathrm{Im}\sum_{n,\boldsymbol{k}}\left[(\mu\rho_n + k_\mathrm{B}Tc_1(\rho_n))\left\langle \frac{\partial u_n}{\partial k_\alpha}\middle|\frac{\partial u_n}{\partial k_\beta}\right\rangle\right], \tag{13}$$

$$(M^B)^{\alpha\beta}_{22} = -(S^B)^{\alpha\beta}_{22} + \frac{2}{\hbar V}\mathrm{Im}\sum_{n,\boldsymbol{k}}\left[\quad \mu^2\rho_n + 2\mu k_\mathrm{B}Tc_1(\rho_n) + (k_\mathrm{B}T)^2 c_2(\rho_n) \quad \left\langle \frac{\partial u_n}{\partial k_\alpha}\middle|\frac{\partial u_n}{\partial k_\beta}\right\rangle\right], \tag{14}$$

where $\rho_n = \rho(\varepsilon_{n,\boldsymbol{k}})$, and the thermal coefficients $L^B(= S^B + M^B)$ are obtained as

$$(L^B)_{11}^{\alpha\beta} = -\frac{1}{\hbar V}\sum_{n,\boldsymbol{k}} \rho_n \Omega_n^z(\boldsymbol{k}), \tag{15}$$

$$(L^B)_{12}^{\alpha\beta} = -\frac{1}{\hbar V}\sum_{n,\boldsymbol{k}} \left(\mu\rho_n + k_{\rm B}Tc_1(\rho_n)\right)\Omega_n^z(\boldsymbol{k}), \tag{16}$$

$$(L^B)_{22}^{\alpha\beta} = -\frac{1}{\hbar V}\sum_{n,\boldsymbol{k}} \ \mu^2\rho_n + 2\mu k_{\rm B}Tc_1(\rho_n) + (k_{\rm B}T)^2 c_2(\rho_n) \ \ \Omega_n^z(\boldsymbol{k}), \tag{17}$$

where $\Omega_n^z(\boldsymbol{k})$ is the Berry curvature in momentum space: $\Omega_n^z(\boldsymbol{k}) = i\left[\langle \frac{\partial u_n}{\partial \boldsymbol{k}} | \times | \frac{\partial u_n}{\partial \boldsymbol{k}} \rangle\right]_z$. Thus we can derive the thermal Hall conductivity $\kappa^{xy} = ((L^B)_{22}^{xy} - 2\mu(L^B)_{12}^{xy} + \mu^2(L^B)_{11}^{xy})/T$ from equation (15)-(17), and obtain

$$\kappa^{xy} = \frac{2k_{\rm B}^2 T}{\hbar V}\sum_{n,\boldsymbol{k}} c_2(\rho_n){\rm Im}\left\langle \frac{\partial u_n}{\partial k_x}\middle| \frac{\partial u_n}{\partial k_y}\right\rangle = -\frac{k_{\rm B}^2 T}{\hbar V}\sum_{n,\boldsymbol{k}} c_2(\rho_n)\Omega_{n,z}(\boldsymbol{k}). \tag{18}$$

As we can see from these equations (15)-(18), the thermal transport coefficients are expressed by the Berry curvature. Therefore, the thermal Hall effect of magnon is totally due to the magnon band structure because $\Omega_n^z(\boldsymbol{k})$ depends only on the magnon band structure. Compared with the previous work [5, 6], which includes only the $(S^B)_{ij}^{\alpha\beta}$, our results include both $(S^B)_{ij}^{\alpha\beta}$ and $(M^B)_{ij}^{\alpha\beta}$.

4. Conclusion

We calculated the thermal transport coefficients in the magnon system, as an example of the boson system, by using the linear response theory in analogy with the electron system. We expressed these coefficients in terms of the Bloch wave function, and found that these coefficients and the corresponding thermal Hall conductivity are described by the Berry curvature in momentum space. Consequently, we conclude that the thermal Hall effect of the magnon in the clean limit is purely due to the magnon band structure.

Acknowledgments

We would like to thank B. I. Halperin and T. Ono for discussions. This work is partly supported in part by Grant-in-Aids from the Ministry of Education, Culture, Sports, Science and Technology of Japan (No. 21000004 and 22540327). We also acknowledge the financial support from the Global Center of Excellence Program by MEXT, Japan through the "Nanoscience and Quantum Physics" Project of the Tokyo Institute of Technology.

References

[1] Kubo R 1957 *J. Phys. Soc. J.* **12** 570
[2] Luttinger L M 1964 *Phys. Rev.* **135** A1505-A1514
[3] Smrčka L and Středa P 1977 *J. Phys.* C **10** 2153-2161
[4] Oji H and Středa P 1985 *Phys. Rev.* B **31** 7291-7295
[5] Katsura H, Nagaosa N and Lee P A 2010 *Phys. Rev. Lett.* **104** 066403
[6] Onose Y, Ideue T, Katsura H, Shiomi Y, Nagaosa N and Tokura Y 2010 *Science* **329** 297
[7] Matsumoto R and Murakami S 2011 Rotating magnon wavepacket in ferromagnets *Preprint* arXiv:1103.1221v1[cond-mat.mes-hall]
[8] Xiao D, Shi J and Niu Q 2005 *Phys. Rev. Lett.* **95** 137204
[9] Thonhauser T, Ceresoli D, Vanderbilt D and Resta R 2005 *Phys. Rev. Lett.* **95** 137205
[10] Xiao D, Yao Y, Fang Z and Niu Q 2006 *Phys. Rev. Lett.* **97** 026603
[11] Resta R 2008 *J. Phys.* C **117** 012024
[12] Bergman D L and Oganesyan V 2010 *Phys. Rev. Lett.* **104** 066601

Thermoelectric figure of merit in topological insulators

Ryuji Takahashi
Tokyo Institute of Technology, Ookayama, Meguro-ku, Tokyo 152-8551, Japan
E-mail: r-takahashi@stat.phys.titech.ac.jp

Shuichi Murakami
Tokyo Institute of Technology, Ookayama, Meguro-ku, Tokyo 152-8551, Japan
PRESTO, Japan Science and Technology Agency (JST), Kawaguchi, Saitama 332-0012, Japan
E-mail: murakami@stat.phys.titech.ac.jp

Abstract. Transport behavior of two-dimensional topological insulators is theoretically studied in narrow ribbon geometry. The system has perfectly conducting edge channels, which are free from backscattering. At high temperature, the bulk states dominate in the transport phenomenon. However, at low temperature the conducting channels along the edges become dominant. It causes a bulk-to-edge crossover. Namely, by lowering temperature, the figure of merit first decreases by a competition between the bulk and the edge transport, and then increases again because the edge transport become larger.

1. introduction

Thermoelectric transport of materials is characterized by the thermoelectric figure of merit $ZT = \frac{\sigma S^2 T}{\kappa}$ [1], where T is the temperature, σ is the electrical conductivity, S is the Seebeck coefficient, and κ is the thermal conductivity from electrons and phonons. To achieve efficient energy conversion, high ZT is favorable. One of the prevailing approaches is the phonon glass and electron crystal (PGEC) [2]. Phonon conduction reduces thermoelectric efficiency because the phonon transport increases κ. A relatively new approach is "low-dimensionality" [3]: a peaked structure in the density of states in low dimensions enhances S.

Recently we proposed that the two-dimensional topological insulators (2D TI) have a potential to enhance thermoelectric figure of merit at low temperature [4]. In this paper we explain the calculations made in [4], and discuss the results. The 2D TIs are qualitatively a new state of matter [5, 6, 7], realized experimentally. The 2D TI is an insulator in the bulk, and has gapless edge states which are stable against nonmagnetic impurities [8, 9]. Therefore we assume that in impure systems, electron conduction contributed from the edge states remains good, while phonon conduction is suppressed, satisfying the PGEC condition. At the same time, the edge currents are one-dimensional (1D), which fulfills the "low-dimensional" criterion. In addition, the 3D TI was observed in good thermoelectric materials such as $Bi_{1-x}Sb_x$ [10], Bi_2Se_3 [11], and Bi_2Te_3 [12]. Although these are 3D TIs, there is a close relationship between 2D and 3D TIs, and the candidate materials are largely in common between 2D and 3D TIs.

Therefore we study thermoelectric transport of 2D TI in this paper [4]. In 2D TIs in ribbon structure, both the bulk states and edge states contribute to the transport. To discuss transport phenomena where the edge states can have comparable or even larger contribution compared with the bulk, we set the ribbon width to be very narrow. We then show that the bulk and edge contributions to the thermoelectric phenomenon compete each other, and this competition gives rise to a bulk-to-edge crossover. The edge contribution to the transport is largely affected by inelastic scattering length ℓ of the edge states because the inelastic scattering breaks their coherence. As the system is cooled, ℓ becomes longer, and the edge states become dominant in thermoelectric transport. We note that at room temperature the bulk transport is completely dominant because ℓ might become very short.

2. Model calculation and results

The transport matrix is given by

$$\begin{pmatrix} j/q \\ w \end{pmatrix} = \begin{pmatrix} L_0 & L_1 \\ L_1 & L_2 \end{pmatrix} \begin{pmatrix} -\frac{d\mu}{dx} \\ -\frac{1}{T}\frac{dT}{dx} \end{pmatrix}, \tag{1}$$

where j is the electric current, w is the thermal current, q is the electron charge $-e$, and μ is the chemical potential. From the matrix, the transport properties are given by

$$\sigma = e^2 L_0, \quad S = -\frac{1}{eT}\frac{L_1}{L_0}, \quad \kappa_e = \frac{1}{T}\frac{L_0 L_2 - L_1^2}{L_0}, \quad ZT = \frac{L_1^2}{L_0 L_2 - L_1^2 + \kappa_L T L_0}, \tag{2}$$

where κ_e is the electron thermal conductivity, κ_L is phonon thermal conductivity, and $\kappa = \kappa_e + \kappa_L$. The total transport matrix elements are given by $L_\nu = L_\nu^{\rm e} + L_\nu^{\rm b}$, where the suffix e and b mean the edge transport and bulk transport, respectively.

Fig. 1 schematically shows the density of states. The energy E is measured from the bottom of the bulk conduction band, and the chemical potential μ is assumed to be close to zero. We neglect the hole band i.e. $-\Delta/2 \ll \mu$ where Δ is the energy gap. In the calculation we used the parameters estimated for Bi_2Te_3 at T =1.8K[4]. We assume that the edge states are perfectly conducting over the whole sample and keep its coherence. In coherent systems, transport properties are calculated by the Landauer formula, with the transmission coefficient being unity when the electron energy is in the bulk gap ($-\Delta < E < 0$). $L_\nu^{\rm e}$ is then given by

$$L_\nu^{\rm e} = \frac{-\ell}{sh}\int dE T(E)(E-\mu)^\nu \frac{\partial f}{\partial E} = c_{\rm e}\int_{-\bar{\Delta}-\bar{\mu}}^{-\bar{\mu}} x^\nu \frac{{\rm e}^x}{({\rm e}^x+1)^2}dx, \quad c_{\rm e} = \frac{2\ell}{sh}, \tag{3}$$

where f is the Fermi distribution function, h is the Planck constant, s is the cross section of the sample, the factor 2 comes from the two gapless channels of the 2D TI, $\bar{\mu} = \mu/(k_B T)$ $\bar{\Delta} = \Delta/(k_B T)$, and k_B is the Boltzmann constant. We use the gap size of Bi_2Te_3: Δ =0.15 eV.

The bulk transport is calculated by the Boltzmann equation. We assume that the relaxation time τ is constant, and the bulk band is parabolic with an effective mass m. We use only the first subband because there is a large gap between the first subband and second subband due to the narrow-ribbon confinement. The transport coefficients are calculated as

$$L_\nu^{\rm b} = \frac{-1}{s}\int_{-\infty}^{\infty}\frac{dk}{2\pi} c\tau v^2(\epsilon-\mu)^\nu \frac{\partial f(\epsilon)}{\partial \epsilon} = c_{\rm b}\int_{-\bar{\mu}}^{\infty}\frac{\sqrt{x+\bar{\mu}}\, x^\nu {\rm e}^x}{({\rm e}^x+1)^2}dx, \quad c_{\rm b} = \frac{4\sqrt{2mk_BT}\mu^* c}{esh} \tag{4}$$

where the coefficient c is the number of the carrier pockets, and μ^* is the mobility.

The results are shown in Fig. 2. μ is assumed to be less than the bottom of the second subbands. The curves "Total" in Fig. 2 (a)(b) mean that both the bulk transport and the edge transport are considered. While in the transport coefficients L_i, the bulk and edge contributions are summed, the Seebeck coefficient and ZT from Eq. (2) are not simply the sums of the bulk and edge contributions. Fig. 2(b) shows that the maximum ZT appears when the chemical potential μ is near the band edge. It comes from a competition between the bulk and the edge transport from the following argument. When the chemical potential μ is near the band edge ($\mu \sim 0$), the bulk charges are holes and edge carriers are electrons. Therefore the bulk and edge transport give positive and negative Seebeck coefficients, respectively. The Seebeck coefficient from the bulk states becomes larger when μ is in the bulk gap, while that from the edge states is larger when μ is in the bulk band. The behavior of the total Seebeck eoefficients can be understood as such bulk-edge competition. This behavior of the Seebeck coefficients also governs the behavior of ZT, which roughly agrees with the Wiedemann-Franz law. This is why maximum of ZT occurs when μ is around the band edge. The maximum ZT in Fig. 2(b) is around 0.4 at 1.8K. For comparison, experimental value of ZT of Bi_2Te_3 at room temperature is around unity [13]. Hence, in ribbon geometry at low temperature, by using the edge states, it might be possible to achieve ZT comparable to the value at room temperature.

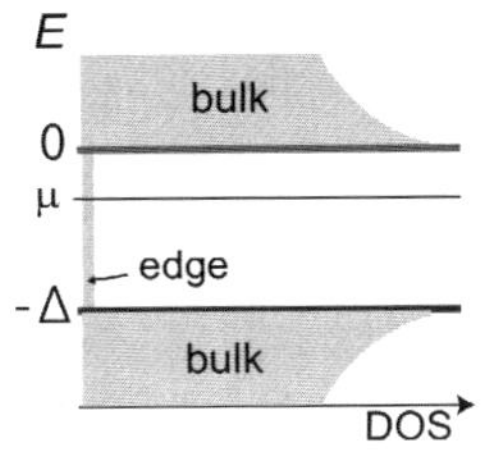

Figure 1. Schematic bands for the bulk and edge states.

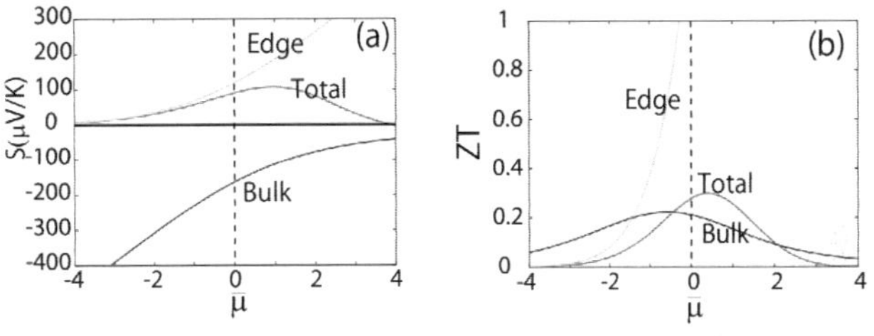

Figure 2. Calculation result of (a) the Seebeck coefficient and (b) ZT as a function of $\bar{\mu}$.

3. Search for maximum ZT

To search for optimal thermoelectric figure of merit in 2D TIs by changing the physical parameters, we define the following dimensionless parameters; $r = c_{\mathrm{e}}/c_{\mathrm{b}} = \frac{e\ell}{2\sqrt{2mk_BT\mu^*c}}$, $g = \left(\frac{\kappa_L}{k_B^2 T}\right)/c_{\mathrm{b}} = \frac{\kappa_L esh}{4\sqrt{2mk_B^5T^3\mu^*c}}$. r represents the ratio of the edge transport compared with the bulk transport, and g represents the ratio between the phonon heat transport and the bulk transport. From these parameters, we calculate ZT as a function of μ, and then study the maximum ZT_{max} when we vary μ. Fig. 3(a) shows that ZT_{max} becomes minimum at $r \sim 2.6$ because of a competition between the edge- and bulk-state transport. The value of $\bar{\mu} = \mu/(k_BT)$ where the ZT becomes maximum as a function of μ is shown in Fig. 3(b). It prominently indicates this competing transport. A jump appears in Fig. 3(b) at around $r \sim 2.6$, because ZT as a function of μ has two peaks, whose relative sizes vary as we change r. The crossover occurs when $\bar{\mu}$ is fixed and r is varied. Fig. 4 shows the value of $r = r_{\mathrm{crossover}}$ when the Seebeck coefficient vanishes as a function of the chemical potential. We see that for larger $\bar{\mu}$, the Fermi energy is deep inside the bulk band, and crossover occurs at larger r.

From Fig. 3(a), when $g = 0 = r$, ZT becomes very large; nevertheless, it cannot be realized because of relatively large phonon transport in real materials. Indeed, by increasing g (phonon transport), ZT is rapidly suppressed. Instead, in our theory we propose that increasing r (edge transport) will be a promising route for enhancing ZT.

Fig. 3(a) shows that in usual material, i.e. for $r = 0$ (no edge transport), ZT_{max} is sensitive to g (phonon transport). Disorder suppresses the heat transport, but it also kill the electric transport, and ZT is not enhanced so much. On the other hand, in the region where the r is

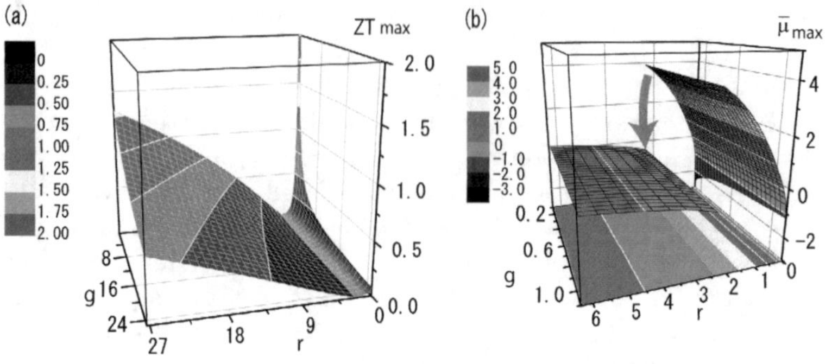

Figure 3. (a) $ZT_{\max}$ and (b) $\bar{\mu}_{\max}$ as a function of r and g.

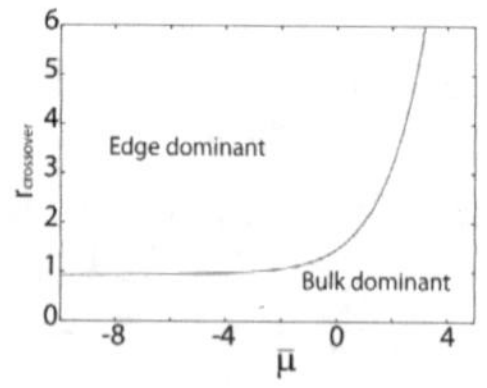

Figure 4. r giving zero Seebeck coefficient.

much larger than the crossover "valley" around $r \sim 2.6$, the figure of merit becomes relatively insensitive to g. Disorder will lower the bulk mobility, and r becomes large. Then the ZT will be enhanced. In ordinary metals, both r and g generally become large as T decreases at low temperature. An estimation on Bi_2Te_3 at $T = 1.8$K gives $r = 9.4$ and $g = 8.2$, which is located in the edge-dominated regime. We estimate the crossover temperature to be 5K-10K for Bi_2Te_3 narrow ribbon by taking into account the temperature dependence μ^*, κ_L, and ℓ. To keep the edge transport robust, the ribbon width w should be much larger than the penetration depth λ of the edge states; if w is less or equal to λ, a hybridization of the gapless edge states at the opposite edges occurs, and opens a sizable gap [14], killing the gapless edge channels.

4. Conclusion

In conclusion, we study thermoelectric properties of two-dimensional topological insulators. Around 5K-10K for narrow ribbons the edge states begin to dominate thermoelectric transport as the temperature is lowered. The bulk-to-edge crossover temperature becomes higher for longer inelastic scattering length of edge states.

Acknowledgements

This work is supported by the "Nanoscience and Quantum Physics" project of Tokyo Institute of Technology through the Global Center of Excellence Program by the Ministry of Education, Culture, Sports, Science and Technology (MEXT), and from Grant-in-Aids from MEXT.

References

[1] H. J. Goldsmid "*Thermoelectric Refrigeration*" (Plenum, New York, 1964).
[2] G.A. Slack, in "*CRC Handbook of Thermoelectrics*" (Ed. D.M. Rowe), CRC Press, Boca Raton, 1995.
[3] L. D. Hicks and M. S. Dresselhaus, Phys. Rev. B**47**, 12727 (1993); *ibid.* **47**, 16631 (1993).
[4] R. Takahashi and S. Murakami, Phys. Rev. B **81**, 161302 (2010).
[5] C. L. Kane and E. J. Mele, Phys. Rev. Lett. **95**, 146802 (2005); *ibid.* **95**, 226801 (2005).
[6] B. A. Bernevig and S.-C. Zhang, Phys. Rev. Lett. **96**, 106802 (2006).
[7] S. Murakami, Phys. Rev. Lett. **97**, 236805 (2006).
[8] C. Wu, B. A. Bernevig and S.-C. Zhang, Phys. Rev. Lett. **96**, 106401 (2006).
[9] C. Xu and J. E. Moore, Phys. Rev. B **73**, 045322 (2006).
[10] D. Hsieh *et al.*, Nature **452**, 970, (2008).
[11] Y. Xia *et al.*, Nature Phys., **5** 398, (2009).
[12] Y. L. Chen *et al.*, Science **325**, 178 (2009).
[13] G. S. Nolas, J. Sharp, H. J. Goldsmid, pp. 128-129 in "*Thermoelectrics – Basic Principles and New Materials Development*", Springer, Berlin, 2001.
[14] B. Zhou *et al.*, Phys. Rev. Lett., **101**, 246807 (2008).

Journal of Physics: Conference Series **302** (2011) 012027 doi:10.1088/1742-6596/302/1/012027

Lattice orientation of driven vortex matter in amorphous superconducting films

D. Shimamoto[1], S. Okuma[1], and N. Kokubo[2]

[1] Department of Physics and Research Center for Low Temperature Physics, Tokyo Institute of Technology, 2-12-1, Ohokayama, Meguro-ku, Tokyo 152-8551, Japan

[2] Center for Research and Advancement in Higher Education, Kyushu University, 4-2-1, Ropponmatsu, Chuoh-ku, Fukuoka 810-0044, Japan

E-mail: `sokuma@o.cc.titech.ac.jp`

Abstract. We present measurements of the mode-locking resonance in the high- eld region prior to the dynamic melting eld, where driven vortex lattices melt into a moving liquid, in an amorphous $Mo_xGe_{1\;x}$ lm. Under increasing eld, a switching of the lattice orientation from a perpendicular to parallel orientation is observed at the characteristic eld B_{ori}. We nd a trend for the perpendicular orientation to change to the parallel orientation with increasing the lattice velocity in elds just prior to B_{ori}, while B_{ori} is not sensitive to the velocity within the range studied. We suggest that the dynamic pinning force moving vortices feel is important for the lattice orientation and is signi cantly reduced around B_{ori} as a precursor of dynamic melting.

1. Introduction

Vortices in type-II superconductors form a triangular lattice by vortex-vortex interactions and its dynamics is readily controlled by a Lorentz force due to an applied current. For isotropic superconductors with random point pinning, such as amorphous lms studied here, the lattice orientation with respect to the ow direction is either parallel or perpendicular, while it is not trivial which orientation the driven vortex lattice favors. A theory showed that at large velocities the motion of vortices driven over the random pinning potential is parallel to their nearest-neighbor direction (i.e., parallel orientation) [1], as schematically illustrated in the top inset of Fig. 1(b). However, recent mode-locking (ML) experiments on amorphous lms conducted both at high and low temperatures T have shown evidence for the perpendicular orientation [bottom inset of Fig. 1(b)] over the broad elds B [2, 3]. As the eld approaches the dynamic melting eld $B_{c,dyn}$, where driven lattices melt into a moving liquid, the moving lattice rotates from the perpendicular to parallel orientation at a T-dependent characteristic eld B_{ori}. The results suggest that the lattice orientation is closely related to lattice softening due to thermal or quantum uctuations, while its mechanism has not been fully understood [3].

In general, the lattice orientation would be dependent on the velocity of driven vortices; namely, dc and ac components of the velocity at ML, which can be controlled by changing the frequency and amplitude of the ac drive, respectively. Previous work has focused on the data measured at moderate velocities, while in this work we present detailed measurements of the ML resonance around B_{ori} including the data measured at larger velocities. We nd a trend for the perpendicular lattice orientation to change to the parallel orientation with increasing the dc and/or ac velocities in elds just prior to B_{ori}. The result is consistent with the recent simulation that the lattice orientation is parallel in the absence of pinning but perpendicular in the presence of pinning [4]. We also nd that B_{ori} extracted from the main signal of the ML resonance is not sensitive to the dc velocity within the range studied, indicating that the dynamic pinning force is signi cantly reduced around B_{ori} prior to dynamic melting. The detailed data and discussion concerning present work have been published elsewhere [3].

2. Experimental

The amorphous $\mathrm{Mo}_x\mathrm{Ge}_{1\quad x}$ lm with thickness of 330 nm was fabricated by rf sputtering onto a silicon substrate mounted on a water cooled rotating copper stage [5]. The mean- eld transition temperatures T_{c0} de ned by a 95 % criterion of the normal-state resistivity and the zero-resistivity temperatures T_c are 6.1 and 6.0 K, respectively. In measuring the ML resonance, the ac current I_{rf} with frequency f_{ext} of 10 and 70 MHz was applied through an rf transformer superimposed with the dc current I [6, 7]. By increasing the frequency f_{ext} and amplitude of I_{rf}, we are able to increase the dc and ac components of the velocity of driven vortex matter at ML, respectively. The lm was attached to the cold plate of our ^{3}He-^{4}He dilution refrigerator. The eld was applied perpendicular to the plane of the lm.

3. Results and discussion

From the $I \quad V$ characteristics, we de ne the depinning current I_c as a threshold current at which the vortices start to move, where a 10 ^{7} V criterion is used [7, 8]. The peak in the $I_c(B)$ curve indicative of the peak e ect [9, 10, 11] is visible over the whole temperature region (T =0.08-5.0 K) studied. The representative data at 1 K is shown in Fig. 1(a). In our previous work it has been shown [8] that the peak eld B_p in $I_c(B)$ marks an order-disorder transition [12, 13, 14] from the ordered vortex lattice (OP)($B < B_p$) to amorphouslike disordered phase (DP) ($B_p < B < B_c$). B_c at which I_c falls to zero re ects a static "melting" eld. We have revealed recently that the dynamic melting eld $B_{c,dyn}$ is close to the static melting eld B_c at high temperature (T >2 K), but it is signi cantly suppressed from B_c at low T(<2 K) due to strong quantum uctuation e ects [6].

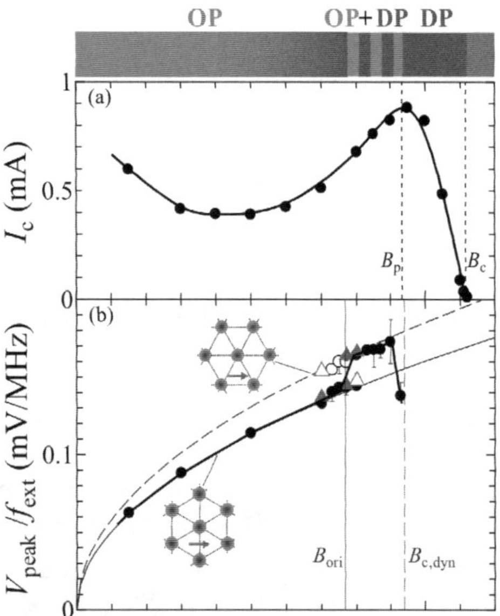

Figure 1. (Color online) (a) I_c vs B at 1.0 K. Vertical dots indicate the location of B_p (left) and B_c (right). (b) V_{peak}/f_{ext} vs B at 1 K measured with I_{rf} of f_{ext} =10 MHz (circles) and 70 MHz (triangles). In the particular B region where two peaks are observed in the dI/dV vs V curves, we plot the positions of the main and secondary peaks with solid and open symbols, respectively. Solid and dashed curves represent $V_{1/2}^{perp}(B)/f_{ext}$ and $V_{1/1}^{para}(B)/f_{ext}$, respectively. Vertical solid and dashed lines indicate B_{ori} and $B_{c,dyn}$, respectively. Other lines are guides for the eye. Insets: schematic diagrams of moving vortex lattices, where arrows indicate the directions of the vortex motion. Static vortex phase diagram is illustrated above (a). OP, OP+DP, and DP denote the ordered phase, coexistence phase of OP and DP, and disordered phase, respectively.

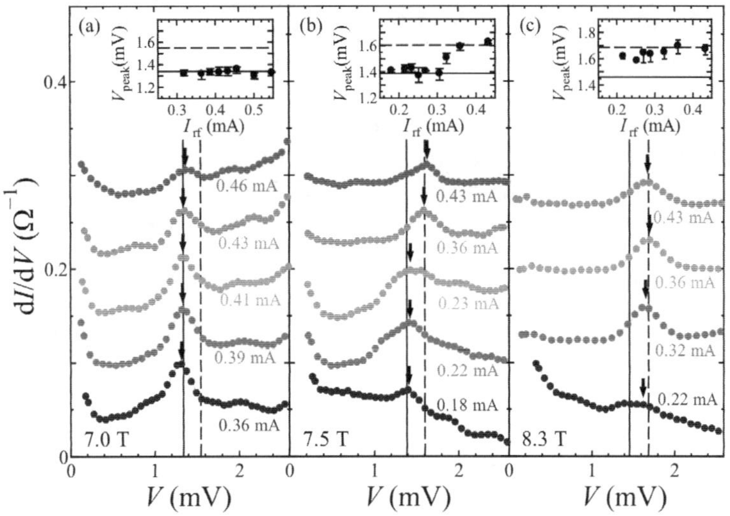

Figure 2. (Color online) The V dependence of dI/dV measured with superimposed 10-MHz I_{rf} of di erent amplitudes, which are shown in each gure, at 1.0 K in (a) 7.0 T, (b) 7.5 T, and (c) 8.3 T. The location of $V_{1/2}^{perp}$ and $V_{1/1}^{para}$ is indicated with vertical solid and dashed lines, respectively. Arrows mark the peak position. The curves are vertically shifted for clarity. Insets: V_{peak} plotted against I_{rf}. Horizontal solid and dashed lines indicate $V_{1/2}^{perp}$ and $V_{1/1}^{para}$, respectively.

The peaks in dI/dV vs V curves, namely, ML steps are visible over the broad elds from the lowest eld measured, 1.5 T, up to 9.3 T. In Figs. 2(a)-2(c) we show dI/dV vs V taken for 7.0, 7.5, and 8.3 T, respectively, at 1.0 K with superimposed 10-MHz I_{rf} of di erent amplitudes, which are shown in each gure. Here, the amplitude of I_{rf} covers the region where the width of the ML current step in the I V curves or the amplitude of the ML peak in the dI/dV vs V curves as a function of I_{rf} takes a rst peak, around which the ML resonance is most clearly observed. Assuming a triangular vortex array moving in the direction perpendicular to one side of the triangles, as schematically shown in the bottom inset of Fig. 1(b), one can calculate a value of the voltage for a given B satisfying the subharmonic resonant condition of $p/q = 1/2$; i.e., $V_{1/2}^{perp} = l(p/q)f_{ext}2a_{\perp}B = lf_{ext}(\sqrt{3}\ \ _0B/2)^{1/2}$, where l is the distance between the voltage contacts, $2a_{\perp}$ is the lattice period in the direction of vortex motion, and $_0$ is the ux quantum [15]. For 7.0 T the rst ML peak voltage V_{peak} marked with an arrow well coincides with the calculation for $V_{1/2}^{perp}$ indicated with a vertical solid line. The inset shows the insensitiveness of V_{peak} to I_{rf} in the I_{rf} range studied, where a horizontal solid line indicates the location of $V_{1/2}^{perp}$. Qualitatively the same behavior is observed in other elds down to 1.5 T [3].

As B increases slightly from 7.0 to 7.5 T, however, the ML peak becomes broad and I_{rf}-dependent, as seen in Fig. 2(b). With increasing I_{rf}, the ML peak voltage V_{peak} increases from a value around $V_{1/2}^{perp}$ indicated with a vertical solid line to a value approximately 15% larger than $V_{1/2}^{perp}$ indicated with a vertical dashed line. The dependence of V_{peak} on I_{rf} is more clearly seen in the inset, where V_{peak} exhibits a jump from $\approx V_{1/2}^{perp}$ (a horizontal solid line) to $\approx 1.15V_{1/2}^{perp}$ (a horizontal dashed line). As the eld increases further (B =8-9 T), V_{peak} takes a value close to $1.15V_{1/2}^{perp}$ for most of I_{rf} measured, as illustrated in the main panel and inset of Fig. 2(c). This resonant voltage, as indicated with dashed lines, is consistent with the condition for the parallel lattice orientation [see the top inset of Fig. 1(b)], which is given by $V_{1/1}^{para} = (2/\sqrt{3})V_{1/2}^{perp}$. On approaching B_p(=9.34 T), V_{peak} drops abruptly below $V_{1/2}^{perp}$ and at 9.4 T all evidence for ML disappears, indicative of dynamic quantum melting of the driven lattice [6].

Thus, the ML data in the low- eld region (B = 1.5 7 T) are explained by a triangular vortex array moving with the perpendicular orientation, whereas those in higher elds (B = 8 9 T) indicate the parallel orientation. In the crossover region (B_{ori} =7-8 T) around which the switching of the lattice orientation occurs, the triangular arrays with either orientation may exist depending on the amplitude of I_{rf}.

When the frequency of I_{rf} is increased from f_{ext} =10 to 70 MHz, which leads to the increase in the dc component of the lattice velocity at ML by seven times, the dI/dV vs V curves exhibit peculiar double peaks at around $V_{1/2}^{perp}$ and $V_{1/1}^{para}$. Since the dc voltage V is proportional to the dc velocity, this result may imply the dc-velocity-induced rotation of the vortex lattice from the perpendicular to parallel orientation. Such a behavior could not be observed for 10 MHz. This is probably because the separation between the dc velocities at $V_{1/2}^{perp}$ and $V_{1/1}^{para}$ for 10 MHz is much smaller compared to that for 70 MHz.

In order to see how the lattice orientation changes with B, in Fig. 1(b) we plot the B dependencies of V_{peak}/f_{ext} at 1 K for f_{ext} =10 and 70 MHz with circles and triangles, respectively. In the particular eld region where the two peaks with di erent amplitudes are observed in the dI/dV vs V curves, we plot the positions of the main and secondary peaks with solid and open symbols, respectively. Over the broad elds the data points for 10 MHz fall onto a solid curve $V_{1/2}^{perp}(B)/f_{ext}$ corresponding to the perpendicular lattice orientation, while for high B, $V_{peak}(B)/f_{ext}$ satis es with the parallel resonant condition given by $V_{1/1}^{para}(B)/f_{ext} = (2/\sqrt{3})V_{1/2}^{perp}(B)/f_{ext}$, as indicated with a dashed curve. Although we do not have enough data for f_{ext} =70 MHz at present, the location of B_{ori} extracted from the main signal of the ML resonance nearly coincides with that for 10 MHz.

It has been shown numerically using the time-dependent Ginzburg-Landau theory that in thin superconductors with random point pinning, as studied in this work, the lattice orientation of driven vortex matter is parallel in the absence of pinning but perpendicular in the presence of pinning [4]. This is consistent with the velocity-induced rotation of the lattice orientation observed in this work. We may also interpret the insensitiveness of B_{ori} to the dc velocity in the range studied as well as the appearance of the parallel orientation in elds exceeding B_{ori} in terms of signi cantly decreased dynamic pinning forces prior to dynamic melting.

Acknowledgments
We thank N. Nakai and X. Hu for useful discussions.

References

[1] Schmid A and Hauger W 1973 *J. Low Temp. Phys.* **11** 667
[2] Kokubo K, Nishizaki T, Shinozaki B and Kes P H 2010 *Physica C* **470** 43.
[3] Okuma S, Imaizumi H, Shimamoto D and Kokubo N 2011 *Phys. Rev. B* **83** 064520
[4] Nakai N, Hayashi N and Machida M 2009 *Physica C* **469** 1106
[5] Kokubo N, Asada T, Kadowaki, Takita K, Sorop T G and Kes P H 2007 *Phys. Rev. B* **75** 184512
[6] Okuma S, Imaizumi H and Kokubo N 2009 *Phys. Rev. B* **80** 132503
[7] Okuma S, Inoue J and Kokubo N 2007 *Phys. Rev. B* **76** 172503
[8] Okuma S, Kashiro K, Suzuki Y and Kokubo N 2008 *Phys. Rev. B* **77** 212505
[9] Kes P H and Tsuei C C 1983 *Phys. Rev. B* **28** 5126
[10] Bhattacharya S and Higgins M J 1993 *Phys. Rev. Lett.* **70** 2617
[11] Xiao Z L, Andrei E Y, Shuk P and Greenblatt M 2000 *Phys. Rev. Lett.* **85** 3265
[12] Paltiel Y *et al.* 2000 *Phys. Rev. Lett.* **85** 3712
[13] van der Beek C J, Colson S, Indenbom M V and Konczykowski M 2000 *Phys. Rev. Lett.* **84** 4196
[14] Xiao Z L, Dogru O, Andrei E Y, Shuk P and Greenblatt M 2004 *Phys. Rev. Lett.* **92** 227004
[15] Kokubo N, Kadowaki K and Takita K 2005 *Phys. Rev. Lett.* **95** 177005

Mode-locking resonance for driven vortex matter in thick and thin superconducting films

H. Sato and S. Okuma

Department of Physics and Research Center for Low Temperature Physics, Tokyo Institute of Technology, 2-12-1, Ohokayama, Meguro-ku, Tokyo 152-8551, Japan

E-mail: sokuma@o.cc.titech.ac.jp

Abstract. We report on measurements of the mode-locking (ML) resonance for the thick and thin films of amorphous Mo_xGe_{1-x} with weak pinning. The clear ML resonance indicative of driven vortex lattices is observed for the thick film, while it is not visible for the thin film. The results suggest that for the thin film the elasticity of driven lattices may be significantly reduced and the lattices may be unstable against small pinning.

1. Introduction

Vortex states in type-II superconductors are largely dependent on the dimensionality (thickness) of superconductors. When the samples contain moderately strong pinning, a continuous transition from the vortex glass to liquid takes place [1]. For thick films the vortex-glass phase occupies a large area of the field-temperature ($B-T$) plane [2], while for thin films only the liquid phase exists at any nonzero T in the mixed states [1, 3, 4]. In the case of weak-pinning samples we focus on in this work, the vortex solid state for the thick samples is somewhat complicated. Under increasing the field, the critical current J_c exhibits a peak just prior to the melting field B_c. This is called a peak effect (PE)[5, 6] and the peak field marks the sharp structural transition from the vortex lattice to amorphouslike glass state. This is actually observed in $NbSe_2$ single crystals [7] and amorphous (a-)Mo_xGe_{1-x} films [8]. For thin films with no pinning the melting line of vortex lattice is predicted to be nearly independent of field at low B [9]. However, to the best of our knowledge, the sharp melting transition as well as PE has not yet been observed in real thin films [10]. These results suggest that the vortex lattice for the thin films may be unstable against small pinning.

The effects of pinning would be much reduced by simply driving the vortices. Here, we conduct a mode-locking (ML) experiment [11, 12, 13, 14], which enables us to detect the coherent motion of driven vortices, for a-Mo_xGe_{1-x} films with thicknesses of 10 and 350 nm. For the thin film both the transition temperature T_c in zero field and the static "melting" field B_c are smaller than those for the thick film, while the difference between the upper critical field $B_{c2}(T)$ and $B_c(T)$ is significantly larger, indicative of the enhanced liquid phase due to the reduced dimensionality [1, 15]. Neither the PE for $J_c(B)$ nor the ML resonance observed for the thick film is visible for the thin film. The results suggest that for the thin film the elasticity of driven lattice may be much reduced, which would account for the "absence" of the sharp melting transition of vortex lattice for the thin films.

2. Experimental

The a-Mo$_x$Ge$_{1-x}$ films with thicknesses of 350 and 10 nm were prepared by rf sputtering on a substrate mounted on a water cooled rotating copper stage. The thickness of the films was controlled by changing the sputtering time. The resistivities ρ in the normal state (at 10 K) are ρ_n =1.63 and 1.98 $\mu\Omega$m for the thick and thin films, respectively. The mean-field transition $[\rho(T_{c0}) = 0.95\rho_n]$ and zero-resistivity temperatures $[\rho(T_c) = 10^{-3}\rho_n]$ are T_{c0} =5.8 K and T_c =5.6 K, and T_{c0} =3.8 K and T_c =3.3 K for the thick and thin films, respectively. The relative widths of the resistive transition, as defined as $(T_{c0} - T_c)/T_{c0}$, are 0.04 and 0.12 for the thick and thin films, respectively. The superconducting coherence length ξ estimated from B_{c2} at $T = 0$ is around 15-20 nm, which is close to or slightly larger (/sufficiently smaller) than the thickness of the thin (/thick) film. Thus, for the thin (/thick) film the dimensionality with respect to ξ is two (/three). Both the lower T_c (T_{c0}) and wider transition width for the thin film than for the thick film are due mainly to the reduced dimensionality. In measuring the ML resonance, ac currents I_{rf} with frequencies f_{ext} of 35 and 50 MHz were applied through an rf transformer for the thick and thin films, respectively [14, 16]. The films were directly immersed in the liquid ^{4}He and the field (up to 6.5 T) was applied perpendicular to the plane of the film.

3. Results and discussion

From the current-voltage ($I-V$) characteristics, we define the critical (depinning) current density J_c as a threshold current at which the vortices start to move, where a 10^{-7} V criterion is used [8, 14]. In Fig. 1 we plot the field dependence of J_c at 3.0 K for the thick film (black solid circles) and at 1.8 K for the thin film (red solid circles). Here, we selected the temperatures T which give rise to the same reduced temperature $T/T_c = 0.54$ for both films. To compare the shape of $J_c(B)$ between the two films, the field is normalized by $B_c(T)$ of each film. For the thick film we observe the peak behavior in the $J_c(B)$ curve indicative of PE [5, 6, 17]. The peak is more pronounced at lower temperatures, which is seen from the typical data at 2.3 K shown in the inset. The peak field B_p in $J_c(B)$ marks the order-disorder transition (ODT) from the ordered (or weakly disordered) vortex-lattice phase (OP) to disordered (amorphouslike) vortex-glass phase (DP) [7, 8].

In the case of the thin film, by contrast, the peak in $J_c(B)$ is not visible at any temperature measured, suggesting the absence of ODT. Furthermore, the value of J_c, which reflects the pinning strength, is comparable to or slightly larger than that for the thick film. These results imply that for the thin film the vortex solid phase is composed of only the vortex-glass phase

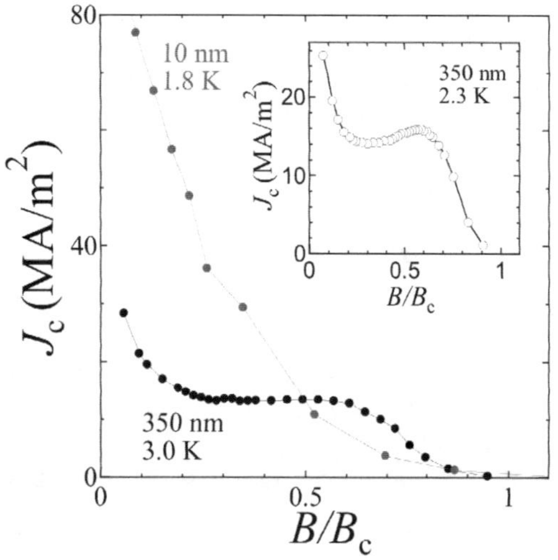

Figure 1. (Color online) The critical current density J_c vs reduced field B/B_c at 3.0 K for the thick (350 nm) film (black solid circles) and at 1.8 K for the thin (10 nm) film (red solid circles). Inset: J_c vs B/B_c at 2.3 K for the thick film. All the lines are guides for the eye.

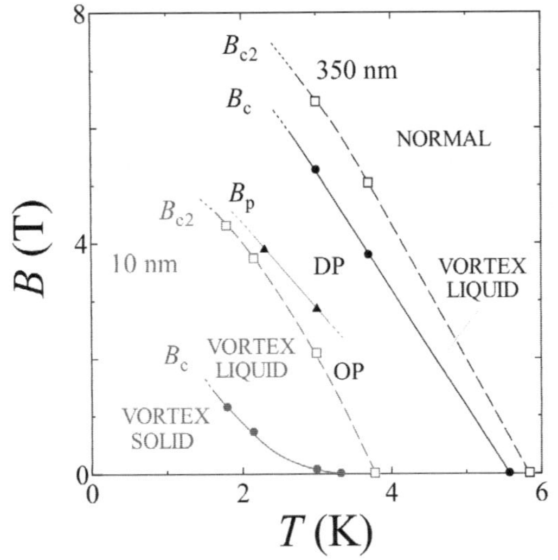

Figure 2. (Color online) B_c vs T (solid circles) and B_{c2} vs T (open squares) for the thick (black symbols) and thin (red symbols) films. B_p vs T (triangles) for the thick film. All the lines are guides for the eye.

(DP), as observed earlier in thick and thin a-Mo_xSi_{1-x} films with stronger pinning.

Figure 2 displays the vortex phase diagram in the $B-T$ plane for the thick (black symbols) and thin (red symbols) films, where we plot the static melting line $B_c(T)$ (solid circles) and upper-critical-field line $B_{c2}(T)$ (open squares) determined from $\rho(B_c) = 10^{-3}\rho_n$ and $\rho(B_{c2}) = 0.95\rho_n$, respectively. We also show $B_p(T)$ (triangles) indicative of ODT for the thick film. One can clearly see that, as compared with the thick film, the $B_c(T)$ line for the thin film is significantly suppressed from the $B_{c2}(T)$ line, resulting in the enhanced liquid phase in a low-field region. The larger liquid phase for the thin film than for the thick film is attributed to stronger fluctuation effects for two dimensions.

In order to explore the possibility of the vortex-lattice phase in the thin film, we perform the measurements of the ML resonance. Figures 3(a) and 3(b) show the V dependencies of the differential conductance dI/dV superimposed with I_{rf} with different amplitudes at 3.0 K in 2.0 T (OP) and at 1.8 K in 0.1 T (vortex-solid phase) for the thick and thin films, respectively. The clear ML peak indicative of driven lattice is detected for the thick film. Assuming a triangular vortex array moving in the direction perpendicular to one side of the triangles (i.e., the perpendicular orientation), we can calculate a value of the voltage ($V_{p/q}^{perp}$) for a given B satisfying the subharmonic resonant condition of $p/q = 1/2$; i.e., $V_{1/2}^{perp} = l(p/q)f_{ext}a_\perp B = lf_{ext}(\sqrt{3}\Phi_0 B/2)^{1/2}$, where l is the distance between the voltage contacts, $a_\perp$ is the lattice period in the direction of vortex motion, and Φ_0 is the flux quantum [18]. The location of $V_{1/2}^{perp}$ is indicated with vertical dashed lines in Fig. 3(a) and 3(b). The results in Fig. 3(a) show that for the thick film the driven vortex matter is the triangular vortex array with perpendicular lattice orientation, consistent with previous work on the similar thick films.

For the thin film, however, evidence for the ML resonance cannot be found for any I_{rf} measured. To reduce the dynamic pinning force, I_{rf} with f_{ext} =50 MHz higher than f_{ext} = 35 MHz for the thick film is superimposed. We also perform the measurements of ML in 0.5 T, which corresponds to the high-B part of the solid phase; however, the results stay unchanged. In our previous work using the thick films we have revealed that at relatively high T ($> 0.4T_{c0}$), as studied in this work, the moving-lattice states are observed over the broad fields in the equilibrium DP as well as OP, while the moving-liquid state is visible only in the equilibrium liquid phase ($B_c < B < B_{c2}$). In other words, even if the equilibrium solid phase is the disordered vortex-glass phase (DP), it can be dynamically ordered when it is driven at high velocity. Based on these facts, we may interpret the absence of ML for the thin film as that

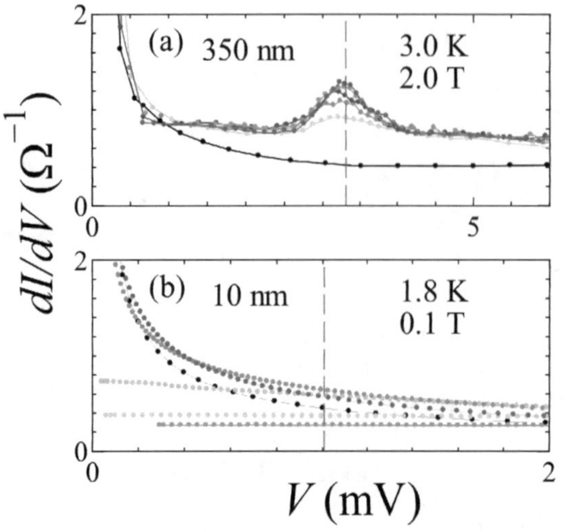

Figure 3. (Color online) (a) V dependencies of dI/dV superimposed with I_{rf} with different amplitudes in the range I_{rf} = 2.9-4.8 mA at 3.0 K in 2.0 T for the thick film and (b) with I_{rf} in the range I_{rf} = 0.4-1.6 mA at 1.8 K in 0.1 T for the thin film. The black symbols correspond to dI/dV without superimposed I_{rf}. Vertical dashed lines represent the location of $V_{1/2}^{perp}$. Other lines are guides for the eye.

for the thin film the elasticity of driven lattice may be significantly reduced and the vortex lattice may be unstable against small pinning as well as thermal/quantum fluctuations. This may account for the "absence" of the sharp melting transition reported so far in various thin superconducting films.

Acknowledgments

We thank D. Shimamoto for his technical assistance. This work was supported by a Grant-in-Aid for Scientific Research from the Ministry of Education, Culture, Sports, Science, and Technology of Japan.

References

[1] Fisher D S, Fisher M P A and Huse D A 1991 *Phys. Rev. B* **43** 130
[2] Okuma S, Imamoto Y and Morita M 2001 *Phys. Rev. Lett.* **86** 3136
[3] Chervenak J A and Valles J M Jr. 1996 *Phys. Rev. B* **54** R15649
[4] Okuma S, Terashima T and Kokubo N 1998 *Phys. Rev. B* **58** 2816
[5] Kes P H and Tsuei C C 1983 *Phys. Rev. B* **28** 5126
[6] Bhattacharya S and Higgins M J 1993 *Phys. Rev. Lett.* **70** 2617; Marley A C, Higgins M J and Bhattacharya S 1995 *ibid.* **74** 3029
[7] Paltiel Y, Myasoedov Y, Zeldov E, Jung G, Rappaport M L, Feldman D E, Higgins M J and Bhattacharya S 2002 *Phys. Rev. B* **66** 060503
[8] Okuma S, Kashiro K, Suzuki Y and Kokubo N 2008 *Phys. Rev. B* **77** 212505
[9] Fisher D S 1980 *Phys. Rev. B* **22** 1190
[10] Saeki T and Ikeda R, preprint
[11] Fiory A T 1971 *Phys. Rev. Lett.* **27** 501
[12] Van Look L, Rosseel E, Van Bael M J, Moshchalkov V V and Bruynseraede Y 1999 *Phys. Rev. B* **60** R6998
[13] Kokubo N, Besseling R, Vinokur V M and Kes P H 2002 *Phys. Rev. Lett.* **88** 247004
[14] Okuma S, Inoue J and Kokubo N 2007 *Phys. Rev. B* **76** 172503
[15] Blatter G, Ivlev B, Kagan Y, Theunissen M, Volokitin Y and Kes P 1994 *Phys. Rev. B* **50** 13013
[16] Okuma S, Imaizumi H and Kokubo N 2009 *Phys. Rev. B* **80** 132503
[17] Ghosh K *et al.* 1996 *Phys. Rev. Lett.* **76** 4600
[18] Kokubo N, Kadowaki K and Takita K 2005 *Phys. Rev. Lett.* **95** 177005

Plastic depinning in superconducting vortices

A. Motohashi and S. Okuma

Department of Physics and Research Center for Low Temperature Physics, Tokyo Institute of Technology, 2-12-1, Ohokayama, Meguro-ku, Tokyo 152-8551, Japan

E-mail: `sokuma@o.cc.titech.ac.jp`

Abstract. We study a plastic depinning phenomenon in a sheared vortex system in a Corbino-disk superconductor, where some vortices are mobile while others remain pinned. We measure the time-dependent voltage $V(t)$ just after the dc drive due to the dc current I with a sharp rise is suddenly applied to the vortex system. We find decaying $V(t)$ toward the steady-state value, indicating that the moving vortices are gradually pinned. The decay time $\tau(I)$ diverges at around a depinning current, as determined from the static $I - V$ characteristics. The value of τ is dependent on the initial vortex configuration, while the critical dynamics is insensitive to it. These results provide a strong support for the plastic depinning transition.

1. Introduction

In type-II superconductors quantized flux lines induced by a magnetic field B repel each other and they are driven by a Lorentz force due to an applied current I. Thus, the vortex system can be used extensively as a general model system to study various nonequilibrium many-body effects [1]. Recently, we have found a novel dynamic transition, indicative of a reversible to irreversible flow transition (RIT), in a periodically sheared vortex system in a Corbino-disk (CD) superconductor [2, 3]. This transition was originally observed in a colloidal particle system, where the irreversible collisions that give rise to chaotic dynamics can also cause a system to self-organize to avoid future collisions [4, 5]. When a displacement of particles per cycle is small, they return to their initial position at the end of each cycle of a periodic shear. This is identified with reversible flow. As the displacement exceeds a critical distance for RIT, the particles do not return to their initial position. This is considered as irreversible flow. This RIT [1,4-8] has also been interpreted in terms of an absorbing transition [9-11]. Within the interpretation, the irreversible regime corresponds to the fluctuating active state where particles are diffusing, while the reversible regime corresponds to the absorbing state where the particles are self-organized and absorbed into a nonfluctuating quiescent state [7]. We have obtained evidence for RIT with the critical behavior similar to that of the absorbing transition in our vortex system [3]. The result suggests the universality of these dynamic transitions.

In the meantime, it has been shown by simulation that the plastic depinning transition would occur in many-particle systems with random pinning and that it exhibits a critical dynamics similar to that of the absorbing transition [12]. When a dc force f is suddenly applied to the particles, they move in the form of complex fluctuating channels, where some particles are mobile while others remain pinned. The moving particles are gradually pinned to the pinning centers. When f is lower than the depinning threshold f_d, all the particles are finally pinned and thus this is identified with an absorbing state. By contrast, for $f > f_d$, nearly a constant number

Published under licence by IOP Publishing Ltd

of particles are flowing in a steady state and hence this is considered to be an active state. The depinning phenomenon is widely observed in nature, but the true nature of the depinning transition is still not fully understood. We have measured the time-dependent voltage $V(t)$ to detect the dynamic response of vortices to the suddenly applied *dc* current I larger than the depinning threshold I_d and observed decaying $V(t)$. This result implies that the moving vortices are pinned gradually and finally $V(t)$ approaches a steady-state voltage, where nearly a constant number of vortices are flowing. The decay time $\tau(I)$ is found to diverge at the depinning current I_d, as determined from static $I-V$ characteristics, suggesting that plastic depinning is a dynamic phase transition. In this work, to get insight into the plastic depinning phenomenon, we conduct detailed measurements of $V(t)$ in the same vortex system. Specifically, we have changed the initial vortex configurations, that is, the number of depinned vortices at $t = 0$ by changing the field-sweep process prior to measurements. It is expected that the difference in the initial vortex configurations would alter the transient vortex dynamics $V(t)$ and the relaxation time $\tau(I)$. To demonstrate the depinning transition, however, it is necessary to verify that the critical behavior is insensitive to the initial vortex configurations.

2. Experimental

The 330-nm-thick a-Mo_xGe_{1-x} film was prepared by rf sputtering on a Si substrate held at room temperature [13]. Mean-field transition and zero-resistivity temperatures are 6.3 and 6.2 K, respectively. The vortex motion can be studied from the voltage V, which is proportional to the average velocity, as a function of the dc drive due to the dc current I. Arrangement of silver electrical contacts is shown schematically in the right inset of Fig. 1(a). The current flows between the contact +C of the center and that -C of the perimeter of the disk, which produces a radial current density that decays as $1/r$, where r is a radius of rotating vortices. The inner radius of CD is 0.8 mm. We used the voltage contacts, +V and -V, to measure the voltage V. The time-dependent voltage $V(t)$ enhanced with a preamplifier was taken and analyzed using a fast-Fourier transform spectrum analyzer with the time-resolution of up to 40 kHz.

3. Results and discussion

All the data presented in this paper were taken at 4.1 K in 3.0 T, which corresponds to the coexistence vortex phase composed of ordered vortex lattice and amorphouslike disordered phase, where pinning is very effective [13, 14]. The main panel and left inset of Fig. 1(a) show the $I-V$ characteristics plotted on linear-linear and log-log scales, respectively. The depinning current I_d, at which the vortices start to move, is determined from the onset of voltage. Precisely speaking, it is difficult to determine the exact value of I_d, because it is dependent on the experimental resolutions of V. In this work, we have measured V with improved resolution ($> 10^{-8}$ V) than before ($> 10^{-7}$ V), which causes a slight decrease in I_d from 0.35 to 0.25 mA, as indicated with an arrow in the main panel and a dashed line in the inset.

Now, we focus on the transient vortex dynamics, that is, $V(t)$, in 3.0 T just after the dc current I with a sharp rise is applied to the vortex system. In our previous experiment [3] we firstly decreased the field from 3.0 to 2.6 T and then increased up to 3.0 T before applying dc current $I(t)$ with a rise time as small as 10^{-8} s at $t = 0$. This initial field-sweep process was necessary to avoid a situation that a large number of vortices are strongly pinned in the initial state, which is most likely realized in field-cooled mode. In fact, when the field of 3.0 T was applied in the normal state and then the film was cooled through T_c with no applied current, decaying $V(t)$ is not clearly visible. In this work, in order to examine a change in transient vortex dynamics that might occur when the initial vortex configuration is changed, we have decreased the field down to zero, instead of 2.6 T, each time before starting the measurements of $V(t)$ in 3.0 T. It is expected that this initial field-sweep process extending to even lower field down to $B = 0$ would induce more depinning and tend to enhance the degree of order of the vortex lattice at $t = 0$.

For $I < I_d$ (=0.25 mA) detectable $V(t)$ is not visible within our experimental resolutions, whereas above I_d we observe the transient voltage $V(t)$ that decays toward the steady-state value $V^\infty [\equiv V(t \to \infty)]$. The decaying behavior looks more pronounced than what was observed in our previous experiment, where more vortices were pinned and immobile at $t = 0$. Figure 2 displays $V(t)/V^\infty$, that is, $V(t)$ divided by the steady-state voltage V^∞, taken at different amplitudes of I, which are shown in the figure. One can see that the relaxation time τ for I =0.8 mA is largest and decreases with increasing I: For I =4.0 mA, τ becomes diminishingly small. We can reproduce the $V(t)$ curves well above I_d using a simple exponential function, while in the vicinity of I_d the functional form approaches a power law. To obtain τ, we fit $V(t)$ to the following relaxation function, which was employed previously [3, 7, 12]:

$$V(t) = V^\infty - (V^\infty - V(0)) \exp(-t/\tau)/t^\alpha.$$

Here, we use α 0.5 near I_d, consistent with the previous simulation [12]. As shown with solid circles in Fig. 1(b), $\tau(I)$ diverges at $I_c = 0.20 \pm 0.10$ mA that is close to I_d 0.25 mA and the data points fall onto a curve expressed as $\tau \propto (I - I_c)^{-\nu}$ with ν 1.33. The same data are plotted on a log-log scale in the inset, giving $\nu = 1.33 \pm 0.20$. This is close to the value ($\nu = 1.26 \pm 0.15$) obtained in our previous experiment [3]. Open circles show $\tau(I)$ in the previous

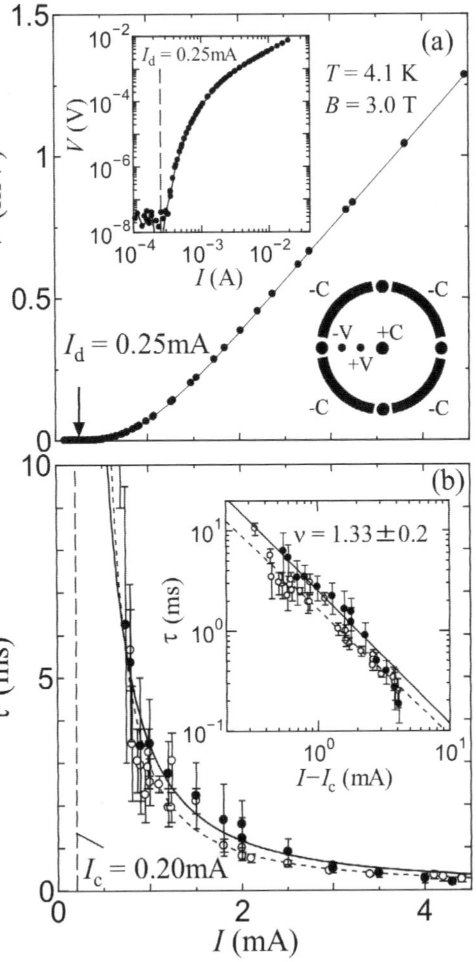

Figure 1. (a) $I - V$ characteristics at 4.1 K in 3.0 T plotted on a linear-linear scale. An arrow indicates the location of the depinning current I_d. Inset: (Left) The same data on a log-log scale. A dashed line marks I_d. (Right) Arrangement of the electrical contacts of CD. (b) τ vs I (solid circles) obtained in the present experiment, where the field was decreased to zero before measuring $V(t)$ in 3.0 T, and that (open circles) obtained from previous work [3], where the field was decreased to 2.6 T before measuring $V(t)$ in 3.0 T. The present data diverge at $I_c = 0.20 \pm 0.10$ mA, as indicated with a vertical dashed line, and previous ones diverge at $I_c = 0.37 \pm 0.15$ mA, both of which are close to I_d 0.25 mA. Solid and dashed curves indicate a power-law fit. Inset: The same data and their fit plotted on a log-log scale. The symbols and lines correspond to those in the main panel.

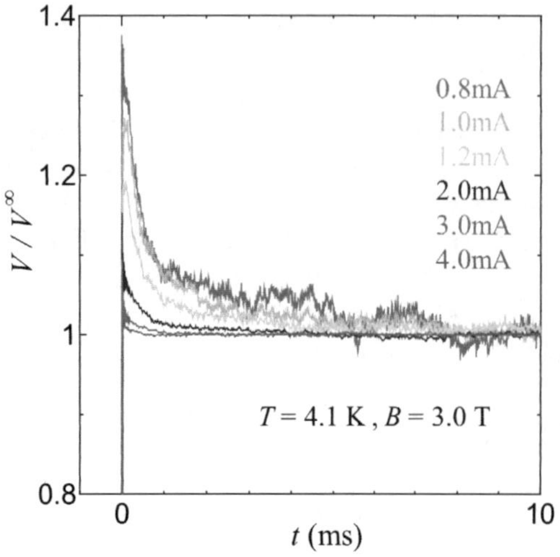

Figure 2. (Color online) Transient $V(t)/V^{\infty}$ at 4.1 K in 3.0 T just after the dc current I with amplitudes of 0.8, 1.0, 1.2, 2.0, 3.0, and 4.0 mA (from top to bottom) was applied at $t = 0$.

experiment, where more vortices were pinned at $t = 0$ [3]. One can see a trend that τ obtained in this work is slightly larger than that in our previous work. This is reasonable considering the fact that, as more vortices are depinned at $t = 0$, more time τ is needed to relax into the steady state. It is important to note that critical dynamics, $\tau(I)$, near I_c, as well as the critical points, $I_c = 0.20 \pm 0.10$ and 0.37 ± 0.15 mA for present and previous work, respectively, is nearly independent of the initial vortex states, providing a strong support for the dynamic transition.

The exponent ν 1.3 obtained in this work is close to $\nu = 1.36 \pm 0.06$ reported in the simulation for the depinning transition in two dimensions (2D) [12] and $\nu_{RIT} = 1.3 \pm 0.3$ for RIT found previously in the periodically sheared vortex system in the same CD superconductor [3]. It is also close to $\nu_{RIT} = 1.1 \pm 0.3$ and 1.33 ± 0.02 found in the colloidal experiment in 3D and simulation for random organization in 2D [7], respectively. All these results strongly suggest that the nonequilibrium RIT, absorbing, and depinning transitions actually occur in our 2D vortex system and exhibit the similar critical behavior.

We thank C. Reichhardt for helpful discussions.

References

[1] Mangan N, Reichhardt C and Olson Reichhardt C J 2008 *Phys. Rev. Lett.* **100** 187002
[2] Okuma S, Suzuki Y and Tsugawa Y 2010 *Physica C* **470** S842
[3] Okuma S, Tsugawa Y and Motohashi A 2011 *Phys. Rev. B* **83** 012503
[4] Pine D J, Gollub J P, Brady J F and Leshansky A M 2005 *Nature* **438** 997
[5] Gollub J and Pine D 2006 *Phys. Today* **59** 8
[6] Moretti P and Miguel M C 2009 *Phys. Rev. B* **80** 224513
[7] Corté L, Chaikin P M, Gollub J P and Pine D J 2008 *Nature Phys.* **4** 420
[8] Zhang W, Zhou W and Luo M 2010 *Phys. Lett. A* **374** 3666
[9] Takeuchi K A, Kuroda M, Chaté H and Sano M 2007 *Phys. Rev. Lett.* **99** 234503
[10] Hinrichsen H 2000 *Adv. Phys.* **49** 815
[11] Dickman R 2002 *Physica A* **306** 90
[12] Reichhardt C and Olson Reichhardt C J 2009 *Phys. Rev. Lett.* **103** 168301
[13] Okuma S, Kashiro K, Suzuki Y and Kokubo N 2008 *Phys. Rev. B* **77** 212505
[14] Paltiel Y, Zeldov E, Myasoedov Y, Rappaport M L, Jung G, Bhattacharya S, Higgins M J, Xiao Z L, Andrei E Y, Gammel P L and Bishop D J 2000 *Phys. Rev. Lett.* **85** 3712

Journal of Physics: Conference Series **302** (2011) 012030 doi:10.1088/1742-6596/302/1/012030

Large-scale electronic-structure calculations for nanomaterials in density functional theory

Atsushi Oshiyama
Department of Applied Phsyics, The University of Tokyo
Hongo, Tokyo 113-8656, Japan
Jun-ichi Iwata
Center for Computational Sciences, University of Tsukuba
Tennodai, Tsukuba 305-8577, Japan

E-mail: oshiyama@ap.t.u-tokyo.ac.jp

Abstract. We present our efforts to develop a real-space scheme in calculating total energies and electron states within the local density approximation and the generalized gradient approximation in the density functional theory, with an intention to perform large-scale computations on massively parallel many-core architecture next-generation computers in order to reveal and predict properties of nanometer scale materials. Our real-space scheme is applied to Si nanodots containing more than 10,000 atoms. The calculated ionization energy, the electron affinity and the energy gap reveal the origin of the underestimate of the energy gap in the local approximation.

1. Introduction

Computational condensed matter science has been initiated in 80s of the previous century and contributed tremendously to the progress of physics and chemistry of materials. Density functional theory (DFT) has been a powerful tool to reveal atom-scale mechanisms for a variety of phenomena in real materials. Yet target systems which DFT has treated in the past are relatively small: calculations for 100-atom systems are typical and those for 1000-atom systems are rare at present. On the other hand, the nanometer-scale system contains more than 10,000 atoms. There the nano-scale shape in the systems is now recognized as an important factor which is decisive in physical properties [1]. Hence it is highly demanded to perform DFT calculations with the levels of local density approximation (LDA) or the generalized gradient approximation (GGA) for 10,000-atom systems to clarify physics and chemistry of nano-structures. In the standard scheme using the plane-wave basis set, Fast Fourier Transform (FFT) is frequently used to compute the necessary quantities in either real space or in k space. On next-generation massively parallel muti-core architecture computers, however, FFT causes a heavy burden in communication tasks so that the large-scale calculations may be formidable. Our choice to overcome this difficulty is to compute the necessary quantities in the real space. This makes the scheme FFT-free and the efficiency of the computation is expected to be high.

Accuracy is another important factor to make simulations reliable. Both LDA and GGA are known to underestimate energy gaps of materials substantially although they predict relatively accurate structural properties. Improvement of the approximation to the exchange-correlation energy is therefore necessary.

Journal of Physics: Conference Series **302** (2011) 012030 doi:10.1088/1742-6596/302/1/012030

In this article, we present our real space DFT (RSDFT) scheme which allows us to perform LDA/GGA calculations for more than 10,000-atom systems [2]. The calculated results for Si nanodots are presented to elucidate the reason for the underestimate of the energy gap.

2. RSDFT Scheme

To numerically minimize the energy functional in DFT, we introduce a 3-dimensional spatial grid and consider the minimization problem within the discrete space of M_L grid points. Then the orbitals, the density, and the potentials are expressed as the corresponding vectors whose elements are the value at each grid point. For example, the Kohn-Sham orbital $\phi(\boldsymbol{r})$ and the charge density $\rho(\boldsymbol{r})$ are expressed as vectors,

$$\vec{\phi} = \begin{pmatrix} \phi^1 \\ \vdots \\ \phi^i \\ \vdots \\ \phi^{M_L} \end{pmatrix}, \quad \text{and} \quad \vec{\rho} = \begin{pmatrix} \rho^1 \\ \vdots \\ \rho^i \\ \vdots \\ \rho^{M_L} \end{pmatrix}, \tag{1}$$

respectively, where the i-th elements are the values at the grid point $\boldsymbol{r}_i$: $\phi^i = \phi(\boldsymbol{r}_i)$ and $\rho^i = \rho(\boldsymbol{r}_i)$. In this discrete scheme, the energy functional can be written as

$$E[\vec{\phi}_1, \vec{\phi}_2, \cdots, \vec{\phi}_{M_B}] = -\frac{1}{2}\sum_{n=1}^{M_B} f_n \sum_{i,j=1}^{M_L} \phi_n^{i*} L_{ij} \phi_n^j \Delta\Omega - 2\pi \sum_{i,j=1}^{M_L} \rho^i L_{ij}^{-1} \rho^j \Delta\Omega + E_{XC}[\rho^1, \rho^2, \cdots, \rho^{M_L}]$$
$$+ \sum_{i=1}^{M_L} v_L^i \rho^i \Delta\Omega + \sum_{i=1}^{M_B} f_n \sum_{i,j=1}^{M_L} \phi_n^{i*} V_{ij}^{NL} \phi_n^j \Delta\Omega \tag{2}$$

where f_n, M_B and $\Delta\Omega$ are the occupation number of n-th Kohn-Sham orbital, the total number of the orbitals and the volume element, respectively. The matrix L_{ij} is the Laplacian in discrete space; i.e., the finite difference (FD) operator. The formula for the higher-order FD coefficients, from which the matrix L_{ij} is constructed, is given in the reference [2]. V^{NL} is the matrix for the non-local pseudopotential. The second term in the right-hand side of Eq. (2) represents the Hartree energy. This form can be derived from that the Hartree potential $\vec{v}_h$ satisfies the Poisson equation (in discrete space): $L\vec{v}_h = -4\pi\vec{\rho}$. The density is defined as $\rho^i = \sum_{n=1}^{M_B} f_n |\phi_n^i|^2$. $E_{XC}[\rho^1, \rho^2, \cdots \rho^{M_L}]$ is the exchange-correlation energy functional. Minimizing the energy functional (2) under the ortho-normalization condition, $\sum_{i=1}^{M_L} \phi_n^{i*} \phi_m^i \Delta\Omega = \delta_{nm}$, we reach the eigenvalue equation,

$$\left[-\frac{1}{2} L + V + V^{NL} \right] \vec{\phi}_n = \varepsilon_n \vec{\phi}_n \tag{3}$$

Where L is a sparse matrix of the FD operator, V^{NL} is the matrix of the non-local operator, and V is the diagonal matrix whose elements are the sum of the local potentials. The Helmann-Feynman forces acting on atoms are also expressed in this real-space scheme.

2.1. A technique: Gram-Schmit orthogonalization

There are a lot of important details which make our RSDFT code run excellently on parallel architecture computers. We here explain just one of such details, Gram-Schmidt (GS) orthonormalization, which is a typical order N^3 operations (N: system size) and a heavy burden in large-scale calculations. In solving (3), we use iterative techniques such as conjugate gradient and residual minimization methods. After the iterative minimization, it is usually necessary to recover the orthonormalization relations among the orbitals through the GS procedure. Since the GS procedure is widely applied in many fields in the computational sciences, several algorithms have been proposed to perform these calculations efficiently. We have developed a GS algorithm suitable for the RSDFT, and the computational cost of the GS is thereby substantially reduced.

The usual GS procedure is as follows:

$$\begin{aligned}
\vec{\varphi}_1 &= \vec{\phi}_1 \quad , \\
\vec{\varphi}_2 &= \vec{\phi}_2 - \vec{\varphi}_1(\vec{\varphi}_1^*, \vec{\phi}_2) \quad , \\
\vec{\varphi}_3 &= \vec{\phi}_3 - \vec{\varphi}_1(\vec{\varphi}_1^*, \vec{\phi}_3) - \vec{\varphi}_2(\vec{\varphi}_2^*, \vec{\phi}_3) \quad , \\
\vec{\varphi}_4 &= \vec{\phi}_4 - \vec{\varphi}_1(\vec{\varphi}_1^*, \vec{\phi}_4) - \vec{\varphi}_2(\vec{\varphi}_2^*, \vec{\phi}_4) - \vec{\varphi}_3(\vec{\varphi}_3^*, \vec{\phi}_4) \quad , \\
\vec{\varphi}_5 &= \vec{\phi}_5 - \vec{\varphi}_1(\vec{\varphi}_1^*, \vec{\phi}_5) - \vec{\varphi}_2(\vec{\varphi}_2^*, \vec{\phi}_5) - \vec{\varphi}_3(\vec{\varphi}_3^*, \vec{\phi}_5) - \vec{\varphi}_4(\vec{\varphi}_4^*, \vec{\phi}_5) \quad , \\
\vec{\varphi}_6 &= \vec{\phi}_6 - \vec{\varphi}_1(\vec{\varphi}_1^*, \vec{\phi}_6) - \vec{\varphi}_2(\vec{\varphi}_2^*, \vec{\phi}_6) - \vec{\varphi}_3(\vec{\varphi}_3^*, \vec{\phi}_6) - \vec{\varphi}_4(\vec{\varphi}_4^*, \vec{\phi}_6) - \vec{\varphi}_5(\vec{\varphi}_5^*, \vec{\phi}_6) \quad , \\
&\cdots\cdots
\end{aligned}$$

This procedure consists of inner products and scalar-by-vector products. These operations could be done using linear-algebra library, e.g., level 1 of Basic Linear Algebra Subprograms (BLAS). BLAS-1 is, however, extremely inefficient when treating large dimension vectors because of high ratio of the data-load tasks to the cpu operations. However, a part of the GS operations can be grouped and regarded as matrix-by-matrix products: e.g., the last three operations in the above are written as ,

$$\left(\vec{\phi}_4, \vec{\phi}_5, \vec{\phi}_6\right) - \left(\vec{\varphi}_1, \vec{\varphi}_2, \vec{\varphi}_3\right) \begin{pmatrix} {}^t\varphi_1^* \\ {}^t\varphi_2^* \\ {}^t\varphi_3^* \end{pmatrix} \left(\vec{\phi}_4, \vec{\phi}_5, \vec{\phi}_6\right) \quad .$$

This is the matrix-matrix product which can be performed efficiently on modern computers by employing a highly tuned linear-algebra library, e.g., level 3 of BLAS. We may tune the size of the matrix block to the computer hardware.

The present RSDFT scheme has been applied to several issues in condensed matter sciences, including microscopic identifications of vacancies in Si crystal [3] and in carbon nanotubes [4], and of dislocation cores at the mismatched interfaces [5]. On the supercomputer developed at Kobe, Japan, electron-state calculations of Si nanowires containing more than 100,000 atoms now become feasible.

2.2. *Energy gap of Si nanodots*

Si nanodots with the dimension of tens of nanometers have been already fabricated and used for electron devices such as memory units. Reading or writing the data on the memory units corresponds to charging or discharging electrons to the nanodots. This technologically fundamental quantity is related to the energy gap (band gap) of the dot. The energy gap is rigorously defined using the ionization energy and the electron affinity as

$$E_g = E(N_e + 1) + E(N_e - 1) - 2E(N_e) \tag{4}$$

where $E(N_e+1)$, $E(N_e-1)$ and $E(N_e)$ represent the total energies of singly negative, singly positive, and neutral charge-state systems, respectively. However, it is impossible to calculate the total energy of a singly charged solid which is of infinite size. Then the energy gaps of solids are usually evaluated by the difference in the orbital energy between the highest occupied state and the lowest unoccupied state. This quantity, occasionally called HOMO-LUMO gap, is obviously different from E_g in (4). The calculated HOMO-LUMO gaps in LDA or in GGA are known to be substantially smaller than the experimental gaps. In order to develop exchange-correlation energy functionals which are capable of describing the energy gaps, it is imperative to examine the difference between E_g and the HOMO-LUMO gap.

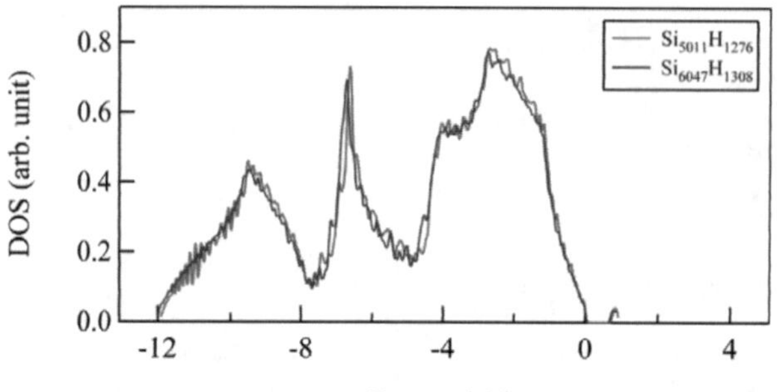

Figure 1: Density of states for $Si_{5011}H_{1276}$ and $Si_{6047}H_{1308}$ nanodots. The zero energy corresponds to the position of the highest-occupied state.

We have examined it by performing large-scale electronic-structure calculations for large enough Si nanodots [6]. Figure 1 shows the calculated density of staets (DOS) of the Si nanodots consisting of about 6000 Si atoms surrounded by 1000 H atoms. The calculated DOSs are essentially identical to each other. Furthermore they are almost same with DOS of bulk crystalline Si. Hence we are able to simulate infinite-size Si crystal by large enough Si dots. In Figure 2, we show the energy gap E_g calculated by (4) and the HOMO-LUMO gap as a function of the dot size. In the infinite size limit, those quantities should converge to the bulk values. As shown in Figure 2, the energy gap decreases as the system size increases. For the largest cluster $Si_{10701}H_{1996}$ the energy gap E_g is 1.07 eV and the HOMO-LUMO gap is 0.66 eV. The former is close to the experimental band gap 1.17 eV in Si crystal, and the latter is close to the HOMO-LUMO gap in bulk Si.

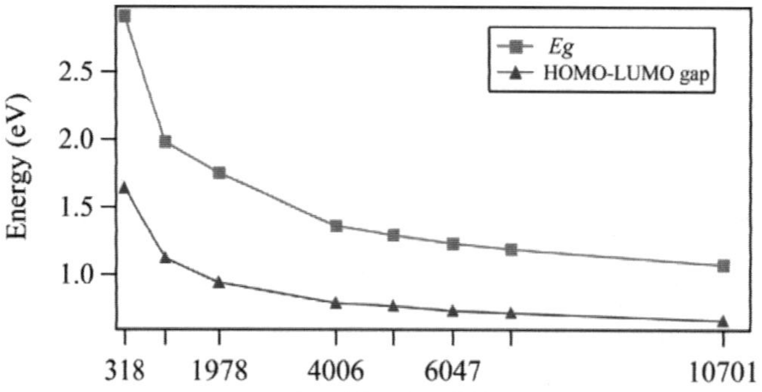

Figure 2: Dot-size dependence of the energy gap E_g and the HOMO-LUMO gap.

In the DFT, the following relation between E_g and the Kohn-Sham levels holds:

$$E_g = \varepsilon_{N_e+1}(N_e+1) - \varepsilon_{N_e}(N_e) \quad . \tag{5}$$

Journal of Physics: Conference Series **302** (2011) 012030 doi:10.1088/1742-6596/302/1/012030

Here $\varepsilon_{N_e}(M)$ depicts the M-th Kohn-Sham level in the N_e-electron system. Hence the difference between E_g (squares in Figure 2) and the HOMO-LUMO gap (triangles in Figure 2) is

$$E_g - (\varepsilon_{N_e}(N_e+1) - \varepsilon_{N_e}(N_e)) = \varepsilon_{N_e+1}(N_e+1) - \varepsilon_{N_e}(N_e+1) \quad . \tag{6}$$

We have analyzed the difference between E_g and the HOMO-LUMO gap computed in LDA (Figure 2) and found that the difference corresponds to Coulomb energy of the additional electron distributed over the dot. Then the difference is likely to disappear in the infinite-size limit. This is consistent with the expression (6). Hence we reach a conclusion that the exact expressions (5) and (6) hold even in LDA and that E_g defined in terms of the total-energy difference for the crystalline Si underestimate the experimental energy gap substantially.

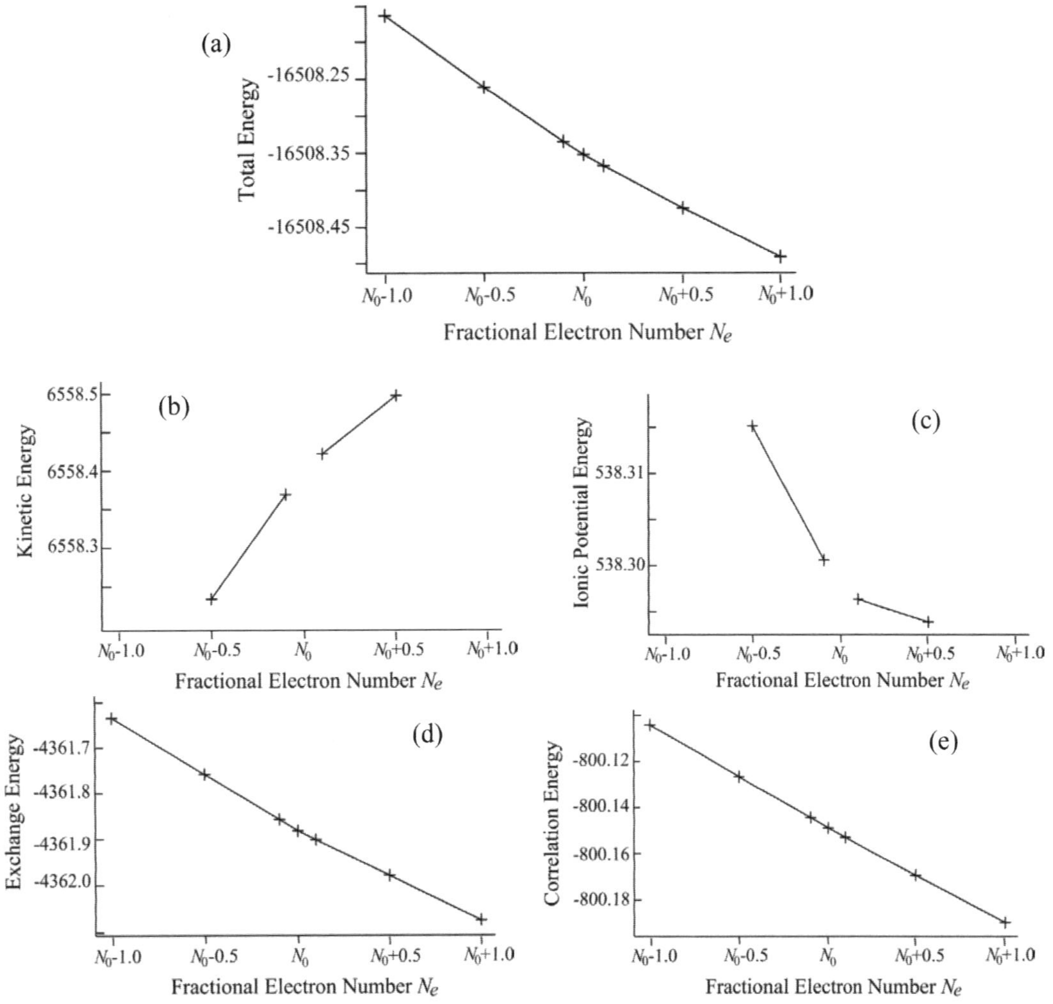

Figure 3: (a) Calculated total energy of $Si_{4006}H_{1038}$ nanodot as a function of fractional number of electrons N_e in LDA. The neutral state corresponds to N_e=N_0=17062. Each contribution to the total energy is also shown as a function of N_e: (b) the kinetic energy, (c) the ionic pseudopotential energy, (d) the exchange energy, and (e) the correlation energy. The unit of the energy is the atomic unit: i.e., 27.21 eV.

From the expression (5) combined with the fractional charge states, it is deduced that the energy gap comes from the discontinuity in the total energy at the integer value of electrons [7-9]. It is thus useful to appropriately define the total energy as a function of fractional electron number N_e[10], and compute it. Figure 3 shows the calculated total energy of $Si_{4006}H_{1038}$ nanodot along with each contribution to the total energy. In LDA or GGA, the Kohn-Sham levels generally shift upwards with increasing the number of fractional electrons (the self interaction error). This causes concave behaviour of the total energy $E_{tot}(N_e)$ as a function of the number of the fractional electrons N_e since Janak theorem $\partial E_{tot} / \partial N_e = \varepsilon(N_e)$ holds [10]. Although the concave behaviour in Figure 3(a) is not very prominent since the dot is so large that the upward shifts of the Kohn-Sham levels are small, the discontinuity of $E_{tot}(N_e)$ at N_e=N_0 (the neutral state) corresponds to the energy gap E_g. As for each contribution to the total energy, the kinetic energy and the ionic pseudopotential energy clearly show the discontinuity at N_e=N_0. This corresponds to the energy gap between the bonding and the anti-bonding states, being referred as the chemical gap. On the other hand, the exchange energy shows faint discontinuity at N_e=N_0 and the correlation energy is almost continuous. This infers that many body effects are not included enough in LDA, causing the underestimate of the energy gap.

3. Summary

We have presented our RSDFT scheme which allows us to perform unprecedentedly large-scale density-functional-theory-based electronic-structure calculations. At present, calculations of electron states of Si nano-structures containing 10,000 – 100,000 atoms are feasible. The calculated energy gaps of Si nanodots using the total energy difference among charged systems have elucidated the reason for well-known underestimate of the energy gaps in the local density approximation in the density functional theory.

Acknowledgments

This work is partly supported by the Grant-in-Aid project from MEXT, Scientific Research on Innovative Areas, ``Materials Design through Computics: Complex Correlation and Non-equilibrium Dynamics" (http://computics-material.jp/).

References

[1] Oshiyama A, and Okada S 2009 , Roles of Shape and Space in Electronic Properties of Carbon Nanomaterials, The Oxford Handbook of Nanoscience and Technology Volumes II (Oxford University Press) pp94 – pp140.
[2] Iwata J.-I, Takahashi D, Oshiyama A, Boku B, Shiraishi K, Okada S and Yabana K 2010, *J. Comp. Phys.* **229** 2339.
[3] Iwata J.-I, Shiraishi K, and Oshiyama A 2008, *Phys. Rev.* B **77**, 115208.
[4] Nishikawa M, Iwata J.-I, and Oshiyama A 2010, Phys. Rev. B **81**, 205542.
[5] Fujimoto Y, and Oshiyama A 2010, Phys. Rev. B **81**, 205309.
[6] Calculational parameters such as grid spacing are carefully examined to assure the necessary accuracy.
[7] Perdew J. P, Parr R. G, Levy M and Balduz Jr. J. L 1982, *Phys. Rev. Lett.* **49**, 1691.
[8] Perdew J. P, Levy M 1983, *Phys. Rev. Lett.* **51**, 1884.
[9] Sham L. J and Schlüter M 1983, *Phys. Rev. Lett.* **51**, 1888.
[10] Janak J. F 1978, *Phys. Rev.* B **18**, 7165.

Nonlinear resistivity in a *d*-wave superconductor $YBa_2Cu_4O_8$ of sub-micron scale grains

H Deguchi[1], T Shoho[1], Y Kato[1], T Ashida[1], M Mito[1], S Takagi[1], M Hagiwara[2] and K Koyama[3]

[1] Faculty of Engineering, Kyushu Institute of Technology, Kitakyushu 804-8550, Japan

[2] Faculty of Engineering and Design, Kyoto Institute of Technology, Kyoto 606-8585, Japan

[3] Faculty of Integrated Arts and Science, The University of Tokushima 770-8502, Japan

E-mail: deguchi@tobata.isc.kyutech.ac.jp

Abstract. The *d*-wave ceramic $YBa_2Cu_4O_8$ superconductor composed of sub-micron size grains is considered as random Josephson-coupled network of 0 and π junctions and shows successive phase transitions. The upper transition occurs inside each grain at T_{c1}=82 K and the lower transition occurs among the grains at T_{c2}=66 K. We measured the temperature dependence of the current-voltage characteristics of the ceramic $YBa_2Cu_4O_8$ and derived the linear and nonlinear resistivity. The nonlinear resistivity ρ_2 and ρ_4 have finite values between T_{c1} and T_{c2} and have the peak at the same temperature T_p=70 K above T_{c2}. The result agrees with the theoretical one obtained by Li and Domínguez. They interpreted T_p as the crossover temperature from the normal state phase to a chiral paramagnetic one.

1. Introduction

Ceramic high-T_c superconductors, depending on the relative orientation of the *d*-wave superconductor grains, may be regarded as random Josephson-coupled networks that contain so-called π junctions. The circulation of a local loop-supercurrent is generated spontaneously in a zero field when there is an odd number of π junctions in a closed loop that consists of some Josephson junctions. The frustration effect due to the random distribution of π junctions should lead to a novel glass state. Kawamura and Li investigated the novel glass phase of ceramic high-T_c superconductors using Monte Carlo simulations.[1] They proposed that the chiral glass phase characterized by a broken time-reversal symmetry may occur in zero external magnetic field. Evidence for the transition to chiral glass has been seen from experimental studies of nonlinear ac magnetic susceptibility, dynamic scaling analysis and of ac resistivity in the *d*-wave ceramic $YBa_2Cu_4O_8$ and $YBa_2Cu_3O_{7-\delta}$ superconductors.[2-6] The two ceramic samples are weak-link coupled systems and show successive phase transitions. With decreases in temperature, the first transition occurs inside each grain at T_{c1} and the second transition occurs among the grains at T_{c2}. The discrepancy between the field-cooled and zero-field-cooled magnetization appears at below T_{c2}, and negative divergence of nonlinear susceptibility is observed at

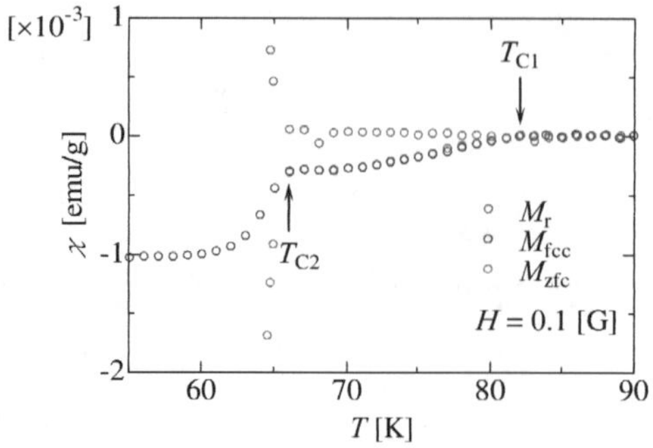

Figure 1. Temperature dependence of zero-field-cooled magnetization (M_{zfc}), field-cooled-magnetization (M_{fcc}), and thermoremanent magnetization (M_r).

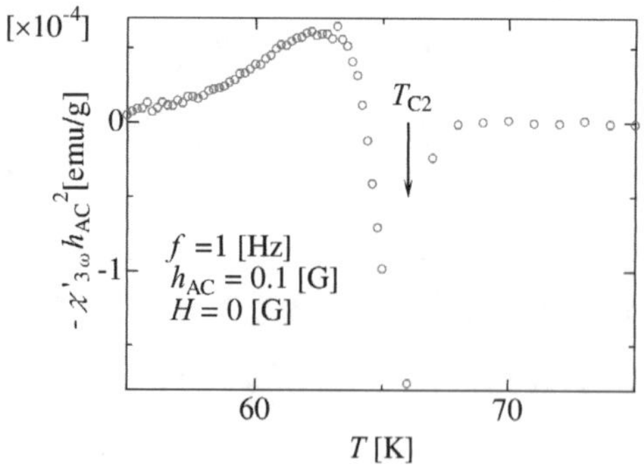

Figure 2. Temperature dependence of the nonlinear susceptibility.

T_{c2}, which reflect a chiral-glass transition.

Yamao *et al.* [7] measured the ac linear resistivity ρ_0 and nonlinear resistivity ρ_2 of the ceramic $YBa_2Cu_4O_8$. Since ρ_0 does not vanish at the peak position of ρ_2, they identify the transition at the peak position of ρ_2 as a transition to the chiral-glass phase. Li and Domínguez [8] calculated the linear and nonlinear resistivity by a Langevin dynamics simulation. They found that ρ_2 has a peak at T_p above T_{c2} and interpreted T_p as the crossover temperature from the normal-state phase to a chiral paramagnet. In order to probe the existence of the crossover point T_p and chiral glass transition T_{c2} we have measured the current-voltage characteristics of the ceramic $YBa_2Cu_4O_8$ composed of sub-micron size grains and derived the linear and nonlinear resistivity with magnetic measurements.

2. Experimental

The sample was prepared using the citrate pyrolysis method.[9] The precursor was calcined for 140 h at 777 ℃ to yield the pure $YBa_2Cu_4O_8$ phase, which was sieved and pressed, and then sintered for 100 h at 778 ℃, to became a pure $YBa_2Cu_4O_8$ ceramic of sub-micron size grains. The dc magnetization and the ac susceptibility were measured with a SQUID magnetometer (Quantum Design MPMS-5) using the ultra-low-field option. The sample space in the magnetometer was shielded with μ -metal. As a result, the residual field was reduced to less than 10 mG. The *E-J* characteristics were measured by low-level pulsed electrical characterization with a current source (KEITHLY 6221) and nanovoltmeter (KEITHLEY 2182A) combination.[6] The delta method for low-voltage measurements was used to eliminate the constant thermoelectric voltage and to minimize the amount of power dissipated in the sample.

3. Results and Discussion

The temperature dependence of the dc magnetization and ac susceptibility were measured in order to determine the transition temperature T_{c1} and T_{c2}. Figure 1 shows the temperature dependence of the zero-field-cooled, field-cooled, and thermoremanent magnetizations at $H = 0.1$ G. The upper transition at T_{c1} = 82 K was identified as the intragrain superconducting ordering, in which small diamagnetism due to the Meissner effect appears in zero-field-cooled and field-cooled-magnetizations. The irreversibility point in zero-field-cooled and field-cooled magnetizations occurs and the thermoremanent magnetization appears at temperatures lower than T_{c2}= 66 K. The nonlinear susceptibility estimated from the first term of the series of in-phase odd-harmonic responses in the frequency of 1 Hz with ac field amplitude of 0.1 G is shown in Figure 2. Negative divergence of

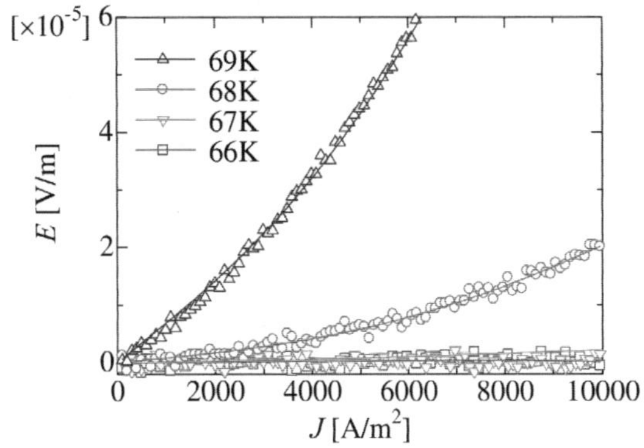

Figure 3. E-J characteristics at different temperatures near T_{c2}.

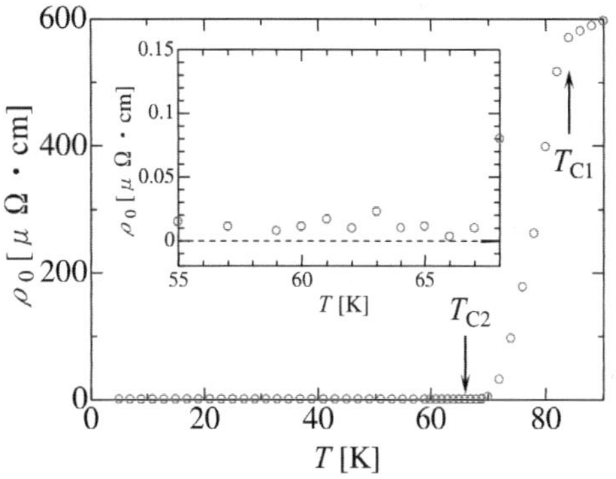

Figure 4. Temperature dependence of the linear resistivity ρ_0

nonlinear susceptibility was observed at T_{c2} = 66 K, at which temperature the ceramic $YBa_2Cu_4O_8$ underwent a chiral-glass transition. In order to investigate the transport property of the ceramic $YBa_2Cu_4O_8$, we measured the E-J curves with zero-magnetic field in the temperature range between 5 and 90 K. Figure 3 shows the E-J curves near the chiral-glass transition temperature T_{c2}. Below T_{c2}, E increased almost linearly with J. At temperatures greater than T_{c2}, rapid increases in E were observed as J increased, perhaps due to the nonlinear resistivity.

Linear resistivity ρ_0 and nonlinear resistivity ρ_2, ρ_4 are defined as the coefficients of the first, third and fifth power term of $E(J)$ expanded in a power series of J.[7] Figure 4 shows the temperature dependence of the linear resistivity ρ_0 estimated from E-J curves. As the temperature was reduced, ρ_0 decreased rapidly at about T_{c1} = 82 K, and then decreased monotonously and almost disappears around a temperature of 70K. Below T_{c2}=66 K, ρ_0 remains finite and has the low value of about 0.01 $\mu\Omega\cdot$cm, which is consistent with the theoretical estimation of chiral-glass ordering [10] and the experimental results.[6,7]

Figures 5 and 6 show the temperature dependence of the nonlinear resistivity ρ_2 and ρ_4 estimated from E-J curves, respectively. As shown in Fig.5, ρ_2 has a positive peak at T = 70 K, at which temperature the chiral-glass transition does not occur, and diminishes continuously toward T_{c2} with decreasing temperature. The behavior of ρ_2 differs from that obtained by Yamao *et al.*[7] The peak position of ρ_2 is concluded at T_{c2} by Yamao *et.al.* but does not agree with T_{c2} in our result. T_{c2} is estimated by both the irreversibility point in the dc magnetization and the divergence point of nonlinear susceptibility in our study, however, Yamao *et.al.* obtained T_{c2} by only the irreversibility point in the dc magnetization. Therefore, we confirm that the peak position of ρ_2 is higher than T_{c2}. Kawamura studied the critical dynamic properties of the chiral-glass transition using Monte Carlo simulations.[11] He investigated transport property of ceramic high-T_c superconductors near the chiral-glass transition. The temperature dependence of ρ_2 in our experiment corresponds to the theoretical prediction by Kawamura with the case of the chiral-glass exponent $z > 5$.

The nonlinear resistivity ρ_4 has the finite value between T_{c2} and T_{c1} and has a negative peak at the same temperature T = 70 K as that for ρ_2. Thus, the nonlinear resistivity ρ_2 and ρ_4 show a great anomaly at T = 70 K above T_{c2}. Li and Domínguez obtained theoretically ρ_2 of d-wave ceramic superconductor by solving the corresponding Langevin dynamical equations.[8] They found that ρ_2 has a peak at T_p above the chiral-glass transition temperature and interpreted T_p as the crossover temperature from the normal-state phase to a chiral paramagnet. They conclude that peak of ρ_2 has no

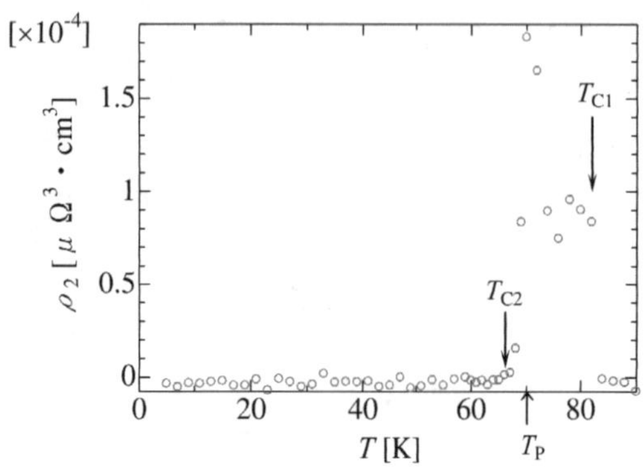

Figure 5. Temperature dependence of the nonlinear resistivity ρ_2.

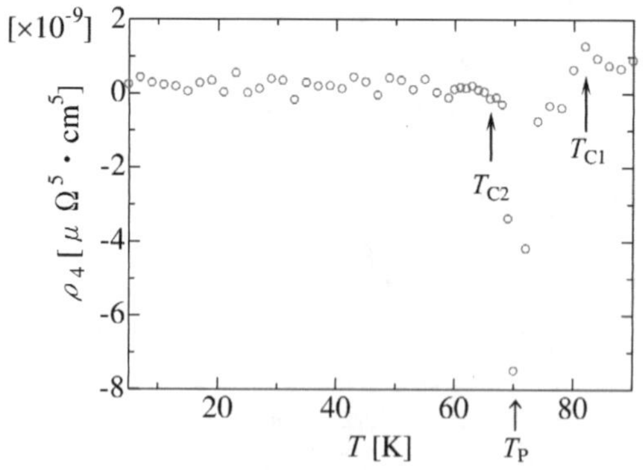

Figure 6. Temperature dependence of the nonlinear resistivity ρ_4.

relation to the chiral-glass phase transition. Thus, T_p just separates the normal-sate phase from a chiral paramagnet where there are local chiral magnetic moments. At a lower temperature, collective phenomena due to the interactions among the chiral moments will start to be important, leading to the transition to the chiral-glass state. Similar
theoretical results for nonlinear resistivity in *d*-wave granular superconductors obtained by Li *et al.*[12] Therefore, the temperature T =70 K of the peak corresponds to the crossover temperature T_p.

In conclusion, our experimental results for the liner and nonlinear resistivity are consistent with those of theoretical studies by Kawamura [11] and Li *et al.*[8,12] The linear resistivity ρ_0 remained at a finite value at temperature less than the chiral-glass transition T_{c2}. The nonlinear resistivity ρ_2 and ρ_4 show a great anomaly at T_p = 70 K above T_{c2}. The temperature T_p does not correspond to the chiral-glass transition but corresponds to the onset of the chiral paramagnet.

Acknowledgments

The authors express their sincere thanks to Prof. H. Kawamura and Prof. M. Matsuura for their continual interest and helpful discussions. The SQUID measurement was carried out at the Center for Instrumental Analysis, Kyushu Institute of Technology. This work was supported by Grant-in-Aid for Scientific Research on Priority Areas (19052006).

References

[1] Kawamura H and Li M S 1997 *Phys. Rev. Lett.* **78** 1556
[2] Kawachi M, Hagiwara M, Koyama K and Matsuura M 1994 *J. Phys. Soc. Jpn.* **63** 3405
[3] Koyama K, Deguchi H, Takagi S and Mizuno K 2003 *J. Low Temp. Phys.* **131** 595
[4] Matsuura M, Kawachi M, Miyoshi K, Hagiwara M and K Koyama K 1995 *J. Phys. Soc. Jpn.* **64** 4540
[5] Deguchi H, Yasunaka M, Oniki R, Takagi S, Koyama K and Mizuno K 2003 *Physica* C **388-389** 269
[6] Deguchi H, Nawa D, Hashimoto Y, Mito M, Takagi S, Hagiwara M and Koyama K 2009 *J. Phys. Conf. Series* **150** 052043
[7] Yamao T, Hagiwara M, Koyama K and Matsuura M 1999 *J. Phys. Soc. Jpn.* **68** 871
[8] Li M S and Domínguez D 2000 *Phys. Rev.* B **62** 14554
[9] Yamao T, Hagiwara M, Shima T and Matsuura M 2004 *Physica* C **412-414** 98
[10] Kawamura H and Li M S 1997 *J. Phys. Soc. Jpn.* **66** 2110
[11] Kawamura H 2000 *J.Phys. Soc. Jpn.* **69** Suppl.A 281
[12] Li M S, Zung H and Domínguez D 2002 *Phys. Rev. Lett.* **88** 257004

Journal of Physics: Conference Series **302** (2011) 012032 doi:10.1088/1742-6596/302/1/012032

Cathodoluminescence Induced by Surface Plasmon Polaritons in 2-Dimensional Plasmonic Crystals

Kengo Takeuchi and Naoki Yamamoto

Department of Physics, Tokyo Institute of Technology, Oh-okayama, Meguro-ku, Tokyo 152-8551, Japan

E-mail: ktakeuch@surface.phys.titech.ac.jp

Abstract. A cathodoluminescence technique using a 200-keV transmission electron microscope revealed the dispersion patterns of surface plasmon polaritons (SPPs) in two-dimensional plasmonic crystals with cylindrical hole arrays. The dispersion curves of the SPP modes involving the Γ point were derived from the angle-resolved spectrum patterns. The contrast along the dispersion curves changed with the polarization direction of the emitted light due to the property of the SPP modes. The SPP modes at the Γ point were identified from the photon maps, which mimicked standing SPP waves in a real space.

1. Introduction

A surface plasmon polariton (SPP) is a transverse magnetic (TM) mode electromagnetic wave propagating at a metal/dielectric interface, which is evanescently confined in the perpendicular direction. SPPs can mutually be converted with light, and metal surfaces with periodic arrangements on a sub-wavelength scale have been utilized as couplers. Similar to electrons in a crystal potential, the dispersion of SPPs on a flat metal surface forms a band structure. An SPP whose energy is within the band gap cannot propagate in the periodic structure. In contrast, the conversion efficiency of an SPP at the band edge energy is enlarged due to the increased density of states. These properties allow devices that manipulate SPPs to be produced, including mirrors [1], wave guides [2], cavities [3], and bio-sensors [4,5]. These types of metal surfaces with periodic structures are called plasmonic crystals, and are recognized as valuable elements in the field of plasmonics.

Spatial distribution of the SPP waves within the band gaps or at the band edges are important for designing plasmonic devices. Direct observation of surface plasmon in real space has been performed by optical microscopy [6], near-field scanning optical microscopy (NSOM) [7] and photon-STM [8], though the resolution is not enough for visualizing details of the field distribution of SPP on surface structures. Electron beam can generate a point source of SPP at any place on a metal surface. Thus cathodoluminescence (CL) technique combined with transmission electron microscopy enables us to study real space distribution of SPPs in plasmonic crystals by scanning electron beam with high spatial resolution [9, 10]. The polarization selected and angle-resolved measurements give important information on the SPP dispersion property [10]. In the present paper we show the dispersion patterns of SPPs in a two-dimensional plasmonic crystal with cylindrical hole arrays taken by the angle-resolved spectrum measurement. We first observed photon maps taken at the resonant energies of SPP modes in the two dimensional plasmonic crystal, which clearly revealed the symmetry of the standing SPP wave modes at the Γ point. From these results we can identify the SPP bands near the Γ point.

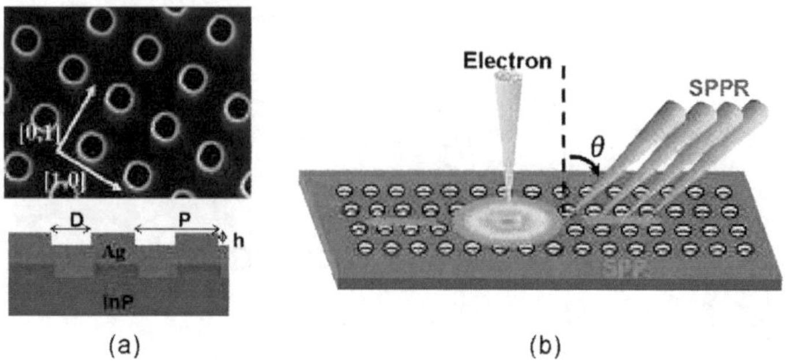

Figure 1 (a) SEM image of a specimen with a periodic array of cylindrical holes and a period of 600 nm, and a diagram of the cross-section. Hole diameter and depth are 240 nm and 200 nm, respectively. InP substrate is coated by a 200-nm thick silver layer. (b) SPP to photon conversion process on the 2D-plasmonic crystal after excitation by an electron beam.

2. Experimental setting

Figure 1(a) shows the SEM image of an array of cylindrical holes on an InP substrate created by electron-beam lithography. A square lattice in the array has a periodicity of 600 nm, and the holes have a diameter and depth of 240 nm and 200 nm, respectively. A 200-nm thick silver layer is evaporated onto the substrate by thermal evaporation in a vacuum. In the present experiment, we employ a TEM combined with a light detection system, which operates at an acceleration voltage of 200 kV. The electron beam has a diameter of 10 nm and a beam current of 1 nA. The incident electron beam excites an SPP, which then propagates into the surroundings as a spherical wave on the specimen surface (Figure 1(b)). SPPs that propagate on a periodic structure induce light emissions. This emitted light is collected by a parabolic mirror, and transformed into a parallel ray, which is then detected by a CCD detector. The polarization direction of light is determined by a polarizer, which is placed between the parabolic mirror and detector. Here we define *p*-polarization as parallel to the emission plane subtended by the axis of parabola and normal to the specimen surface, and *s*-polarization is normal to the emission plane. For an angle-resolved measurement, a mask with a small hole is inserted between the polarizer and the CCD detector. The mask position selects the emission direction of the detected light, which is specified by angles θ and φ. The mask is moved vertically when the azimuth angle $\varphi = 0°$ and the emission spectra are successively measured to determine the θ-dependence. This procedure provides an emission intensity distribution of an angle-resolved spectrum (ARS), which shows variations with angle θ and forms an ARS pattern, $I(\theta,E)$ [10]. After the measurement, the variations in the observed spectrum due to the variability in the detection efficiency are corrected and the θ-dependence due to the solid angle subtended by the hole is taken into account.

3. Results and discussion

We observed the ARS patterns in arrangements where the [1,0] directions of the specimen are parallel to the parabolic mirror axis. During the measurement, the electron beam is scanned over a $5 \times 5\ \mu m^2$ area. The observed ARS patterns are then transformed into the dispersion patterns, $I(k_{//},E)$, by changing the emission angle to the wave vector component parallel to the surface, $k_{//}$, using the relation

$$k_{//} = |\boldsymbol{k}| \sin\theta = \frac{E_{ph}}{\hbar c} \sin\theta \qquad (1)$$

where $\boldsymbol{k}$ is the wave vector for the emitted light, $\hbar$ is Plank constant divided by 2π, and c is the velocity of light. Figures 2(a) and (b) show the dispersion patterns along the Γ–X direction for (a) *p*-polarized and (b) *s*-polarized emissions. The emission near the Γ point disappears because the hole in the mirror where the electron beam passes is incident to the specimen surface. The green lines in Figure 2 indicate SPP dispersion curves along the Γ–X directions. Here we treat the SPP states at the Γ point on the basis of group theory. An SPP in a plasmonic crystal can be expressed by the Bloch wave function, and can be solved from Maxwell equation under the boundary condition of the periodic surface structure. Group theory indicates four eigenvectors and eigenvalues, although two of them are

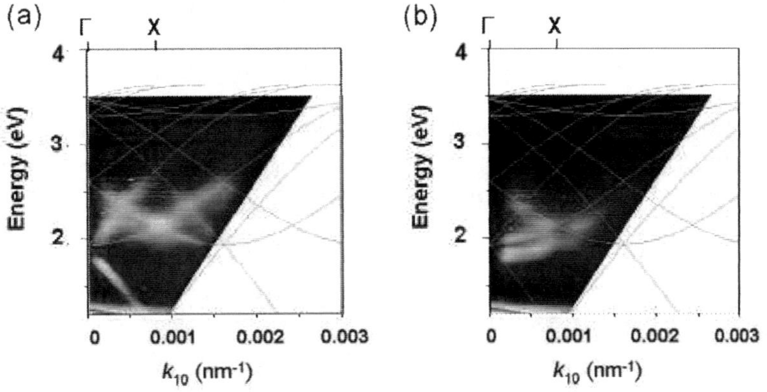

Figure 2 Dispersion patterns along the *Γ–X* direction derived from the ARS patterns for (a) *p*-polarized and (b) *s*-polarized emissions. Green lines indicate the SPP dispersion curves of the plasmonic crystal in the empty lattice approximation.

degenerate. The field distributions of the four eigenstates are shown in Figure 3(a)–(d) where the red and blue colors denote opposite signs. These patterns can also be interpreted as the surface charge distributions because the induced surface charge is proportional to the electric field component normal to the surface. Figure 3(a) corresponds to the total symmetry mode (*A* mode), which has a maximum at the lattice point and the center of the unit cell. In contrast, the *B* mode in Figure 3(b) has minimum at these points. The E_1 and E_2 modes in Figures 3(c) and (d), respectively, have inversion symmetry around the lattice point and are energetically degenerate.

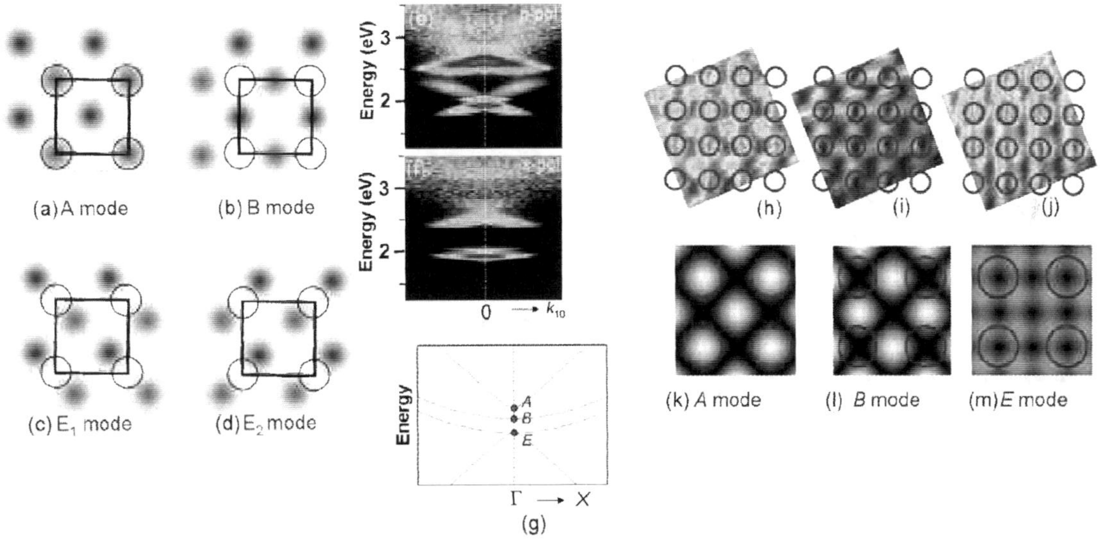

Figure 3 (a)–(d) represent the Bloch wave patterns of the four SPP modes at the *Γ* point. Solid circles indicate holes and squares indicate a unit cell. ARS patterns near the *Γ* point taken by (e) *p*-polarized and (f) *s*-polarized emissions. (g) Schematic diagram of the dispersion curves along the *Γ–X* direction. Photon maps taken by the *p*-polarized emission at peak energies of (h) 2.02 eV, (i) 1.97 eV, (j) 1.89 eV. (k)–(m) show patterns of the time-averaged field strength associated with the SPP modes at the *Γ* point. Red circles indicate holes.

Figures 3(e) and (f) show the ARS patterns of (e) *p*-polarized and (f) *s*-polarized emissions near the surface normal direction acquired by tilting the specimen in the [1,0] direction with respect to the incident beam direction. These reveal the dispersion pattern of the SPP bands around the Γ point. The contrast disappears in the larger k_{10} due to the hole in the mirror; the dispersion line contrasts are actually related to those in Figures 2(a) and (b). The four dispersion lines gather at the Γ point, as illustrated in Figure 3(g). The line contrast of the highest energy mode appears only in *p*-polarization, although it disappears at the Γ point (Figure 3(e)). Therefore, this mode should correspond to the *A* mode, which does not emit light, due to the symmetrical charge distribution around the hole. The dispersion line contrast of the lowest energy mode in *p*-polarization intersects with the energy axis at the same position as the one of the line contrasts that extends in the horizontal direction in *s*-polarization. This suggests that these lines are connected via the *E* mode at the Γ point.

Figures 3(h)–(j) are photon maps taken with light emitted near the surface normal direction at the peak energies of the Γ point for the *p*-polarization where the red circles denote hole positions. Figures 3(k)–(m) show patterns of the time-averaged field strength associated with the SPP modes at the Γ point. There is a clear correlation between the photon maps and calculated patterns; (h) corresponds to (k), (i) corresponds to (l), and (j) correspond to (m). Thus, we can identify the modes at the Γ point, as indicated in Figure 3(g).

4. Conclusion

The dispersion relation of SPPs in 2D-PLC with arrays of cylindrical holes is investigated using a 200-keV TEM combined with the angle-resolved CL system. In this technique, a focused electron beam provides a point source of an SPP at any location on the surface, and photon maps can be measured with a high spatial resolution on the nanometer scale. The photon map acquired at the peak energy at the Γ point mimics a standing SPP wave, which allows the SPP mode to be identified. The dispersion patterns of SPPs propagating in special directions are derived from the ARS pattern measured along select emission directions. After considering the excitation probability of an SPP by a 200-keV electron, the dispersion pattern can yield the efficiency of the SPP to photon conversion. These results provide useful information for designing SPP–photon coupling devices using a plasmonic crystal.

Acknowledgment

This work has been supported by a Grant-in-Aid for Scientific Research (Nos.19101004 and 21340080) from the MEXT of Japan, and MEXT-Nanotech NP.

References

[1] Ditlbacher H, Galler N, Koller D M, Hohenau A, Leitner A, Aussenegg F R and Krenn J R 2008 *Opt. Express.* **16** 10455–10464
[2] Bozhevolnyi S I, Erland J, Leosson K, Skovgaard P M W and Hvam J M 2001 *Phys. Rev. Lett.* **86** 3008–3011
[3] Weeber J C, Bouhelier A, Colas des Francs G, Markey L and Dereux A 2007 *Nano Lett.* **7** 1352-1359
[4] Homola J 2008 *Chem. Rev.* **108** 462–93
[5] Truong T T, Maria J, Yao J, Stewart M E, Lee T-W, Gray S K, Nuzzo R G and Rogers J A 2009 *Nanotechnology* **20** 434011
[6] Mock J J, Barbic M, Smith D R, Schultz and Schultz S 2002 *J Chem. Phys.* **116**, 6755
[7] Zayats A V, Smolyaninov I I, Maradudin A A 2004 *Phys. Rep.* **11**, 001-184
[8] Weeber J C, Krenn J R, Dereux A, Lamplecht B, Lacroute Y, and Goudonnet J P 2001 *Phys. Rev. B*, **64**, 045411
[9] Yamamoto N and Suzuki T 2008 *Appl. Phys. Lett.* **93** 093114
[10] Suzuki T and Yamamoto N 2009 *Opt. Express* **17** 23664-23671

Tip-Enhancement Effect on the PL of Localized Excitons in an InGaN Thin Film

K. Katayama[1], Y. Ogawa[1], F. Minami[1], T. V. Shubina[2], and A. A. Toropov[2]

[1]Department of Physics, Tokyo Institute of Technology, Meguro, Tokyo, 152-8551, Japan
[2]Ioffe Physical-Technical Institute, Russian Academy of Sciences, 26 Polytekhnicheskaya, St. Petersburg, 194021, Russia

E-mail: katayama.k.ac@m.titech.ac.jp

Abstract. Simultaneous measurements of the morphology, μ-PL, and tip-enhanced PL (TEPL) images of an InGaN thin film have been done. From a comparison between the μ-PL and TEPL images, the appearance of the bright spots in the TEPL image should be determined by the emission yield, permittivity of the sample and dipole orientation. The tip-enhancement effect on the temporal behaviour of the PL of the InGaN thin film has been measured. The temporal behaviour is altered by the tip-enhancement effect from a multi-exponential to single-exponential shape. This shows that the influx of the excitons from other potential minima to the luminescent energy level is suppressed by the existence of the gold-nano-particle attached to the tip.

1. Introduction

The coupling of photo-excited dipoles and surface-plasmons yields the photoluminescence (PL) enhancement [1]. To obtain the PL enhancement by surface-plasmons, researchers coated their samples by metals [1, 2] or dispersed metallic nano-particles on their samples [3]. These methods are appropriate for practical applications. In contrast to these methods, the tip-enhancement method is an effective tool to study the effect of surface-plasmons. In tip-enhanced PL measurements, a metal nano-particle is attached onto the tip of an atomic force microscope (AFM) probe and located in nano-meter scale proximity to the photo-excited dipole. Thus we can turn the effect of surface-plasmons on and off by adjusting the position of the metal nano-particle. Recently, we have reported the tip-enhanced PL (TEPL) of the InGaN thin film [4] and discussed the mechanism of TEPL from topographical, μ-PL, and tip-enhanced PL images, temperature dependence of the TEPL effect and PL temporal behaviour of the coated and non-coated samples. In the present paper, with an improved measurement system, topographical, μ-PL, and TEPL images of an InGaN thin film were measured simultaneously and the conditions of TEPL in the InGaN thin film are discussed. We have also measured the tip-enhancement effect on the PL dynamics of the InGaN thin film for the first time and found the suppression of the phonon-assisted tunneling process by the tip-enhancement effect.

2. Sample and measurement set-up

The sample was an $In_{0.25}Ga_{0.75}N$ thin film that was grown by plasma-assisted molecular beam epitaxy on c-plane sapphire above a GaN buffer [4]. The thickness of the sample was 200 nm [5]. A confocal microscope measurement system integrated with an AFM was applied for the TEPL measurement. The diameter of the gold nano-particle on the AFM probe and the laser spot were about 200 nm and 1.7 μm, respectively. The excitation wavelength was 532 nm and the light sources were CW and ML (76 MHz, 70ps) -Nd:YAG lasers for PL mapping and decay measurements, respectively. The integrated intensity of whole PL spectra has been recorded by an APD for the PL mapping measurement. The PL spectra were recorded by a nitrogen cooled CCD equipped with a 275 mm spectrometer. The dynamics of the PL was measured by a time-correlated single photon counting system with an APD equipped with a 100 mm spectrometer. Every experiment was done at room temperature.

3. TEPL mapping

Figure 1 (a) shows a topographic image of the sample. The surface roughness of the sample is several tens of nanometers and this roughness arises from the N-rich growth conditions [4]. Fig. 1 (b) and (c) show μ-PL and TEPL images of the same area of the sample, respectively. The PL originates from the recombination of excitons captured at potential local minima created by concentration fluctuation of In [4]. The localized excitons migrate to another potential minimum by phonon-assisted tunnelling. The probe is about 4 μm above the sample surface in Fig. 1 (b) and on the sample surface within the focused laser spot in Fig. 1 (c). The integrated intensity of a whole PL spectrum has been recorded by an APD at each point of the sample. During each scan, the relative position of the gold nano-particle with respect to the laser spot was fixed, while the sample was scanned laterally by a piezoelectric transducer scanner. In Fig. 1 (c), the enhanced PL intensity is about three times larger than that of the μ-PL image.

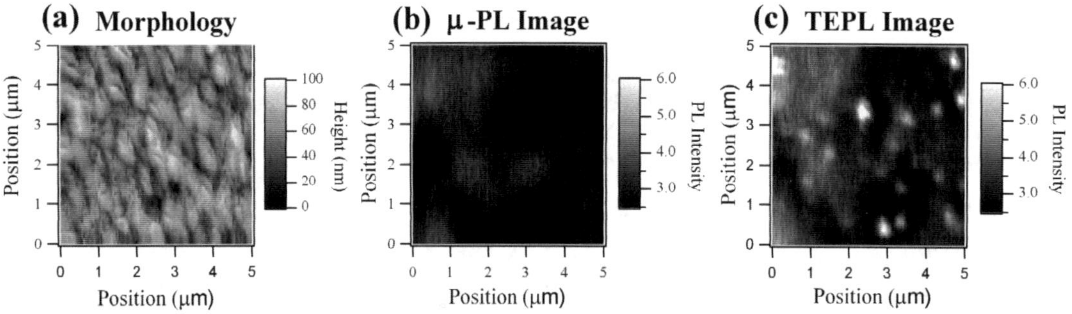

Figure 1. Topographic (a), μ-PL (b), and tip-enhanced PL (c) images of the $In_{0.25}Ga_{0.75}N$ thin film. The probe is about 4 μm above the sample surface (b) and on the sample surface within the focused laser spot (c).

Recently, the topographic and TEPL images have been measured simultaneously and it has been proposed that segregated metallic In droplets exist on the sample surface and the coupling between surface plasmons in the metallic In droplets and in the gold nano-particle on the tip plays an important role for the occurrence of the tip-enhancement effect [6]. If the metallic In droplets yield the enhancement, they should exist at the bright spots of the TEPL image. Furthermore, the μ-PL image should reflect the In droplets distribution. However, the bright spots in the TEPL image in Fig. 1(c) do not correspond to the bright places in μ-PL image in Fig. 1(b). Therefore, metallic In droplets do not contribute to the tip-enhancement effect in this area of the sample. Previously, we have showed the enhancement up to 20 times at the pit of the sample surface [4] and described the mechanism as below. The radial component of the exciton dipole increases gradually from bottom to the peripheral border of the pit. The radial component is more efficient than that of the tangential component for the tip-enhancement. Therefore, the enhancement occurs at the pit of the sample surface. In the TEPL image (Fig 1. (c)), the bright spots do not correspond exactly to the pit of the sample surface morphology

(Fig .1(a)). At a glance, the dipole orientation effect seems to be less effective in this result. However, the bright spots in Fig. 1(c) are at the slopes where the ratio of the radial component is larger than at the crest. Also, the emission yield and permittivity of the sample are important for the tip-enhancement [7]. Therefore, the appearance of the bright spots in the TEPL image (Fig 1. (c)) should be determined by the dipole orientation, emission yield and permittivity.

4. TEPL decay

Figure 2 (a) shows PL spectra of the sample at room temperature. In the case of "Tip-On" and "Tip-off" in Fig. 2 (a), the gold nano-particle on the AFM probe was inside and outside of the focused laser spot on the sample surface. The PL intensity is enhanced about 3 times by the tip-enhancement effect. The temporal behaviour of the PL is depicted in Fig. 2 (b). The integrated energy range for the time-resolved measurement is the area between two lines depicted in Fig. 2 (a). In the case of "Tip-off" in Fig. 2 (b), the temporal behaviour of excitons shows a multi-exponential shape. The decay of excitons by emitting a photon, by non-radiative recombination, by outflow due to the phonon-assisted tunnelling are expressed as product of each rate and the exciton population at the observed energy level. These decay processes yield a single exponential decay. On the other hand, the influx of excitons from other potential minima by phonon assisted tunnelling results in the multi-exponential shape. Therefore, the temporal behaviour without tip-enhancement effect shows various decay processes and the influx of the excitons from other potential minima. The temporal behaviour is altered by the tip-enhancement effect in the case of "Tip-on" in Fig. 2(b). To reveal the tip-enhancement effect on the temporal behaviour, the difference between "Tip-on" and "Tip-off" is shown in Fig. 2 (c). The enhanced PL shows single exponential decay (1.3 ns). This result shows that the influx of the excitons from other potential minima is suppressed by the existence of the gold-nano particle.

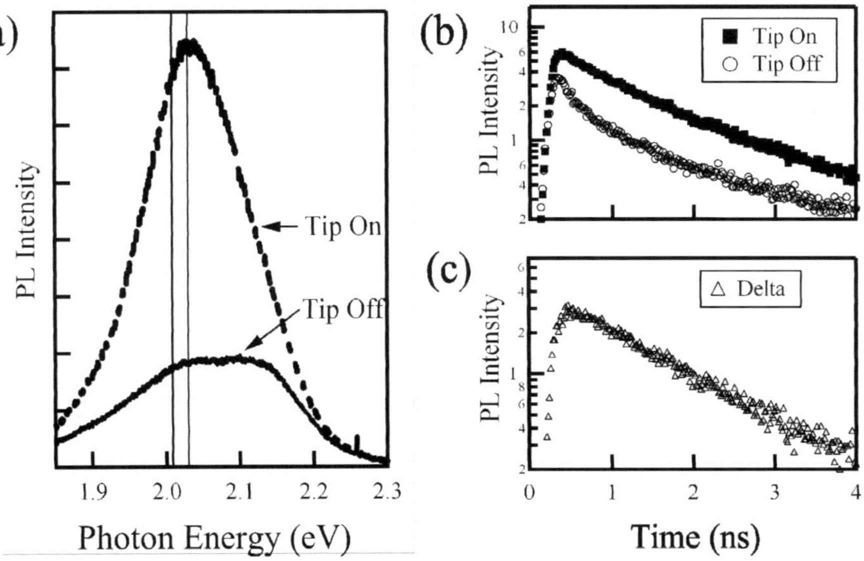

Figure 2. (a) PL spectra of the sample at room temperature. The dashed and continuous lines are in the cases of "Tip-On" and "Tip-off". (b) The dynamics of the PL in the cases of "Tip-On" (filled rectangle) and "Tip-off" (open circle). (c) The difference between "Tip-On" and "Tip-off" (open triangle).

5. Conclusions

Simultaneous measurements of the topographical, μ-PL, and TEPL images of the InGaN thin film show that the appearance of the bright spots in the TEPL image should be determined by the emission yield and the permittivity of the sample and the dipole orientation. The temporal behaviour of the PL of the InGaN thin film is altered by the tip-enhancement effect from a multi-exponential to single-exponential shape. These results show that the influx of the excitons from other potential minima to the luminescent energy level is suppressed by the existence of the gold-nano particle.

Acknowledgments
The researchers at Tokyo Institute of Technology acknowledge the financial support from the Global Center of Excellence Program by MEXT, Japan through the Nanoscience and Quantum Physics Project of the Tokyo Institute of Technology. The researchers from Ioffe Institute acknowledge the financial support of the Russian Foundation for Basic Research.

References

[1] Okamoto K, Niki I, Shvartser A, Narukawa Y, Mukai T and Shere A 2004 Nat. Mat. **3** 601
[2] Okamoto K, Vyawahare S and Sherer A 2006 J. Opt. Soc. Am. B **23** 1674
[3] Matsuda K, Ito Y, Kanemitsu Y 2008 Appl. Phys. Lett 211911
[4] Toropov A A, Shubina V T, Jmerik N V, Ivanov V S, Ogawa Y and Minami F, 2009 Phys. Rev. Lett. **103** 037403
[5] Toropov A A, Shubina V T, Jmerik N V, Ivanov V S, Ogawa Y and Minami F, 2009 Phys. Stat. Solidi. **6** S619
[6] Katayama K, Ogawa Y and Minami F 2011 Physica E **43** 808
[7] Toropov A A, Shubina V T, Belyaev G K, Ivanov V S, Kop'ev S P, Ogawa Y and Minami F, in preparation

Impurities Effects on the Electronic Structure of Titanium Dioxide

Yuta Aoki and Susumu Saito

Department of Physics, Tokyo Institute of Technology, 2-12-1 Oh-okayama, Meguro-ku, Tokyo 152-8551, Japan

E-mail: y-aoki@stat.phys.titech.ac.jp

Abstract. Effects of substitutional doping of nitrogen to TiO_2 are studied within the framework of the density-functional theory (DFT). We compare the total energy and the electronic structure of pristine TiO_2 with those of nitrogen-doped TiO_2. It is found that the smaller the concentration of doped nitrogen is, the more energy is required for the substitutional doping of the nitrogen atom in both rutile and anatase phases. This result suggests that the doped nitrogen atoms tend to be clustered. A sizable reduction of the photoexcitation energy is also suggested in the rutile structure.

1. Introduction

Titanium dioxide (TiO_2) are attracting a wide attention because of its remarkable photocatalytic properties. Especially, photo-induced water decomposition utilizing its photocatalysis [1] is thought to be a potential method for a sustainable hydrogen production using solar light [2].

The mechanism of semiconductor photocatalytic water decomposition is schematically shown in Figure 1. A semiconductor can be excited by the photons with energy equal to or larger than the bandgap E_g, and photo-excited electrons and holes are generated in the conduction band and the valence band, respectively. A water decomposition reaction

$$H_2O \longrightarrow H_2 + \frac{1}{2}O_2 \qquad (1)$$

consists of following two half-reactions.

$$H_2O + e^- \longrightarrow \frac{1}{2}H_2 + OH^- \qquad (2)$$

$$\frac{1}{2}H_2O + h^+ \longrightarrow \frac{1}{4}O_2 + H^+ \qquad (3)$$

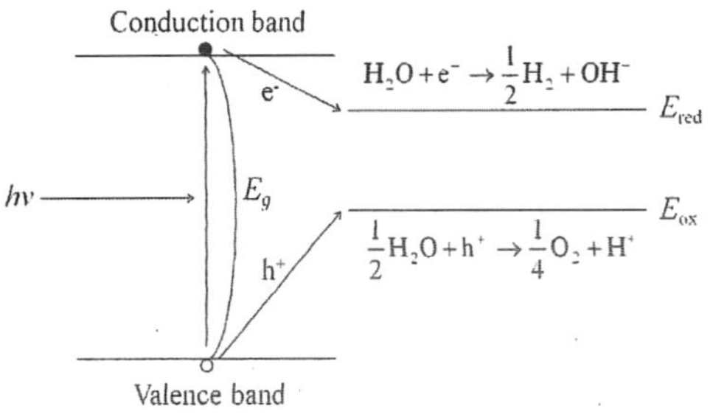

Figure 1. The schematic mechanism of semiconductor photocatalytic reaction.

When the redox potential of the reaction (2) E_{red} is lower than the conduction-band bottom and the redox potential of the reaction (3) E_{ox} is higher than the valence-band top, photo-excited electron-hole pairs in the semiconductor can initiate the reaction (1). The electronic structure of TiO_2 is known to satisfy the above-mentioned requirements.

However, some di culties still remain in its practical application. One of these di culties is that the bandgap of TiO_2 is so wide that TiO_2 only absorbs light with energy of 3.0 eV or larger. This energy is just around the borderline between ultraviolet ray and visible ray although visible ray accounts for as much as around 50 percent of the energy of sunlight [2]. The wide bandgap of TiO_2 greatly lowers the solar-to-hydrogen energy conversion e ciency. In order to overcome this di cuty, several kinds of methods to narrow the bandgap are widely studied.

The rst successful study of shifting photoresponse into visible spectrum was done by Asahi and co-workers [3]. They investigated the e ects of substitutional doping of nitrogen both theoretically and experimentally. They calculated the densities of states (DOSs) of TiO_2 with interstitial doping of nitrogen and substitutional doping of nitrogen, uorine, carbon, sulfur and phosphorus in the anatase structure. The result shows that the nitrogen is the most e ective element to narrow the bandgap among the dopants studied. They also prepared nitrogen-doped TiO_2 lms, which were the mixtures of anatase and rutile phases. These lms absorbed the light with the wavelength less than 500 nm, corresponding to the photon energy larger than 2.48 eV.

In the present study, the e ects of substitutional doping of nitrogen to TiO_2 are studied in more detail within the framework of the density-functional theory (DFT). We compare the total energy of nitrogen-doped TiO_2 with that of pristine TiO_2 for two representative solid phases of TiO_2, rutile and anatase, and for di erent concentrations of substitution sites. Also, we calculate the electronic structures of nitrogen-doped TiO_2 and pristine TiO_2 with the rutile structure.

2. Research Objects and Calculational Methods

Our research objects are $[TiO_2]_{2n\ 1}TiON$ and $[TiO_2]_{2n}$ ($n = 1, 2, 4$) both for rutile and anatase structures. The pre-optimized crystal structures of $[TiO_2]_{2n\ 1}TiON$ is shown in Figure 2. It should be noted that $1/4n$ indicates the concentration of substitution sites in these systems.

In the present work, we perform the electronic-structure calculations based on the DFT [4,5] with the local density approximation (LDA) using the Quantum ESPRESSO package [6].

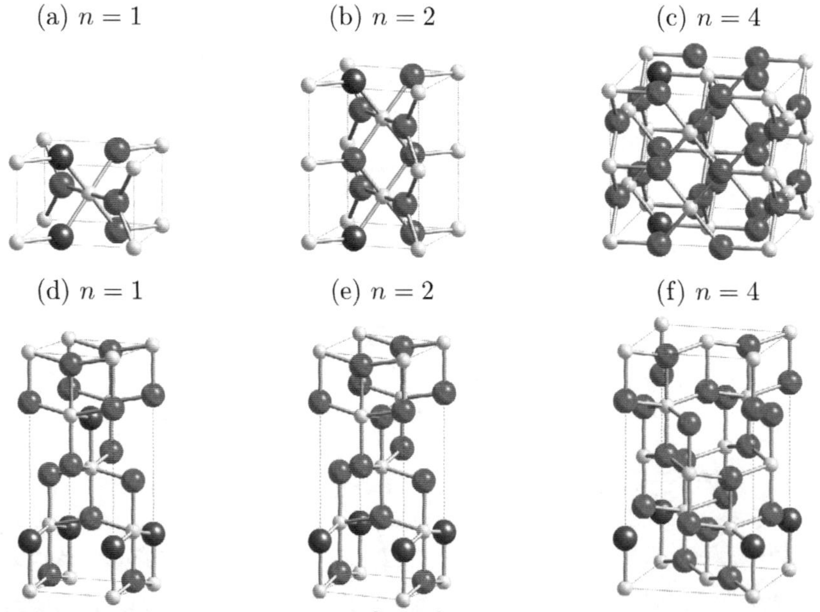

Figure 2. The pre-optimized structures of $[TiO_2]_{2n\ 1}TiON$. Sky-blue ball is titanium, red ball is oxygen, and purple ball is nitrogen. (a), (b), and (c) are rutile-structured, and (d), (e), and (f) are anatase-structured materials.

The exchange-correlation functional of Ceperley and Alder [7] parameterized by Perdew and Zunger [8] and the Vanderbilt ultrasoft pseudopotential [9] are used. The energy cuto for the plane-wave basis is set to be 70 Ry. The structural optimization is performed using the damped dynamics of the Wentzcovitch Lagrangian [10].

3. Energetics

In the present work, we compare the relative tendencies for the nitrogen substitution to occur in di erent crystalline geometries with di erent nitrogen concentrations. So, we compare the energies required for nitrogen substitution per nitrogen atom both for rutile and anatase structures in di erent nitrogen concentrations, with respect to the highest nitrogen concentration case considered for the rutile structure in the present work ($n = 1$). Assuming that the doped nitrogen atoms are derived from nitrogen molecules and the replaced oxygen atoms desorb as oxgen molecules, the relative energies required for nitrogen substitution is given by $\mathscr{E}_n \quad \mathscr{E}_1(rutile)$ where

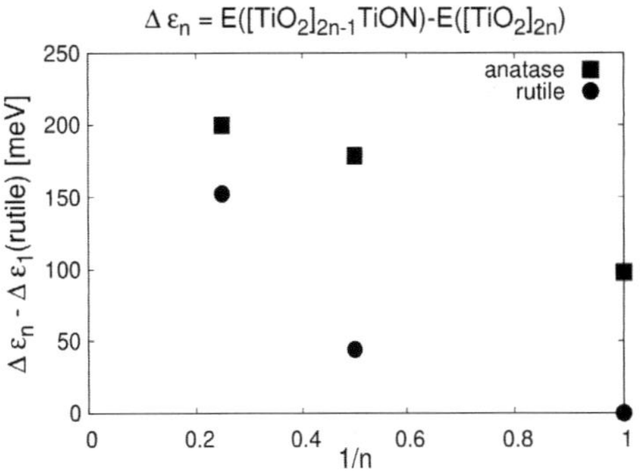

Figure 3. ε_n vs 1/n. ε_1 of the rutile structure is taken as zero.

$$\mathscr{E}_n \quad E([\mathrm{TiO_2}]_{2n\ \ 1}\mathrm{TiON}) + \frac{1}{2}E(\mathrm{O_2}) \quad E([\mathrm{TiO_2}]_{2n}) \quad \frac{1}{2}E(\mathrm{N_2}). \tag{4}$$

Here, each E is the total energy of the structural unit given in the parenthesis. The value to be compared $\mathscr{E}_n \quad \mathscr{E}_1(rutile)$ is, on the other hand, equal to $\varepsilon_n \quad \varepsilon_1(rutile)$ where

$$\varepsilon_n \quad E([\mathrm{TiO_2}]_{2n\ \ 1}\mathrm{TiON}) \quad E([\mathrm{TiO_2}]_{2n}). \tag{5}$$

Therefore, $E(\mathrm{O_2})$ and $E(\mathrm{N_2})$ are not neccesary for the present purpose.

The result is shown in Figure 3. It is found that the smaller the concentration of substitution sites is, the more energy is required for substitution per atom in both rutile and anatase phases. This suggests that, when a certain number of nitrogen atoms are doped into $\mathrm{TiO_2}$, the doped nitrogen atoms should tend to be concentrated in small domains. This "clustering" is known to take place in doping into semiconductor device materials, and is attracting more attentions since it can become a serious obstacle in the future nanometer-scale technology [11–13]. Also, the present result suggests that it is easier to dope nitrogen atoms into rutile than into anatase.

4. Electronic Structure

We also calculate the electronic structure of pristine $\mathrm{TiO_2}$ rutile and that of rutile-structured $\mathrm{TiO_2TiON}$ (Figure 4). Here, the optimized geometry of rutile-structured $\mathrm{TiO_2TiON}$ is primitive monoclinic while that of the pristine $\mathrm{TiO_2}$ is primitive tetragonal. However, symmetry-point names usually used for primitive monoclinic Brillouin zones are used for the corresponding symmetry points of the primitive tetragonal Brillouin zones (Figure 4(a)) in order to compare directly their electronic structures.

As for rutile-structured $\mathrm{TiO_2TiON}$, the bottom of the original conduction band and the top of the original valence band are on the A point and on the Γ point, respectively, and the gap is 1.11 eV. However, since $\mathrm{TiO_2TiON}$ has odd-numbered electrons per unit cell, it should possess the metallic band structure and the Fermi level is not within the fundamental gap anymore. Therefore, the energy di erence between the conduction-band bottom and the Fermi energy is also important to discuss photoexcitation. It is found that the energy di erence between the conduction bottom and the Fermi level is 2.02 eV. This value should be compared not to the

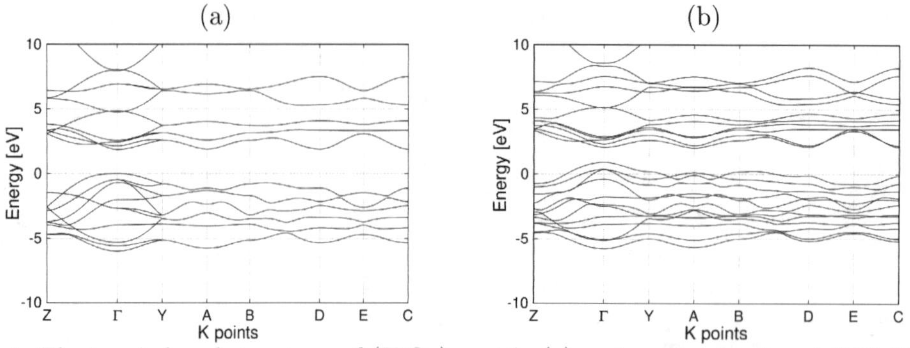

Figure 4. Electronic band structure of $(TiO_2)_2$ rutile (a) and rutile-structured TiO_2TiON (b). (a): The valence top is taken as zero. (b): The Fermi level is taken as zero.

LDA fundamental-gap value of rutile (1.85 eV) but to the experimental bandgap value of 3.0 eV [14], because the LDA gives electronic structures of metallic systems quantitatively well while it underestimates the bandgap values of insulators and semiconductors [15, 16]. Therefore, the present result suggests that the photoexcitation energy reduces by 0.98 eV with nitrogen doping.

5. Summary and Conclusion

We have compared the total energy and the electronic structures between nitrogen-doped TiO_2 and pristine TiO_2 both for rutile and anatase structures. The calculations have been based on the DFT with the LDA using the ultrasoft pseudopotential. Result suggest that the doped nitrogen atoms may tend to be clustered. In addition, it is shown that the conduction-band bottom shifts from the Γ point to the A point, and the photoexcitation energy can reduce considerably with nitrogen substitutional doping.

Acknowledgements

Numerical calculations were mainly performed on TSUBAME 2.0 at the Global Scienti c Information and Computing Center, Tokyo Institute of Technology. We also acknowledge the nancial supports from the Global Center-of-Excellence Program by Ministry of Education, Culture, Sports, Science and Technology of Japan (MEXT) through the Nanoscience and Quantum Physics Project of Tokyo Institute of Technology, and the Elements Science and Technology Project by MEXT.

References

[1] Fujishima A and Honda K 1972 *Nature* **238** 37
[2] Ni M, Leung M K H, Leung D Y C and Sumathy K 2007 *Renew. Sust. Energy. Rev.* **11** 401
[3] Asahi R, Morikawa T, Ohwaki T, Aoki K and Taga Y 2001 *Science* **293** 269
[4] Hohenberg P and Kohn W 1964 *Phys. Rev.* **136** B864
[5] Kohn W and Sham L J 1965 *Phys. Rev.* **140** A1133
[6] Giannozzi P *et al.* 2009 *J. Phys.: Condens. Matter* **21** 395502
[7] Ceperly D M and Alder B J 1980 *Phys. Rev. Lett.* **45** 566
[8] Perdew J P and Zunger A 1981 *Phys. Rev.* B **23** 5048
[9] Vanderbilt D 1990 *Phys. Rev.* B **41** 7892
[10] Wentzcovitch R M 1991 *Phys. Rev.* B **44** 2358
[11] Chadi D J *et al.* 1997 *Phys. Rev. Lett.* **79** 4834
[12] Kim Y S, Lee E C and Chang K J 2003 *Phys. Rev. Lett.* **91** 125503
[13] Mueller D C and Fichtner W 2004 *Phys. Rev.* B **70** 245207
[14] Kavan L, Gratzel M, Gilbert S E, Klemenz C and Scheel H J 1996 *J. Am. Chem. Soc.* **118** 6716
[15] Wang C S and Pickett W E 1983 *Phys. Rev. Lett.* **51** 597
[16] Perdew J P and Levy M 1983 *Phys. Rev. Lett.* **51** 1884

Sequential Photon Emissions from Quad-excitons in Single GaAs Quantum Dots

Y. Arashida, T. Hanazawa, Y. Ogawa, and F. Minami

Department of Physics, Tokyo Institute of Technology, Oh-okayama 2-12-1, Tokyo, Japan

E-mail: arashida.y.aa@m.titech.ac.jp

Abstract. We studied photoluminescence (PL) spectra of a single self-assembled GaAs quantum dot (QD) grown by modified droplet epitaxy. The single QD exhibits several sharp peaks in the PL spectrum indicating the formation of multi-excitonic complexes. Most of these peaks are originated from the recombinations of electron-hole pairs in the lowest level of the QD. We observed several emission peaks arising from recombinations from the second level of the QD at the condition of strong excitation. The difference of the recombination energy between the lowest and second levels is about 30 meV. We identified the emissions of tri-excitons and quad-excitons in the PL spectrum by single photon correlation spectroscopy. Quad-exciton, which is the four-exciton complex, emits correlated four photons in the cascade recombination. We observed bunching of photons emitted from the cascade by cross-correlation measurements.

1. Introduction

Semiconductor quantum dots (QDs) exhibit various interesting properties, which can not appear in bulk materials [1, 2]. Discretization of the energy structure and formation of multi-excitonic complexes, due to strong confinement effect, are one of the most attractive features for application into nano-scale electronic devices [3-5]. Multi-excitonic complexes are expected for a candidate of non-classical light sources [5]. The quantum features of light are essential to realize quantum computation and quantum cryptography [6]. In the single QD spectroscopy, the production of single photons has been performed by the recombination of excitons [7]. Biexciton cascade are demonstrated to generate correlated photon pairs [8]. Especially, it is shown that the photon pairs of biexcitons have entanglement as for their polarization in symmetric QDs [9, 10]. In practice, larger number of correlated photons are required for the quantum information processing [6]. Multi-excitonic complexes are suitable sources of large number of photons because they can emit photons depending on the number of excitons included in themselves. However, the spectra of QDs are complicated with increasing carriers because many-body effects, especially the exchange interaction, become pronounced [11, 12]. Therefore, understanding of the properties of few-particle states are essential to improve insight into the energy structure and the electronic properties of QDs.

In this report, we observed photoluminescence (PL) spectrum of multi-excitonic complexes recombining an electron and a hole in the second levels of a QD. The emission peaks of tri-excitons, which is the bound state of three excitons, and quad-excitons, which is the bound state of four excitons, are identified in the PL spectrum by single photon correlation measurements. The energy structure of the quad-exciton is discussed by considering the exchange interaction.

2. Experimental

Micro-PL measurements were performed on GaAs QDs grown by modified droplet epitaxy (MDE) [13, 14]. The GaAs QDs were formed self-assembly on a $Al_{0.3}Ga_{0.7}As$ layer deposited on a GaAs substrate with [001] growth direction and capped by a $Al_{0.3}Ga_{0.7}As$ layer. Atomic force microscopy and high resolution scanning electron microscopy demonstrated the formation of QDs with low density ($\sim 7 \times 10^8$ cm^{-2}), 20 nm in height, and 30 nm in base size before the capping. The size of the QDs is reduced due to interdiffusion of Ga and Al during post-growth annealing [15].

The QDs were kept at 8 K in a cryostat and excited by a diode laser, which produced 60 ps pulses of 407 nm in center wavelength and with 100 MHz repetition rate. A confocal microscope setup with an objective lens ($NA = 0.55$) was used for excitation and collection. The luminescence was separated by a beam splitter in a Hambury-Brown & Twiss (HBT) interferometer. Each beam was detected by an avalanche photodiode (APD) through a spectrometer equipping a N_2-cooled charge coupled device (CCD) camera. Outputs of the two APDs acted as start or stop events for a time-correlated coincidence counter, constructing a histogram of the measured time separation τ. This setup can measure unnormalized second-order correlation function $g^{(2)}(\tau)$, which is useful to investigate the mechanism of the radiative decay in single QDs. When two states are excited exclusively each other in the same QD by the same excitation pulses, the cross-correlation function should show suppressed coincidence counts at $\tau = 0$ (antibunching), whereas if two photons are generated in a cascade process, the coincidence counts increase at $\tau = 0$ (bunching).

3. Results and Discussion

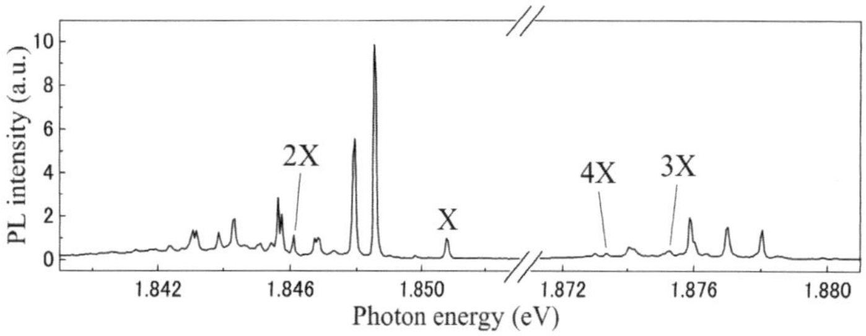

Figure 1. Photoluminescence spectrum of a single GaAs QD. Two regions of the spectrum are displayed.

Figure 1 shows a typical PL spectrum of a single QD. The spectrum has two groups of peaks separated by ~30 meV. In extremely low excitation power, no peaks are observed at the higher energy band 1.872 - 1.879 eV, while the peak X appears at the lower energy band 1.842 - 1.852 eV. X can be identified to originate from excitons, which emit photons by the recombination of electron-hole pairs in the lowest energy levels of the QD ($1e$ and $1h$). As the excitation power is increased, the intensity of X increases linearly and several peaks appear at the low energy side of X. Their intensities increase superlinearly, indicating the formation of multi-excitonic complexes in the QD. The energy differences of each peak depend on their binding energies. In previous works, we have studied the origin of PL peaks of the QDs by polarized PL, time resolved PL, excitation power dependence, and photon correlation measurements. It is shown that neutral and positively charged excitonic complexes can be excited in the QDs, and the recombinations of $1e$-$1h$ pairs contribute to the peaks in the lower energy band [16, 17]. On the other hand, increasing the excitation power furthermore, several peaks appear at the higher energy band. Their intensities increase superlinearly. By considering the excitation processes, the origin of the higher energy band can be interpreted as the emissions from the second levels in the QD ($2e$ and $2h$). In this experiment, carriers are excited in the barrier at 3.05 eV and captured into the QD at 1.88 - 1.85 eV with energy relaxation. When the excitation power is

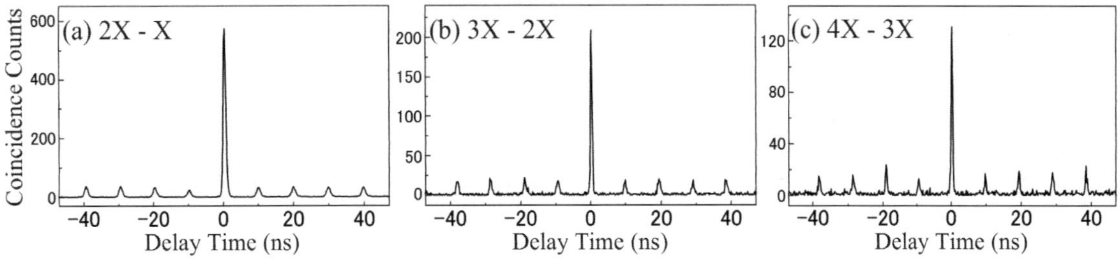

Figure 2. Results of cross-correlation measurements for neutral excitonic complexes.

low, the probability to find one or two excitons in the QD is dominant so that the recombinations of $1e$-$1h$ pairs mainly occur. In such condition, carriers immediately relax into the lowest levels even if they occupy the second levels. In contrast, because the probability to find more than two excitons in the QD has a large value with increasing the excitation power, the lowest levels are fully occupied and the recombinations of $2e$-$2h$ pairs begin to take place. Thus, the higher energy band should be attributed to highly multi-excitonic complexes.

We assumed $3X$ and $4X$ in Fig. 1 as the emissions from $2e$-$2h$ pairs in tri-excitons and quad-excitons, respectively. The tri-exciton consists of two electrons and two holes in the lowest levels and one electron and one hole in the second levels ($1e^2 1h^2 2e^1 2h^1$), where the superscripts denote the numbers of carriers occupied in the levels. The quad-exciton has four electrons and four holes, and the lowest and the second levels are fully occupied ($1e^2 1h^2 2e^2 2h^2$). If the peak $4X$ results from the quad-excitons, correlated four photons should be emitted by the cascade recombination which involves tri-excitons, bi-excitons, and excitons. To confirm this assignment, we performed single photon correlation measurements.

There are many peaks without labels in the PL spectrum. These peaks are originated from charged excitonic states because each of them shows antibunching in the cross-correlation with each neutral state. We do not mention furthermore about the charged states in this report. The intensity of the charged states are much larger than that of neutral states because the sample has slightly p-doped structure introduced not intentionally by the growth process [18].

The results of cross-correlation measurements are displayed in Fig. 2. As shown in Fig. 2(a), the cross-correlation between $2X$ and X shows apparent bunching in the central peak at $\tau = 0$, which corresponds to the probability that both X and $2X$ are emitted during the same laser pulse. The visibility of the bunching in $2X$-X is a higher value than other cross-correlation measurements associated with X. Therefore, $2X$ can be assigned as the biexciton emission.

The cross-correlation function between $3X$ and $2X$ is presented in Fig. 2(b). Since bunching is observed at $\tau = 0$, $3X$ and $2X$ are emitted in the same cascade process. This indicates that $3X$ is the tri-exciton emission. When the $2e$-$2h$ emission of the tri-exciton occurs, the biexciton is left behind in the QD and then $2X$ and X can be emitted in order. The cross-correlation between $3X$ and X also shows bunching as supporting this assignment (not shown here).

Figure 2(c) is the result of cross-correlation between $4X$ and $3X$, which shows bunching at $\tau = 0$. In addition, $4X$ exhibits bunching in the cross-correlations function with other neutral states (not shown here). Therefore, $4X$ can be identified to the emission of quad-excitons. When the $2e$-$2h$ emission of the quad-exciton occurs, the tri-exciton is left behind in the QD and the cascade of the tri-exciton should take place. It is demonstrated from the above data that the temporally correlated four photons are emitted from the QD.

The energy structure of the second levels of the QD is important to understand the PL spectrum of the quad-excitons because the second levels have p-orbital-like envelop functions, where p_x, p_y, p_z are degenerate in a fully symmetric QD. If the degeneracy of the p_x, p_y, p_z is retained or the energy spacings between them are smaller than the exchange energy between $2e$-$2e$ or $2h$-$2h$, the ground state of the quad-exciton contains two electrons (holes) with spin-parallel in the lower two $p^{e(h)}$-orbitals. (The accurate value of the $2e$-$2e$ or $2h$-$2h$ exchange energy could not be obtained, though these values would have a few meV because the exchange energy between $1e$-$2h$ is ~ 1.8 meV in the QD measured by an energy separation of emission peaks of the positively charged biexcitons [19].) On the other hand, if the energy spacing of lower two $p^{e(h)}$ orbitals is larger than the $2e$-$2e$ ($2h$-$2h$) exchange energy due to the anisotropic shape or strain of the QD, the stable condition for the quad-exciton is that two electrons (holes) occupy the lowest $p^{e(h)}$ orbital with anti-parallel spins. In the former case, $4X$ should have several peaks in the PL spectrum due to the exchange interaction [20]. However, we can not find any other peaks satisfying the properties of the quad-exciton: it have only one peak in the higher energy band. Therefore we adopt the latter case for the QD.

In conclusion, we observed the PL from the single GaAs QD arising from the emissions from the lowest levels and the second levels. It is shown that the quad-excitons are excited in the QD and several peaks emitted in the cascade recombination of the quad-excitons are identified in the PL spectrum by using the single photon correlation spectroscopy. The results demonstrate that the correlated four photons are emitted from the quad-excitons, where the two electron-hole pairs in the second levels recombine at first and then the recombinations of the two electron-hole pairs in the lowest levels take place.

4. Acknowledgments

The authors thank Professor N. Koguchi for providing the samples. This work was supported by the financial support from the Global Center of Excellence Program by MEXT, Japan through the Nanoscience and Quantum Physics Project of the Tokyo Institute of Technology.

References

[1] Marzin J -Y, Gérard J -M, Izraël A, Barrier D and Bastard G 1994 *Phys. Rev. Lett.* **73** 716-19
[2] Dekel E, Gershoni D, Ehrenfreund E, Spektor D, Garcia J M and Petroff P M 1998 *Phys. Rev. Lett.* **80** 4991–94
[3] Huffaker D L, Park G, Zou Z, Shchekin O B and Deppe D G 1998 *Appl. Phys. Lett.* **73** 2564–66
[4] Klein D L, Roth R, Lim A K L, Alivisatos A P and McEuen P L 1997 *Nature (London)* **389** 699–701
[5] Michler P, Imamoğlu A, Mason M D, Carson P J, Strouse G F and Buratto S K 2000 *Nature* **406**, 968–70
[6] Bouwmeester D, Ekert A and Zeilinger A 2000 *The Physics of Quantum Information (Springer, Berlin)*
[7] Lounis B, Bechtel H A, Gerion D, Alivisatos P and Moerner W E 2000 *Chem. Phys. Lett.* **329** 399–404
[8] Moreau E, Robert I, Manin L, Thierry-Mieg V, Gerard J M and Abram I 2001 *Phys. Rev. Lett.* **87** 183601
[9] Benson O, Santori C, Pelton M and Yamamoto Y 2000 *Phys. Rev. Lett.* **84** 2513–16
[10] Stevenson R M, Young R J, Atkinson P, Cooper K, Ritchie D A and Shields A J 2006 *Nature* **439** 179–82
[11] Kavokin K V 2003 *phys. stat. sol.* (a) **195** 592–95
[12] Akimov I A, Kavokin K V, Hundt A and Henneberger F 2005 *Phys. Rev. Lett.* **71** 075326
[13] Koguchi N, Takahashi S and Chikyow T 1993 *J. Cryst. Growth* **32** 688–92
[14] Watanabe K, Koguchi N and Gotoh Y 2000 *Jpn. J. Appl. Phys.* **39** 79–81
[15] Sanguinetti S, Watanabe K, Kuroda T, Minami F, Gotoh Y and Koguchi N 2002 *J. Cryst. Growth* **242** 321–31
[16] Kihira T, Arashida Y, Yamagiwa M, Ogawa Y, Minami F and Koguchi N 2008 *J. Phys.: Conf. Series* **109** 012019
[17] Arashida Y, Ogawa Y and Minami F 2010 *Superlatt. Microst.* **47** 93–97
[18] Abbarchi M *et al.* 2008 *Appl. Phys. Lett.* **93** 162101
[19] Poem E, Kodriano Y, Tradonsky C, Gerardot B D, Petroff P M and Gershoni D 2010 *Phys. Rev.* B **81** 085306
[20] Hawrylak P 1999 *Phys. Rev.* B **60** 5597–608

Spectral narrowing of four-wave mixing signals for excitons in the layered semiconductor GaSe

H. Tahara, Y. Ogawa, and F. Minami

Department of Physics, Tokyo Institute of Technology, Oh-Okayama 2-12-1, Tokyo 152-8551, Japan

E-mail: tahara.h.aa@m.titech.ac.jp

Abstract. The spectral narrowing has been observed using time-resolved and spectrally resolved four-wave mixing measurements on a layered semiconductor GaSe. The narrowing of four-wave mixing spectra is one of the evidences of the non-Markovian dephasing. From the calculation of the non-Markovian theory for exciton-phonon interaction, it has been found that the excitons in GaSe are confined to 7 layers and that the correlation time of the interaction is 1.1 ps, which defines the limit of the non-Markovian time regime.

1. Introduction

Electronic dynamics can be observed as coherent transient phenomena by employing femtosecond pump-probe spectroscopy. By measuring this transient phenomenon, electronic properties are revealed in terms of time domain. Four-wave mixing (FWM) measurement is the method for measuring this dephasing process by using two incident pulses. Exciton polarization induced in a semiconductor by a coherent light pulse is lost by several interactions. Dephasing of exciton polarization is expressed as exponential decay for Markovian dynamics, which is basically treated by the optical Bloch equations. In contrast, exciton polarization of some semiconductors exhibits complicated decay in the initial dephasing. This complicated dynamics are characterized by non-Markovian dynamics of several interactions such as exciton-phonon and exciton-exciton interactions. Beating signals generated by the exciton-phonon interaction have been observed in some semiconductors [1-3]. For exciton-exciton interaction, memory time induced by two pair correlations has been estimated [4]. Because non-Markovian dephasing is determined by the time correlation of interactions, investigations of non-Markovian dephasing are necessary to understand the microscopic behavior of electron dynamics.

As non-Markovian feature, the exciton wave-packet motion has been observed in the layered semiconductor GaSe. It has been found that the wave-packet motion is caused by the exciton-phonon interaction along the c-axis direction, as reported in our previous study [5]. In order to understand the non-Markovian dephasing process, changes of the shape of exciton spectra should be measured in detail. In this study, we report the spectral narrowing of non-Markovian dephasing signals for excitons in the layered semiconductor GaSe. The spectral narrowing was observed using time-resolved (TR) and spectrally resolved (SR) FWM spectroscopy.

2. Experimental details

The sample was a layered semiconductor GaSe, and it was set at 3.4 K in a closed cycle refrigerator. Excitons in GaSe are confined to about 10 layers due to a stacking disorder [6, 7]. Because of the confinement, the dephasing of exciton polarization is caused by the acoustic deformation potential coupling [5]. The excitation source was a frequency-doubled optical parametric oscillator pumped by a mode-locked Ti:sapphire laser tuned to the exciton resonant energy of 2.109 eV. The optical pulse had a duration of 200 fs and a repetition rate of 76 MHz. The first and second excitation pulses were sent into the sample with wave vectors $\boldsymbol{k}_1$ and $\boldsymbol{k}_2$, respectively. The diffracted signal in the direction of $2\boldsymbol{k}_2-\boldsymbol{k}_1$ was measured at various delay times τ, which is the time interval between the first and second excitation pulses. The TR-FWM signals were detected by using the heterodyne technique with acousto-optic modulators, which were used to shift the frequency of the reference pulse and that of the second excitation pulse. In this technique, the interference of the reference pulse and the FWM signals was detected by a photo detector to which a spectrum analyzer was connected. For the SR-FWM measurements, the signals were spectrally resolved by a monochromator and detected by a photomultiplier tube.

3. Results

The TR-FWM signals are shown in Fig. 1 for various delay times τ. Fast damping is observed only in the initial dephasing at the delay times of 0.0 and 0.4 ps. After the delay time of 0.4 ps, the fast damping gradually disappears and then only the Gaussian signal is observed. The Gaussian signal is created by the distribution of the exciton resonances, i.e., so-called inhomogeneous broadening. However, the fast damping cannot be explained by the phenomenological introduction of inhomogeneous broadening. In order to show the spectral linewidth, the inverse of half-decay time $1/\tau_{1/2}$ is plotted in the inset of Fig. 1, where $\tau_{1/2}$ is measured from the corresponding signal peak. In the inset, $1/\tau_{1/2}$ decreases from 2.9 to 1.5 ps^{-1} with increasing the delay times from 0.0 to 1.0 ps, respectively. This decrease of $1/\tau_{1/2}$ expresses the spectral narrowing.

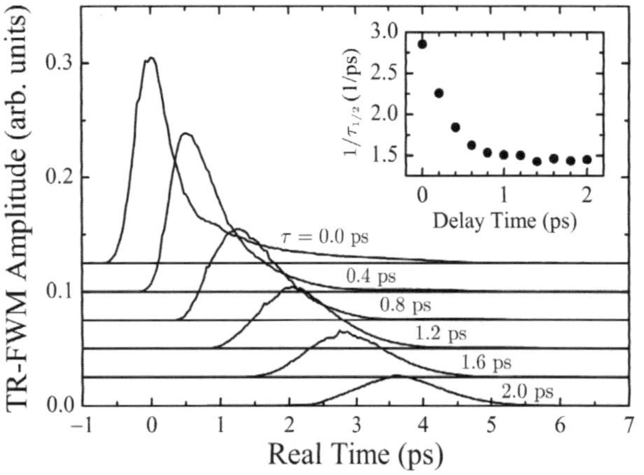

Figure 1. Time-resolved four-wave mixing signals for delay times τ. The delay time varies from 0.0 to 2.0 ps in 0.4 ps steps, as indicated. The inverse of half-decay time $1/\tau_{1/2}$ is plotted as a function of the delay time in the inset.

The Fourier transformation of the TR-FWM signal is performed for each delay time as shown in Fig. 2 (a). It is clearly observed that the FWM spectra become narrower with increasing the delay time. The FWM spectrum for the delay time of 0.0 ps extends up to ±5 meV from the center of the spectrum as indicated by the arrows. However, for the delay time of 2.0 ps, the spectrum extends only up to ±2 meV. The spectral narrowing is also observed in SR-FWM measurements as shown in Fig. 2 (b), which is consistent with the Fourier transformation of the TR-FWM signal in terms of spectral narrowing. The asymmetric shape of the SR-FWM signal is caused by the wave-packet motion of exciton coupled to acoustic phonons [5]. It is not

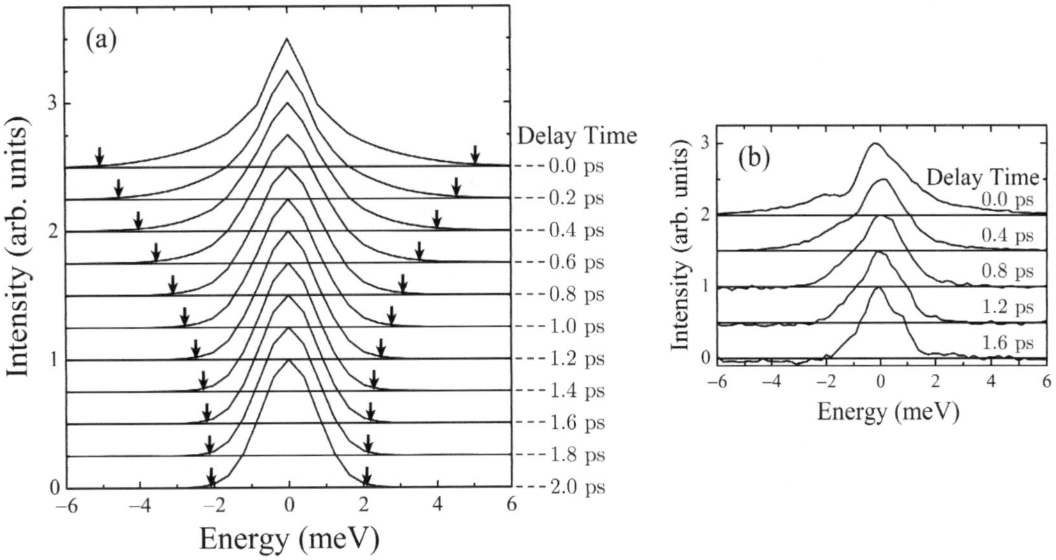

Figure 2. (a) Fourier transformation of the time-resolved four-wave mixing signal. The arrows point to the edges of spectra. (b) Spectrally resolved four-wave mixing intensities for the indicated delay times. The centers of the peaks are at an energy of 0 meV.

obtained from the Fourier transformation of the TR-FWM signals, because the time evolution of the phase factor cannot be determined from the TR-FWM measurements.

4. Discussion

In order to understand the spectral narrowing of the FWM signals, the dephasing process is discussed on the basis of a microscopic treatment of acoustic phonons, instead of the phenomenological introduction of exponential decay term. In this theoretical treatment, which is characterized as non-Markovian theory [8, 9], exciton polarization is lost by the interaction between excitons and the reservoir of acoustic phonons. The non-Markovian dephasing exhibits only within the correlation time of exciton-phonon interactions. In this time regime, the FWM spectrum is observed as the exciton spectrum involving the spectral broadening due to the correlation. After the loss of the correlation, the linewidth of the FWM spectrum is decreased to the linewidth determined by inhomogeneous broadening. The interaction without the correlation also causes spectral linewidth related to the dephasing, but it is narrower than that of inhomogeneous broadening. The narrowing of the FWM spectra expresses the loss of the correlation.

The full width at quarter maximum (FWQM) of the FWM spectrum is measured for comparing the experimental data with the non-Markovian theory. The FWQM of the SR-FWM signals and the fitting result are shown in Fig. 3. In the fitting, the spectral density of the interaction, which has been analyzed from the wave-packet motion observed in GaSe [5], is applied to the non-Markovian theory. The localization length of excitons is found to be 7 layers, which is consistent with the previous investigations [5-7]. In addition, the correlation time of the interaction is obtained as 1.1 ps, which defines the switching time from the non-Markovian to Markovian dynamics. Because the interaction behaves as inhomogeneous broadening only for the non-Markovian dynamics [8], this result indicates that the dephasing can be suppressed if the second excitation pulse arrives within a few picoseconds. In fact, the decoherence suppression of the excitons in GaSe has been reported within the non-Markovian regime [10].

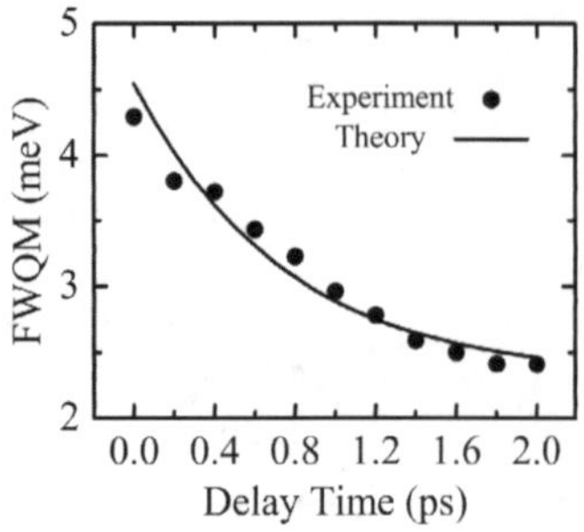

Figure 3. Full width at quarter maximum (FWQM) of the four-wave mixing spectrum as a function of the delay time. The solid line represents the calculation results based on the non-Markovian theory.

5. Conclusions

We have observed the spectral narrowing of four-wave mixing signals for excitons in the layered semiconductor GaSe. The narrowing, which arises in the non-Markovian dephasing, has been measured by time-resolved and spectrally resolved four-wave mixing spectroscopies. From the comparison between the experimental data and the calculation based on the non-Markovian theory, the correlation time of the exciton-phonon interaction defining the non-Markovian regime is obtained as 1.1 ps.

Acknowledgments

This work was supported by the Global Center of Excellence Program by MEXT, Japan through the Nanoscience and Quantum Physics Project of the Tokyo Institute of Technology.

References

[1] Bányai L, Tran Thoai D B, Reitsamer E, Haug H, Steinbach D, Wehner M U, Wegener M, Marschner T and Stolz W 1995 *Phys. Rev. Lett.* **75** 2188
[2] Steinbach D, Kocherscheidt G, Wehner M U, Kalt H, Wegener M, Ohkawa K, Hommel D and Axt V M 1999 *Phys. Rev.* B **60** 12079
[3] Woggon U, Gindele F, Langbein W and Hvam J M 2000 *Phys. Rev.* B **61** 1935
[4] Axt V M, Kuhn T, Haase B, Neukirch U and Gutowski J 2004 *Phys. Rev. Lett.* **93** 127402
[5] Tahara H, Ogawa Y and Minami F 2010 *Phys. Rev.* B **82** 113201
[6] Forney J J, Maschke K and Mooser E 1977 *J. Phys.* C **10** 1887
[7] Capozzi V and Maschke K 1986 *Phys. Rev.* B **34** 3924
[8] Aihara M 1982 *Phys. Rev.* B **25** 53
[9] Mukamel S 1995 *Principles of Nonlinear Optical Spectroscopy* (New York: Oxford University Press)
[10] Kishimoto T, Hasegawa A, Mitsumori Y, Ishi-Hayase J, Sasaki M and Minami F 2006 *Phys. Rev.* B **74** 073202

Quantum coherence in biological systems

Seth Lloyd
Department of Mechanical Engineering, MIT

Abstract. This paper reviews the role of quantum mechanics in biological systems, and shows how the interplay between coherence and decoherence can strongly enhance quantum transport in photosynthesis.

Nature is the great nano-technologist. The chemical machinery that powers biological systems consists of complicated molecules structured at the nanoscale and sub-nanoscale. At these small scales, the dynamics of the chemical machinery is governed by the laws of quantum mechanics. Quantum mechanics is well known to exhibit strange and counterintuitive effects. Accordingly, it makes sense to investigate the extent to which peculiarly quantum effects such as coherence and entanglement play an important role in living systems.

Quantum mechanics and quantum coherence play a central role in chemistry. Quantum coherence and entanglement determine the valence structure of atoms and the form of covalent bonds. Quantum mechanics fixes the set of allowed chemical compounds and sets the parameters of chemical reactions. Indeed, the very fact that there are only a countable, discrete set of possible chemical compounds arises from the fundamentally discrete nature of quantum mechanics.

Chemistry, in turn, lays down the rules for what biological structures are possible and for how they function. Biomolecules can contain many atoms (billions in the case of DNA). As molecules become larger and more complex, quantum coherence becomes harder to maintain. Vibrational modes and interactions with the environment tend to decohere quantum superpositions. Consequently, most biomolecular mechanisms have traditionally been modeled as essentially classical processes. For example, classical 'ball and spring' models of molecules can capture accurately many aspects of molecular dynamics. As a result, for many years, the conventional answer to the question of whether quantum coherence and entanglement are important for biological processes has been, not really.

Recently, however, strong evidence has emerged that quantum coherence is playing an important role in certain selected biological processes [1-3]. The strongest evidence for quantum coherence occurs in photosynthesis, which will form the main topic of this paper (a review of the mechanisms for photosynthesis can be found in [4]. Indirect evidence for quantum coherence also appears in bird navigation [5-6] – the so-called 'avian compass' – and in the sense of smell [7-8]. Before turning to photosynthesis, I will review the evidence for quantum effects in these biological mechanisms.

(1) The avian compass. Some birds, such as homing pigeons, possess a small piece of magnetite in their beaks which functions as a compass, allowing them to tell North from South. Other birds, such as the European robin, seem to possess a different mechanism for sensing the earth's magnetic field [5-6]. This mechanism (a) allows the birds to identify the angle the magnetic

Published under licence by IOP Publishing Ltd

field lines make with the earth, but does not distinguish North from South; (b) is only activated in blue or green light; and (c) is disrupted by oscillating fields a bit above one megahertz, at the electron's Larmor frequency in the earth's magnetic field. These features are difficult to explain using classical mechanisms. The only known mechanism consistent with these features is quantum mechanical.

The proposed mechanism postulates that absorption of blue or green light creates a radical pair of electrons, separated within an oriented molecule (perhaps a cryptochrome) within the bird's eye. The racial pair undergoes coherent quantum oscillations between entangled singlet and triplet states, at a rate that depends on the orientation of the host molecule with respect to the earth's magnetic field. A spin-dependent electron transition then allows current to flow only if the spins in the pair are in a particular symmetry state. The rate of current flow then depends on the molecule's relative orientation to the field. This purely quantum mechanical mechanism explains all peculiar observed features of the birds' behavior, including their inability to tell North from South. Ongoing experimental and theoretical investigations are attempting to identify the molecules or molecular complexes within which this mechanism can function, and to devise detailed and testable models for its operation.

(2) Sense of smell. The conventional mechanism postulated for smell is a lock and key model, in which different types of odorant molecule bind to different types of olfactory sensors. The different types of olfactory sensors in the human nose have been identified. In this model, the particular smell of an odorant molecule depends only on its affinity to different sensors. There are several issues with this model. First of all, the smell of a molecule depends at most weakly on shape, the primary determinant of the sensors to which the molecule binds in a lock and key model. Second, smell correlates well with the vibrational spectrum of the molecule: for example, a sulphur-less molecule whose vibrational spectrum exhibits a resonance at the same frequency as the sulphur-hydrogen stretch mode, smells of sulphur. The nose is evidently a vibrational spectrometer.

Again, the only known mechanism that can explain this vibrational sensitivity of the sense of smell is quantum mechanical [7-8]. The mechanism relies on inelastic tunneling. Once an odorant molecule docks into the olfactory sensor, electrons can pass through the molecule, and current can flow, only by emitting a phonon of specific frequency to one of the molecule's dominant vibrational modes. The strongest confirming evidence for this mechanism is the ability of fruit flies (drosophila) to smell the difference between an organic molecule and a deuterated version of the same molecule. A molecule whose hydrogens have been replaced by deuterium should bind to the same receptors as the original molecule, with similar affinities. So if the shape-based lock and key mechanism were true, it should smell the same. By contrast, deuterium-carbon bonds vibrate at a rate $\sqrt{2}$ slower than hydrogen-carbon bonds: deuteration significantly changes the vibrational spectrum of the molecule. Ongoing work to validate further this purely quantum mechanical mechanism includes performing X-ray crystallography to identify the precise geometry of the receptor molecules, and creating detailed quantum mechanical models to test and elucidate the specific mechanisms of inelastic tunneling.

Photosynthesis. I now turn to quantum coherence in photosynthesis, the primary topic of this article. In the cases of the avian compass and sense of smell just described, the primary evidence for quantum effects is observation of animal behavior, combined with a lack of plausible classical models. By contrast, in photosynthesis, the evidence for the effects of quantum coherence is direct and overwhelming [1-3].

In photosynthesis, a particle of light or photon is absorbed by a photocomplex, a set of molecular structures embedded in the membrane of a plant or bacterial cell. Part of the energy of the photon is converted into heat in the form of molecular vibrations, but most of it is captured as an exciton, a bound electron-hole pair that resides in a chromophore, a 'light-carrying' molecule such as a plant or bacterial chlorophyll. The exciton is so-called Frenkel exciton, tightly bound within a region a few angstroms on each side. Energy transfer occurs via the dipolar interaction between chromophores: a so-called Förster coupling between induced or transition dipoles de-excites one chromophore while exciting a neighboring one. The exciton must propagate through hundreds or thousands of chromophores in order to reach the reaction center, a special molecular complex in which charge separation takes place.

The exciton transport process is, of course, governed by the laws of quantum mechanics. For decades, however, this process was thought to be minimally quantum mechanical, in the sense that while quantum mechanics determined couplings between chromophores, the actual transition rates could be calculated semi-classically using Fermi's golden rule. In 2007, this picture changed dramatically when the Fleming group in Berkeley used femtosecond two-dimensional spectroscopy based on four-wave mixing to demonstrate that quantum coherence was at work the transfer process. The spectroscopic signature exhibited clear evidence of quantum 'beating' at 77K, a result that was later confirmed at room temperature. The evidence for quantum coherence in photosynthetic transport was essentially incontrovertible.

My own background is in quantum computation, not photochemistry. I became involved in this field completely by accident. When the Fleming paper came out, the New York Times reported it as claiming that green sulphur bacteria were performing a quantum computation. Quantum computers are devices that use quantum coherence to process information in ways that classical computers can not. Specific methods for performing a quantum computation are called quantum algorithms. The Fleming paper speculated that the demonstrated coherence in excitonic transport meant that bacteria were performing a quantum algorithm called 'quantum search' to allow the exciton to travel from the place and to find the reaction center. This claim caused great hilarity in our quantum information theory group at MIT, as it seemed implausible that a quantum algorithm such as quantum search could be performed in the hot, wet, and noisy conditions inside a living cell. Despite our collective skepticism, conscientiousness obliged us to take a look. Moreover, Alan Aspuru-Guzik, an assistant professor of chemistry at Harvard, had interacted with the Fleming group and could vouch for the seriousness and accuracy of their results. So I read the paper and discussed it with Alan.

The paper was a masterwork. The evidence for quantum coherence was exceptionally strong. The line about quantum search algorithms was essentially an afterthought. Quantum search algorithms, invented by Lov Grover in 1996, take a unique form [9]: we were able to show right away that the bacteria were not performing a quantum search algorithm. To our surprise, however, the bacteria were in fact performing a different type of quantum algorithm, called a quantum walk.

A random walk is a transport mechanism in which the walker takes steps in random directions through the network or structure to be traversed. In a classical random walk, the position of the walker gradually diffuses throughout the network. Because the walker is equally likely to move in any direction, after time t the walker has moved from its original position by an distance proportional to $\sqrt{t}$. In a quantum walk, by contrast, the walker takes a quantum superposition of paths through the network [10]. These paths can exhibit quantum interferences. If the interference is destructive, the walker becomes stuck or 'localized' within the network. By contrast, if the interference is constructive, the walker moves from its original position by a distance proportional to t, a considerable improvement over the classical random walk. The coherence demonstrated by the Fleming group showed that the excitons were performing a quantum walk, at least over short distances.

Excited, we decided to make a detailed quantum mechanical model of the excitonic transport process, a model that took into account the strong role of quantum coherence. We were aided by the fact that the quantum structure of the particular photosynthetic complex on which

Fleming had performed his experiment was well-known. The Fenna-Matthews-Olsen complex, or FMO, had been exhaustively studied using both spectroscopy and by X-ray crystallography [1-3]. Consequently, we were able to make a full quantum model of the dynamics of excitonic transport through the complex, including its interaction with the environment. Because of the well-studied nature of the complex, the only free parameter in our model was the temperature, which we could vary at will. All other parameters in the model were determined by pre-existing experiments.

When Alan's postdoc, Masoud Mohseni, and graduate student Patrick Rebentrost coded the model into a computer and fired it up, we found that the efficiency of excitonic transport had a remarkable dependence on temperature [11-12]. (Subsequently but independently, Martin Plenio and co-workers at Imperial College developed a similar model which confirmed this behavior [13].) At low temperatures, the interference between paths in the excitonic walk was predominantly destructive, localizing the exciton to within a few sites of its initial position. At higher temperatures, however, the interaction with the environment decohered the quantum walk sufficiently to remove that destructive interference, allowing the exciton to diffuse throughout the FMO complex towards the reaction center. At very high temperatures, the decoherence induced by the environment was so destructive that it froze the exciton in place, reducing the efficiency of the transport. The highest efficiency for transport was 290K, the average temperature of the water in which the bacteria lived. Moreover, the transport was robust with respect to variation about this optimal temperature, exhibiting almost 100% efficiency for a range of many tens of degrees K in both directions.

Our model gave a clear picture of the role of quantum coherence in excitonic transfer in FMO: quantum coherence allowed the exciton to spread rapidly and coherently over a small number of sites, then environmentally induced decoherence kicked in to prevent the exciton from being localized. The transport efficiency exhibited a distinctive signature as a function of temperature – low at low temperature, high for a range around some optimum temperature, then low again at high temperature. Because this particular signature was a function not of the individual system under investigated, but of any partially coherent transport process, we gave it the name ENAQT for Environmentally Assisted Quantum Transport. ENAQT applies to essentially any type of partially coherent quantum transport through a disordered medium.

In addition, our model gave insight into the evolutionary mechanism underlying a phenomenon that had puzzled photochemists and biologists for decades – the convergence of time scales in photosynthesis. Separation of time scales is a common phenomenon in physical systems. The initial absorption of a photon and creation of an exciton in a chromophore is a fast process, taking only a few femtoseconds. By contrast, the exciton lives for several nanoseconds before decaying. So there is a six order of magnitude separation of time scales between absorption and decay. The physiological benefits of this particular separation of time scales is easy to see: fast absorption arises from strong coupling of the chromophore to light, and to absorption over a broad band of frequencies, both desirable characteristics. Long lifetimes increase the chance that the exciton will make it to the reaction center. All other time scales in the excitonic transport process, however – coupling constants, energy differences, decoherence rates, and environmental correlation times – converged to the scale of a picosecond or so. Why?

The convergence of time scales makes excitonic transport in photosynthesis hard to model. When time scales are separated, different processes at different time scales can be modeled accurately as perturbation to an underlying process. When time scales converge, no simple model is possible. So one explanation of why time scales converge is that Nature is modest and prefers not to reveal her secrets to scientists. A more physical explanation is that when time scales for different processes converge, the strong interaction between the processes can either help transport, or hurt it. In systems that have undergone a billion years of natural selection, it should come as little surprise that these temporally convergent processes help each other out.

To test this hypothesis, we varied the parameters in our model to test the efficiency of transport in non-physiological regimes where the time scales for different processes were not the same [14]. These studies showed that when time scales failed to converge, the efficiency of transport was lower. We call this effect the 'quantum Goldilocks effect' after the little girl in the folk tale who enters a house in which there is three of everything – three chairs, three beds, three bowls of porridge on the table. One of the chairs is too large, one too small, and one is just the right size. So Goldilocks sits on the chair that is just right. One of the bowls of porridge is too cold, one too hot, and the third is just the right temperature. So she eats that one. Finally, she falls asleep on the bed that has just the right degree of softness. (At this point, the three bears who live in the house return home and become upset, but that is another story.)

The Goldilocks effect – natural selection chooses just the right degree for each effect – is well known in biology. There is also a Goldilocks principle for complexity of evolved or designed systems [15]: too little complexity compromises function, while too much complexity diminishes robustness. Well-designed or well-evolved systems typically exhibit just the right degree of complexity. In the quantum Goldilocks effect, natural selection has made time scales converge, adding between quantum processes in photosynthesis until photosynthetic systems attain just the right degree of quantum complexity. Our simulations that vary the different parameters of the FMO complex show that the FMO complex has attained just the right degree of quantum complexity to give essentially 100% efficiency while remaining robust.

Conclusion: Quantum coherence plays a strong role in photosynthetic energy transport, and may also play a role in the avian compass and sense of smell. In retrospect, it should come as no surprise that quantum coherence enters into biology. Biological processes are based on chemistry, and chemistry is based on quantum mechanics. If an organism can attain an advantage in reproduction, however slight, by putting quantum coherence to use, then over time natural selection has a chance to engineer the necessary biochemical mechanisms to exploit that coherence. Different types of quantum processes that operate at the same time scale can interact strongly either to assist or to impede one another. In photosynthetic energy transfer, the convergence of quantum time scales gives rise to more efficient and robust transport. Evolved biological systems exhibit the quantum Goldilocks effect: natural selection pushes together time scales to allow quantum processes to help each other out.

References

[1] Engel G S, Calhoun T R, Read E L, Ahn T K, Mancal T, Cheng Y C, Blankenship R E and Fleming G R 2007 *Nature* **446** 782
[2] Collini E, Wong C Y, Wilk K E, Curmi P M, Brumer P and Scholes G D 2010 *Nature* **463** 644
[3] Panitchayangkoon G, Hayes D, Fransted K A, Caram J R, Harel E, Wen J, Blankenship R E and Engel G S 2010 *Proc. Nat. Acad. Sci.* **107** 12766
[4] Blankenship R E 2002 *Molecular Mechanisms of Photosynthesis,* (London: Blackwell Science)
[5] Ritz T, Adem S and Schulten K 2000 *Biophys. J.* **78** 707
[6] Rodgers T and Hore P J 2009 *Proc. Nat. Acad. Sci. USA* **106** 353
[7] Turin L 1996 *Chem. Senses* **21** 773
[8] Franco M I, Turin L, Mershin A and Skoulakis E M C 2011 *Proc. Nat. Acad. Sci. USA* **108** 3797
[9] Grover L K 1996 *Proc. 28th Ann. ACM Sym. Th. Comp.* p 212
[10] Farhi E and Gutmann S 1998 *Phys. Rev. A* **58** 915
[11] Mohseni M, Rebentrost P, Lloyd S and Aspuru-Guzik A 2008 *J. Chem. Phys.* **129** 174106
[12] Rebentrost P, Mohseni M, Kassal I, Lloyd S and Aspuru-Guzik A 2009 *New J. Phys.* **11** 033003
[13] Plenio M B and Huelga S F 2008 *New J. Phys.* **10** 113019
[14] Shabani A, Mohseni M, Rabitz H and Lloyd S 2011 (in preparation)
[15] Lloyd S, lecture at Santa Fe Institute, 1990, as reported by Gell-Mann M *The Quark and the Jaguar,* 1994 (New York: A. Knopf)

Ultra Slow Muon Microscopy for Nano-science

Y. Miyake[1,2], N Nishida[3], J. Yoshino[3], W. Higemoto[2,4], E. Torikai[5], K. Shimomura[1,2],Y. Ikedo[1,2], N. Kawamura[1,2], P. Strasser[1,2], S. Makimura[1,2],H. Fujimori[1,2],K. Nakahara[1,2,8], A. Koda[1,2], Y. Kobayashi[1,2], K. Nishiyama[1,2], R. Kadono[1,2],T. Ogitsu[2,6], Y. Makida[2,7], K. Sasaki[2,6], T. Adachi[6], and K.Nagamine[1,9]

[1] Muon Science Laboratory, High Energy Accelerator Research Organization,Tokai, Ibaraki 319-1106, Japan

[2] Muon section, Materials and Life Science division, J-PARC Center,Tokai, Ibaraki 319-1106, Japan

[3]Department of Physics, Faculty of Science, Tokyo Institute of Technology, 201201 O-okayama Meguro-ku, Tokyo 152, Japan

[4]Advanced Science Research Center, Japan Atomic Energy Agency, Tokai, Ibaraki 319-1195, Japan

[5]Faculty of Engineering, Yamanashi University, Kofu, Yamanashi, 4008511, Japan

[6]Cryogenic Science center, High Energy Accelerator Research Organization, Tsukuba, Ibaraki 319-1106, Japan

[7] Institute of Particle and Nuclear Studies, High Energy Accelerator Research Organization,Tokai, Ibaraki 319-1106, Japan

Abstract. The" surface" muon beam which has been used for the studies of condensed matter physics or chemistry is conventionally obtained from the decay of positive pions (π^+) stopped near the surface of the pion production target in the proton beam line and has large energy broadening with an implantation depth of 0.1 to 1 mm. Despite the name of" surface" muon, it is used as a probe of bulk phenomena rather than surface phenomena. In these two decades, the new method to generate ultra-slow muon beam with energy 0.2 eV has been developed and successfully obtained by KEK and RIKEN group. When the production of intense ultra-slow muon source will be realized, the use of its short-range penetration depth will allow muon science to be expanded towards a variety of new nano-scienti c elds, which we call "Ultra Slow Muon Microscope" such as, 1) Surface/boundary magnetism utilizing its spin polarization and unique time-window. 2) Surface chemistry, utilizing a feature of a light isotope of hydrogen; such as catalysis reactions. 3) Muon Microscopy, utilizing a feature of micron meter beam size, when ultra slow muon is accelerated. 4) Precise atomic physics testing QED, since Mu is the simplest lepton pair consisting μ^+ and e . 5) Ion sources for" g-2" experiment, and towards $\mu^+\mu$ collider experiments in high-energy physics.
Int this paper, the latest status of the intense low-emittance ultra slow muon source and its scienti c prospects will be reported.

1. Introduction

A muon is a kind of lepton. It has a spin, 1/2 and a mass about 1/9 of proton (207 times heavier than electron). There are two kind of muons, positively charged muon ($^+$) and negatively

[8] Present address: Physics Department, University of Maryland, College Park, MD 20742, USA

[9] Department of Physics & Astronomy, University of California, Riverside, California 92521, USA

charged muon ( ). $^+$ is a decay product of $^+$, associated with . $^+$ is emitted towards the opposite direction from and hence, $^+$ is spin polarized inherently, since has negative helicity. When $^+$ is implanted into the materials, $^+$ stops between atoms, and there, $^+$ starts Lamour precession according to the local magnetic eld. $^+$ decays to e^+ and two neutrino, $_e$ and anti- in a lifetime of 2.2 s. $^+$ emits positron preferentially to the direction when their spins were pointing upon decay. Therefore, spin evolution of $^+$ can be observed in the time histogram, by detecting e^+ events by scintillation counters. This experimental technique is called as SR.

The conventional muon beams are obtained from proton accelerators with proton energies higher than 400 MeV. According to the momentum of the generated muon beam these beams are usually referred to either as" decay muon" beam or" surface muon" beam. The decay muons are obtained through the in- ight decay of $^+$/ , con ned by a strong longitudinal eld of several T from a superconducting solenoid magnet. Muons obtained from the in- ight decay of $^+$/ have high momentum, between 40 to several hundreds of MeV/c. Not only positive but also negative muon beam are available[1]. The surface muons, as the naming suggests, are obtained from the decay of positive pions ($^+$) stopped near the surface of the pion production target in the proton beam line[2]. Compared to the decay muons which have a high momentum due to the in- ight decay of pions, the surface muons obtained from the decay of $^+$ at rest have a low and unique kinetic energy of 4.1 MeV (corresponding to a momentum of 29.8 MeV/c). Since the negative pions ( ) are easily captured by the nucleus, only positive surface muon beams ($^+$) are available. The spin polarized surface muons are widely used as a magnetic microprobe in material science [3]. The surface muon beam has a lower energy compared to the decay muon beam and when injected into a bulk sample, they have, therefore, an implantation depth of just 0.1 to 1 mm. Despite the name' surface' muons, it is typically used as a probe to bulk phenomena rather than surface phenomena.

Lately, the studies of surface/subsurface, nano materials and multi layered thin lms, are becoming increasingly important. However, for that purpose, we must have muon beam, that has su ciently low energy to stop on or near the surface of the sample. To perform such studies, so called slow (low energy) muons are required with energy that is of the order of several eV to a few tens of keV, far lower than the energies available from the conventional muon beams.

2. J-PARC Facility

The accelerator complex of the Japan Proton Accelerator Complex (J-PARC) has been constructed in the south part of Tokai-JAEA site, and consists of a 181 MeV (400 MeV at Phase2) LINAC, and 3 GeV and 50 GeV proton synchrotron rings. About 90 % of the 3 GeV, 333 A (1.0 MW) is sent to the Materials and Life Science facility (MLF) for the production of intense pulsed neutron and muon beams, while the remaining 10 % will be sent to the 50 GeV ring for further acceleration in order to produce a 15 A, 50 GeV proton beam to be used for the kaon or neutrino physics programs[4].

3. J-PARC Muon Facility, MUSE

The muon science experimental area (MUSE) is located in the integrated building of MLF for the materials and life science program. The 3 GeV proton beam from the 3 GeV proton synchrotron ring will be transported through the 300 m long beam transport line into the MLF building. The beam will be focused on the muon target (installed upstream of the neutron target) with a size of approximately 24 mm in diameter. The repetition rate of the 3 GeV synchrotron is 25 Hz, with a pulse-separation of 40 ms. One proton pulse consists of two bunches which are separated by 600 ns with a full width of 100-120 ns. Construction of the MLF building was started in the beginning of 2004, and rst muon beam was delivered in the autumn of 2008.

For Phase 1, we installed a superconducting decay/surface channel with a modest-acceptance (about 40 msr) pion injector, where both surface muons, and decay muons can be extracted.

In November 2009, the surface muons ($^{+}$) extraction rate was recorded to be 1.8 10^6/s with use of a 120 kW proton beam from the RCS. The achieved intensity is equivalent or even larger than that at the RIKEN-RAL muon facility [5]. This is the reason why that the world strongest pulsed muon source was claimed to be achieved at MUSE, even if with use of a 120 kW proton beam intensity. This intensity would correspond to 1.5 10^7/s surface muons when in the future the proton beam will reach a power of 1MW [6] .

In addition to Phase 1, as one of the principal muon beam channel at MUSE, we are planning to install the Super Omega Muon beam channel, which consists of a large acceptance solenoid made of mineral insulation cables (MIC), a superconducting curved transport solenoid and a superconducting axial focusing magnets . There, we can collect either surface or cloud muons with a large acceptance of 400 mSr. Compared to conventional beamlines, the large acceptance of the front-end solenoid will allow for the capture of the highest intensity pulsed muon beam in the world. With a muon capture rate of 5 10^8 surface muons/s and an approximate transport e ciency of 80%, we can expect 4 10^8 /s surface muons and 10^7/s negative cloud muons in the experimental hall [7]. At the super omega channel, we are aiming to create a new type of muon source; the intense ultra-slow muon source. Fig. 1 shows a schematic view of the super omegamuon beam line.

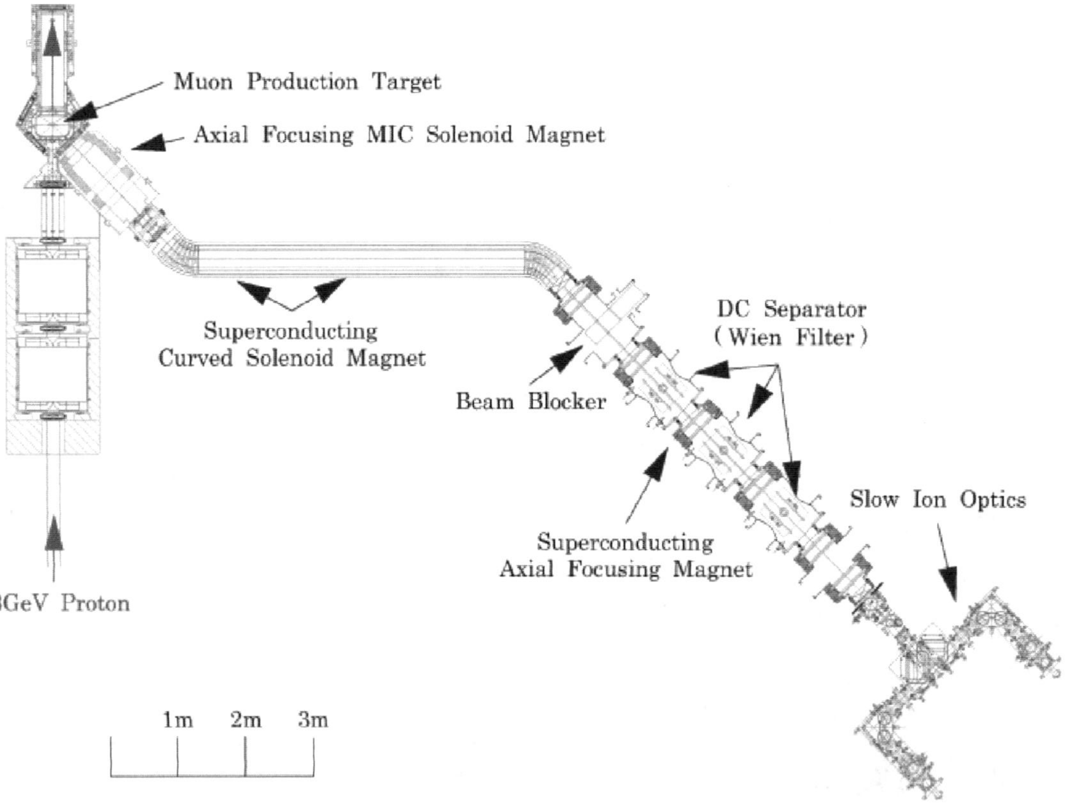

Figure 1. A schematic view of the superomega muon beamline consisting of a axial focusing MIC solenoid, a superconducting curved solenoid magnet, a superconducting axial focusing magnets and a slow ion optics.

4. Slow muon

There are proposed two methods to generate slow muon beam. The one is the generation of low energy $^+$ through the cryogenic moderation method using Van der Waals solids such as solid Ar or N_2, has been initiated at TRIUMF and is developed at PSI [8]. Another method is the Ultra slow muon generation through the resonant ionization of thermal Mu atoms, developed at KEK and RIKEN RAL pulsed muon facilities[9]. At J-PARC, we are proceeding the latter method to generate intense ultra slow muon source at J-PARC [10].

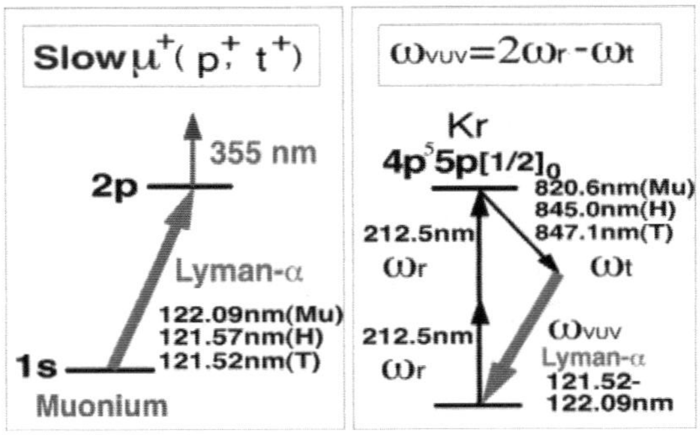

Figure 2. A resonant ionization scheme via the 1S→ 2P→ unbound transition for Mu and the scheme for the Lyman- generation via the four-wave frequency mixing method using two 212.55 nm (ω_r) photons for the two-photon resonant excitation of the $4P^5 5P[1/2]_0$ state in krypton, subtracted by a photon with a tunable di erence wavelength ω_t.

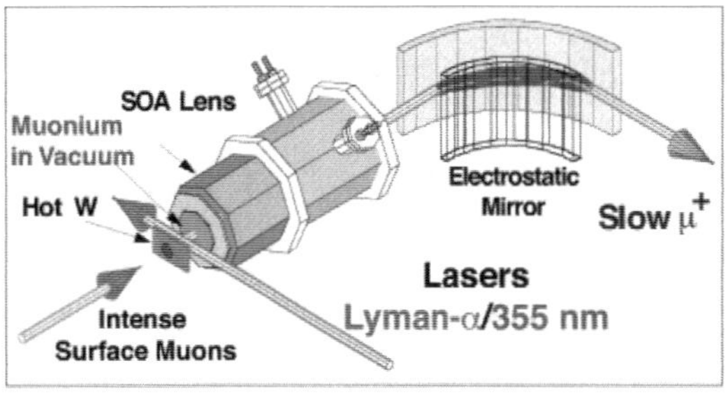

Figure 3. A schematic view of generation of the ultra slow muon beam. Ultra slow muons generated by the resonant ionization of Mu are accelerated and transported by " Slow-Ion Optics " consisting of an immersion lens (SOA), an electrostatic mirror, etc.

4.1. EXPERIMENTAL TECHNIQUE OF THE ULTRA SLOW $^+$ GENERATION BY RESONANT IONIZATION OF MU

The ultra slow pulsed muons are generated by the resonant ionization of thermal Muonium (resighnated by Mu) atoms generated from the surface of a hot tungsten foil, placed at the intense surface muon beam line. In order to e ciently ionize the Mu near the W surface, a resonant ionization scheme via the 1S→2P→unbound transition was adopted. This requires a laser system to generate Lyman- photon (in the vacuum ultraviolet region around 120 nm) and 355 nm photon (for the ionization step from the 2P state). Lyman- photon is generated by the resonant four-wave frequency mixing, through a parametric process of adding energies of two 212.55 nm photons (ω_r) and subtracting energy of one tunable infrared photon (ω_t). (ω_{VUV} = 2 ω_r- ω_t). Fig. 2 (left) shows a resonant ionization scheme via the 1S→2P→unbound transition for the hydrogen atom isotopes and the scheme for the Lyman- generation via the four-wave frequency mixing method using two 212.55 nm (ω_r) photons for the two-photon resonant excitation of the $4P^5 5P[1/2, 0]$ state in krypton, subtracted by a photon with a tunable di erence wavelength. Fig. 2 (right) shows how to generate Lyman- light by the four wave mixing in Kr. Ultra Slow muons which is generated by ionization of Mu are accelerated and transported by "Slow-Ion Optics" , which consists of an immersion lens (SOA), a magnetic bend for mass separation, an electrostatic de ector, and ve sets of electrostatic quadrupoles. Fig. 3 shows a schematic view how to generate and transport ultra slow muon beam [9].

Journal of Physics: Conference Series **302** (2011) 012038 doi:10.1088/1742-6596/302/1/012038

4.2. UNIQUE FEATURES OF THE PULSED ULTRA SLOW $^+$ BEAM

One of the most important features of the ultra slow $^+$ beam generated by the resonant laser ionization of Mu is its short pulse width. Regardless the temporal pro le of the surface muon pulse with its double bunch structure consisting of 100 ns (FWHM) pulses separated by 320 ns at RIKEN-RAL, the pulse duration of the ultra slow muon beam is determined only by the duration of the laser pulse and by the transport properties of the low energy ion transport beamline. Consequently, at RIKEN-RAL the duration of the ultra slow $^+$ pulse in the time-of- ight spectrum at the end of the transport beamline is just 7.5 ns (FWHM). In principle, the pulse duration can be reduced further to 1 ns by using a laser with shorter pulse duration and optimizing the beamline transport [11].

Since the Mu atoms evaporate from the hot W (2300K) at thermal energies, the initial kinetic energy of the ultra slow $^+$ generated by the laser resonant ionization is only 0.2 eV. Therefore the beam emittance of the ultra slow $^+$ beam accelerated by SOA lens to 9 keV is so small that the beam can be focused onto a tiny spot on the sample. Compared to the size of the incident surface muon beam of 40 mm (FWHM), the resulting ultra slow $^+$ beam it could be focused to a spot of just 3.3 mm (FWHM, horizontally) by 4.1 mm (FWHM, vertically) presently at RIKEN-RAL. In principle, the beam size is not limited by the beam emittance but by the transport characteristics of the beamline and by better optimization of the beamline spots as small as 1 mm (FWHM) can be potentially obtained, perhaps at a cost of reducing the transport e ciency.

4.3. Yield of ultra slow Muon by the laser resonant ionization method

At the RIKEN/RAL muon facility, 20 ultra slow $^+$/s [11] at the sample out of 1.2 10^6 /s surface muons can be extracted. At the J-PARC MUSE Super Omega muon channel 4 10^8 surface muons/s can be extracted from a large acceptance of 400 msr. From the viewpoint of the intensity of the surface muon, we can gain a factor of 300. Taking into account the repetition rate of the pulsed laser system and the muon beam, we can gain a factor of two, since at J-PARC, both the laser and muon beams can be synchronized to 25 Hz, whereas at the RIKEN-RAL facility, the muon beam (50Hz) is synchronized to every second laser pulse. Therefore, we can expect 1.3 10^4 /s of the ultra slow $^+$/s without any additional laser development at J-PARC. Although more than 71 J/cm2 of the Lyman- light is needed in order to saturate the transition of Mu from the 1S state to the 2P state, we are, at present, producing about 1 J/cm^2 of the Lyman- light through the Resonant Four-wave Frequency Mixing. With su cient laser development, we can expect about 100 times gain. Finally, we are expecting a maximum rate of 1.3 10^6 /s ultra slow $^+$/s [10].

5. Conclusion

As one of the principal muon beam line at the J-PARC muon facility (MUSE), we are now constructing the Super Omega Muon beam line, which consists of a large acceptance solenoid made of mineral insulation cables (MIC) and a superconducting curved transport solenoid. There, we can extract 4 10^8 /s surface muons towards the experimental hall. At the super omega line, we are aiming to create a new type of muon source; the intense ultra-slow muon source by resonant ionization of thermal Muonium (designated as Mu; consisting of a $^+$ and an e) atoms generated from the surface of a hot tungsten foil. When the production of intense ultra-slow muon source is realized, the use of its short-range penetration depth (eg. 1 nm resolution at a penetration of 1 nm, and 10 nm at a penetration of 6 nm in Gold) will allow muon science to be expanded towards a variety of new nano-scienti c elds such as, 1) Surface/boundary magnetism utilizing its spin polarization and unique time-window. 2) Surface chemistry, utilizing a feature of a light isotope of hydrogen; such as catalysis reactions. 3) Muon Microscopy, utilizing a feature of micron meter beam size, when ultra slow muon is accelerated.

4) Precise atomic physics testing QED, since Mu is the simplest lepton pair consisting $^{+}$ and e . 5) Ion sources for" g-2" experiment, and towards $^{+}$ collider experiments in high-energy physics.

Table 1 shows a comparison of features between conventional surface muon and ultra slow muon.

Table 1. A comparison of features between conventional surface muon and ultra slow muon.

	Ultra Slow Muon	Surface Muon ($ConventionalMuon$)
Muon Beam Line	J-PARC U-Line	J-PARC D-Line
Implantation Energy	0.05 - 30 keV	4.1 MeV
Monochromacity	14 eV	0.4 MeV
Beam Size	ϕ 1 - 4	ϕ 50 - 70
Time Resolution	sub ns - ns	100 - 150 ns
Polarization	50 %	100 %
Intensity	10^6/s (@1MW)	3 10^7/s (@1MW)

References

[1] Nagamine, K. 1981. Pulsed μ SR facility at the KEK BOOSTER. Hyper ne Intera. 8: 787-796.
[2] Pifer, A.E., Bowen, T., Kendall, K.R. 1976. A high stopping density μ + beam. Nucl. Instru. Metho. 135:39-46.
[3] Schenck, 1985. A. Muon Spin Rotation Spectroscopy. Adam Hilger Press, London.
[4] S. Nagamiya, T. Nagae, Y. Ooyama, Y. Miyake, H. Takano, J.R. Helliwell, J.C. Peng, Energy Review 19, 12(1999) 4-23
[5] K. Nagamine, T. Matsuzaki, K. Ishida, I. Watanabe, S.N. Nakamura, R. Kadono, N. Kawamura, S. Sakamoto, M. Iwasaki, M. Tanase, M. Kato, K. Kurosawa, G.H. Eaton, H.J. Jones, G. Thomas, and W.G. Williams, Hyper ne Interactions 101/102(1996)521
[6] Y. Miyake, K. Nishiyama, N. Kawamura, P Strasser, S. Makimura, A. Koda, K. Shimomura, H Fujimori, N. Nakahara, R. Kadono, M. Kato, S Takeshita, W. Higemoto, K. Ishida, T. Matsuzaki, Y. Matsuda, and K. Nagamine, Nucl. Instr. and Meth. in Phys. Res. A 600(2009) 22-24
[7] K. Nakahara, Y. Miyake, K. Shimomura, P. Strasser, K. Nishiyama, N. Kawamura, H. Fujimori, S. Makimura, A. Koda, K. Nagamine, T. Ogitsu,A. Yamamoto, T. Adachi, K Sasaki, K. Tanaka, N. Kimura, Y. Makida, Y. Ajima, K. Ishida, Y. Matsuda, AIP conference Proceedings, 981(2007)312-314
[8] Morenzoni, E., Kottmann, F., Maden, D., Matthias, B., Meyberg, M., Prokscha, Th., Wutzke, Th., Zimmermann, 1994. U. Generation of Very Slow Polarized Positive Muons.Phys. Rev. Lett., 72,17: 2793-2796
[9] Nagamine, K., Miyake, Y., Shimomura, K., Birrer, P., Marangos, J.P., Iwasaki, M, Strasser, P., Kuga, T., Ultra slow Positive-Muon Generation by Laser Ionization of Thermal Muonium from Hot tungstem at Primary Proton Beam. 1995. Phys. Rev. Lett., 74,24: 4811-4814.
[10] " Ultra Slow Muon Project at J-PARC, MUSE" Y. Miyake, K. Nakahara, K. Shimomura, P. Strasser, N. Kawamura, A. Koda, S. Makimura, H. Fujimori, K. Nishiyama, Y. Matsuda, P. Bakule, T. Adachi, and T. Ogitsu, AIP Conference Proceedings1104 (2009) 47-52
[11] P. Bakule, Y. Matsuda, Y. Miyake, K. Nagamine, M. Iwasaki, Y. Ikedo, K. Shimomura, P. Strasser, S. Makimura, Nucl. Instr. and Meth. in Phys. Res. B266 (2008) 335-346

Search for an Electric Dipole Moment in ^{129}Xe Atom with Nuclear Spin Oscillator Technique

K. Asahi[1], T. Furukawa[1], T. Inoue[1], A. Yoshimi[2], T. Nanao[1], M. Chikamori[1], K. Suzuki[1], M. Tsuchiya[1], H. Hayashi[1], M. Uchida[1], H. Ueno[2], Y. Matsuo[2], and T. Fukuyama[3]

[1] Tokyo Institute of Technology, Meguro-ku, Tokyo 152-8551, Japan
[2] RIKEN, 2-1 Hirosawa, Wako-shi, Saitama 351-0198, Japan
[3] Ritsumeikan University, Nojihigashi, Kusatsu, Shiga 525-8577, Japan

E-mail: asahi@phys.titech.ac.jp

Abstract. Development of a nuclear spin oscillator intended for the experimental search for an electric dipole moment in ^{129}Xe atom is reported. The spin oscillation is brought about by polarizing ^{129}Xe nuclear spins by spin exchange with optically polarized Rb atoms, detecting the spin precession through the transmission of a circularly polarized laser light, and applying a feedback field according to the detected spins. Until present, a frequency precision, in a measurement time of $T = 45{,}000$ s, of $\delta\nu = 5.2$ nHz has been achieved.

1. Introduction

Constituents of the matter world and their interactions known from laboratory experiments are surprisingly well described by the standard model (SM) of elementary particles, major part of which was established already in 1980's. Presently, however, most of us cannot avoid feeling dissatisfaction with this model, since there are numbers of questions the model is unable to answer, such as why quarks and leptons exist in three generations, why there are three kinds of interactions, and what is the 95 % of energy content of our universe which seemingly exist but cannot be explained by the SM [1]. The permanent electric dipole moment (EDM) serves as an exclusively suited observable to detect flaws, if any, of the standard model. This prominent role allotted to the EDM stems from the fact that this flavor-diagonal observable violates T and hence CP invariances. Thus, EDMs of particles are predicted to be almost undetectably small in the SM, while most of theories proposed beyond the SM predict their sizes in the regions reachable in not very distant future.

EDMs are being searched for in various sites: in neutron, paramagnetic atoms, diamagnetic atoms, molecules, and even in charged particles such as muon, proton, deuteron and ions. In the present report we describe the present status of our experimental attempt to detect an EDM in the diamagnetic atom ^{129}Xe.

2. EDM in a diamagnetic atom

Total angular momentum (the atomic spin) of a diamagnetic atom with a singlet cloud of atomic electrons comes solely from its nuclear spin I. Since the electron spins are totally paired off, EDMs of electrons give no contribution to the atomic EDM, and thus the EDM of a diamagnetic atom is induced solely from the nucleus. While the Schiff's theorem [2] prohibits direct contribution of a nuclear EDM

Journal of Physics: Conference Series **302** (2011) 012039 doi:10.1088/1742-6596/302/1/012039

to the atomic EDM, there remains a route in which a CP-violating higher-order moment of nucleus, the Schiff moment S as defined shortly below, can give rise to the atomic EDM.

2.1 Atomic EDM and the Schiff moment

Consider a nucleus of point size and charge Ze but with no EDM, in a neutral atom. The nucleus locates itself at the center of the atom, $\boldsymbol{r}_{\rm nucl} = 0$, because the electrostatic potential takes minimum there. Once an EDM $\boldsymbol{d}_{\rm nucl}$ is attached to it, the nucleus shifts its position by a displacement $\Delta\boldsymbol{r}_{\rm nucl} = -\boldsymbol{d}_{\rm nucl}/Ze$ so that its effective position of charge

$$\begin{aligned}\langle \boldsymbol{r}\rangle_{\rm ch} &\equiv \frac{1}{Ze}\int \rho_{\rm ch}(\boldsymbol{r}-\boldsymbol{r}_{\rm nucl})\boldsymbol{r}\,{\rm d}^3\boldsymbol{r}\\ &= \frac{1}{Ze}\int\left[Ze\delta(\boldsymbol{r}-\boldsymbol{r}_{\rm nucl}) + q\delta(\boldsymbol{r}-\boldsymbol{r}_{\rm nucl}-\frac{1}{2}\boldsymbol{a}) - q\delta(\boldsymbol{r}-\boldsymbol{r}_{\rm nucl}+\frac{1}{2}\boldsymbol{a})\right]\boldsymbol{r}\,{\rm d}^3\boldsymbol{r}\\ &= \boldsymbol{r}_{\rm nucl} + \frac{1}{Ze}\boldsymbol{d}_{\rm nucl}\end{aligned} \tag{1}$$

becomes zero (the center of the atom). Here $\boldsymbol{r}_{\rm nucl}$, $\boldsymbol{d}_{\rm nucl} = q\boldsymbol{a}$, and $\rho_{\rm ch}(\boldsymbol{r})$ denote the position, EDM, and charge distribution of the nucleus, respectively. Since the effective position of charge remains at the center anyhow, the atom does not "know" that the nucleus possesses an EDM $\boldsymbol{d}_{\rm nucl}$ (Schiff's theorem). Now we take a finite nuclear size into account. The effect of having $\boldsymbol{d}_{\rm nucl}$ is again simply a position shift $\Delta\boldsymbol{r}_{\rm nucl} = -\boldsymbol{d}_{\rm nucl}/Ze$, but in this case atomic electrons "feel" that the electrostatic interaction with the nucleus $\Phi(\boldsymbol{R})$ has changed [3]:

$$\Delta\Phi(\boldsymbol{R}) \equiv \int \frac{\rho_{\rm ch}(\boldsymbol{r})}{|\boldsymbol{R}-\boldsymbol{r}|}{\rm d}^3\boldsymbol{r} - \int \frac{\rho_{\rm ch}(\boldsymbol{r}-\Delta\boldsymbol{r}_{\rm nucl})}{|\boldsymbol{R}-(\boldsymbol{r}-\Delta\boldsymbol{r}_{\rm nucl})|}{\rm d}^3\boldsymbol{r} = \frac{1}{Ze}(\boldsymbol{d}_{\rm nucl}\cdot\nabla_{\boldsymbol{R}})\int\frac{\rho_{\rm ch}(\boldsymbol{r})}{|\boldsymbol{R}-\boldsymbol{r}|}{\rm d}^3\boldsymbol{r} \tag{2}$$

where $\boldsymbol{R}$ is the position of electron. Taking the lowest order term (setting an octupole term aside), Eq. (2) is equivalent to an electric field

$$\boldsymbol{E}_{\rm Schiff} \approx \frac{15}{R_{\rm nucl}^{5}}\boldsymbol{S} \tag{3}$$

acting inside the nuclear volume to the electrons [3], where $R_{\rm nucl}$ is the nuclear radius, and the Schiff moment $\boldsymbol{S}$ is defined as

$$\boldsymbol{S} \equiv \frac{1}{10}\int_{\rm nucleus}\rho_{\rm ch}(\boldsymbol{r})\boldsymbol{r}\left(r^2 - \frac{5}{3}\langle r^2\rangle_{\rm ch}\right){\rm d}^3\boldsymbol{r} \tag{4}$$

and $\langle r^2\rangle_{\rm ch}$ is the mean square charge radius of the nucleus. Electric field (2) polarizes the cloud of atomic electrons, leading to a finite EDM of atom

$$d_{\rm atom} = \alpha S\,, \tag{5}$$

where the constant of proportionality α is obtained from atomic physics calculations [4,5] for individual atoms.

2.2 Schiff moment from fundamental processes

The Schiff moment is induced by CP violating processes in the nucleus. In a nuclear physics terminology, the relevant processes would be the mixing of small CP violating amplitude into the nuclear ground state wave function, and CP violating nucleon-nucleon interactions that produce the mixings. Thus, nuclear structure calculations give the Schiff moment S as a function of isoscaler, isovector, and isotensor components of CP violating πNN coupling constants $\tilde{g}^{(0)}_{\pi NN}$, $\tilde{g}^{(1)}_{\pi NN}$ and $\tilde{g}^{(2)}_{\pi NN}$ through a perturbation expression

$$S \approx \sum_i \frac{\langle \text{g.s.}|S|i\rangle\langle i|V_{\text{CPV}}|\text{g.s.}\rangle}{E_0 - E_i} \tag{6}$$

and CP violating one-pion exchange NN interactions

$$\begin{aligned} V_{\text{CPV}} = \frac{g_{\pi NN}}{8\pi m_N}\sum\Big\{\Big[\tilde{g}^{(0)}_{\pi NN}\left(\boldsymbol{\tau}_i\cdot\boldsymbol{\tau}_j\right) - \frac{1}{2}\tilde{g}^{(1)}_{\pi NN}\left(\tau_i^z + \tau_j^z\right) + \tilde{g}^{(2)}_{\pi NN}\left(3\tau_i^z\tau_j^z - \boldsymbol{\tau}_i\cdot\boldsymbol{\tau}_j\right)\Big]\left(\boldsymbol{\sigma}_i - \boldsymbol{\sigma}_j\right) \\ -\frac{1}{2}\tilde{g}^{(1)}_{\pi NN}\left(\tau_i^z - \tau_j^z\right)\left(\boldsymbol{\sigma}_i + \boldsymbol{\sigma}_j\right)\Big\}\cdot\nabla_i \frac{\exp\left(-m_\pi|\boldsymbol{r}_i - \boldsymbol{r}_j|\right)}{|\boldsymbol{r}_i - \boldsymbol{r}_j|}, \end{aligned} \tag{7}$$

while particle physics analyses [6,7] provide the latter constants in terms of quark chromoelectric dipole moment $\tilde{d}_{\text{u}}$, $\tilde{d}_{\text{d}}$, and $\tilde{d}_{\text{s}}$ of up, down, and strange quarks. Analyses in Ref. [6] give (assuming the Peccei-Quinn symmetry [8] and inserting numerical values therein) $\tilde{g}^{(0)}_{\pi NN} \approx -1.02\left(\tilde{d}_{\text{u}} + \tilde{d}_{\text{d}}\right)$, $\tilde{g}^{(1)}_{\pi NN} \approx -9.2\left(\tilde{d}_{\text{u}} - \tilde{d}_{\text{d}}\right)$ and $\tilde{g}^{(2)}_{\pi NN} \approx 0$, where $\tilde{d}_{\text{q}}$ are in GeV.

2.3 Recent status

EDMs of diamagnetic atoms have been studied for ^{129}Xe and ^{199}Hg. Also, there are experimental plans for radioactive ^{209}Rn and ^{225}Ra isotopes [9,10], in which enhancements of EDM due to octupole deformation are actively discussed [11]. The most studied theoretically is the case of ^{199}Hg, on which the recent experiment [12] has placed very stringent limit, $d(^{199}\text{Hg}) < 3.1\times10^{-29}$ $e$cm. Atomic physics calculation [13] gives $d(^{199}\text{Hg}) = -2.8\times10^{-17}S$ [in $e$fm^3] $e$cm. As for the Schiff moment $S(^{199}\text{Hg})$, the earlier calculation [5] provided the relation

$$S(^{199}\text{Hg}) = g_{\pi NN}\left(-0.0010\tilde{g}^{(0)}_{\pi NN} + 0.074\tilde{g}^{(1)}_{\pi NN} + 0.018\tilde{g}^{(2)}_{\pi NN}\right)\, e\text{fm}^3. \tag{8}$$

The above experimental limit thus has been supposed to give an upper limit on a combination of the quark chromoelectric dipole moment, $\tilde{d}_{\text{u}} - \tilde{d}_{\text{d}} < 6\times10^{-27}$ cm . Very recently, however, fully self-consistent calculations [14] of the Schiff moment has given a result which turns out quite different from the earlier one:

$$S^{\text{New}}(^{199}\text{Hg}) = g_{\pi NN}\left(0.016\tilde{g}^{(0)}_{\pi NN} - 0.006\tilde{g}^{(1)}_{\pi NN} + 0.019\tilde{g}^{(2)}_{\pi NN}\right)\, e\text{fm}^3. \tag{9}$$

Note that the previously perceived dominance of the isovector component $\tilde{g}^{(1)}_{\pi NN}$ has now disappeared, weakening the above obtained constraint on $\tilde{d}_{\text{u}} - \tilde{d}_{\text{d}}$ by an order of magnitude, and even its sign has changed.

In the case of ^{129}Xe atom, atomic physics calculations [13,15] give $d(^{129}\text{Xe}) = 0.38\times10^{-17}S$ [in e fm^3] $e$cm, whereas the relation of $S(^{129}\text{Xe})$ to the CP violating πNN coupling constants $\tilde{g}^{(0)}_{\pi NN}$, $\tilde{g}^{(1)}_{\pi NN}$

and $\tilde{g}^{(2)}_{\pi NN}$ was calculated more than two decades ago with a rather simple account of CP violating *NN* interactions and core polarization effects. Thus, new calculation of $S(^{129}\mathrm{Xe})$ is strongly awaited.

3. Experiment

We aim to measure the atomic EDM of ^{129}Xe. ^{129}Xe is a stable nucleus so that a macroscopic number of particles can be subjected to the measurement. It has spin $I = 1/2$ and exists in a form of a rare gas, and there is a well established method to obtain a high degree of spin polarization by using the spin-exchange with optically pumped alkali atoms. Furthermore, the ^{129}Xe spins show quite long transverse relaxation times (spin coherent times) [16]. These advantageous features of ^{129}Xe have made us choose it as a physical system where an EDM is looked for, even with supposed disadvantages that it is not as heavy as the other candidates ^{199}Hg, ^{209}Rn, and ^{225}Ra and there is no enhancement due to octupole collectivity expected.

3.1 EDM measurement with a nuclear spin oscillator

The detection of an EDM is, in most cases, nothing other than a detection of extremely small change in frequency of the spin precession upon a reversal of the applied electric field. The key issue for a sensitive EDM detection is therefore a high precision determination of frequency with well stabilized (or monitored) strength of the applied magnetic field. Thus, the EDM d is deduced from the observed frequency change $\Delta\nu = \nu_+ - \nu_-$, as

$$d = \frac{h\Delta\nu}{4E}. \tag{10}$$

Usually the observation time for a single shot of continued spin precession is limited by the transverse spin relaxation time T_2. The measurement is made by repeating n times the shot of precession for a

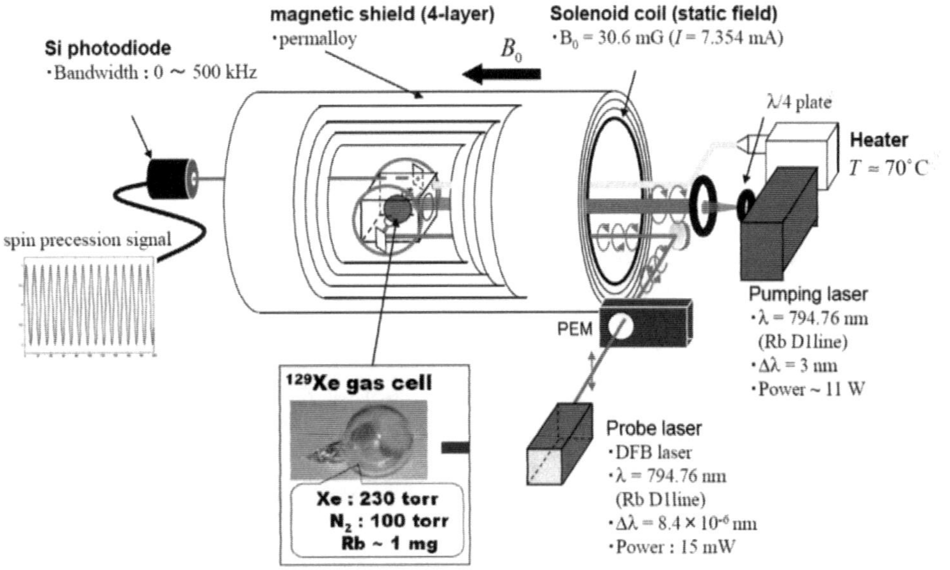

Figure 1. Setup for the spin oscillator experiment.

time period T, having the total measurement time to be $T_{\rm total} = nT$, with $T \sim T_2$. The frequency precision for such a measurement is given by

$$\delta\nu = \delta\nu_{\rm single} / \sqrt{n} \propto (T_{\rm total})^{-1/2}. \qquad (11)$$

If, on the other hand, the spin coherence time is elongated up to the full measurement time $T_{\rm total}$, the precision $\delta\nu$ should go as

$$\delta\nu \propto (T_{\rm total})^{-3/2}, \qquad (12)$$

enabling a much faster improve-ment in $\delta\nu$. In order to realize such an elongation of the spin precession time, we employ the technique of a nuclear spin oscillator with feedback by optical spin detection [17].

Figure 1 shows the experimental setup for the spin oscillator experiment. A ^{129}Xe gas contained in a 18 mm diameter spherical glass cell together with a small amount of Rb vapor is illuminated by a circularly polarized laser light of 794.7 nm wave length. By spin exchange with the optically pumped Rb atoms, the ^{129}Xe spins are polarized in the direction of a static magnetic field $B_0 = 30.6$ mG. Once the ^{129}Xe spins are tilted, they start precession about B_0, which is observed through transmission of a probe laser light traversing the cell in the direction perpendicular to B_0. Thus, the precession signal obtained at a Si photodiode is lock-in amplified, digitalized, processed, and used for generating a feedback field $B_{\rm FB}$ applied on the ^{129}Xe cell. The $B_{\rm FB}$ field is so arranged that it is kept 90 degree in advance from the transverse component of the precessing ^{129}Xe spins, thus acting to increase transverse component of ^{129}Xe spins to maintain the precession. Detailed explanation of the operation of the spin oscillator is given in Ref. [17].

3.2 Frequency characteristics

After succeeding in the realization of steady state oscillation, detailed studies of operation

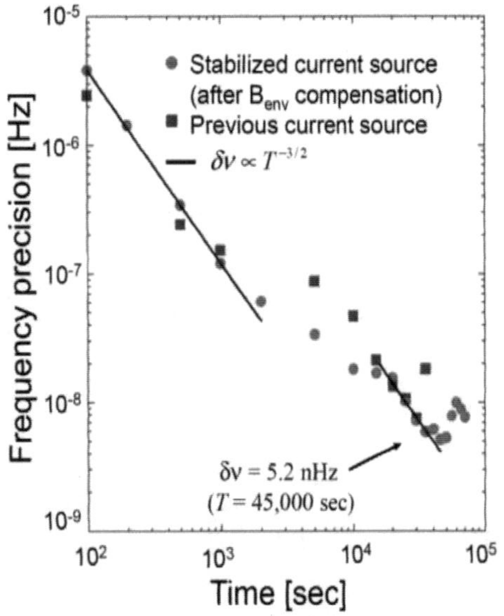

Fig. 2: Frequency precision attained by operating the spin maser for time periods T. Blue squares and red circles represent, respectively, data before and after the current source was replaced by a new one.

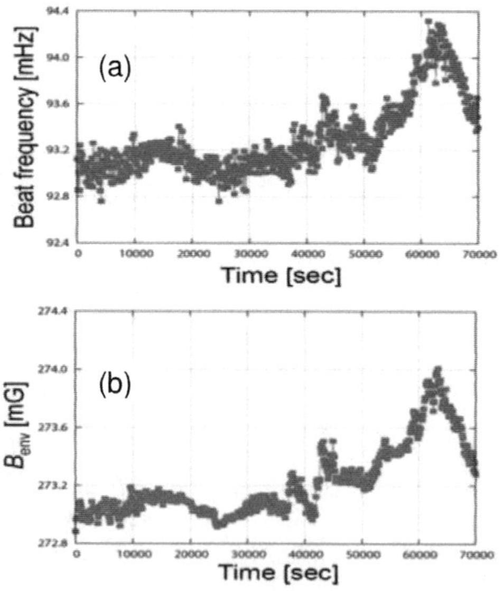

Fig. 3: Time variations of the maser frequency (a), and the environmental field (b). A clear correlation between (a) and (b) is seen.

characteristics such as frequency and amplitude stabilities, their dependences on the cell temperature, pumping laser power, environmental field, and strength of the B_0 and B_{FB} fields were conducted. In particular, the frequency drifts and fluctuations were found to occur from several different sources. Some part of drifts appears to be increased regularly in the midday when the ambient temperature raises, while some part shows a clear dependence on drifts in the current of the solenoid coil for applying B_0. Figure 2 shows the frequency precision, $\delta\nu$, attained in long-time operations of the spin oscillator. The blue squares represent the data in the measurements before the current source for the B_0 solenoid was replaced by a new one. Although the expected improvement of frequency precision, $\delta\nu \propto (T_{\rm total})^{-3/2}$ was observed in the region up to 1000 s, it became flattened for T = 1,000 - 5,000 s, and then started to be improved faster. In about 30,000 s, $\delta\nu$ reached the best value of $\delta\nu$ = 9.3 nHz which corresponded to an EDM precision of 9×10^{-28} *e*cm provided that an electric field of $E = \pm 10$ kV/cm was applied. Detailed inspection revealed that in the region of time T = 1,000 - 5,000 s frequency drifts were largest. Since there seemed several sources of drifts and fluctuations as stated earlier, we made several measurements using different setups: i) the current source was replaced with a new one which was equipped with specially designed feedback stabilizer, ii) a lower cell temperature of 50°C was chosen at which Rb density in the cell was calculated to be low, and iii) the frequency was corrected for the drift in the environmental field measured with a flux gate magnetometer. The result is shown in Fig. 2 by the red filled circles. Now $\delta\nu$ remains almost in the $(T_{\rm total})^{-3/2}$ dependence until 10,000 s, and reached an improved limit of $\delta\nu$ = 5.2 nHz corresponding to a 5×10^{-28} *e*cm EDM precision. In Fig. 3, the observed frequency is plotted as a function of time, in comparison with the simultaneously monitored environmental field $B_{\rm env}$. A close resemblance between the behaviors of $\delta\nu$ and $B_{\rm env}$ is clearly seen. We are now preparing for an active compensation of the environmental field.

4. Outlook

After spending a rather long term of "groping in the dark" to find out sources of frequency drifts and fluctuations, we have reached a point to establish the final setup for the EDM search with the best tuned nuclear spin oscillator. To this end, we are preparing for cells with electrodes for the E field application, a reinforced pumping laser, a fluid circulation system for the cell temperature stabilization, and high precision magnetometry using nonlinear magneto-optical rotation [18].

References

[1] Trodden M 1999 *Rev. Mod. Phys.* **71** 1463
[2] Schiff L I 1963 *Phys. Rev.* **132** 2194
[3] Ginges J S M and Flambaum V V 2004 *Phys. Reports* **397** 63
[4] Flambaum V V, Khriplovich I B and Sushkov O P 1986 *Nucl. Phys.* A **449** 750
[5] de Jesus J H and Engel J 2005 *Phys. Rev.* C **72** 045503
[6] Hisano J and Shimizu Y 2004 *Phys. Rev.* D **70** 093001
[7] Faessler A *et al.* 2006 *Phys. Rev.* D **74** 074013
[8] Peccei R D and Quinn H R 1977 *Phys. Rev. Lett.* **38** 1440
[9] Chupp T 2009 *Nucl. Phys.* A **827** 428c
[10] Guest J R *et al.* 2007 *Phys. Rev. Lett.* **98** 093001
[11] Dobaczewski J and Engel J 2005 *Phys. Rev. Lett.* **94** 232502
[12] Griffith W C *et al.* 2010 *Phys. Rev.* C **82** 015501
[13] Dzuba V A *et al.* 2002 *Phys. Rev.* A **66** 012111
[14] Ban S, Dobaczewski J, Engel J and A. Shula 2010 *Phys. Rev.* C **82** 01550182
[15] Dzuba V A *et al.* 2009 *Phys. Rev.* A **80** 032120
[16] Gemmel C *et al.* 2010 *Eur. Phys. J.* D **57** 303
[17] Yoshimi A *et al.* 2002 *Phys. Lett.* A **304** 13
[18] Budker D *et al.* 2002 *Rev. Mod. Phys.* **74** 1153

International Symposium "Nanoscience and Quantum Physics 2011" IOP Publishing
Journal of Physics: Conference Series **302** (2011) 012040 doi:10.1088/1742-6596/302/1/012040

Superfluid state in the periodic Anderson model with attractive interactions

Akihisa Koga
Department of Physics, Tokyo Institute of Technology, Tokyo 152-8551, Japan
E-mail: koga@phys.titech.ac.jp

Philipp Werner
Theoretische Physik, ETH Zurich, 8093 Zürich, Switzerland

Abstract. We investigate the periodic Anderson model with attractive interactions. By combining dynamical mean-field theory (DMFT) with a continuous-time quantum Monte Carlo impurity solver, we study the competition between the superfluid state and the paramagnetic Kondo insulating state. It is found that both the particle and hole doping of the Kondo insulating state yields a phase transition to the superfluid state, and induces low energy in-gap states in the density of states. The phase diagram of the model is determined.

1. Introduction

Ultracold atoms have attracted considerable interest [1, 2] since the successful realization of Bose-Einstein condensation in a bosonic ^{87}Rb system [3]. One of the most active topics in this field is the study of fermionic optical lattice systems, which are formed by loading ultracold fermions into a periodic potential [4, 5, 6, 7]. Due to the high controllability of the lattice structure and onsite interactions, fermionic optical lattice systems are sometimes regarded as quantum simulators of the Hubbard model. Recent papers have suggested the possible realization of optical lattice systems described by a multi-component model [8, 9, 10] and the Kondo lattice model [11, 12, 13]. This motivates us to investigate the periodic Anderson model (PAM) with itinerant and localized bands [14]. The PAM may provide a stage to discuss quantum phase transitions in optical lattice systems, since quantum critical behavior is expected to occur in both the repulsive and attractive models. In our previous paper [15], we have investigated the PAM with attractive interactions to clarify how the chemical potential affects the competition between the Kondo insulating state and the superfluid state at low temperatures. It is also instructive to clarify the doping dependence of low temperature properties in the PAM model. This topic was beyond the scope of our previous paper, but it may be important for the investigation of quantum critical behavior in optical lattice systems.

We study here the PAM with attractive interactions by combining dynamical mean-field theory (DMFT) [16, 17, 18, 19] with the continuous-time quantum Monte Carlo (CTQMC) approach [20]. We then discuss the competition between the superfluid state and the Kondo insulating state and determine the phase diagram. By examining dynamical properties, we clarify that the particle or hole doping of the Kondo insulating state yields a phase transition to the superfluid state, which induces low energy in-gap states in the density of states.

Journal of Physics: Conference Series **302** (2011) 012040 doi:10.1088/1742-6596/302/1/012040

The paper is organized as follows: In Sec. 2, we introduce the model Hamiltonian and briefly summarize our theoretical approach. We demonstrate how the superfluid state competes with the Kondo insulating state at low temperatures in Sec. 3. A brief summary is given in the last section.

2. Model and Method

We consider the periodic Anderson model with conduction and localized bands, which is described by the following Hamiltonian,

$$\begin{aligned} H &= -t \sum_{\langle i,j \rangle \sigma} c_{i\sigma}^\dagger c_{j\sigma} + V \sum_{i\sigma} \left(c_{i\sigma}^\dagger f_{i\sigma} + f_{i\sigma}^\dagger c_{i\sigma} \right) + E_f \sum_{i\sigma} n_{i\sigma}^f \\ & -U \sum_i \left[n_{i\uparrow}^f n_{i\downarrow}^f - \frac{1}{2} \left(n_{i\uparrow}^f + n_{i\downarrow}^f \right) \right], \end{aligned} \tag{1}$$

where $c_{i\sigma}$ ($f_{i\sigma}$) annihilates a fermion in the conduction (localized) band on the ith site with spin σ, and $n_{i\sigma}^a = a_{i\sigma}^\dagger a_{i\sigma}$ ($a = c$ and f). t is the nearest-neighbor hopping matrix for the conduction band, V is the hybridization between the two bands, U is the attractive interaction, and E_f is the energy level for the localized band. In this model, the hybridization between the conduction and localized bands favors the formation of a local Kondo singlet state, and stabilizes the Kondo insulating state around half filling. On the other hand, the attractive interaction enhances pairing correlations. This tends to stabilize the superfluid state, which should be characterized by a pair potential $\Delta_a = \frac{1}{N} \sum_i \langle a_{i\uparrow} a_{i\downarrow} \rangle$, where $a = c, f$, and N is the total number of sites. Therefore, it is expected that the Kondo insulating state competes with the superfluid state at low temperatures.

To discuss how the superfluid state competes with the Kondo insulating state, we make use of DMFT, where local particle correlations can be taken into account precisely. In DMFT, the lattice Green's function is obtained via a self-consistency condition imposed on the impurity problem. The non-interacting Green's function for the lattice model is given as

$$\hat{G}_0^{-1}(k, i\omega_n) = \begin{pmatrix} (i\omega_n + h_c)\sigma_0 + (\mu_c - \epsilon_k)\sigma_z & -V\sigma_z \\ -V\sigma_z & (i\omega_n + h_f)\sigma_0 + (\mu_f - E_f)\sigma_z \end{pmatrix}, \tag{2}$$

where ϵ_k is the dispersion relation of the conduction band, σ_z is the z component of the Pauli matrix, σ_0 is the identity matrix, $\omega_n = (2n+1)\pi T$ is the Matsubara frequency, and T is the temperature. Here, the Green's function is represented in the Nambu formalism to enable a description of the superfluid state. The lattice Green's function is then given in terms of the self-energy $\hat{\Sigma}(i\omega_n)$ as,

$$\begin{aligned} \hat{G}(i\omega_n) &= \begin{pmatrix} G_{cc}(i\omega_n) & G_{cf}(i\omega_n) \\ G_{fc}(i\omega_n) & G_{ff}(i\omega_n) \end{pmatrix} = \int dk \left[\hat{G}_0^{-1}(k, i\omega_n) - \hat{\Sigma}(i\omega_n) \right]^{-1}, \\ \hat{\Sigma}(i\omega_n) &= \begin{pmatrix} 0 & 0 \\ 0 & \Sigma_f(i\omega_n) \end{pmatrix}. \end{aligned} \tag{3}$$

The self-energy for the localized band Σ_f is given as,

$$\Sigma_f(i\omega_n) = \begin{pmatrix} \Sigma_{f\uparrow}(i\omega_n) & S_f(i\omega_n) \\ S_f(i\omega_n) & -\Sigma_{f\downarrow}^*(i\omega_n) \end{pmatrix}, \tag{4}$$

where $\Sigma_{f\sigma}(i\omega_n)$ $[S_f(i\omega_n)]$ is the normal (anomalous) part of the self-energy for the localized band. In DMFT, the self-consistency condition is given by $G_{ff}(i\omega_n) = G_{imp}(i\omega_n)$, where G_{imp}

is the Green's function of the effective impurity model. The effective medium for each site is given by $\mathcal{G}^{-1}(i\omega_n) = [G_{imp}(i\omega_n)]^{-1} + \Sigma_f(i\omega_n)$.

In the PAM with attractive interactions, it is necessary to accurately treat different energy scales such as the hybridization gap and the superfluid gap. To this end, we make use of the CTQMC method, which has recently been developed [20, 21] and has successfully been applied to general classes of models such as the Hubbard model [22, 23, 24, 25, 26, 27], the PAM [15, 28], the Kondo lattice model [29], and the Holstein-Hubbard model [30]. Here, we use the continuous-time auxiliary field version of the weak-coupling CTQMC method [31] extended to the Nambu formalism, which allows us to directly access the superfluid state at low temperatures [26]. In our CTQMC simulations, we measure normal and anomalous Green's functions on a grid of one thousand points. In the following, we set $\mu = \mu_c = \mu_f$, $h = h_c = h_f = 0$, and $E_f = 0$ for simplicity.

3. Results

We consider the PAM on the hypercubic lattice. The bare density of states for the conduction band is given by a Gaussian with bandwidth t^*: $\rho(x) = \exp[-(x/t^*)^2]/\sqrt{\pi t^{*2}}$. In the paper, we use t^* as the unit of energy. Here, we focus on a system with $V = 0.4$ and $U = 1.0$, which is in the Kondo insulating state at half filling, at least, down to $T = 0.005$. To clarify how

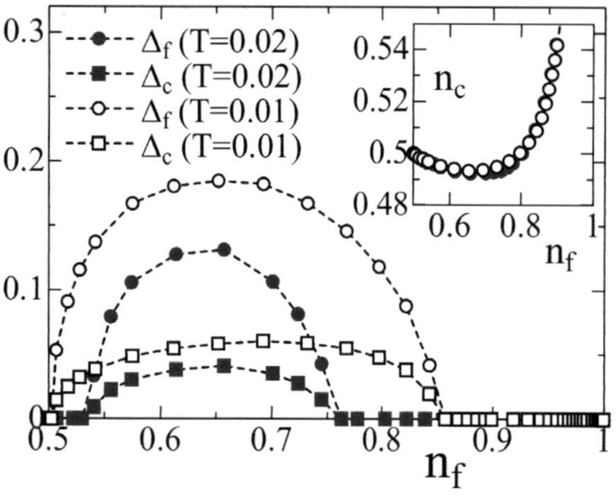

Figure 1. Pair potentials as a function of n_f when $U = 1.0$ and $V = 0.4$ at $T = 0.02$ (solid symbols) and $T = 0.01$ (open symbols). Squares (circles) represent the results for the conduction (localized) band. The inset shows the particle number for the conduction band at $T = 0.02$ (solid circles) and $T = 0.01$ (open circles). The results are almost identical.

the superfluid state is realized by the particle doping, we calculate the pair potential for each band, as shown in Fig. 1. It is found that when $n_f = 0.5$, the pair potentials are zero and the hybridization stabilizes the Kondo insulating state. In fact, the density of states is symmetric, with a hybridization gap, as shown in Fig. 2. The introduction of the chemical potential leads to an increase in the number of the particles. At a certain critical value $n_f = n_{fc1}$, a phase transition to the superfluid state occurs and the pair potential is induced. The simulation gives $n_{fc1} \sim 0.54$ ($T = 0.02$) and 0.503 ($T = 0.01$). An in-gap state is simultaneously induced below the Fermi level inside the hybridization gap, as shown in Fig. 2. This low energy peak is associated with the superfluid state, since it is always located in the vicinity of the Fermi level as long as the system is in the superfluid state. If we further increase the total particle number, the pair potential goes through a maximum and finally vanishes at $n_f = n_{fc2}$, where $n_{fc2} \sim 0.75$

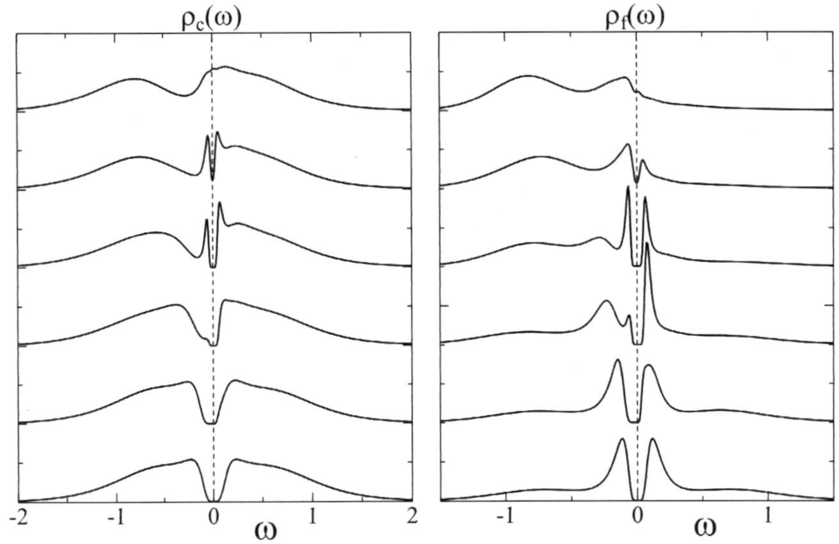

Figure 2. Density of states for the conduction (left) and localized (right) band of the system with $U = 1$ and $V = 0.4$ at $T = 0.01$. The data are for the following particle numbers in the localized band: $n_f = 0.5, 0.503, 0.575, 0.732, 0.823$, and 0.858 (from the bottom to the top).

($T = 0.02$) and 0.85 ($T = 0.01$). The second phase transition is to the normal metallic state and marked by the collapse of the superfluid gap, as shown in Fig. 2. We also note that the small increase in the total number of particles (the chemical potential) is associated with an increase in the number of particles in the localized band n_f, while a slight decrease is observed in the other band, as shown in the inset of Fig. 1. This behavior is not sensitive to the temperature, in contrast to the the pair potential and the density of states, as shown in the inset of Fig. 1. Therefore, we conclude that this charge transfer does not originate from the realization of the superfluid state, but is simply a consequence of the attractive interactions, which favor doubly occupied states at each site.

By performing similar calculations, we obtained the phase diagram shown in Fig. 3. At low temperatures, the introduction of holes or particles in the Kondo insulating state immediately induces a phase transition to the superfluid state. Therefore, we find that the Kondo insulating state is realized only around $n_f \sim 0.5$, while the superfluid state is stable in a wide region of parameter space. This phase diagram may be useful in understanding the low temperature properties of optical lattice systems with a confining potential, where the local particle number depends on the location of the sites.

4. Summary

We have investigated the periodic Anderson model with attractive interactions on the hypercubic lattice. By combining DMFT with the CTQMC method based on the Nambu formalism, we have studied quantitatively how the superfluid state is stabilized at low temperatures. It has been found that a low-energy state characteristic of the superfluid phase appears in the hybridization gap when particles are doped into the Kondo insulating state.

Acknowledgement

We would like to thank H. Kusunose for fruitful discussions during the Japan-Swiss Joint Workshop "New Trends in the Theory of Correlated Materials". This work was partly supported

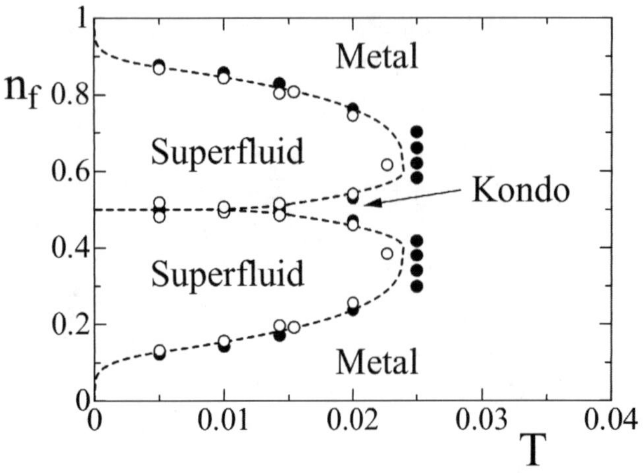

Figure 3. Phase diagram for the system with $U = 1.0$ and $V = 0.4$. Open (solid) circles indicate the state with (without) pair potentials. The phase boundary is a guide to eyes.

by the Grant-in-Aid for Scientific Research 20740194 (A.K.) and the Global COE Program "Nanoscience and Quantum Physics" from the Ministry of Education, Culture, Sports, Science and Technology (MEXT) of Japan. PW acknowledges support from SNF Grant PP002-118866.

References

[1] Pethick C J and Smith H 2002 *Bose-Einstein Condensation in Dilute Gases* (Cambridge University Press, Cambridge)
[2] Pitaevskii L and Stringari S 2003 *Bose-Einstein Condensation* (Clarendon Press, Oxford)
[3] Anderson, M H Ensher J R, Matthews M R, Wieman C E, and Cornell E A 1995 *Science* **269** 198
[4] Bloch I and Greiner M 2005 *Advances in Atomic, Molecular, and Optical Physics*, edited by P. Berman and C. Lin (Academic Press, New York), Vol. 52 p. 1
[5] Bloch I 2005 *Nature Physics* **1** 23
[6] Jaksch D and Zoller P 2005 *Ann. Phys. (NY)* **315** 52
[7] Morsch O and Oberhaler M 2006 *Rev. Mod. Phys.* **78** 179
[8] Ottenstein T B, Lompe T, Kohnen M, Wenz A N, and Jochim S 2008 *Phys. Rev. Lett.* **101** 203202
[9] Huckans J H, Williams J R, Hazlett E L, Stites R W, and O'Hara K M 2009 *Phys. Rev. Lett.* **102** 165302
[10] Fukuhara T, Sugawa S, Takasu Y, and Takahashi Y 2009 *Phys. Rev.* A **79** 021601
[11] Gorshkov A V, Hermele M, Gurarie V, Xu C, Julienne P S, Ye J, Zoller P, Demler E, Lukin M D, and Rey A M 2010 *Nat. Phys.* **6** 289
[12] Foss-Feig M, Hermele M, and Rey A M 2010 *Phys. Rev.* A **81** 051603
[13] Foss-Feig M, Hermele M, Gurarie V, and Rey A M 2010 *Preprint* arXiv:1007.5083
[14] Hewson A C 1993 *The Kondo Problem to Heave Fermions*, Cambridge University Press (Cambridge)
[15] Koga A and Werner P 2010 *J. Phys. Soc. Jpn.* **79** 114401
[16] Metzner W and Vollhardt D 1989 *Phys. Rev. Lett.* **62** 324
[17] Müller-Hartmann E 1989 *Z. Phys.* B **74** 507
[18] Georges A, Kotliar G, Krauth W, and Rozenberg M J 1996 *Rev. Mod. Phys.* **68** 13
[19] Pruschke T, Jarrell M, and Freericks J K 1995 *Adv. Phys.* **44** 187
[20] Rubtsov A N, Savkin V V and Lichtenstein A I 2005 *Phys. Rev.* B **72** 035122
[21] Werner P, Comanac A, de'Medici L, Troyer M, and Millis A J 2006 *Phys. Rev. Lett.* **97** 076405
[22] Werner P and Millis A J 2007 *Phys. Rev.* B **75** 085108
[23] Werner P and Millis A J 2006 *Phys. Rev.* B **74** 155107
[24] Werner P and Millis A J 2007 *Phys. Rev. Lett.* **99** 126405; Werner P, Gull E, and Millis A J 2009 *Phys. Rev.* B **79** 115119
[25] Dao T-L, Ferrero M, Georges A, Capone M, and Parcollet O 2008 *Phys. Rev. Lett.* **101** 236405
[26] Koga A and Werner P 2010 *J. Phys. Soc. Jpn.* **79** 064401
[27] Koga A, Bauer J, Werner P, and Pruschke Th *Physica* E **43** 697

[28] Luitz D J and Assaad F F 2010 *Phys. Rev.* B **81** 024509
[29] Otsuki J, Kusunose H, and Kuramoto Y 2009 *Phys. Rev. Lett.* **102** 017202
[30] Assaad F F and Lang T C 2007 *Phys. Rev.* B **76** 035116; Werner P and Millis A J 2007 *Phys. Rev. Lett.* **99** 146404
[31] Gull E, Werner P, Parcollet O and Troyer M 2008 *Europhys. Lett.* **82** 57008

Quantum Betatron

Yosuke Kayanuma

Research Institute for the 21st Century, Osaka Prefecture University, Sakai, 599-8531 Japan, and CREST Japan Science and Technology Agency, Kawaguchi, Saitama 332-0012, Japan

E-mail: kayanuma@pe.osakafu-u.ac.jp

Abstract. A quantum mechanical model of the betatron is proposed for two-dimensional electron gases, in which the betatron condition is satisfied globally under a special type of axially symmetric inhomogeneous magnetic field named beta-field. The eigenvalue problem for the electron in the beta-field is solved analytically. It is shown that, for electrons undergoing the cyclotron motion in a uniform magnetic field, the adiabatic increase of the beta-field accelerates the speed of the rotation without changing the orbital radius.

1. Introduction

In 1941, Kerst[1] succeeded in constructing an accelerator of electrons to the energy as high as 2.3MeV by using the principle of electromagnetic induction. This appratus, now called a betatron, is operated based on the theory of stability orbit in an inhomogeneous time-dependent magnetic field. A bunch of electrons rotating in a vacuum chamber is accelerated by the induced electric field due to the increase of the magnetic flux linking the electron orbit, while the magnetic field at the orbit provides the central force (Lorentz force) that maintains the radius of the circular motion constant. For that purpose, the so-called 2:1 rule should be satisfied[1],

$$\Phi(r,t) = 2\pi r^2 B(r,t), \tag{1}$$

where $\Phi(r,t)$ is the magnetic flux penetrating the orbit with radius r, and $B(r,t)$ is the magnetic field at the orbit. Thus, the magnetic flux within the orbit must be twice the value obtained if the flx density is uniform and is equal to the field at the orbit. The betatron has advantages that it can accelerate electrons avoiding the insulation breakdown of the direct voltage generators, and that the size of the apparatus is relatively small.

In the present work, I study the quantum mechanical properties of the betatron downsized to the mesoscopic scale. Such a small size of apparatus may be realized by the nano fabrication technique applied to the two-dimensional electron gases (2DEG), or quantum dots. It will be shown below that, for a special type of inhomogeneous time-dependent magnetic field, the electrons can be accelerated coherently in their circular motion without changing the orbital radius, just like the classical betatron. It will provide a new tool of controlling electronic systems, and also a new source of radiation, in mesoscopic scale.

2. Beta-field and stationary states

Let the coordinates of an electron confined in a plane be $\vec{r} = (x, y, 0)$. We assume that a time-dependent axially symmetric magnetic field is applied perpendicularly to the plane,

Published under licence by IOP Publishing Ltd

$\vec{B}(r,t) = B(r,t)\vec{e}_z$, where $\vec{e}_z$ is the unit vector in z-direction and $r = \sqrt{x^2+y^2}$. In the cylindrical coordinate system (r,θ,z) defined as usual, the corresponding vector potential $\vec{A}$ has only the azimuth angle component $A_\theta(r,t)$ given by $B(r,t) = r^{-1}\partial\left(rA_\theta(r,t)\right)/\partial r$.

The Hamiltonian is given by

$$H(t) = -\frac{\hbar^2}{2m_e}\left[\frac{\partial^2}{\partial r^2} + \frac{1}{r}\frac{\partial}{\partial r} - \frac{1}{r^2}\left\{-i\frac{\partial}{\partial\theta} + \frac{er}{\hbar}A_\theta(r,t)\right\}^2\right], \tag{2}$$

in which m_e is the effective mass of the electron, and $e(>0)$ is the magnitude of the electric charge. Here and hereafter we suppress the argument z. In this work, we neglect the interaction of the magnetic field with the spin of electrons. It will give rise to an additional spin-dependent force because of the inhomogeneity of the magnetic field like the Stern-Gerlach effect. Since the angular momentum $L_z \equiv -i\hbar\partial/\partial\theta$ is a constant of motion, the time-dependent Schrödinger equation $i\hbar\partial\psi(\vec{r},t) = H(t)\psi(\vec{r},t)$ is reduced to that for the radial function by the separation of the variable, $\psi(\vec{r},t) = \exp(im\theta)R(r,t)$.

In considering the quantum betatron, we assume that the magnetic field $\vec{B}(r,t)$ satisfies the betatron condition Eq.(1) everywhere in the plane. This is natural since, in quantum mechanics, the wave function depends globally on the field profile. Furthermore, this is desirable from the experimental as well as practical point of view, since a substantial part of the electrons in the mesoscopic scale of 2DEG can be accelerated coherently without changing the orbital size. It can be easily shown that the profile of the magnetic field satisfying this requirement is given by

$$B(r,t) = C(t)/r, \tag{3}$$

where $C(t)(>0)$ is a function of time. The vector potential is given by $A_\theta = C(t)$. We will call such a magnetic field a *beta-field.*

First we solve the eigenvalue problem for the time-*in*dependent beta-field, $C(t) = C$. For the angular momentum quantum number m, the Schrödinger equation for the radial wave function $R(r)$ is given by

$$\left\{\frac{d^2}{dr^2} + \frac{1}{r}\frac{d}{dr} - \left(\frac{m}{r}+\beta\right)^2 + \epsilon\right\}R(r) = 0, \tag{4}$$

where $\beta = eC/\hbar$ and $\epsilon = 2m_eE/\hbar^2$ with E being the eigenenergy. If one neglects the second term, the problem becomes analogous to to that in one-dimension with the effective potential $V(r) = (m/r+\beta)^2$. For $m<0$, $V(r)$ takes a minimum value at $r_0 = |m|/\beta$, and tends to $V(r\to\infty) = \beta^2$, $V(r\to 0)\to\infty$. Therefore, there are bound states with ϵ satisfying $0<\epsilon<\beta^2$. The eigenstates with $\epsilon \geq \beta^2$ form a continuum of scattering states. For $m>0$, the effective potential becomes repulsive, so that all eigenstates become scattering states with $\epsilon \geq \beta^2$.

For $m<0$ and $\epsilon \leq \beta^2$, we transform Eq.(4) into the form,

$$\rho\frac{d^2}{d\rho^2}L + (2|m|+1-2\rho)\frac{dL}{d\rho} - (2\lambda m + 2|m|+1)L = 0, \tag{5}$$

by the change of variables $\rho = \alpha r$ with $\alpha = \sqrt{\beta^2-\epsilon}$, $\lambda = \beta/\alpha$, and by setting $R(r) = \rho^{|m|}\exp(-\rho)L(\rho)$. This is Laguerre's equation. The quantization condition is derived by the requirement that $L(\rho)$ is a polynomial as usual. Thus we find

$$\lambda|m| - |m| - \frac{1}{2} = n, \quad n = 0,1,2,\cdots, \tag{6}$$

and the eigenenergy as

$$E_{n,m} = \frac{\hbar^2\beta^2}{2m_e}\left(1 - \frac{|m|^2}{(n+|m|+\frac{1}{2})^2}\right). \tag{7}$$

The solution to Eq.(5) is given by the associated Laguerre polynomial $L_n^{(|2|m|)}(2\rho)$. The eigenenergies depend both on the principal quantum number n and the magnetic quantum number m. For each m, the eigenstates form an infinite ladder of eigenvalues from $n = 0$ to $n \to \infty$.

The scattering states for the energy $\epsilon \geq \beta^2$ can be treated simultaneously both for $m < 0$ and $m > 0$. It can be easily shown that, by setting $\rho = i\alpha r$ with $\alpha = \sqrt{\epsilon - \beta^2}$, and $R(r) = \rho^m \exp(-\rho) L(\rho)$ in this case, the solution is written by a hypergeometric function explicitly. The case of zero angular momentum requires a special attention. For $m = 0$, Eq.(4) reduces essentially to that for a free particle, with the change of the energy from ϵ to $\epsilon + \beta^2$. The Schrödinger equation is solved trivially, and the solution is written by the Bessel function of order 0, as is well known.

3. Betatron

Feldman and Kahn[2] showed that a quantum mechanical description of the cyclotron motion under a uniform magnetic field B_0 is given by the two-dimensional coherent state representation. The wave function is characterized by two complex-valued parameters, δ and ξ. It is shown that the minimum uncertainty state with width of the quantum mechanical cyclotron radius $l = \sqrt{\hbar/eB_0}$ undergoes a cyclotron motion around $x = \mathrm{Re}\ \xi$, $y = \mathrm{Im}\ \xi$ with radius $|\delta|$ without changing the Gaussian wave form. The frequency of this cyclic motion is the cyclotron frequency $\omega_c = eB_0/m_e$. The classical picture is recovered in the limit $|\delta|/l >> 1$.

Suppose that the beta-field $B(r,t)$ is switched on and increased adiabatically, superimposing on the static uniform field B_0. The beta-field breaks the translational symmetry. We consider such an orbit encircling the origin with radius $|\delta|$. The corresponding wave function is given by the superposition of the eigenstates of the angular momentum m, with the maximum value of the weight taken at $m \simeq |\delta|^2/2l^2$[2]. Under the coexistence of B_0 and $B(r,t)$, the Hamiltonian is given by

$$H(t) = -\frac{\hbar^2}{2m_e}\left\{\frac{d^2}{dr^2} + \frac{1}{r}\frac{d}{dr} - \left(\frac{m}{r} + \frac{r}{2l^2} + \beta(t)\right)^2\right\}, \tag{8}$$

in the subspace of the angular momentum m. A crucial observation is that the effective potential

$$V(r) = \left(\frac{m}{r} + \frac{r}{2l^2} + \beta(t)\right)^2$$

takes the minimum value at $r_m = \sqrt{2m}l$ $(m > 0)$ independently of $\beta(t)$. This is a unique feature of the beta-field. For $m >> 1$, the lowest adiabatic eigenvalue of $H(t)$ is approximately given by

$$E_0^{(m)}(t) \simeq V(r_m) = \frac{\hbar^2}{2m_e}\left(\frac{\sqrt{2m}}{l} + \beta(t)\right)^2, \tag{9}$$

where we neglect the zero point energy estimated by the curvature of $V(r)$ at $r = r_m$. For those states with $|\delta|/l >> 1$, $E_0^{(m)}(t)$ is approximated as $E_0^{(m)}(t) \simeq m\hbar\tilde{\omega}_c(t)$ with $\tilde{\omega}_c(t) = \omega_c(1 + l^2\beta(t)/|\delta|)^2$. Since the initial wave packet of the cyclotron motion is a coherent state given by the sum of the eigenstates of L_z, the wave packet does not change the wave form nor the orbital radius, with only its frequency of circulation increasing as $\beta(t)$ is increased adiabatically.

4. Discussion

It has been shown that the velocity of the cyclotron motion of the quantum mechanical electrons can be increased adiabatically, with its orbital radius kept nearly constant, by the induction of a magnetic field with a specially designed profile. This means that the Hamiltonian (2) has an approximate adiabatic invariant $\langle r \rangle$ for such type of magnetic field. At first sight, this seems paradoxical since the angular momentum is a strictly conserved quantity. It should be noted that the velocity of the electron is given by $\vec{v} = (\vec{p} + e\vec{A})/m_e$ with $\vec{p}$ being the momentum. Therefore, the *kinetic part* of the angular momentum $\tilde{L}_z$ is given by

$$\tilde{L}_z = (\vec{r} \times m_e \vec{v})_z = L_z + erA_\theta. \tag{10}$$

Thus, the increase of the angular momentum of the electromagnetic field is transferred to that of the kinetic part of the angular momentum.

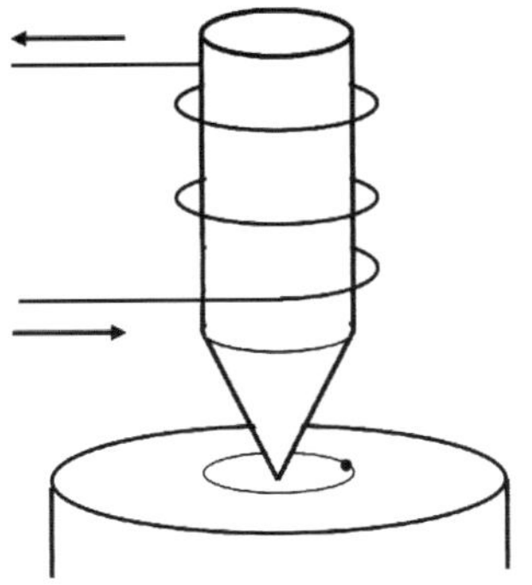

Figure 1. A possible design of the electromagnet that produces the beta-field.

For actual implementation of the quantum betatron, a possible way of producing the beta-field may be to use an electromagnet with a cone-shaped core as a magnetic pole, with its tip of cone nearly touching the sample, as shown in Fig.1. An alternative and more interesting way will be to irradiate a circularly polarized laser pulse focused to the sample which is put on an appropriate magneto-optically active material. The inverse Faraday effect[3] will produce a beta-field with a designed spatio-temporal profile of the magnetic field.

Like classical betatrons, the accelerated electrons in the quantum betatron will emit radiations. It was shown[4] that in the nonrelativistic regime, the emission spectrum has a single peak at the cyclotron frequency. The accelerated frequency $\tilde{\omega}_c$ is given by $\tilde{\omega}_c = \omega_c F$ with the enhancement factor F given by $F = (1 + B_\delta/B_0)^2$, where B_δ is the intensity of the beta-field at the orbit, $B_\delta = C/|\delta|$. Therefore, $\tilde{\omega}_c$ is higher for the electrons circulating closer to the origin. The quantum betatron will be a source of broad band radiation in mesoscopic scale.

Acknowledgments

The author thanks Prof. Suemoto for valuable information on the inverse Faraday effect. This work was partially supported by the Grant-in-Aid for Scientific Research (B) from Japan Society for the Promotion of Science (Grant No. 20340079).

References

[1] Kerst W D *Phys. Rev.* **60**, 47 (1941).
[2] Feldman A and Kahn A *Phys. Rev.* B **1**, 4584 (1970).
[3] Kimel A V, Kirilyuk A, Usachev P A, Pisarev R V, Balbashov A M and Rasing Th *Nature* **435**, 655 (2005).
[4] Schwinger J *Phys. Rev.* **75**, 1912 (1949).

$\mathcal{N}=2$ Instanton Effective action in Ω-background and D3/D(-1)-brane System in R-R background

K Ito[1], H Nakajima[2,3], T Saka[1] and S Sasaki[1]

[1] Department of Physics, Tokyo Institute of Technology, Tokyo, 152-8551, Japan
[2] School of Physics, Korea Institute for Advanced Study, Seoul, 130-722, Korea
[3] Department of Physics, Kyungpook National University, Teagu, 702-701, Korea

E-mail: ito@th.phys.titech.ac.jp, nakajima@kias.re.kr, bstsaka@th.phys.titech.ac.jp, shin-s@th.phys.titech.ac.jp

Abstract. We study the relation between the ADHM construction of instantons in the Ω-background and the fractional D3/D(-1)-branes at the orbifold singularity of $\mathbb{C}\times\mathbb{C}^2/\mathbb{Z}_2$ in Ramond-Ramond (R-R) 3-form field strength background. We calculate disk amplitudes of open strings connecting the D3/D(-1)-branes in certain R-R background to obtain the D(-1)-brane effective action deformed by the R-R background. We show that the deformed D(-1)-brane effective action agrees with the instanton effective action in the Ω-background.

1. Introduction

On the study of the instanton effects for the gauge theory, it is shown in [1] that the deformations of the theory are very useful in the calculation of the instanton partition function for $\mathcal{N}=2$ supersymmetric Yang-Mills theory using the localization technique, which is called the Nekrasov's formula. The noncommutativity of the gauge theory resolves the the small instanton singularity of the instanton moduli space [2], and the Ω-background isolates the fixed points of BRST nilpotent operator, which is obtained by the topological twist of the theory [3].

In order to apply Nekrasov's formula to various gauge theories, it would be important to realize the Ω-background in superstring theory. Recently, the low-enegy effective theories on D-branes in the constant R-R 3-form field strength background have been attracted much attentions in study of non-perturbative effects in supersymmetric gauge theories. For example, the field theory on D-branes with NS-NS B-field background is described by the noncommutative gauge theory [4, 5]. The noncommutativity of the gauge theory resolves the small instanton singularity of the instanton moduli space [2], which helps the calculation of the instanton partition function via localization technique [1]. In [6], the authors studied the low-energy effective action of D(-1)-branes in the D3-D(-1) system located at the singularity of the orbifold $\mathbb{C}^2/\mathbb{Z}_2$ in some constant self-dual R-R 3-form background and found that the effective action of the D(-1)-branes coincides with the effective action for the instanton moduli, which is called the "instanton effective action", of $\mathcal{N}=2$ super Yang-Mills theory in the self-dual Ω-background.

Follwoing these results, in this work [7] we will study the instanton effective action in general (non-(anti-)self-dual) Ω-background and its relations to the D3/D(-1)-brane system in certain R-R 3-form background. First we we introduce four-dimensional $\mathcal{N}=2$ super Yang-Mills theory in the Ω-background and study the ADHM construction of instantons. We compute the instanton effective action in the Ω-background and show that the action is exact under the deformed supersymmetry. Next we investigate general R-R background and calculate the disk amplitudes in the presence of the R-R 3-form background. Based on the amplitudes, we construct the effective action of D(−1)-branes in the fractional D3/D(−1)-brane system.

2. $\mathcal{N}=2$ instanton effective action in Ω-background

The Lagrangian for $\mathcal{N}=2$ super Yang-Mills theory is given by

$$\mathcal{L}_0 = \frac{1}{\kappa}\mathrm{Tr}\Big[\frac{1}{4}F_{mn}F^{mn} - \frac{i\theta g^2}{32\pi^2}F_{mn}\tilde{F}^{mn} + \Lambda^I\sigma^m D_m\bar{\Lambda}_I + (D_m\varphi)D^m\bar{\varphi} \\ - i\frac{g}{\sqrt{2}}\Lambda^I[\bar{\varphi},\Lambda_I] + i\frac{g}{\sqrt{2}}\bar{\Lambda}_I[\varphi,\bar{\Lambda}^I] + \frac{g^2}{2}[\varphi,\bar{\varphi}]^2\Big], \tag{1}$$

where $F_{mn} = \partial_m A_n - \partial_n A_m + ig[A_m, A_n]$ is the gauge field strength of $U(N)$ gauge field A_m ($m = 1,2,3,4$), Λ^I_α, $\bar{\Lambda}^I_{\dot{\alpha}}$ are Weyl fermions , φ and $\bar{\varphi}$ are complex scalars, g is the gauge coupling constant and $D_m = \partial_m + ig[A_m, *]$ is the gauge covariant derivative. We also define the Dirac matrices $\sigma_m = (i\tau^1, i\tau^2, i\tau^3, 1)$ and $\bar{\sigma}_m = (-i\tau^1, -i\tau^2, -i\tau^3, 1)$, where τ^c ($c = 1,2,3$) are the Pauli matrices. σ^{mn} and $\bar{\sigma}^{mn}$ are the Lorentz generators defined by

$$(\sigma^{mn})_\alpha{}^\beta = \frac{1}{4}\big(\sigma^m_{\alpha\dot{\alpha}}\bar{\sigma}^{n\dot{\alpha}\beta} - \sigma^n_{\alpha\dot{\alpha}}\bar{\sigma}^{m\dot{\alpha}\beta}\big), \quad (\bar{\sigma}^{mn})^{\dot{\alpha}}{}_{\dot{\beta}} = \frac{1}{4}\big(\bar{\sigma}^{m\dot{\alpha}\alpha}\sigma^n_{\alpha\dot{\beta}} - \bar{\sigma}^{n\dot{\alpha}\alpha}\sigma^m_{\alpha\dot{\beta}}\big). \tag{2}$$

θ is the theta-angle and $\tilde{F}_{mn} = \frac{1}{2}\epsilon_{mnpq}F^{pq}$ is the dual of F_{mn}. The projection of the field strength into the (anti-)self-dual part is given by $F^{\pm}_{mn} = \frac{1}{2}(F_{mn} \pm \tilde{F}_{mn})$. The second term of (1) is topological and the instanton number is defined by

$$k = \int d^4x \, \frac{1}{\kappa}\mathrm{Tr}\Big[\frac{g^2}{32\pi^2}F_{mn}\tilde{F}^{mn}\Big]. \tag{3}$$

Four-dimensional $\mathcal{N}=2$ super Yang-Mills theory in Ω-background is obtained by dimensional reduction of six-dimensional $\mathcal{N}=1$ super Yang-Mills theory. The Ω-background [8, 9, 1, 10, 11] is defined by a nontrivial fibration of $\mathbb{R}^4$ over T^2. We denote the complex coordinates of a two-dimensional torus T^2 by $(z, \bar{z})$, The six-dimensional metric is given by

$$ds^2_{6\mathrm{D}} = 2d\bar{z}dz + (dx^m + \Omega^m d\bar{z} + \bar{\Omega}^m dz)^2, \tag{4}$$

where $\Omega^m = \Omega^{mn}x_n$, $\bar{\Omega}^m = \bar{\Omega}^{mn}x_n$. Ω^{mn} and $\bar{\Omega}^{mn}$ are constant antisymmetric matrices. We require that the metric (4) is flat. This leads to the condition that Ω^{mn} and $\bar{\Omega}^{mn}$ commute with each other:

$$\Omega^{mn}\bar{\Omega}_{np} - \bar{\Omega}^{mn}\Omega_{np} = 0. \tag{5}$$

We also introduce the R-symmetry Wilson line by gauging $SU(2)_I$ R-symmetry as

$$\boldsymbol{A}^I{}_J = \mathcal{A}^I{}_J d\bar{z} + \bar{\mathcal{A}}^I{}_J dz, \tag{6}$$

where we consider the case such that the R-symmetry gauge fields $\mathcal{A}^I{}_J$, $\bar{\mathcal{A}}^I{}_J$ are constant.

The instanton effective action for the action (1) in the Ω-background is obtained by deriving the instanton solution for the equations of motion for the action using the ADHM construction [12, 13] and substituting the solution to the action. In order to solve the equations of motion, we developed the deformed ADHM construction of the instanton solution. The result of the instanton effective action is

$$
\begin{aligned}
S^{(0)}_{\text{eff}} = \frac{2\pi^2}{\kappa}\text{tr}_k\Big[& -2\Big([\bar{\chi}, a'_m] - i\bar{\Omega}_{mn}a'^n\Big)\Big([\chi, a'^m] - i\Omega^{mp}a'_p\Big) \\
& + (\bar{\chi}\bar{w}^{\dot{\alpha}} - \bar{w}^{\dot{\alpha}}\bar{\phi}^0)(w_{\dot{\alpha}}\chi - \phi^0 w_{\dot{\alpha}}) + (\chi\bar{w}^{\dot{\alpha}} - \bar{w}^{\dot{\alpha}}\phi^0)(w_{\dot{\alpha}}\bar{\chi} - \bar{\phi}^0 w_{\dot{\alpha}}) \\
& + \frac{i}{2\sqrt{2}}\mathcal{M}'^{\alpha I}\epsilon_{IJ}\Big([\bar{\chi}, \mathcal{M}'^J_\alpha] - \frac{i}{2}\bar{\Omega}^+_{mn}(\sigma^{mn})_\alpha{}^\beta\mathcal{M}'^J_\beta\Big) \\
& - \frac{i}{\sqrt{2}}\bar{\mu}^I\epsilon_{IJ}(\mu^J\bar{\chi} - \bar{\phi}^0\mu^J) + \frac{1}{\sqrt{2}}\bar{A}^J{}_I\Big(\bar{\mu}^I\mu_J + \frac{1}{2}\mathcal{M}'^{\alpha I}\mathcal{M}'_{\alpha J}\Big) \\
& + i\Omega^-_{mn}(\bar{\sigma}^{mn})^{\dot{\alpha}}{}_{\dot{\beta}}\bar{w}^{\dot{\beta}}\bar{\phi}^0 w_{\dot{\alpha}} + i\bar{\Omega}^-_{mn}(\bar{\sigma}^{mn})^{\dot{\alpha}}{}_{\dot{\beta}}\bar{w}^{\dot{\beta}}\phi^0 w_{\dot{\alpha}} \\
& - 2i\Omega^{-mn}\bar{\chi}[a'_m, a'_n] - 2i\bar{\Omega}^{-mn}\chi[a'_m, a'_n] + \frac{1}{2}\Omega^{-mn}\bar{\Omega}^-_{mn}\bar{w}^{\dot{\alpha}}w_{\dot{\alpha}} \\
& - i\bar{\psi}^{\dot{\alpha}}_I(\bar{\mu}^I w_{\dot{\alpha}} + \bar{w}_{\dot{\alpha}}\mu^I + [\mathcal{M}'^{\alpha I}, a'_{\alpha\dot{\alpha}}]) \\
& + iD^c(\tau^c)^{\dot{\alpha}}{}_{\dot{\beta}}(\bar{w}^{\dot{\beta}}w_{\dot{\alpha}} + \bar{a}'^{\dot{\beta}\alpha}a'_{\alpha\dot{\alpha}})\Big].
\end{aligned} \tag{7}
$$

This instanton effective action has an equivariantly nilpotent operator, which is needed to obtain the instanton partition function using the localization formula.

3. Effective action of D3/D(−1)-brane system in R-R background

In the context of the string theory, the instanton effective action of 4d $\mathcal{N}=2$ $U(N)$ super Yang-Mills theory is equivalent to the effective action of D3/D(−1)-brane system [14, 15] located at the fixed point of the orbifold $\mathbb{C}\times\mathbb{C}^2/\mathbb{Z}_2$ [6]. In order to consider deformations to the D(−1) effective action, we incorporate R-R 3-form flux. In [6], the case of the constant self-dual R-R 3-form background $\mathcal{F}_{mna}$ $(m,n=1,\cdots,4, a=5,6)$ in type IIB string theory has been studied. By identifying the self-dual R-R backgrounds with the self-dual Ω-background parameters with $\Omega_{mn}=\Omega^+_{mn}, \bar{\Omega}_{mn}=\bar{\Omega}^+_{mn}$, the authors showed that the deformed low-energy effective action of D3/D(−1)-branes coincides with the instanton effective action in $\mathcal{N}=2$ super Yang-Mills theory in the self-dual Ω-background. In order to realize the deformed instanton effective action in the general Ω-background, we need to turn on non-(anti-)self-dual R-R 3-form flux with indices $\mathcal{F}_{mna}$ and $\mathcal{F}_{abc}$. Computing the disk amplitudes of strings connecting the branes and collecting the non-vanishing amplitudes under the field theory limit, we obtain the following effective action :

$$
\begin{aligned}
S_{\text{str}}(C,\bar{C},m) = \frac{2\pi^2}{\kappa}\text{tr}_k\Big[& -2\left([\bar{\chi}, a'_m] + \bar{C}_{mn}a'^n\right)\left([\chi, a'^m] + C^{mk}a'_k\right) \\
& + \frac{i}{2\sqrt{2}}\mathcal{M}'^{\alpha I}\left([\bar{\chi}, \mathcal{M}'_{\alpha I}] + \frac{1}{2}\bar{C}^{+mn}(\sigma_{mn})_\alpha{}^\beta\mathcal{M}'_{\beta I} + 2\sqrt{2}im_{(IJ)}\mathcal{M}'_\alpha{}^J\right) \\
& + \left(\bar{\chi}\bar{w}^{\dot{\alpha}} - \bar{w}^{\dot{\alpha}}\bar{\phi}^0 + \frac{1}{2}\bar{C}^{-mn}(\bar{\sigma}_{mn})^{\dot{\alpha}}{}_{\dot{\beta}}\bar{w}^{\dot{\beta}}\right)\left(w_{\dot{\alpha}}\chi - \phi^0 w_{\dot{\alpha}} + \frac{1}{2}C^{-mn}(\bar{\sigma}_{mn})^{\dot{\gamma}}{}_{\dot{\alpha}}w_{\dot{\gamma}}\right) \\
& + \left(\chi\bar{w}^{\dot{\alpha}} - \bar{w}^{\dot{\alpha}}\bar{\phi}^0 + \frac{1}{2}C^{-mn}(\bar{\sigma}_{mn})^{\dot{\alpha}}{}_{\dot{\beta}}\bar{w}^{\dot{\beta}}\right)\left(w_{\dot{\alpha}}\bar{\chi} - \bar{\phi}^0 w_{\dot{\alpha}} + \frac{1}{2}\bar{C}^{-mn}(\bar{\sigma}_{mn})^{\dot{\gamma}}{}_{\dot{\alpha}}w_{\dot{\gamma}}\right) \\
& + \frac{i}{\sqrt{2}}\bar{\mu}^I\left(-\mu_I\bar{\chi} + \bar{\phi}^0\mu_I + 2\sqrt{2}i\mu^J m_{(IJ)}\right) \\
& - i\bar{\psi}^{\dot{\alpha}}_I(\bar{\mu}^I w_{\dot{\alpha}} + \bar{w}_{\dot{\alpha}}\mu^I + [\mathcal{M}'^{\alpha I}, a'_{\alpha\dot{\alpha}}]) \\
& + iD^c(\tau^c)^{\dot{\alpha}}{}_{\dot{\beta}}(\bar{w}^{\dot{\beta}}w_{\dot{\alpha}} + \bar{a}'^{\dot{\beta}\alpha}a'_{\alpha\dot{\alpha}})\Big],
\end{aligned} \tag{8}
$$

where we have defined $C_{mn} = C^{+}_{mn} + C^{-}_{mn}$, $\bar{C}_{mn} = \bar{C}^{+}_{mn} + \bar{C}^{-}_{mn}$ and

$$\begin{aligned} C^{+mn} &= -2\pi\sqrt{2}(2\pi\alpha')^{\frac{1}{2}}\epsilon_{\beta\gamma}(\sigma^{mn})_{\alpha}{}^{\gamma}\epsilon_{IJ}\mathcal{F}^{(\alpha\beta)[IJ]}, \\ C^{-mn} &= -2\pi\sqrt{2}(2\pi\alpha')^{\frac{1}{2}}\epsilon^{\dot{\beta}\dot{\gamma}}(\bar{\sigma}^{mn})^{\dot{\alpha}}{}_{\dot{\gamma}}\epsilon^{I'J'}\mathcal{F}_{(\dot{\alpha}\dot{\beta})[I'J']}, \\ \bar{C}^{+mn} &= -2\pi\sqrt{2}(2\pi\alpha')^{\frac{1}{2}}\epsilon_{\beta\gamma}(\sigma^{mn})_{\alpha}{}^{\gamma}\epsilon_{I'J'}\overline{\mathcal{F}}^{(\alpha\beta)[I'J']}, \\ \bar{C}^{-mn} &= -2\pi\sqrt{2}(2\pi\alpha')^{\frac{1}{2}}\epsilon^{\dot{\beta}\dot{\gamma}}(\bar{\sigma}^{mn})^{\dot{\alpha}}{}_{\dot{\gamma}}\epsilon^{IJ}\overline{\mathcal{F}}_{(\dot{\alpha}\dot{\beta})[IJ]}, \\ m_{(IJ)} &= \pi i(2\pi\alpha')^{\frac{1}{2}}\bar{\mathcal{F}}_{[\dot{\alpha}\dot{\beta}](IJ)}\epsilon^{\dot{\alpha}\dot{\beta}}. \end{aligned} \tag{9}$$

Comparing the actions (8) with (7), they coincide if we identify $C^{mn} = -i\Omega^{mn}, \bar{C}^{mn} = -i\bar{\Omega}^{mn}$ and

$$m_{(IJ)} = -\frac{1}{2\sqrt{2}}\epsilon_{IK}\bar{A}^{K}{}_{J} \tag{10}$$

then redefine D^c in (7) as

$$D^c \to D^c - \frac{i}{2}\bar{\eta}^{c}_{mn}(\bar{\Omega}^{-mn}\chi + \Omega^{-mn}\bar{\chi}). \tag{11}$$

4. Conclusion

In this paper, we have studied general (non-(anti-)self-dual) Ω-background deformation of the instanton effective action of $\mathcal{N} = 2$ super Yang-Mills theory and the R-R 3-form background deformation of the D(-1)-brane effective action. We have found that these two effective actions coincide under the appropriate identification of the deformation parameters.

Acknowledgments
The work of H. N. is supported by Mid-career Researcher Program through the National Research Foundation of Korea(NRF) grant funded by the Korea government(MEST)(No. 2009-0084601). The work of T. S. is supported by the Global Center of Excellence Program by MEXT, Japan through the "Nanoscience and Quantum Physics" Project of the Tokyo Institute of Technology, and by Iwanami Fujukai Foundation. The work of S. S. is supported by the Japan Society for the Promotion of Science (JSPS) Research Fellowship.

References

[1] Nekrasov N A 2004 *Adv. Theor. Math. Phys.* **7** 831
[2] Nekrasov N and Schwarz A S 1998 *Commun. Math. Phys.* **198** 689
[3] Witten E 1988 *Commun. Math. Phys.* **117** 353
[4] Chu C S and Ho P M 1999 *Nucl. Phys.* B **550** 151
[5] Seiberg N and Witten E 1999 String theory and noncommutative geometry *J. High Energy Phys.* JHEP09(1999)032
[6] Billó M, Frau M, Fucito F and Lerda A 2006 Instanton calculus in R-R background and the topological string *J. High Energy Phys.* JHEP11(2006)012
[7] Ito K, Nakajima H, Saka T and Sasaki S 2010 $\mathcal{N} = 2$ Instanton Effective action in Ω-background and D3/D(-1)-brane System in R-R background *J. High Energy Phys.* JHEP11(2010)093
[8] Moore G W, Nekrasov N and Shatashvili S 2007 *Commun. Math. Phys.* **209** 97
[9] Losev A, Nekrasov N and Shatashvili S L 1998 Testing Seiberg-Witten Solution *Preprint* hep-th/9801061
[10] Losev A S, Marshakov A and Nekrasov N A 2003 Small Instantons, Little Strings and Free Fermions *Preprint* hep-th/0302191
[11] Nekrasov N and Okounkov A 2003 Seiberg-Witten Theory and Random Partitions *Preprint* hep-th/0306238
[12] Atiyah M F, Hitchin N J, Drinfeld V G and Manin Y I 1978 *Phys. Lett.* A **65** 185
[13] Dorey N, Hollowood T J, Khoze V V and Mattis M P 2002 *Phys. Rept.* **371** 231
[14] Douglas M R 1995 Branes within Branes *Preprint* hep-th/9512077
[15] Witten E 1995 *J. Geom. Phys.* **15** *(1995) 215*

Bifurcation phenomena of mean-field coupled self-sustained oscillators with two different time scales under the influence of external white noise

Keiji Okumura and Masatoshi Shiino

Department of Physics, Faculty of Science, Tokyo Institute of Technology, Meguro-ku, Tokyo 152-8551, Japan

E-mail: kokumura@mikan.ap.titech.ac.jp

Abstract. We report stochastic phenomena of inter-cluster synchronization in ensembles of coupled self-sustained oscillators with different time scales. A mean-field model of two inter-connected clusters of coupled limit cycle oscillators having same trajectories with two native frequencies is considered to *analytically* derive the time evolution of the order parameters in the thermodynamic limit. A set of the order parameters obtained for the system exhibits noise induced bifurcations, including those from torus type attractors to fixed point ones for the case with two different heat baths in accordance with the introduction of the time scale.

1. Introduction

Synchronization phenomena are ubiquitous ones observed in many fields of natural sciences [1]. While effects of noise on synchronization phenomena in oscillatory systems are breakdown of synchronization, they are widely reported to exhibit opposite effects on synchronization [2, 3]. As far as we know, there are few papers which study the outcome of considering same trajectories with *different* native frequencies within the frame work of noise induced synchronization. In this paper, we study the relationships between noise influence and synchronization phenomena in ensembles of two clusters of globally coupled limit cycle oscillators with different time scales. A nonlinear Fokker-Planck equation (NFPE) approach to our model, which turns out to be noise level-free analysis, is applied to derive the time evolution equation of the order parameters in the thermodynamic limit. Noting that our basic oscillator model includes an excitable element model with changes in the parameters, observations of noise induced synchronization in our system will be of practical interest.

2. Model and nonlinear Fokker-Planck equation approach

We present a simple model where the ensemble consists of two inter-connected clusters of mean-field coupled limit cycle oscillators having same trajectories with two native frequencies [4, 5, 6]:

$$\frac{\mathrm{d}z_{1i}^{(x)}}{\mathrm{d}t} = -a^{(x)}z_{1i}^{(x)} + \frac{J^{(x)}}{N}\sum_{j=1}^{N} F_1^{(x)}(b^{(x,x)}z_{1j}^{(x)} + b^{(x,y)}z_{1j}^{(y)}) + I$$

Published under licence by IOP Publishing Ltd

$$+\epsilon\frac{K^{(x)}}{N}\sum_{k=1}^{N}F_2^{(x)}(b^{(x,x)}z_{2k}^{(x)}+b^{(x,y)}z_{2k}^{(y)})+\eta_{1i}^{(x)}(t), \tag{1}$$

$$\frac{dz_{1i}^{(y)}}{dt} = \kappa(-a^{(y)}z_{1i}^{(y)}+\frac{J^{(y)}}{N}\sum_{j=1}^{N}F_1^{(y)}(b^{(y,x)}z_{1j}^{(x)}+b^{(y,y)}z_{1j}^{(y)})$$

$$+\epsilon\frac{K^{(y)}}{N}\sum_{k=1}^{N}F_2^{(y)}(b^{(y,x)}z_{2k}^{(x)}+b^{(y,y)}z_{2k}^{(y)}))+\eta_{1i}^{(y)}(t), \tag{2}$$

$$\frac{dz_{2i}^{(x)}}{dt} = \tau(-a^{(x)}z_{2i}^{(x)}+\frac{J^{(x)}}{N}\sum_{j=1}^{N}F_2^{(x)}(b^{(x,x)}z_{2j}^{(x)}+b^{(x,y)}z_{2j}^{(y)})+I$$

$$+\epsilon\frac{K^{(x)}}{N}\sum_{k=1}^{N}F_1^{(x)}(b^{(x,x)}z_{1k}^{(x)}+b^{(x,y)}z_{1k}^{(y)}))+\eta_{2i}^{(x)}(t), \tag{3}$$

$$\frac{dz_{2i}^{(y)}}{dt} = \tau\kappa(-a^{(y)}z_{2i}^{(y)}+\frac{J^{(y)}}{N}\sum_{j=1}^{N}F_2^{(y)}(b^{(y,x)}z_{2j}^{(x)}+b^{(y,y)}z_{2j}^{(y)})$$

$$+\epsilon\frac{K^{(y)}}{N}\sum_{k=1}^{N}F_1^{(y)}(b^{(y,x)}z_{1k}^{(x)}+b^{(y,y)}z_{1k}^{(y)}))+\eta_{2i}^{(y)}(t), \tag{4}$$

where $z_{\alpha i}^{(\mu)}$ $(\mu = x, y)$ $(\alpha = 1, 2)$ are the dynamical variables of the two dimensional oscillators at site i $(i = 1, \cdots, N)$ of cluster α, $a^{(\mu)}$ and $b^{(\mu,\nu)}$ are constants, $J^{(\mu)}$ and $K^{(\mu)}$ are intra- and inter-cluster coupling strengths, $F_\alpha^{(\mu)}(\cdot)$ are nonlinear coupling functions, and $\eta_{\alpha i}^{(\mu)}(t)$ are independent external noise, respectively. Due to the introduction of the time scale parameter τ, the native frequency of each oscillator in cluster 2 can be different from that of each oscillator in cluster 1, in spite of their same trajectories. The inter-cluster coupling strength ϵ may take any real constant value, not limited to weak connections. The external noise are of independent white Gaussian type, $\langle\eta_{\alpha i}^{(\mu)}(t)\rangle = 0$, $\langle\eta_{\alpha i}^{(\mu)}(t)\eta_{\beta j}^{(\nu)}(t')\rangle = 2D_\alpha^{(\mu)}\delta_{ij}\delta_{\alpha\beta}\delta_{\mu\nu}\delta(t-t')$.

Applying an NFPE approach to this model [4, 5, 6], we derive the time evolution of each moment and obtain a set of closed ordinary differential equations involving at most second moments of the Gaussian probability density ensured for large times by the H theorem:

$$\frac{d\langle z_1^{(x)}\rangle_G}{dt} = -a^{(x)}\langle z_1^{(x)}\rangle_G + J^{(x)}\langle F_1^{(x)}\rangle_G + I + \epsilon K^{(x)}\langle F_2^{(x)}\rangle_G, \tag{5}$$

$$\frac{d\langle z_1^{(y)}\rangle_G}{dt} = \kappa(-a^{(y)}\langle z_1^{(y)}\rangle_G + J^{(y)}\langle F_1^{(y)}\rangle_G + \epsilon K^{(y)}\langle F_2^{(y)}\rangle_G), \tag{6}$$

$$\frac{d\langle z_2^{(x)}\rangle_G}{dt} = \tau(-a^{(x)}\langle z_2^{(x)}\rangle_G + J^{(x)}\langle F_2^{(x)}\rangle_G + I + \epsilon K^{(x)}\langle F_1^{(x)}\rangle_G), \tag{7}$$

$$\frac{d\langle z_2^{(y)}\rangle_G}{dt} = \tau\kappa(-a^{(y)}\langle z_2^{(y)}\rangle_G + J^{(y)}\langle F_2^{(y)}\rangle_G + \epsilon K^{(y)}\langle F_1^{(y)}\rangle_G). \tag{8}$$

We note that $\langle u_\alpha^{(\mu)^2}\rangle_G \to D_\alpha^{(\mu)}/\{a^{(\mu)}[(\tau-1)\delta_{2\alpha}+1][(\kappa-1)\delta_{y\mu}+1]\}$ and $\langle u_\alpha^{(\mu)}u_\beta^{(\nu)}\rangle_G \to 0\,(t\to\infty)$.

To investigate qualitative dynamical behavior of the system, we specify the coupling functions as $F_\alpha^{(x)}(z) = z\exp\left(-\frac{z^2}{2}\right)$ and $F_\alpha^{(y)}(z) = z$, which reflects excitable features of neural oscillators [5]. Then, one has

$$\langle F_\alpha^{(x)}\rangle_G = \frac{m_\alpha^{(x)}}{(\sigma_\alpha^2+1)^{3/2}}\exp\left[-\frac{m_\alpha^{(x)^2}}{2(\sigma_\alpha^2+1)}\right],\quad \langle F_\alpha^{(y)}\rangle_G = m_\alpha^{(y)}, \tag{9}$$

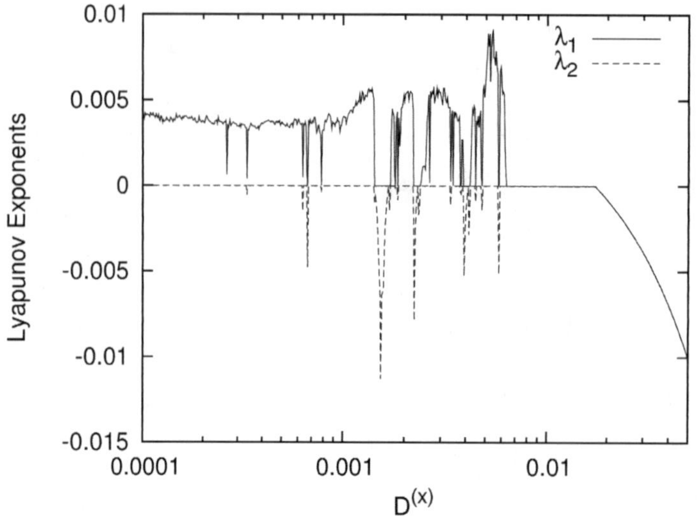

Figure 1. Dependence of the first and second Lyapunov exponents λ_1, λ_2 on the Langevin noise intensity $D_1^{(x)} = D_2^{(x)}/\tau = D^{(x)}$. With an increase of noise intensity, the qualitative changes in the attractors formed in the order parameter system are observed, involving the appearance and disappearance of the inter-cluster synchronization phenomena. In this system, the occurrence of the bifurcations from torus type attractors to fixed point ones is worth noting. The model parameter values are $a^{(x)} = 1.6875$, $a^{(y)} = 2.0925$, $b^{(x,x)} = 0.87750$, $b^{(x,y)} = -0.33750$, $b^{(y,x)} = 2.7000$, $b^{(y,y)} = -0.67500$, $J^{(x)} = K^{(x)} = 2.5650$, $J^{(y)} = K^{(y)} = 2.835$, $\kappa = 0.0068000$, $I = -2.0$, $D^{(y)} = 0$, $\epsilon = 0.0012$, $\tau = 0.9$, $D_\alpha^{(y)} = 0$.

where $m_\alpha^{(\mu)} = b^{(\mu,x)}\langle z_\alpha^{(x)}\rangle_G + b^{(\mu,y)}\langle z_\alpha^{(y)}\rangle_G, \sigma_\alpha^2 = b^{(x,x)^2}\langle u_\alpha^{(x)^2}\rangle_G + b^{(x,y)^2}\langle u_\alpha^{(y)^2}\rangle_G$.

3. Nonequilibrium phase transitions including stochastic inter-cluster synchronization

Using an NFPE approach, we have reduced the system of two clusters of mean-field coupled limit cycle oscillators from the set of the Langevin equations (1) - (4) to the moment equations (5) - (8). Investigating nonlinear aspects of the obtained reduced system, we can reveal the effects of noise, frequency differences, and coupling strengths on inter-cluster synchronization.

We performed numerical calculations with the fourth-order Runge-Kutta method. Stochastic phenomena of inter-cluster synchronization associated with nonequilibrium phase transitions are systematically investigated with changes in the inter-cluster coupling strength ϵ, the time scale τ , and the Langevin noise intensity $D_1^{(x)} = D_2^{(x)}/\tau = D^{(x)}$ which corresponds to the introduction of the two different heat baths based on the time scale τ.

The most interesting result of stochastic phenomena in this system is the noise induced bifurcations from torus type attractors to fixed point ones as shown in Figs. 1 and 2. Conducting a linear stability analysis of fixed points near the bifurcation point $D^{(x)} = 0.017378$, we show that eigenvalues of the Jacobian matrix of the order parameter equations (5) - (8) are two pairs of purely imaginary numbers. Since our model setting of the mean-field coupled oscillators having same trajectories with two native frequencies and two heat baths in accordance with the time scale contribute to this kind of bifurcation structure, it may not be frequently observed in other systems. The typical case of noise induced inter-cluster synchronization phenomena is also observed, as were shown by previously reported work [6].

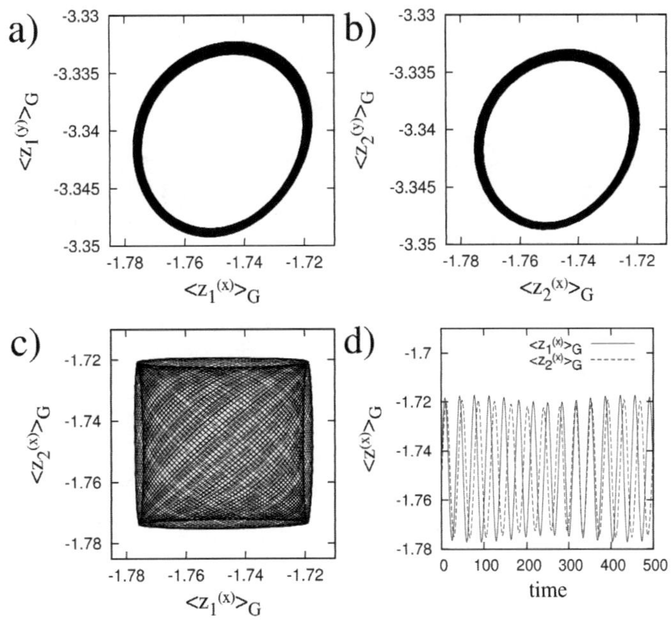

Figure 2. The behavior of a torus type oscillation in the order parameter system corresponding to Fig. 1 with $D^{(x)} = 0.017060$. a), b), c) The trajectories in phase space projected to the $\langle z_1^{(x)}\rangle_G$-$\langle z_1^{(y)}\rangle_G$, $\langle z_2^{(x)}\rangle_G$-$\langle z_2^{(y)}\rangle_G$, and $\langle z_1^{(x)}\rangle_G$-$\langle z_2^{(x)}\rangle_G$ planes, respectively. d) The time evolution of the order parameters of the system. Sizes of the torus type attractors are very small and shapes of those are almost circular near the bifurcation point.

4. Concluding remarks

We have shown the effects of the independent white Gaussian noise on synchronization phenomena in ensembles of limit cycle oscillators having same trajectories with different native frequencies. We have considered a mean-field model of two clusters of coupled limit cycle oscillators with different time scales. Using an NFPE approach, we have analytically reduced the system from a set of the Langevin equations to a set of ordinary differential equations of the order parameters. This noise level-free analysis has allowed us to reveal various interesting bifurcations including the inter-cluster synchronization and torus type attractors by external noise.

Acknowledgments

K.O. acknowledges the financial support from the Global Center of Excellence Program by MEXT, Japan through the "Nanoscience and Quantum Physics" Project of the Tokyo Institute of Technology.

References

[1] Pikovsky A, Rosenblum M, Kurths J 2001 *Synchronization: A universal concept in nonlinear sciences* (University Press, Cambridge)
[2] Marella S, Ermentrout G B 2008 *Phys. Rev. E* **77** 041918
[3] Nagai K H, Kori H 2010 *Phys. Rev. E* **81** 065202
[4] Shiino M, Yoshida K 2001 *Phys. Rev. E* **63** 026210
[5] Okumura K, Shiino M 2010 *17th International Conference on Neural Information Processing* **6443** 486
[6] Okumura K, Ichiki A, Shiino M 2010 *Europhys. Lett.* **92** 50009

Boundary Conditions in One-dimensional Tunneling Junction

Yutaka Shikano[1,2,*] **and Masao Hirokawa**[3,†]

[1]Department of Physics, Tokyo Institute of Technology, 2-12-1 Oh-Okayama, Meguro, Tokyo 152-8551 Japan
[2]Department of Mechanical Engineering, Massachusetts Institute of Technology, 77 Massachusetts Avenue, Cambridge, MA 02139 USA
[3]Department of Mathematics, Okayama University, 3-1-1 Tsushima-Naka, Kita, Okayama 700-8530, Japan

E-mail: *`shikano@th.phys.titech.ac.jp`, †`hirokawa@math.okayama-u.ac.jp`

Abstract. The tunneling effect is first characterized in quantum mechanics. However, the mathematical analysis for the tunneling effect is so complicated. In the conventional sense, the WKB analysis is useful to analyze such effect. Here, we consider the mathematical analysis for the tunneling effect in the one-dimensional Schrödinger system from the viewpoint of functional analysis. From the rigorous analysis, we obtain the interference pattern of the tunneling particle.

1. Introduction

The tunneling effect is first appeared in quantum mechanics but cannot be characterized in classical mechanics. This effect is one of the purely quantum-mechanical effects. However, almost all cases cannot be analytically solved. On considering the tunneling effect, the WKB analysis is often used. This analysis is not rigorous in mathematics since almost all solutions are not converged but this can be well described in physical phenomena. According to the WKB analysis, the interference pattern has been appeared for the tunneling particle. To give the analytical meaning in mathematics, the exact WKB analysis was proposed to be applied the Borel summation method to the WKB solution, which is the solution in the case that the Planck constant is a sufficient small parameter [1]. Even in the exact WKB analysis, the phase factor for the WKB solution is in general appeared on the tunneling. In Ref. [2], we have considered the role of the phase factor from the viewpoint of mathematical physics.

The aim of this paper is to make clear the physical meaning of the results of Ref. [2]. Therefore, some mathematical definitions are not strictly defined in this paper.

2. Preliminaries

2.1. Our setup

In this subsection, our setup is assumed. The Schrödinger particle on the line is to be tunneled in the junction $[-\Lambda, \Lambda]$. Our question is what behavior of the tunneling particle is from the mathematical constraint. In the case of the non-tunneling particle, this only gets the global phase, see more details in Ref. [2]. The potential energy is infinity on the junction and free on the other region. Therefore, the Hamiltonian is mathematically defined as

$$\hat{H} := -\frac{d^2}{dx^2} \text{ on } \Omega_\Lambda := (-\infty, -\Lambda) \cup (\Lambda, \infty). \tag{1}$$

The degrees of the freedoms at the edges of the junction are the wavefunction and its derivative defined on the edges, that is, $U(2) \times U(2)$. However, the tunneling effect can be represented as the unitary transform $U(1)$. Therefore, the total degree of freedom is $U(2) \times U(2) \setminus U(1)$. This schematic figure is illustrated in Fig. 1.

From the viewpoint of physics, the tunneling effect cannot be occurred in this situation since the potential energy on the junction is infinity. However, our situation is also satisfied in the case of the point junction, $\Lambda \to 0+$. This case corresponds to Ref. [3] to be analyzed in the quantum graph theory. It should be noted that our result is consistent with one of Ref. [3] in the case of $\Lambda \to 0+$. The meaning of the mathematical generalization from the point junction is not only mathematical beauty but also to give the analytic solution for the simple model to be extended to the more realistic models, *e.g.*, see quantum wires and fractional quantum Hall devices in Ref. [4]. From the conventional approach without the WKB analysis, we cannot obtain the interference pattern after the tunneling.

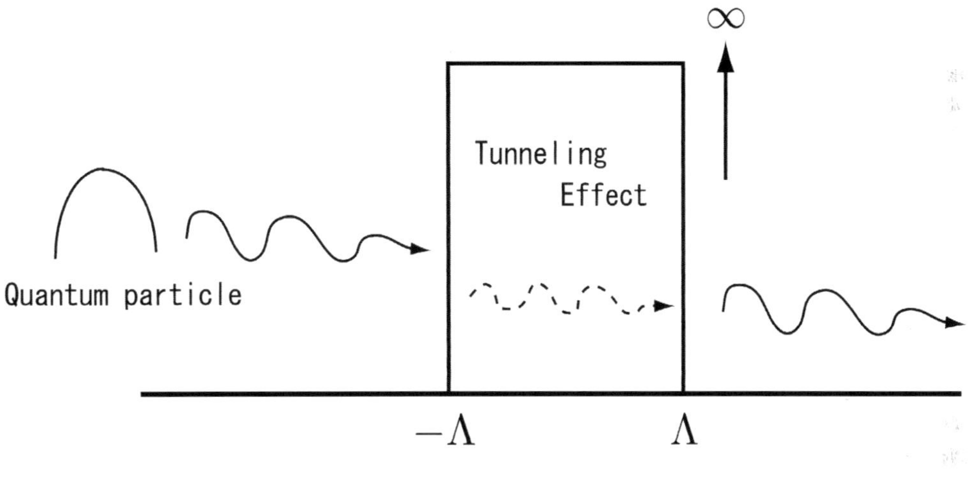

Figure 1. The schematic figure is our considered setup. We only consider tunneling the junction $[-\Lambda, \Lambda]$ of the quantum particle, especially the Schrödinger particle. The potential energy of the junction is infinity. In the case of $\Lambda \to 0+$, the junction can be represented as the Dirac delta function.

2.2. Mathematical notations

Due to the physical constraint that the energy spectrum is real, the Hamiltonian has to be the *self-adjoint operator*. It is remarked that in the case that the Hilbert space is the finite dimensional complex vector field, the Hermitian matrix is self-adjoint. This is because the Hermitian matrix must have the real eigenvalue. However, in general, in the case that the Hilbert space is the square integrable space[1], the Hermite operator is completely different from the self-adjoint operator. Since the self-adjoint operator must have real spectrum, this can be taken as the observable in the von Neumann mathematical axioms of quantum mechanics [5]. The counterexamples that the Hamiltonian is not self-adjoint have been well studied, *e.g.*, see

[1] Our setup is considered in this class.

Ref. [6]. Furthermore, from the viewpoint of statistical physics, the case that the Hamiltonian has the complex eigenvalue were studied to be related with the resonance pole [7]. The symmetric operator is classified with three classes; the operator is self-adjoint, self-adjoint extendable, and the non-self-adjoint extendable, see the pedagogical explanation in Ref. [8, Appendix B]. In the case that the operator is self-adjoint extendable, the operator can be taken as the self-adjoint operator since it can be mapped by the unitary operator.

In our setup, the Hamiltonian should mathematically defined by

$$\begin{aligned} \hat{H}_0 &= -\frac{d^2}{dx^2} \\ &\text{with} \quad Dom(\hat{H}_0) = \{\psi \in Dom(\hat{H}_0^*) \mid \psi(-\Lambda) = \psi(\Lambda) = \psi'(-\Lambda) = \psi'(\Lambda) = 0\}, \end{aligned} \tag{2}$$

where

$$\begin{aligned} \hat{H}_0^* &:= -\frac{d^2}{dx^2} \\ &\text{with} \quad Dom(\hat{H}_0^*) = \{\psi \in \mathcal{L}^2(\Omega_\Lambda) \mid \psi \in AC^2(\overline{\Omega_\Lambda})\}, \end{aligned} \tag{3}$$

and $Dom(\hat{A})$ expresses is the domain of the operator $\hat{A}$. Here, $\overline{S}$ is the closure of the set S, $\mathcal{L}^2(S)$ is the square integrable space on the space S, and $AC^2(S)$ is that every function defined on the space S is absolutely contentious on the space S. Now, the *deficiency subspaces*[2] for the Hamiltonian H_0 are defined by

$$\mathcal{H}_+(\hat{H}_0) := \{\psi \in Dom(\hat{H}_0^*) \mid \hat{H}_0^*\psi = i\psi\}, \tag{4}$$

$$\mathcal{H}_-(\hat{H}_0) := \{\psi \in Dom(\hat{H}_0^*) \mid \hat{H}_0^*\psi = -i\psi\}. \tag{5}$$

Roughly speaking, any wavefunction $\psi \in \mathcal{L}^2(\Omega_\Lambda)$ can be decomposed to three components,

$$\psi = \psi_0 + c_+\psi_+ + c_-\psi_-, \text{ with } \psi_+ \in \mathcal{H}_+(\hat{H}_0) \text{ and } \psi_- \in \mathcal{H}_-(\hat{H}_0), \tag{6}$$

where $c_+, c_- \in \mathbb{C}$. The wavefunction ψ_0 has the eigenfunction with the real eigenvalue for the operator $\hat{H}_0$.

3. Main result

Theorem 1 (Furuhashi, Hirokawa, Nakahara, Shikano [2]) *From the physical constraint that the Hamiltonian $\hat{H}_0$ should be self-adjoint extendable, we obtain*

$$\psi_+(\Lambda) = e^{i(\theta+3\pi/4)}\psi_+(-\Lambda), \tag{7}$$

$$\psi_-(\Lambda) = e^{i(\theta-3\pi/4)}\psi_-(-\Lambda), \tag{8}$$

where θ is the arbitrary real number.

This means that the tunneling Schrödinger particle always leads to the interference pattern after the tunneling by the effect of $\psi_\pm(x)$. Also, this phase is independent of the width of the junction 2Λ. While we simply set $\psi = \psi_0$ and $\psi_0(\pm\Lambda) = \psi_0'(\pm\Lambda) = 0$ as the boundary condition in the standard approach, the more general boundary condition should be considered in the quantum-mechanical system by the von Neumann axiom. It should be noted that our result is globally defined. Therefore, the incoming Schrödinger particle for the tunneling cannot be dealt

[2] The deficiency subspaces are useful to classify the symmetric operator, which is called a *deficiency theorem*. See more details in Ref. [8, Appendix B].

from the viewpoint of mathematical physics. This interference pattern is not always occurred. As the future problem, the concrete example of our system should be constructed. However, the tunneling effect is often analyzed by the WKB analysis. The tunneling effect can be taken as one kind of the scattering problem. Therefore, in the approximate solution for the quantum mechanical system, the interference after the tunneling can be considered. However, our result is not approximated while the potential energy on the junction is infinity. Our result may help us consider the relationship between the self-adjointness and the scattering process in quantum mechanics.

4. Conclusion

We have considered the phase factor of the one-dimensional Schrödinger particle with the junction. The tunneling particle leads to the interference from the viewpoint of the mathematical rigorous analysis. In this paper, we have explained its meaning and the comparison to the conventional setting from the viewpoint of physics.

Acknowledgments

The authors thank Yoshiyuki Furuhashi and Kazumitsu Nakahara for useful collaboration [2]. One of the authors (YS) thanks Yakir Aharonov and Ian Affleck for useful discussion. YS acknowledges the great hospitality and great support, especially from Margaret Sullivan, at Massachusetts Institute of Technology hosted by Seth Lloyd partially supported by JSPS Excellent Young Researcher Overseas Visit Program. YS is supported by JSPS Research Fellowships for Young Scientists (No. 21008624) and Global Center of Excellence Program "Nanoscience and Quantum Physics" at Tokyo Institute of Technology. MH is also supported by JSPS, Grant-in-Aid for Scientific Research (C) 20540171.

References

[1] Voros A 1983 *Ann. Inst. H. Poincaré* **39** 211
[2] Furuhashi Y, Hirokawa M, Nakahara K, and Shikano Y 2010 *J. Phys. A: Math. Theor.* **43** 354010
[3] Exner P and Grosse H 1999 arXiv:math-ph/9910029
[4] Fendley P, Ludwig A W W, and Saleur H 1995 *Phys. Rev. B* **52** 8934
[5] von Neumann J 1932 *Mathematische Grundlagen der Quantumechanik* (Berlin: Springer)
Beyer R T (trans) 1955 *Mathematical Foundations of Quantum Mechanics* (Princeton, NJ: Princeton University Press) (Engl. Transl.)
[6] Kodaira K 1949 *Sugaku* **2** 113 (in Japanese)
[7] Nakano R, Hatano N, and Petrosky T 2010 *Int. J. Theor. Phys.* **50** 1134
[8] Shikano Y and Hosoya A 2008 *J. Math. Phys.* **49** 052104

Superconducting Qubit Measurement and Information Conversion from Quantum to Classical

Hayato Nakano
NTT Basic Research Laboratories, Atsugi-shi, Kanagawa 243-0198, JAPAN
E-mail: nakano.hayato@lab.ntt.co.jp

Abstract. The Josephson bifurcation amplifier (JBA) readout process of a superconducting qubit is analyzed quantum mechanically. We examined the dynamics of the density operator of a detector (a driven nonlinear oscillator) and a qubit coupled system during the measurement process. We have observed the qubit-detector (JBA) entangled state and it is divided into two separable states at the moment the JBA transition begins. Moreover, we discuss the process from the viewpoint of time variations in information quantities, such as mutual information between the qubit and detector. We show that the measurement process corresponds to an information conversion from quantum to classical one.

1. Introduction

A qubit is the most elementary unit of quantum information processing [1]. Qubit readout is one of simple quantum measurements. According to a simple explanation of quantum measurements, it is described as if we can directly obtain an observable eigenvalue of the quantum system by a measurement. In a real measurement, however, we must always use a detector that interacts with the quantum system. A detector is often a macroscopic object. We actually measure a macroscopic quantity in the detector, and merely *postulate* the physical quantity of the quantum system we want to measure. To discuss what we really measure when we measure a qubit, we have to take account of the detector system. A measurement process is a quantum dynamics of the total system consisting of a qubit and a detector.

A good qubit readout gives a probabilistic projection of a qubit state $|\psi_{\rm q}\rangle = a|\uparrow\rangle + b|\downarrow\rangle$ into $|\uparrow\rangle$ or $|\downarrow\rangle$ with a probability $|a|^2$ or $|b|^2$, respectively. Here, $|\uparrow\rangle$ and $|\downarrow\rangle$ are two of the eigenstates of the *observable* in the qubit. The observable is mainly determined by the operator of the qubit appearing in the interaction Hamiltonian between the qubit and the detector. However, the observable is sometimes different from the operator because it is determined by the entire dynamics of the total system during the readout process.

Below, we discuss the readout process for a superconducting qubit. We use a Josephson Bifurcation Amplifier (JBA) as the detector in the measurement [2]. The JBA is an AC-driven SQUID that behaves as a nonlinear oscillator. We find many important quantum characters by investigating the dynamics during the readout process [3].

Usually, the total system is expected to evolve as follows, when the qubit is in a superposition of two of observable eigenstates $|\psi(0)_{\rm q}\rangle = a|\uparrow\rangle + b|\downarrow\rangle$. When the qubit-detector interaction becomes on, an entanglement starts to be formed: $|\psi(0)_{\rm q}\rangle\otimes|{\rm C}_0\rangle \rightarrow a|\uparrow\rangle\otimes|{\rm A}_1\rangle+b|\downarrow\rangle\otimes|{\rm B}_1\rangle$. Since the detector is a macroscopic system where a finite decoherence is inevitable, dephasing occurs.

Journal of Physics: Conference Series **302** (2011) 012045 doi:10.1088/1742-6596/302/1/012045

Then, the total system becomes a classical mixture like $|a|^2| \uparrow\rangle|A_1\rangle\langle\uparrow|\langle A_1| + |b|^2| \downarrow\rangle|B_1\rangle\langle\downarrow|\langle B_1|$. Here, $|C_0\rangle$, $|A_1\rangle$, $|B_1\rangle$ are macroscopic states of the detector, and we can easily distinguish $|A_1\rangle$ or $|B_1\rangle$. Therefore, when we measure the detector state, we can postulate whether the qubit state is $|\uparrow\rangle$ or $|\downarrow\rangle$.

2. Superconducting Qubit and Measurement with a DC-SQUID: Setup

There are three types of superconducting qubits; charge, phase, and flux qubits [4, 5, 6, 7, 8, 9, 10]. A superconducting flux qubit is a superconducting ring interrupted by a few Josephson junctions [10, 11, 12, 13, 14, 15]. When the external magnetic field piercing the ring is in the vicinity of the half flux quantum $\Phi_0/2$, the ring becomes a superposition of two macroscopically distinctive states. Supercurrent flows clockwise around the ring in one state $|R\rangle$, and it flows counterclockwise in the other state $|L\rangle$. The qubit Hamiltonian is given by [16, 17, 18] $H_q = \frac{1}{2}(\varepsilon\sigma_z + \Delta\sigma_x)$, where σ_z and σ_x are two Pauli matrices, and $\sigma_z = |L\rangle\langle L| - |R\rangle\langle R|$. We can control ε by changing the externally applied magnetic flux, but Δ is fixed. The energy-eigenstates of the qubit are given by $|g\rangle = \cos(\theta/2)|L\rangle + \sin(\theta/2)|R\rangle$, and $|e\rangle = \sin(\theta/2)|L\rangle - \cos(\theta/2)|R\rangle$, where $\tan\theta = \Delta/\varepsilon$.

In order to readout superconducting flux qubits, a Superconducting Quantum Interference Device (SQUID) is used as the detector. The SQUID detects a very small magnetic flux induced by supercurrent flowing along the qubit ring, which is only one observable in a flux qubit we can measure from the outside (see, Fig. 1). A SQUID can be approximated as a single Josephson junction whose Josephson energy depends on the magnetic flux Φ_{SQ} piercing the SQUID ring. The Hamiltonian for the SQUID phase γ(=the Josephson phase of the effective single junction) is given by [19], $(\hbar/2e)^2\dot{\gamma}^2/(2C) - E_J(\Phi_{SQ})\cos\gamma - (\hbar/2e)I_b\gamma = (\hbar/2e)^2\dot{\gamma}^2/(2C) - E_J(\Phi_{SQ})(1-\gamma^2/2+\gamma^4/24-\cdots) - (\hbar/2e)I_b\gamma$, where C is the effective capacitance of the junction and I_b is the bias current applied to the SQUID.

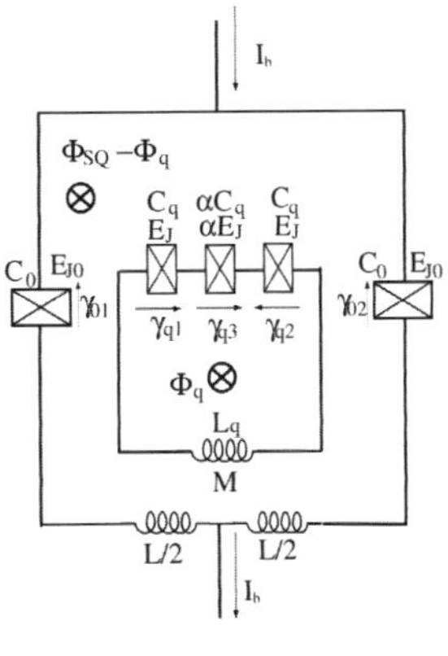

Figure 1. Readout of a flux-qubit system with a detector SQUID that has a continuous variable $\gamma = (\gamma_{01} + \gamma_{02})/2$.

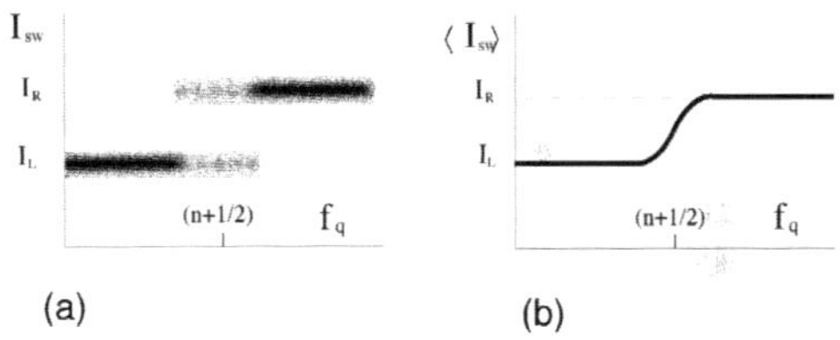

Figure 2. Expected behavior of readout data as a function of ε if the measurement is σ_z projection (schematic). (a) Raw data before averaging. (b) Average values over huge number of detections.

A half decade ago, our experimental results, where raw data before averaging behaved as in the right panel of Fig. 2, clearly showed that our flux-qubit measurement with a DC-SQUID is not a σ_z projection measurement [20, 21]. Moreover, we theoretically clarified that the raw data behave in the qubit measurement with a SQUID when we increase the SQUID bias current slowly, from a detailed calculation of the dynamics of the qubit-SQUID coupled system [19]. The measurement projects the coupled system to one of its energy-eigenstates. In the state, the qubit is approximately in an energy-eigenstate of H_q, and the value (signal) obtained from a single measurement approximately corresponds to $\langle g|\sigma_z|g\rangle$ or $\langle e|\sigma_z|e\rangle$.

3. Quantum dynamics of a Josephson Bifurcation Amplifier

When the amplitude of γ is large, the oscillator becomes bistable under appropriate operation parameters [22]. One stable state has a small amplitude (low-amplitude state), and the other has a larger amplitude (high-amplitude state).

The critical driving force f_c or the critical detuning δ_c for the transition between these two states is very sensitive to small changes in the operational parameters of the oscillator. For example, when we increase or decrease the driving force continuously, the amplitude of the oscillation behaves hysteretically as shown in Fig. 3. When using a JBA as a qubit state readout detector, the JBA detects a small change that depends on the qubit state. The bifurcation phenomenon can be discussed only for classical oscillators, and is impossible from the viewpoint of pure quantum mechanics for an isolated system [23].

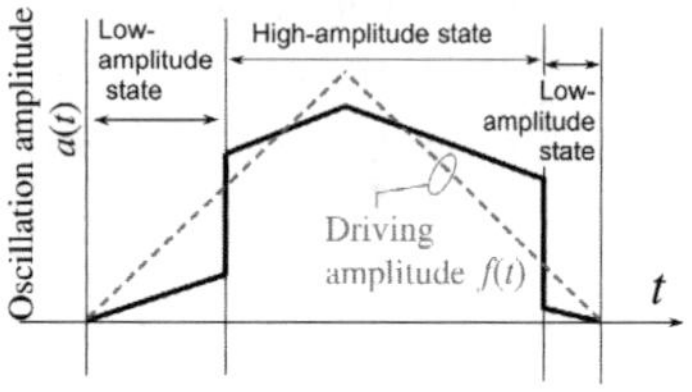

Figure 3. Hysteretic behavior of the oscillation amplitude $a(t)$ of a classical JBA. When the driving force $f(t)$ (broken line) is changed with time t, the amplitude (thick line) varies as shown (schematic).

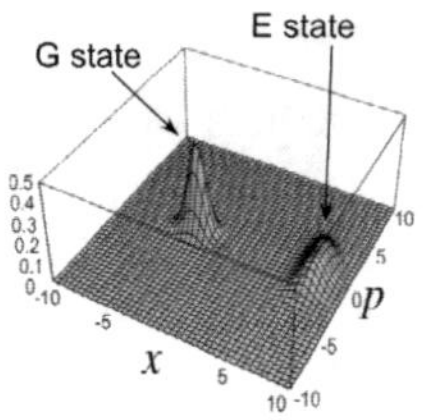

Figure 4. Two macroscopic states of JBA (quasi-distribution representation): p, x are the momentum and amplitude of the oscillation, respectively.

A quantum mechanical analysis is indispensable if we are to understand the readout process [3, 24]. A JBA can be modeled as an anharmonic oscillator in a rotating frame approximation with a Hamiltonian;

$$H_J = (\Omega - \omega)n_a - \alpha {n_a}^2 - \frac{1}{2}f(a^\dagger + a) \tag{1}$$

where, $a^\dagger(a)$ is the creation (annihilation) operator of the Josephson plasma oscillation. The normalized amplitude and momentum of an oscillation are expressed by $x = (a^\dagger + a)/2$ and $p = (a - a^\dagger)/(2i)$, respectively. $n_a = a^\dagger a$, and Ω is the linear resonant frequency of the JBA oscillator. ω is the driving frequency, which is slightly smaller than Ω by the detuning $\delta \equiv \Omega - \omega$. f is the driving strength, and $\alpha(>0)$ is the nonlinearity. In a classical approximation, this model shows the bifurcation in an appropriate parameter region. However, for a quantum-mechanical junction, the transition from $|G\rangle_J$ (low-amplitude state) to $|E\rangle_J$ (high-amplitude state) or, from $|E\rangle_J$ to $|G\rangle_J$ is impossible.

As an example、 here, we introduce linear loss in the oscillator (JBA). The time evolution of the JBA is governed by a Liouville equation [25]:

$$\frac{d\rho}{dt} = \frac{1}{i}[H_J, \rho] + \frac{\Gamma}{2}(2a\rho a^\dagger - a^\dagger a\rho - \rho a^\dagger a), \tag{2}$$

where ρ is the density operator of the system, and Γ is the relaxation rate due to the linear loss in the JBA.

We carried out a numerical calculations [3]. We set $\delta = 0.007\Omega$, nonlinearity $\alpha = 8 \times 10^{-5}\Omega$, quality factor $Q = 2500$, and the driving force f is operated as $0 \to 0.025\Omega \to 0$. These parameters are similar to those used in actual experiments [2]. To order to emphasize quantumness, however, δ and Γ are a factor of 10^{-2} times smaller than in real cases. More than one hundred basis states are prepared in order to avoid numerical artifacts. Then we can reproduce hysteretic bifurcation phenomena of our JBA. We found that the purity $\mathrm{Tr}[\rho^2]$ decreases abruptly. This corresponds to the beginning of the transition from $|G\rangle_J$ to $|E\rangle_J$ of the JBA.

The qubit-JBA composite system is approximately expressed by the Hamiltonian

$$H = H_{\mathrm{J}} + k\sigma_z n_a + H_{\mathrm{q}}, \qquad H_{\mathrm{q}} = \frac{1}{2}(\varepsilon\sigma_z + \Delta\sigma_x) \tag{3}$$

k is the interaction constant between the qubit and the JBA.

4. Quantum Correlation and Quantum Discord

To prepare later discussions, we introduce mutual information in classical information theory and the quantum counterpart.

When there are two information sources A and B, the correlation between A and B is quantified by the mutual information defined by

$$I_{\mathrm{C}}(A, B) = H(\rho_A) + H(\rho_B) - H(\rho_{AB}),$$

where ρ_A and ρ_B are the probability sets for A, B, respectively. ρ_{AB} is the joint probability set of A and B. $H(\rho)$ is the information (Shannon) entropy for ρ. This definition implicitly contains the conditional entropy $K_{\mathrm{C}}(\rho_{B|A}) = \sum_j p_j H(\rho_{B|A=j}) = H(\rho_{AB}) - H(\rho_A)$. The Bayes rule makes the equality satisfied.

On the other hand, the quantum counterparts are given as follows. The density operator of a quantum system $\rho = T^\dagger \begin{pmatrix} p_1 & 0 & 0 & \cdots & 0 \\ 0 & p_2 & 0 & \cdots & 0 \\ 0 & 0 & p_3 & \cdots & 0 \\ \vdots & \vdots & \vdots & \ddots & \vdots \\ 0 & 0 & 0 & \cdots & p_L \end{pmatrix} T$ is defined, where T is a unitary matrix that diagonalizes ρ.

For quantum information, we use von Neumann entropy, $S(\rho) = -\mathrm{Tr}[\rho \log_2 \rho] = -\sum_{l=1}^{L} p_l \log_2 p_l$, instead of Shannon entropy. The quantum mutual information I_{Q} of a two-body (A and B) system is defined by [26]

$$I_{\mathrm{Q}}(A, B) = S(\rho_A) + S(\rho_B) - S(\rho_{AB}).$$

This agrees with the classical mutual information I_{C} if there is only a classical correlation between A and B.

I_{Q} is different from I_{C} because the Bayes rule is not always valid for quantum events. For quantum events, the conditional probability should be expressed by [27] $K(\rho_{B|A}) = \min_{\Pi_j^A} \sum_j^m p_j S(\rho_{B|\Pi_j^A})$, where Π_j^A is an operator that projects the state of A into j and $p_j = \mathrm{Tr}_B[\Pi_j^A \otimes \boldsymbol{I}^B \rho_{AB}]$, $\rho_{B|\Pi_j^A} = \mathrm{Tr}_A[\Pi_j^A \otimes \boldsymbol{I}^B \rho_{AB} \Pi_j^A \otimes \boldsymbol{I}^B]/p_j$, etc.. Then, we obtain, $K(\rho_{B|A}) \geq S(\rho_{AB}) - S(\rho_A)$ and the equality is not always satisfied. Using the conditional entropy K, we can define another type of mutual information, $J_{\mathrm{Q}}(A, B) = S(\rho_B) - K(\rho_{B|A})$, which agrees with

I_{Q} if there is only a classical correlation between A and B. The difference between these two types of mutual information

$$D(\rho_{AB}) = I_{\mathrm{Q}}(A, B) - J_{\mathrm{Q}}(A, B) = K(\rho_{B|A}) - S(\rho_{AB}) + S(\rho_A)$$

is called "Quantum Discord" [27].

Most well known quantum correlation between A and B is entanglement, which is quantified by a measure called "concurrence" [28, 29] as

$$C(\rho_{AB}) = \min_{\{p_k|\psi_k\rangle\}} \sum_k p_k C(|\psi_k\rangle),$$

where $|\psi_k\rangle$and p_k are a pure state and corresponding eigenvalue, with which ρ_{AB} is decomposed as $\rho_{AB} = \sum_k p_k |\psi_k\rangle\langle\psi_k|$, $(0 \le p_k \le 1, \sum_k p_k = 1)$, and $C(|\psi_k\rangle) = \sqrt{2(1 - \mathrm{Tr}[\mathrm{Tr}_A[|\psi_k\rangle\langle\psi_k|]^2]}$ is the concurrence for the pure state $|\psi_k\rangle$. Since ρ_{AB} can be decomposed in many ways, $C(\rho_{AB})$ is defined as the realizable minimum value. Hereafter, we adopt an approximate value for $C(\rho_{AB})$ with its lower bound given by K. Chen et al. [29, 30], instead of an exact $C(\rho_{AB})$ because the minimization procedure is too difficult. Then,

$$C(\rho_{AB}) \ge \sqrt{2/M(M-1)}(\max\{||\rho_{AB}^{T_A}||, ||R(\rho_{AB})||\} - 1).$$

where $\rho_{AB}^{T_A}$is the transposed ρ_{AB} about the indices with A, and $R(\rho_{AB})$ is the transposed ρ_{AB} with both A and B in a cross exchange manner. $||G|| = \mathrm{Tr}[(GG^\dagger)^{\frac{1}{2}}]$ is the trace norm, and we set $M \le N$.

5. Qubit Readout with a JBA

The qubit readout process is well understood by employing knowledge of the quantum behavior in the time evolution of the JBA discussed above.

In Fig. 5, we show the dynamics during the qubit readout process (H_{J} in Eq. (2) replaced with the Hamiltonian H (Eq. (1)) of the qubit-JBA composite system). The JBA parameters are the same as for the above example. The initial state is a separable state; $(\frac{1}{\sqrt{2}}|g\rangle_{\mathrm{q}} + \frac{1}{\sqrt{2}}|e\rangle_{\mathrm{q}}) \otimes |G\rangle_{\mathrm{J}}$.Here, $|g\rangle_{\mathrm{q}}$ and $|e\rangle_{\mathrm{q}}$ are the ground and excited states of the qubit, respectively. The qubit parameters are $\varepsilon = 0.2\Omega$, $\Delta/\varepsilon = 1/2$. The coupling between the qubit and the JBA is set at $k = 0.001\Omega$. The driving force f is increased from 0 to 0.012Ω (slightly larger than the f_c of the JBA) and maintained. This parameter set provides a typical behavior for a successful qubit readout.

The quasi-distribution representations of the JBA state $\mathrm{Tr}_{\mathrm{q}}[\rho]$ are shown in Fig. 5, where ρ is the density operator of the qubit-JBA coupled system, and $\mathrm{Tr}_{\mathrm{q}}[\cdots]$ denotes taking partial trace about qubit degrees of freedom. These peaks constitute an incoherent mixture, so they correspond to two possibilities in the measurement result.

This process can be expressed schematically as

$$\begin{aligned}
& \rho(0) = |G\rangle_{\mathrm{JJ}}\langle G| \otimes \left(\tfrac{1}{\sqrt{2}}|g\rangle_{\mathrm{q}} + \tfrac{1}{\sqrt{2}}|e\rangle_{\mathrm{q}}\right)\left(\tfrac{1}{\sqrt{2}}{}_{\mathrm{q}}\langle g| + \tfrac{1}{\sqrt{2}}{}_{\mathrm{q}}\langle e|\right) \\
\rightarrow\quad & \rho(\tau_1) = \tfrac{1}{2}\left(|G\rangle_{\mathrm{J}}|e\rangle_{\mathrm{q}} + |G'\rangle_{\mathrm{J}}|g\rangle_{\mathrm{q}}\right)\left({}_{\mathrm{J}}\langle G|_{\mathrm{q}}\langle e| +{}_{\mathrm{J}}\langle G'|_{\mathrm{q}}\langle g|\right) \\
\rightarrow\quad & \rho(\tau_2) = \tfrac{1}{2}|G\rangle_{\mathrm{J}}|e\rangle_{\mathrm{qJ}}\langle G|_{\mathrm{q}}\langle e| + \tfrac{1}{2}|G'\rangle_{\mathrm{J}}|g\rangle_{\mathrm{qJ}}\langle G'|_{\mathrm{q}}\langle g| \\
\rightarrow\quad & \rho(\tau_3) = \tfrac{1}{2}|G\rangle_{\mathrm{J}}|e\rangle_{\mathrm{qJ}}\langle G|_{\mathrm{q}}\langle e| + \tfrac{1}{2}|E\rangle_{\mathrm{J}}|g\rangle_{\mathrm{qJ}}\langle E|_{\mathrm{q}}\langle g|.
\end{aligned} \tag{4}$$

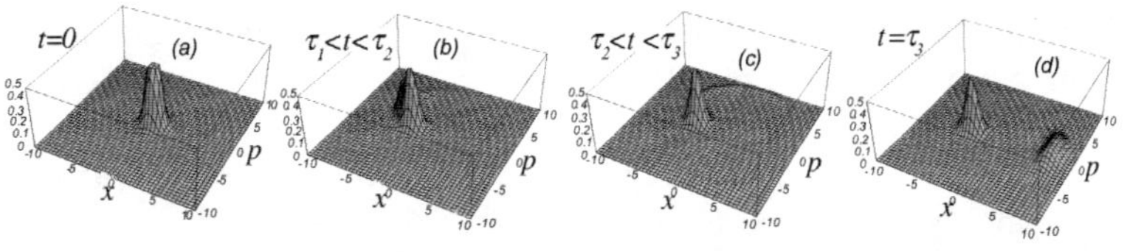

Figure 5. Time evolution of JBA during readout. The figures show quasi-distribution representations ($\langle\alpha|\mathrm{Tr_q}[\rho]|\alpha\rangle$) of the JBA oscillator states, where $|\alpha\rangle$ is the coherent state of a complex amplitude $\alpha = x + ip$ (in the rotating frame). (a) Beginning of the readout. The state of the total system is (schematically) $\rho = |\psi_0\rangle\langle\psi_0|$, with $|\psi_0\rangle = 1/\sqrt{2}(|g\rangle_q + |e\rangle_q) \otimes |G\rangle_J$. (b) Starting the transition. Entanglement formation and projection are carried out during this period. (c) During the transition. Entanglement has already been destroyed. (d) The entire system has become a mixture of classically correlated states.

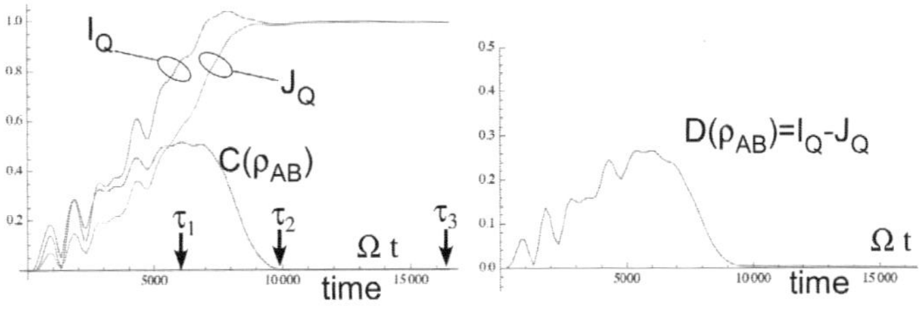

Figure 6. Time evolution of information quantities during the qubit measurement process. I_Q: Quantum mutual information. J_Q: Classical mututal information. C: Concurrence. D: Quantum discord

In Fig. 6, we show the time variations of some of the information quantities introduced in the previous section. The horizontal line corresponds to the time from the onset of the driving, normalized by the linear resonant frequency Ω, the vertical line is in bit or qubit units. The concurrence $C(\rho_{AB})$ grows as the driving strength increases. I_Q and J_Q behave in the same manner. J_Q is kept at almost unity after the concurrence C vanishes. If we measure the JBA state classically within this time span, we can obtain maximum information about the qubit state. After τ_2, the quantum discord ($I_Q - J_Q$) disappears. That is, after τ_2, a measurement on A or B never destroys the correlation between A and B.

Anyway, this time evolution behavior in information quantities shows that the correlation initially invoked as a quantum correlation is converted into a classical correlation during the process. Then, then we can obtain information about the projected qubit via a classical measurement on the JBA. Moreover, at time $t = \tau_3$, the required measurement merely involves distinguishing two peaks in Fig. 5($t = \tau_3$). This is a very easy classical measurement. This situation is established via the separation of the two peaks during the time from τ_2 to τ_3. The entanglement has already been destroyed at τ_2. Therefore, this time variation corresponds to classical information amplification.

6. Conclusion

We analyzed the dynamics of a composite system consisting of a superconducting qubit and a JBA in order to understand the quantum characters in qubit readout process with a JBA, and clarified what happens during the process and how we obtain the qubit information by performing measurements with the detector (JBA). Moreover, we discussed the measurement process from the viewpoint of time variations in such information quantities as quantum and classical mutual information and entanglement concurrence. We clarified that entanglement formation corresponds to the generation of a quantum correlation between the qubit and the detector and the destruction of the entanglement caused by decoherence constitutes the information conversion from quantum to classical.

Acknowledgement
We thank the members of superconducting qubit experiment group in NTT Basic Research Laboratories for fruitful discussions. We also thank Prof. W. Zurek and his colleagues for providing many insights into quantum discord. This work is partially supported by MEXT-KAKENHI, ans JSPS through FIRST program.

[1] Nielsen M and Chuang I 2000 *Quantum Computation and Quantum Information*, (Cambridge: Cambridge Univ. Press)
[2] Siddiqi I *et al.* 2004 *Phys. Rev. Lett.* **93** 207002; Siddiqi I *et al.* 2005 *ibid.* **94** 027005; Lupascu A *et al.* 2006 *ibid.* **96** 127003; Siddiqi I *et al.* 2006 *Phys. Rev.* **B 73** 054510; Lupascu A *et al.* 2007 *Nature Physics* **3** 119; Boulant N *et al.* 2007 *Phys. Rev.* **B 76** 014525
[3] Nakano H, Saito S, Semba K, Takayanagi H 2009 *Phys. Rev. Lett.* **102** 257003
[4] Devoret M and Martinis J 2004 *Quantum Information Processing* **3** 163, and references therein
[5] Friedman J, Patel V, Chen W, Tolpygo S, and Lukens J 2000, *Nature* **406** 43
[6] Vion D, *et al.* 2002 *Science* **296**, 886
[7] Yu Y, Han S, Chu X, Chu S, and Wang Z 2002 *Science* **296** 889
[8] Nakamura Y, Pashkin Yu, and Tsai J-S 1999 *Nature* **398**, 786
[9] Martinis J, Nam S, Aumentado J, and Urbina C 2002 *Phys. Rev. Lett.* **89** 117901
[10] Mooij J *et al.* 1999 *Science* **285**, 1036
[11] Tian L , Lloyd S, and Orlando T 2002 *Phys. Rev.* **B 65** 144516
[12] van der Wal C *et al.* 2000 *Science* **290** 773; van der Wal C 20001 *Quantum superpositions of persistent Josephson currents* Ph.D. thesis, (Delft: Delft Univ. Press)
[13] Chiorescu I, Nakamura Y, Harmans C, and Mooij J 2004 *Nature* **431** 159
[14] Chiorescu I, Bertet P, Semba K, Nakamura Y, Harmans C, and Mooij J 2003 *Science* **299** 1869
[15] Bertet P *et al.* 2004 cond-mat/0412485.
[16] Crankshaw D and Orlando T 2001*IEEE Trans. Appl. Superconductivity* **11** 1006; You J, Nakamura Y, and Nori F 2003 *cond- mat/0309491*; van Brink A *cond- mat/0310425*
[17] van der Wal C, Wilhelm F, Harmans C, Mooij E 2003*Eur. Phys. J.* **B 31** 111 (2003)
[18] Orlando T, *et al.* 1999 *Phys. Rev.* **B 60** 15398
[19] Nakano H and Takayanagi H 2003 *J. Phys Soc. Jpn.* **72** Suppl. **A** 1; Nakano H *et al.* 2004 *arXiv:cond-mat/0406622*; Nakano H and Takayanagi H 2003, in *Toward the Controllable Quantum States* (Singapore :World Scientific) p 359
[20] For review, Semba K, Johansson K, Kakuyanagi K, Nakano H, Saito S, Tanaka H and Takayanagi H 2009 *Quantum Information Processing*: **8, Issue 2** 199.
[21] Tanaka H, Sekine Y, Saito S, Takayanagi H 2002 *Physica* **C 368** 300; Takayanagi H, Tanaka H, Saito S, and Nakano H 2002 *Physica Scripta* **T102** 95; Tanaka H, Saito S and Takayanagi H 2003 *Toward the Controllable Quantum States* (Singapore: World Scientific) p 366
[22] Jordan W and Smith P 1999 *Nonlinear Ordinary Differential Equations* third edition (Oxford: Oxford Univ. Press)
[23] Dykman M and Fistul M 2005 *Phys. Rev.* **B 71** 140508 (R); Dykman M 2007*Phys. Rev.* **E 75** 011101; Peano V and Thorwart M 2006 *Chem. Phys.* **322** 135; Peano V and Thorwart M 2006 *New J. Phys.* **8** 21
[24] Makhlin Y, Shnirman A, and Schön G 2001 *Rev. Mod. Phys.* **73** 357
[25] Lindblad G 1976 *Commun. Math. Phys.* **48** 119
[26] Schumacher B and Westmoreland 2010 *Quantum Processes, Systems & Information* (Cambridge: Cambridge University Press)
[27] Ollivier H and Zurek W 2001 *Phys. Rev. Lett.* **88** 017901
[28] Uhlmann A *Phys. Rev.* **A 62** 32307
[29] Horodecki R, Horodecki P, Horodecki M, and Horodecki K 2009 *Rev. Mod. Phys.* **81** 865
[30] Chen K, Albeverio S, and Fei S-M *et al.* 2005 *Phys. Rev. Lett.* **95** 40504; Chen K, Albeverio S, and Fei S-M *et al.* 2005 *Phys. Rev. Lett.* **95** 210501

International Symposium "Nanoscience and Quantum Physics 2011" IOP Publishing
Journal of Physics: Conference Series **302** (2011) 012046 doi:10.1088/1742-6596/302/1/012046

Kibble-Zurek mechanism in simulated annealing and quantum annealing

Sei Suzuki

Department of Physics and Mathematics, Aoyama Gakuin University, Fuchinobe, Sagamihara 252-5258, Japan

E-mail: `sei @ phys.aoyama.ac.jp`

Abstract. We study errors of quantum annealing and simulated annealing to highlight better performance of quantum annealing over simulated annealing. Quantum annealing and simulated annealing perform optimization through a quantum adiabatic evolution and a quasi-static evolution respectively. In both methods, dynamics across a phase transition plays a crucial role. The Kibble-Zurek mechanism is known as an underlying physics of defect formation during a time-evolution across a phase transition. We apply an argument for the Kibble-Zurek mechanism to the error generation of quantum annealing and simulated annealing. We show that, for the disordered Ising chain, the kink density and residual energy per spin of quantum annealing decay as $(\ln\ )^{\ 2}$ and $(\ln\ )^{\ 4}$ respectively, whereas those of simulated annealing decay as $(\ln\ )^{\ 1}$ and $(\ln\ )^{\ 2}$ with the annealing rate $\ 1/\ $. These results imply better performance of quantum annealing. We also develop our theory for a two-dimensional spin-glass model.

1. Introduction

Solving optimization problems is not only the main topic in computer science but also an important subject in condensed matter physics [1]. Several methods have been proposed for optimization problems from the physicist's point of view. Simulated annealing is an algorithm which is applicable to generic optimization problems and uses the notion of statistical mechanics [2, 3]. Quantum annealing is another generic algorithm [4, 5, 6, 7, 8, 9]. Quantum annealing applies the quantum mechanics for classical problems. A great question is whether the quantum mechanics is able to make speed-up from the classical algorithm in computation. In this paper, we focus on a comparison of quantum annealing and simulated annealing.

Let us consider an Ising model: $H_{\rm Ising} = \sum_{i,j} J_{ij}\ {}^{z}_{i}\ {}^{z}_{j}$, where ${}^{z}_{i}$ is the Ising spin and J_{ij} is the coupling constant. If J_{ij} is disordered, the ground state of this Hamiltonian can be non-trivial and di cult to nd. For instance, there is no known algorithm to nd the ground state of the three-dimensional spin-glass model in a polynomial time with system size [10]. A lot of optimization problems, including the travelling salesman problem, satis ability problem, and so on, can be mapped to such ground-state search problems of a disordered Ising model. Simulated annealing o ers an approximate solution to these problems. In simulated annealing, one introduces a ctitous temperature and makes a thermal equilibrium state by means of the Monte-Carlo simulation or molecular dynamics simulation. When T is su ciently high, it is easy to realize the equilibrium state, but it is not for low T. At rst we make the equilibrium state at high T and then decrease T with time. If the decreasing speed of T is su ciently slow, the distribution of spin states changes with keeping the equilibrium distribution at each temperature and the ground state is obtained with high probability in the end when the temperature becomes zero. Quantum annealing, on the other hand, introduce the transverse eld, $\Gamma \sum_i {}^x_i$, instead of the temperature, where ${}^x_i$ is the x-component of the Pauli's spin operator and Γ stands for the strength of the eld. At rst we set Γ at a su ciently high value and make the ground state. This ground state is easily made since all spins are aligned along the transverse eld. Then we decrease Γ with time. If the decreasing speed of Γ is su ciently slow, the spin state evolves adiabatically and reaches the ground state of H_{Ising} when Γ vanishes. Quantum annealing is naturally implemented in the quantum computer. However, it is possible to implement in the classical computer if one uses the quantum Monte-Carlo method. In this case, the Schrodinger dynamics is replaced by the stochastic dynamics of the Monte-Carlo method.

Both in simulated annealing and quantum annealing, the phase transition is an obstacle to the success. Consider a thermodynamic phase transition. In the vicinity of phase transition, the relaxation time diverges, and hence the distribution of spin states cannot keep track of the Gibbs-Boltzmann's distribution as far as the temperature is decreased with a nite rate. As for a quantum phase transition, the energy gap between the ground state and the rst excited state vanishes at the critical point. Since the adiabatic condition with changing Γ is given by $|\dot{\Gamma}|/ {}^2 \quad 1$ [9, 11], the adiabaticity breaks at the critical point. The Kibble-Zurek mechanism is known as an underlying physics of defect formation during such a dynamics across a phase transition [12, 13]. By applying the Kibble-Zurek argument to simulated annealing and quantum annealing, we explicitly derive scaling formulas with respect to the error after annealing for one-dimensional disordered Ising model. In fact, we show that the densities of kinks after simulated annealing and quantum annealing are scaled as ${}^{\text{SA}} \quad (\ln \ )^{\ 1}$ and ${}^{\text{QA}} \quad (\ln \ )^{\ 2}$ respectively, where $1/$ is the annealing rate. We also show that residual energies per spin are $\varepsilon^{\text{SA}}_{\text{res}} \quad (\ln \ )^{\ 2}$ and $\varepsilon^{\text{QA}}_{\text{res}} \quad (\ln \ )^{\ 4}$. These results reveal that quantum annealing reduces the error faster than simulated annealing at least for the disordered Ising chain.

We organize this paper as follows. In the next section, we explain the Kibble-Zurek mechanism in detail. We then apply the argument of section 2 to the disordered Ising chain in section 3. The main results of this paper is presented here. In section 4, we discuss a two-dimensional system. Although we have not reached de nite results on a two-dimensional system, we provide a hint which may be useful in future work. This paper is concluded in section 5.

2. Kibble-Zurek mechanism

Let us consider a ferromagnet with the critical temperature T_c. In the standard second order phase transition, the correlation length ξ and the relaxation time ${}_{\text{r}}$ increase with decreasing temperature toward T_c as ξ and ${}_{\text{r}} \quad \xi^z \quad {}^{z}$, where $= (T \quad T_c)/T_c$ is the dimensionless temperature, and z and are critical exponents.

Now we consider quenching temperature with time t as $T(t) = T_c(1 \quad t/ \ )$, namely, $(t) = \quad t/$, where $1/$ is the quench rate and we assume that the time moves from ∞ to . We assume that the system is in its equilibrium state initially. As far as the temperature is su ciently high and the relaxation time is su ciently short, the system can maintain its equilibrium with decreasing temperature. However, when the temperature becomes close to T_c, the temperature decreases further before the system attains the equilibrium. Thus the system cannot achieve the ferromagnetic long-range order and contains ferromagnetic domains when the temperature goes below T_c. Once the domain structure forms, it would not vanish since a large energy barrier lies between this state and the uniformly ordered state. The size of the domain is estimated as follows. We focus on two scales of time in our system. One is the remaining time $|t|$ to the critical point, and the other is the relaxation time ${}_{\text{r}}$. We then equate them:

$$ {}_{\text{r}} = |t|. \tag{1}$$

This equality de nes the time $\hat{t}$ at which the relaxation time is equal to the remaining time to the critical point. Since $_{\mathrm{r}} = \quad ^{z} = (|t|/ \quad) \quad ^{z}$, one obtains $|\hat{t}| = \quad ^{z\ /(z\ +1)}$. After the time passes $\hat{t}$, the system never attains the equilibrium since the relaxation time is longer than the remaining time. Suppose here that the system stays in the equilibrium at $\hat{}\quad (\hat{t}) = \quad ^{1/(z\ +1)}$ and does not evolve any more after $\hat{t}$. Then the correlation length of the state after $\hat{t}$ is approximated by $\hat{\xi} \quad \xi(\hat{}) = \quad ^{/(z\ +1)}$. This gives an estimate of the size of the domain produced by quenching temperature. The density of defects can be estimated by $\quad \hat{\xi} \quad ^{d}$, where d is the dimension of the system. Therefore one obtains $\quad ^{d\ /(z\ +1)}$.

3. Disordered Ising chain

In this section, we focus on simulated annealing and quantum annealing of the disordered Ising chain. The Hamiltonian of the disordered Ising chain is given by

$$\mathcal{H}_{\mathrm{Ising}} = \quad \sum_{i} J_i \quad ^{z}_{i} \quad ^{z}_{i+1}. \tag{2}$$

In the present study, we assume that J_i obeys the uniform distribution between 0 and 1, namely, $P(J_i) = 1$ for $0 \quad J_i \quad 1$ and $P(J_i) = 0$ otherwise. The ground state of this model is obviously the state with the complete ferromagnetic order. However the excited states re ect the randomness in the coupling constants, which in uence annealing dynamics.

3.1. Simulated annealing

The disordered Ising chain does not show the phase transition at any nite temperature. The critical temperature is regarded as $T_c = 0$. We assume that the temperature is decreased as $T = \quad t/ \quad$ with time moving from $t = \quad \infty$ to 0. We employ the Glauber's model [14] for the spin dynamics.

Let us apply the argument in section 2 to this system. At rst, the correlation function between two spins is obtained as $[\langle \quad _i \quad _{i+r}\rangle] = (\ln\cosh \quad / \quad)^r$ in the thermodynamic limit, where $[\quad]$ stands for the average with respect to the distribution of J_i and is the inverse of the temperature. From this, the correlation length for $T \quad 1$ is obtained as

$$\xi \approx \quad / \ln 2. \tag{3}$$

Next, the relaxation time under the Glauber's model is given by [15]

$$_{\mathrm{r}} \approx \frac{1}{2}\mathrm{e}^{4} \quad \approx \frac{1}{2}\mathrm{e}^{(4\ln 2)\xi}. \tag{4}$$

The scaling properties of this zero-temperature phase transition is di erent from those of the standard second order phase transition. Hence we return to equation (1). From equations (3) and (4) with $\quad = \quad /t$, one obtains an equation of ξ: $\frac{1}{2}\exp[(4\ln 2)\hat{\xi}] = \quad /(\hat{\xi}\ln 2)$ which yields

$$\hat{\xi} = \frac{1}{4\ln 2}\left(\ln \quad \ln\frac{\hat{\xi}\ln 2}{2}\right). \tag{5}$$

Note that we write the solution of the equation of ξ as $\hat{\xi}$. For $\quad \gg 1$, this reduces to $\hat{\xi} \approx \ln \quad /(4\ln 2)$. The inverse of the correlation length provides an estimate of the density of kinks in the Ising chain:

$$^{\mathrm{SA}} \approx \frac{1}{\hat{\xi}} \approx \frac{4\ln 2}{\ln \quad \ln(\hat{\xi}\ln 2/2)}, \tag{6}$$

which reduces to $^{\mathrm{SA}} \quad (\ln \quad)^{\ 1}$ for $\quad \gg 1$. Since the kink is seen as the error against the ferromagnetic ground state, this result implies that the error of simulated annealing decays as $1/\ln \quad$ with increasing $\quad$.

The residual energy is also an important quantity in measuring the error. The energy expectation value per spin at low temperature is given by $\varepsilon \approx \quad \frac{1}{2} + \frac{\pi^2}{24} \quad ^2$, where the rst term corresponds to the ground-state energy. This expression with equations (3) and (5) leads us to an estimate of the residual energy per spin:

$$\varepsilon_{\mathrm{res}}^{\mathrm{SA}} \approx \frac{2 \quad ^2}{3} \left(\ln \qquad \ln \frac{\hat{\xi} \ln 2}{2} \right)^{2} , \tag{7}$$

which reduces to $\varepsilon_{\mathrm{res}}^{\mathrm{SA}} \quad (\ln \quad)^{\ 2}$ for $\quad \gg 1$.

The analytic results on $\quad^{\mathrm{SA}}$ and $\varepsilon_{\mathrm{res}}^{\mathrm{SA}}$ obtained above have been supported by numerical simulation [16].

3.2. Quantum annealing

To study quantum annealing of the disordered Ising chain, we consider the following Hamiltonian:

$$\mathcal{H} = \mathcal{H}_{\mathrm{Ising}} \quad \Gamma \sum_i h_i \quad {}_i^x , \tag{8}$$

where h_i is a random constant which is introduced for calculational convenience. h_i obeys the same probability distribution as J_i. This model exhibits a quantum phase transition at $\Gamma = 1$ [17]. For $0 \quad \Gamma < 1$, the ground state is ferromagnetic, i.e., $[\langle \ {}_i^z \rangle] > 0$. On the other hand, the paramagnetic ground state is realized for $\Gamma > 1$. The quantum critical point at $\Gamma = 1$ of this model is called the in nite randomness xed point. Its property has been revealed by means of the strong disorder renormalization group [17, 18, 19, 20]. Let us write the distance from the critical point as $\quad = \Gamma \quad 1$. The correlation length is scaled with as

$$\xi \quad | \quad | \quad . \tag{9}$$

What makes this quantum phase transition unusual is that the distribution of the energy gap above the ground state at the critical point is scaled by $\quad \ln \quad /L \quad$ using the system size L and an exponent . Hence the coherence time $\quad_{\mathrm{coh}}$, de ned by the inverse of the averaged energy gap, is scaled with ξ as

$$\ln \quad _{\mathrm{coh}} \quad \xi \quad . \tag{10}$$

The exponents and have been obtained as $\quad = 2$ and $\quad = \frac{1}{2}$ [17, 18]. These scaling properties are the same in the case with $h_i = 1$ for all i.

Now we assume the time-dependent transverse eld: $\quad = \quad t/ \quad$, where t moves from $\quad \infty$ to . We follow the argument of section 2 starting from equation (1). Taking the logarithm of equation (1) and using (9), (10) and $|t| = \quad$, one has

$$\hat{\xi} \quad \ln(\quad \hat{\xi} \quad ^{1/} \quad), \tag{11}$$

which reduces to $\hat{\xi} \quad (\ln \quad)^{1/} \quad = (\ln \quad)^2$ for $\quad \gg 1$. Therefore one obtains the density of kinks after quantum annealing as $\quad^{\mathrm{QA}} \quad (\ln \quad)^{\ 2}$ [21, 22].

The scaling of residual energy after quantum annealing is estimated as follows. Suppose that there is one kink per length L in the chain. The smallest coupling constant in the chain of length L is estimated as $1/L$. Hence the energy cost per spin coming from the kink is given by $1/L^2$. Now the state after quantum annealing should be the state with correlation length $\hat{\xi}$. This implies that there is one kink per length $\hat{\xi}$ on average. Therefore the residual energy per spin is estimated as $\varepsilon_{\mathrm{res}}^{\mathrm{QA}} \quad 1/\hat{\xi}^2 \quad (\ln \quad)^{\ 4}$.

The result on the density of kinks has been supported by numerical simulation [22]. As for the residual energy, numerical simulation has produced $\varepsilon_{\mathrm{res}}^{\mathrm{QA}} \quad (\ln \quad)^{\ 3.4}$ on the basis of results for sizes up to 512 spins [22]. We expect that this discrepancy is due to nite size e ects.

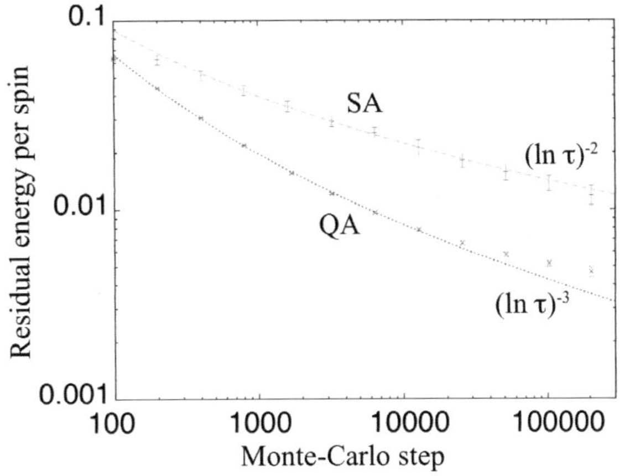

Figure 1. Residual energies after simulated annealing and quantum annealing of the two-dimensional spin-glass model with 99 99 spins. Simulated annealing and quantum annealing are performed by the classical and quantum Monte-Carlo methods [28] respectively. Simulated annealing is started from an initial temperature $T = 5.0$ and quantum annealing is done from $\Gamma = 5.0$ at $T = 0.01$. The uniform transverse eld is applied in quantum annealing. The random average is taken over 100 runs for a single sample for simulated annealing and single run for 16 samples for quantum annealing. The decay of the residual energy of simulated annealing is well tted by (ln)^{2}. For quantum annealing, the residual energy decays approximately as (ln)^{3} except for long where it has a slower decay rate. The decay rate for quantum annealing is faster than simulated annealing. However, its rate is di erent from (ln)$^{4.47}$ obtained theoretically.

4. Spin-glass in two dimension

The disordered Ising chain we studied in the previous section has the trivial ground state. It is signi cant to discuss the case with a non-trivial ground state. Here we consider the two-dimensional spin-glass model: $\mathcal{H}_{\text{Ising}} = \sum_{<ij>} J_{ij} \; {}^{z}_{i} \; {}^{z}_{j}$, where J_{ij} obeys the uniform distribution between 2 and 2. For simulated annealing of spin-glass systems, the Huse-Fisher law provides the upper bound (ln)^{2} for the decay rate of the residual energy [23]. In fact, as shown in gure 1, numerical results are well tted by $\varepsilon_{\text{res}}^{\text{SA}}$ (ln)^{2}. The Huse-Fisher law is independent of the Kibble-Zurek mechanism and originates from spin dynamics of a small spin cluster. Hence this fact implies local spin dynamics dominates the error of simulated annealing. As for quantum annealing, there is no known mechanism like the Huse-Fisher law. In the presence of the transverse eld, the quantum phase transition of this model should be characterized by the in nite-randomness xed point [24], like in one-dimension. Hence, the same argument as the previous section yields $\hat{\xi}$ (ln)$^{1/}$. In order to estimate the residual energy, we assume that a lowlying droplet excitation from the ground state in a system with size L in the absence of the transverse eld costs an energy L^{θ} [25], namely $L^{\theta \;\; d}$ per spin, where is the sti ness exponent and d is the dimension of the system. Then, with the correlation length $\hat{\xi}$ after annealing, the residuel energy per spin should be scaled as $\varepsilon_{\text{res}}^{\text{QA}}$ $\hat{\xi}^{\theta \;\; d}$ (ln)$^{(\theta \;\; d)/}$. Unfortunately we have no estimates for and so far. However, if one substitute the value $\approx$ 0.29 of the Gaussian spin-glass [26] and ≈ 0.48 of the random transverse Ising ferromagnet [27], one reaches ($d)/$ ≈ 4.77. In gure 1, we show results of the quantum Monte-Carlo simulation. It is clear that the decay of residual energy after quantum annealing is faster than simulated annealing. However one nds that $\varepsilon_{\text{res}}^{\text{QA}}$ behaves almost as (ln)^{3} which is slower than (ln)$^{4.77}$ except for long where it has a still slower decay rate. One of the reason of this disagreement might be the di erence between the Schrodinger dynamics and the Monte-Carlo dynamics. However, since our scaling analysis is independent of the detail of the rule of dynamics, our theoretical result may be applicable to the Monte-Carlo dynamics. Next, one can attribute the disagreement to the values of and we used. The uniform distribution of J_{ij} may produce di erent from the Gaussian distribution. Moreover it is not clear whether the frustration is irrelevant or not at the in nite-randomness xed point. Finally, there is a possibility that another unknown mechanism dominates the decay of residual energy. To clarify the nature of quantum annealing in non-trivial systems, we need analytic and numerical studies further.

5. Conclusion

We studied residual errors after simulated annealing and quantum annealing with the particular attention to the Kibble-Zurek mechanism. For the disordered Ising chain, simulated annealing produces the density of kinks and residual energy per spin which are scaled as $^{\rm SA}$ (ln) ^{1} and $\varepsilon_{\rm res}^{\rm SA}$ (ln) ^{2}, whereas quantum annealing produces $^{\rm QA}$ (ln) ^{2} and $\varepsilon_{\rm res}^{\rm QA}$ (ln) ^{4}. These scaling relations are con rmed numerically except $\varepsilon_{\rm res}^{\rm QA}$ for which we ascribed the discrepancy to nite size e ects of simulation. Our results on decay rates of residual errors reveal better performance of quantum annealing in comparison with simulated annealing. We also discussed the two-dimensional spin-glass model. Although we presented theoretical and numerical results, further studies are needed to clarify the discrepancy between them.

Acknowledgments

The author acknowledges support by Grant-in-Aid for Scienti c Research (No. 20740225) from MEXT, Japan.

References

[1] Mezard M, Parisi G and Virasoro M 1987 *Spin Glass Theory and Beyond* (Singapore: World Scienti c)
[2] Kirkpatric S, Gelatt Jr. C D and Vecchi M P 1983 *Science* **220** 671
[3] Cerny V 1985 *J. Optim. Theory Appl.* **45** 41
[4] Finnila A B, Gomez M A, Sebenik C, Stenson C and Doll J D 1994 *Chem. Phys. Lett.* **219** 343
[5] Kadowaki T and Nishimori H 1998 *Phys. Rev.* E **58** 5355
[6] *Quantum Annealing and Related Optimization Methods* ed A Das and B K Chakrabarti 2005 (Heidelberg: Springer)
[7] Santoro G E and Tosatti E 2006 *J. Phys.* A **39** R393
[8] Das A and Chakrabarti B K 2008 *Rev. Mod. Phys.* **80** 1061
[9] Morita S and Nishimori H 2008 *J. Math. Phys.* **49** 125210
[10] Barahona F 1982 *J. Phys.* A **15** 3241
[11] A Messiah 1962 *Quantum Mechanics* Vol. 2 (Amsterdam: North-Holland)
[12] Kibble T W B 1976 *J. Phys.* A **9** 1387
[13] Zurek W H 1985 *Nature* (London) **317** 505
[14] Glauber R J 1963 *J. Math. Phys.* **4** 294
[15] Dhar D and Barma M 1980 *J. Stat. Phys.* **22** 259
[16] Suzuki S 2009 *J. Stat. Mech.* P030032
[17] Fisher D S 1992 *Phys. Rev. Lett.* **69** 534
[18] Fisher D S 1995 *Phys. Rev.* B **51** 6411
[19] Igloi F and Monthus C 2005 *Phys. Rep.* **412** 277
[20] Vojta T 2006 *J. Phys.* A: Math. Gen. **39** R143
[21] Dziarmaga J 2006 *Phys. Rev.* B **74** 064416
[22] Caneva T, Fazio R and Santoro G E 2007 *Phys. Rev.* B **76** 144427
[23] Huse D A and Fisher D S 1986 *Phys. Rev. Lett.* **57** 2203
[24] Pich C, Young A P, Rieger H and Kawashima N 1998 *Phys. Rev. Lett.* **81** 5916
[25] Fisher D S and Huse D A 1988 *Phys. Rev. B* **38** 386
[26] Katzgraber H G Lee L W and Young A. P. 2004 *Phys. Rev.* B **70** 014417
[27] Kovacs I A and Igloi F 2010 *Phys. Rev.* B **82** 054437
[28] Morita S, Suzuki S and Nakamura T 2009 *Phys. Rev.* E **79** 065701(R)

Journal of Physics: Conference Series **302** (2011) 012047 doi:10.1088/1742-6596/302/1/012047

Nonequilibrium work performed in quantum annealing

Masayuki Ohzeki, and Hidetoshi Nishimori

Department of Systems Science, Graduate School of Informatics, Kyoto University, Yoshida Honmachi, Sakyo-ku, Kyoto, 606-8501, Japan

Department of Physics, Tokyo Institute of Technology, Oh-okayama, Meguro-ku, Tokyo, 152-8551, Japan

E-mail: mohzeki@i.kyoto-u.ac.jp

Abstract. Quantum annealing is a generic solver of classical optimization problems that makes full use of quantum fluctuations. We consider work statistics given by a repetition of quantum annealing processes by employing the Jarzynski equality proposed in nonequilibrium statistical physics. In particular, we analyze a nonequilibrium average of the exponentiated work performed by a transverse field. A special symmetry, gauge symmetry, leads to a non-trivial relationship between quantum annealing toward different targets in the theory of spin glasses. We believe that our results will be a step toward an alternative realization of efficient quantum computation as well as our better understanding of nonequilibrium behavior of systems under quantum control.

1. Introduction

We should often solve difficult problems in a reasonable time, which is closely related to the efficiency in cost and time in industry and distribution systems. A well-known instance is to find the minimum path for a person to visit all the cities, called the traveling salesman problem, which is a typical optimization problem [1, 2]. The goal of quantum computation is to solve problems hard to solve quickly on classical computer in a moderate time by use of superpositions, tunneling effects, and entanglement of quantum nature. Quantum annealing is a method to realize quantum computation [3, 4, 5, 6]. Conventional quantum computation uses successive, discrete unitary operations to enhance the probability of the desired state representing the answer of a hard optimization problem. In contrast, quantum annealing uses continuous dynamics represented by a time-dependent Schrödinger equation. Dynamical behavior of quantum systems during such processes is still poorly understood, and an adiabatic control of quantum dynamics is often employed, called the quantum adiabatic computation [7]. Such a quantum adiabatic computation, however, has a bottleneck for a certain type of the optimization problems [8, 9].

A recent study of quantum annealing indicates a possible deviation from adiabatic controls [10]. In the present paper, we proceed along this line and analyze the work performed by quantum fluctuations in a non-adiabatic control of system parameter. The analysis makes use of several remarkable properties found in statistical physics. One is an exact relationship between nonequilibrium processes and equilibrium states at the beginning and end, the Jarzynski equality [11, 12]. Another is a special symmetry, called the gauge symmetry, found in spin glasses

[13, 14]. As detailed below, a combination of these theories enables us to obtain exact relations between equilibrium and nonequilibrium quantities and to evaluate several quantities related to dynamical processes in quantum systems [15].

After an introduction of the Jarzynski equality, we develop our analyses by use of gauge symmetry to show the derivation of an exact identity related with the performed work during quantum annealing.

2. Quantum annealing and spin glasses

We consider the following time-dependent Hamiltonian in quantum annealing

$$H(t) = -f(t)J\sum_{\langle ij\rangle} \tau_{ij}\sigma_i^z\sigma_j^z - \Gamma\left(1 - f(t)\right)\sum_i \sigma_i^x, \tag{1}$$

where $f(t) = t/T$, and t runs continuously from 0 to T. The initial Hamiltonian $H(0)$ is composed only of the transverse-field term $-\Gamma\sum_i \sigma_i^x$. The final Hamiltonian $H(T)$ is chosen to describe the optimization problem to be solved. We take a simple spin glass Hamiltonian as $H(T)$, the $\pm J$ Ising model, on an arbitrary lattice throughout this paper for convenience. The sign of the interaction τ_{ij} follows the distribution function

$$P(\tau_{ij}) = p\delta(1-\tau_{ij}) + (1-p)\delta(1+\tau_{ij}) = \frac{e^{\beta_p J\tau_{ij}}}{2\cosh\beta_p J}, \tag{2}$$

where p is the concentration of the ferromagnetic interaction $J > 0$, and $\exp(-2\beta_p J) = (1-p)/p$. A sufficiently slow decrease of the strength of the transverse field changes the trivial initial state to the nontrivial ground state of the target Hamiltonian $H(T)$. This is the idea of quantum adiabatic computation realized in $T \to \infty$. Instead of the adiabatic control, in the present study, let us consider the repetition of fast quantum annealing (T small) starting from an equilibrium ensemble. We may not be able to always reach the desired state since the system does not trace the instantaneous ground state when the adiabatic condition is not satisfied. Therefore we need to repeat the process to hit the correct ground state in such a non-adiabatic realization of quantum annealing.

We next point out that a class of models of spin glasses has a special symmetry known as the gauge symmetry. We define the gauge transformation as the following local unitary operator for Pauli spin matrices, and the change of the sign of the interaction [16]:

$$\sigma_i^\alpha \to G\sigma_i^\alpha G^{-1}, G = \prod_i G_i, \quad G_i = \begin{cases} 1 & (\xi_i = +1) \\ \exp(i\pi\sigma_i^x/2) & (\xi_i = -1) \end{cases} \tag{3}$$

where $\alpha = x, y, z$ and $\tau_{ij} = \tau_{ij}\xi_i\xi_j$, where ξ is a classical gauge variable taking only ± 1. After the above gauge transformation, the time-dependent Hamiltonian (1) is invariant, but the distribution function (2) is modified into $P(\tau_{ij}) = \exp(\beta_p J\tau_{ij}\xi_i\xi_j)/2\cosh\beta_p J$. This property helps us to derive the following results.

3. Analysis

3.1. Jarzynski equality

Let us introduce another piece of theoretical tool, the Jarzynski equality, which is a generalization of the second law of thermodynamics. We here use the quantum version of the Jarzynski equality for the $\pm J$ Ising model, not the original version for the classical case [11, 12], with a specific configuration $\{\tau_{ij}\}$ as follows [17, 18],

$$\langle e^{-\beta W}\rangle_{0\to T} = \frac{Z_\beta(T;\{\tau_{ij}\})}{Z_\beta(0;\{\tau_{ij}\})}, \tag{4}$$

where $Z_\beta(t;\{\tau_{ij}\})$ is the partition function for the instantaneous Hamiltonian, and β is the inverse temperature. The performed work during a nonequilibrium process is given by the difference between the outputs of the measurements of the initial and final energies as $W = E_m(T) - E_n(0)$, where m and n denote the indices of the eigenstates as $H(T)|m(T)\rangle = E_m(T)|m(T)\rangle$ and $H(0)|n(0)\rangle = E_n(0)|n(0)\rangle$. The left-hand side of equation (4) expresses the average of the exponentiated work over all the realizations of nonequilibrium processes starting from the equilibrium ensemble, and is evaluated as

$$\begin{aligned}\langle \mathrm{e}^{-\beta W}\rangle_{0\to T} &= \sum_{m,n} \mathrm{e}^{-\beta(E_m(T)-E_n(0))} P(m|n) \frac{\mathrm{e}^{-\beta E_n(0)}}{Z_\beta(0;\{\tau_{ij}\})} \\ &= \frac{1}{Z_\beta(0;\{\tau_{ij}\})} \sum_{m,n} \mathrm{e}^{-\beta E_m(T)} P(m|n) = \frac{Z_\beta(T;\{\tau_{ij}\})}{Z_\beta(0;\{\tau_{ij}\})},\end{aligned} \tag{5}$$

where we use the path probability for a nonequilibrium process as, $P(m|n) = |\langle m(T)|U_{0\to T}|n(0)\rangle|^2$. Here $U_{0\to T}$ is the time evolution operator.

3.2. *Jarzynski equality for quantum annealing : Symmetric distribution*

The partition function for the initial Hamiltonian $H(0)$ does not depend on the configuration $\{\tau_{ij}\}$ since it is given only by the transverse field, $Z_\beta(0;\{\tau_{ij}\}) = (2\cosh\beta\Gamma)^N$. Here N is the number of spins. Let us now consider the configurational average of the Jarzynski equality for quantum annealing by summation over all the possible configurations $\{\tau_{ij}\}$ as

$$\left[\langle \mathrm{e}^{-\beta W}\rangle_{0\to T}\right]_{\beta_p} = \frac{[Z_\beta(T;\{\tau_{ij}\})]_{\beta_p}}{(2\cosh\beta\Gamma)^N}. \tag{6}$$

The explicit expression of the quantity on the right-hand side is

$$\frac{[Z_\beta(T;\{\tau_{ij}\})]_{\beta_p}}{(2\cosh\beta\Gamma)^N} = \frac{1}{(2\cosh\beta\Gamma)^N (2\cosh\beta_p J)^{N_B}} \sum_{\tau_{ij}} \prod_{\langle ij\rangle} \mathrm{e}^{\beta_p J\tau_{ij}} Z_\beta(T;\{\tau_{ij}\}), \tag{7}$$

where N_B is the number of interactions. If we set $\beta_p = 0$, which corresponds to the case with the symmetric distribution of the interaction ($p = 1/2$), the above quantity can be evaluated as

$$\left[\langle \mathrm{e}^{-\beta W}\rangle_{0\to T}\right]_{\beta_p=0} = \frac{2^N (2\cosh\beta J)^{N_B}}{2^{N_B} (2\cosh\beta\Gamma)^N}. \tag{8}$$

3.3. *Inverse statistics on special subspace*

Let us show that the gauge transformation reveals a non-trivial property of the nonequilibrium average of the exponentiated work during quantum annealing. We take the configurational average of the inverse of the nonequilibrium-averaged exponentiated work as

$$\left[\frac{1}{\langle \mathrm{e}^{-\beta W}\rangle_{0\to T}}\right]_{\beta_p} = \frac{(2\cosh\beta\Gamma)^N}{(2\cosh\beta_p J)^{N_B}} \sum_{\tau_{ij}} \frac{\prod_{\langle ij\rangle} \mathrm{e}^{\beta_p J\tau_{ij}}}{Z_\beta(T;\{\tau_{ij}\})}. \tag{9}$$

Here we apply the gauge transformation to the above equality. The quantity on the left-hand side does not change, since the Hamiltonian and the time-evolution operator are invariant. On the other hand, the quantity on the right-hand side becomes

$$\left[\frac{1}{\langle \mathrm{e}^{-\beta W}\rangle_{0\to T}}\right]_{\beta_p} = \frac{(2\cosh\beta\Gamma)^N}{(2\cosh\beta_p J)^{N_B}} \sum_{\tau_{ij}} \frac{\prod_{\langle ij\rangle} \mathrm{e}^{\beta_p J\tau_{ij}\xi_i\xi_j}}{Z_\beta(T;\{\tau_{ij}\})}. \tag{10}$$

Setting $\beta = \beta_p$ and taking the summation over all configurations of ξ_i leads us to

$$\left[\frac{1}{\langle \mathrm{e}^{-\beta W}\rangle_{0\to T}}\right]_{\beta} = \frac{2^{N_B}\left(2\cosh\beta\Gamma\right)^{N}}{2^{N}\left(2\cosh\beta J\right)^{N_B}}, \tag{11}$$

which implies that

$$\left[\langle \mathrm{e}^{-\beta W}\rangle_{0\to T}\right]_{\beta_p=0} = \left(\left[\frac{1}{\langle \mathrm{e}^{-\beta W}\rangle_{0\to T}}\right]_{\beta_p=\beta}\right)^{-1}. \tag{12}$$

This equality represents a non-trivial relationship in quantum annealing driven toward two different states in spin glasses $\beta_p = 0$ and $\beta_p = \beta$. It is important to remark that we do not use any approximations in the above calculation, and we deal with the nonequilibrium average during quantum annealing, which does not assume to be quasi-static, unlike quantum adiabatic computation.

4. Summary

We can obtain several further exact equalities for the repetition of non-adiabatic quantum annealing and establish non-trivial relations that hold between quantum annealing toward different spin glasses, as will be announced shortly in a separate paper. Exactness of our results would be useful as a benchmark to check experimental conditions of nonequilibrium process in quantum spin dynamics. Notice that our results are scalable, which means independence of the system size, since the Jarzynski equality holds for any size if we take all fluctuations into account in estimation of the expectation value on the left-hand side of equation (4). Finite errors on the ratio of the partition functions may be attributed to the rare event on nonequilibrium processes. The present study will help us to understand and manipulate quantum dynamics efficiently.

References

[1] Garey M R and Johnson D S 1979 Computers and Intractability: A Guide to the Theory of NP-Completeness (San Francisco: Freeman)
[2] Hartmann A K and Weigt M 2005 Phase Transitions in Combinatorial Optimization Problems: Basics, Algorithms and Statistical Mechanics (Wiley-VCH, Weinheim)
[3] Kadowaki T and Nishimori H 1998 *Phys. Rev. E* **58** 5355
[4] Das A and Chakrabarti B K 2008 *Rev. Mod. Phys.* **80** 1061
[5] Morita S and Nishimori H 2008 *J. Math. Phys.* **49** 125210
[6] Ohzeki M and Nishimori H 2011 to appear in J. Comp. Theor. Nanoscience
[7] Farhi E, Goldstone J, Gutmann S and Sipser M, arXiv:0001106
[8] Jörg T, Krzakala F, Kurchan J and Maggs A C 2008 *Phys. Rev. Lett.* **101** 147204
[9] Young A P, Knysh S and Smelyanskiy V N 2010 *Phys. Rev. Lett.* **104** 020502
[10] Ohzeki M 2010 *Phys. Rev. Lett.* **105** 050401
[11] Jarzynski C 1997 *Phys. Rev. Lett.* **78** 2690
[12] Jarzynski C 1997 *Phys. Rev. E* **56** 5018
[13] Nishimori H 1981 *Prog. Theor. Phys.* **66** 1169
[14] Nishimori H *Statistical Physics of Spin Glasses and Information Processing: An Introduction* (Oxford Univ. Press, Oxford, 2001)
[15] Ohzeki M and Nishimori H 2010 *J. Phys. Soc, Jpn.* **79** 084003
[16] Morita S, Nishimori H and Ozeki Y 2006 *J. Phys. Soc. Jpn.* **75** 014001
[17] Tasaki H arXiv:0009244
[18] Campisi M, Talkner P and Hänggi P 2009 *Phys. Rev. Lett.* **102** 210401

Jarzynski Equality for an Energy-Controlled System

Hitoshi Katsuda[1] and Masayuki Ohzeki[2]

[1] Department of Physics, Tokyo Institute of Technology, Oh-okayama, Meguro-ku, Tokyo 152-8551, Japan

[2] Department of System Science, Graduate School of Informatics, Kyoto University, Yoshida-Honmachi, Sakyo-ku, Kyoto 606-8501, Japan

E-mail: [1] `katsuda@stat.phys.titech.ac.jp`

Abstract. We extend the Jarzynski equality, which is an exact identity between the equilibrium and nonequilibrium averages, to be useful to compute the value of the entropy difference by changing the Hamiltonian. To derive our result, we introduce artificial dynamics where the instantaneous value of the energy can be arbitrarily controlled during a nonequilibrium process. We establish an exact identity on such a process corresponding to the so-called Jarzynski equality. It is suggested that our formulation is valuable in a practical application as in optimization problems.

1. Introduction

To decrease the power loss, we often design the shortest paths to connect components in integrated circuits. Not only the electric loss but also any kind of cost is demanded to be minimized as much as possible in industrial products. Such problems can be formulated into a more generic task to minimize or maximize a real single-valued function of multivariables, which is called optimization problems [1, 2]. Several solvers for optimization problems have been invented in the context of the dynamical process in statistical physics [3, 4]. In these methods, the system is driven to be trapped at the global minimum of the complicated valley structure.

In the present study, we approach this issue in a non-standard way with the recent progress in statistical mechanics. We propose a method to measure the difference between the approximate result given by solvers of optimization problems and the true answer. In practice, we do not always know the ground state energy, however the minimum value of the entropy is known trivially. Therefore the entropy can be an indicator for the deviation of the approximate solution from the accurate answer.

To facilitate our study, we extend the identity proposed by Adib, which is useful to estimate the entropy difference in changing the Hamiltonian in artificial dynamics [5]. This identity is inspired by the Jarzynski equality, which plays a key role to connect equilibrium states at beginning and end with a nonequilibrium process [6]. However the original formulation given by Adib is considered only for the isoenergy process, in which the energy is fixed to be a constant value, while the Hamiltonian changes. We extend the identity for the entropy difference in the isoenergy process to an energy-controlled process [7].

2. Formulation

In order to arbitrarily control the energy, we introduce an artificial field $\mathbf{F} = (\mathbf{F}_x, \mathbf{F}_p)$ to the Hamilton dynamics. The equations of the modified dynamics are

$$\dot{\mathbf{x}} = \frac{\partial H}{\partial \mathbf{p}} + \mathbf{F}_x(\mathbf{\Gamma}), \quad \dot{\mathbf{p}} = -\frac{\partial H}{\partial \mathbf{x}} + \mathbf{F}_p(\mathbf{\Gamma}). \tag{1}$$

The above $\mathbf{\Gamma} = (\mathbf{x}, \mathbf{p})$ describes a point in the phase space. The energy follows an arbitrary function of time $E(t)$, while the Hamiltonian changes from $t = 0$ to τ, if we choose the functional form of $\mathbf{F} = (\mathbf{F}_x, \mathbf{F}_p)$ as

$$\mathbf{F}(\mathbf{\Gamma}) = \frac{\mathbf{X}}{\mathbf{X} \cdot \partial_{\mathbf{\Gamma}} H} \left(\frac{\mathrm{d}E}{\mathrm{d}t} - \frac{\partial H}{\partial t} \right). \tag{2}$$

Here $\mathbf{X}$ is an arbitrary vector on the phase space satisfying $\mathbf{X} \cdot \partial_{\mathbf{\Gamma}} H \neq 0$.

From equations (1) and (2), we can easily confirm that the energy is accurately controlled as $E(t)$ since we have

$$\frac{\mathrm{d}H}{\mathrm{d}t} = \frac{\partial H}{\partial \mathbf{\Gamma}} \cdot \dot{\mathbf{\Gamma}} + \frac{\partial H}{\partial t} = \frac{\mathrm{d}E}{\mathrm{d}t}. \tag{3}$$

Under the special dynamics (1), the ensemble density $\rho_t(\mathbf{\Gamma})$ evolves following the Liouville equation:

$$\rho_t(\mathbf{\Gamma}_t) = \rho_0(\mathbf{\Gamma}_0) \mathrm{e}^{-t\overline{\Lambda_t}(\mathbf{\Gamma}_t)}, \tag{4}$$

where

$$\overline{\Lambda_t}(\mathbf{\Gamma}_t) = \frac{1}{t} \int_0^t \mathrm{d}t' \Lambda(\mathbf{\Gamma}_{t'}). \tag{5}$$

Equation (5) is the time average of the "phase space compression factor" $\Lambda(\mathbf{\Gamma}_t) = \partial_{\mathbf{\Gamma}_t} \cdot \dot{\mathbf{\Gamma}}_t$ along the trajectory that connects $\mathbf{\Gamma}_0$ to $\mathbf{\Gamma}_t$ [8]. This factor appearing in the dynamics of the ensemble density $\rho_t(\mathbf{\Gamma}_t)$ will play the central role to estimate the entropy as shown below. Similarly to the original Jarzynski equality, the system is assumed to be in an equilibrium state at the initial time $t = 0$. The distribution $\rho_0(\mathbf{\Gamma}_0)$ at the initial time is set to be the microcanonical distribution at $E = E(0)$:

$$\rho_0(\mathbf{\Gamma}_0) = \frac{\delta(H(\mathbf{\Gamma}_0) - E(0))}{\Omega_0}, \tag{6}$$

where Ω_0 is the number of states at $t = 0$.

Let us consider the average of $\mathrm{e}^{\tau\overline{\Lambda_\tau}}$ over all possible realizations from $t = 0$ to τ:

$$\langle \mathrm{e}^{\tau\overline{\Lambda_\tau}} \rangle = \int \mathrm{d}\mathbf{\Gamma}_\tau \rho_\tau(\mathbf{\Gamma}_\tau) \mathrm{e}^{\tau\overline{\Lambda_\tau}} = \int \mathrm{d}\mathbf{\Gamma}_\tau \frac{\delta(H(\mathbf{\Gamma}_0) - E(0))}{\Omega_0}. \tag{7}$$

In the second transformation, equations (4) and (6) have been used. Note that we control the energy of the system as $H(\mathbf{\Gamma}_t) = E(t)$, and we thus find

$$H(\mathbf{\Gamma}_t) - E(t) = H(\mathbf{\Gamma}_0) - E(0) \tag{8}$$

for any $t \in [0, \tau]$. Therefore $H(\mathbf{\Gamma}_0) - E(0)$ in the Dirac delta function of equation (7) can be replaced by $H(\mathbf{\Gamma}_\tau) - E(\tau)$. Equation (7) can be reduced to

$$\langle \mathrm{e}^{\tau\overline{\Lambda_\tau}} \rangle = \mathrm{e}^{\Delta S}. \tag{9}$$

The above $\Delta S = \ln(\Omega_\tau / \Omega_0)$ is the entropy difference between the equilibrium states at the different energy values $E(0)$ and $E(\tau)$. Therefore we can obtain the entropy difference after the nonequilibrium procedure following the schedule of the energy $E(t)$. If we estimate the entropy difference with such a naive method as directly calculating the entropy at the initial and final times, we have to investigate the entropy twice. On the other hand, our formula (9) gives the entropy difference by taking just a single average of $\mathrm{e}^{\tau\overline{\Lambda_\tau}}$. This means that we can examine the entropy difference by the direct use of the resulting ensemble after the nonequilibrium process.

3. Application to optimization problems

We here show that our equality can be used for a quantitative estimation, which indicates how much an approximate solution differs from the true solution. Let us consider an arbitrary potential energy $V(\mathbf{x})$ with continuous variables $\mathbf{x}$ which has no degeneracy at the ground state, and assume that a solution with the energy $\mathcal{E}$ has been obtained. We consider the following Hamiltonian

$$H = \mathbf{p}^2 + \frac{t}{\tau}V(\mathbf{x}). \tag{10}$$

At the first stage of the dynamics, the system can visit all the locations since the potential energy is small. By increasing the coefficient of the potential energy, the particle can recognize the energy barriers and be trapped into local minima. Let us consider the schedule of the energy from $E(0) = \mathcal{E}'$ to $E(\tau) = \mathcal{E}$, where $\mathcal{E}' > \mathcal{E}$. The entropy at the initial time $S(0)$ can be calculated easily since the system equals to non-interacting particles with the mass $m = 1/2$ at $t = 0$. On the other hand, the entropy at the final time $S(\tau) = S(0) + \Delta S$ can be estimated by calculating ΔS from our identity.

The detailed protocol is described below. First, we randomly choose an initial condition $(\mathbf{x}_0, \mathbf{p}_0)$ from the set $\{(\mathbf{x},\ \mathbf{p})|H(\mathbf{x}, \mathbf{p}; t = 0) = \mathcal{E}'\}$. Note that the initial Hamiltonian depends only on $\mathbf{p}$. We obtain the path toward a phase point with the lower energy following the energy-controlled dynamics (1) under the given initial condition. Some initial conditions result in the divergence of the factor appearing on the right-hand side of equation (2) when the system is trapped in a local minimum with $\partial_{\mathbf{\Gamma}_t} H = 0$. Such a divergence means that the energy can not decrease toward $\mathcal{E}$ and the system evolving from such initial conditions are unable to reach any points on the phase space at $t = \tau$. Therefore such samples should be excluded in taking the average of $\exp \tau\overline{\Lambda_\tau}$. Finally, we take the average of $\exp \tau\overline{\Lambda_\tau}$ only in the case of the absence of such divergences, and we obtain ΔS appearing on the right-hand side of our equality (9).

In the case of optimization problems, the value of the energy at the ground state cannot be known in advance. On the other hand, the minimum value of the entropy is trivial as $-\infty$ in the case of classical systems. Therefore $S(\tau) \to -\infty$ implies that the obtained solutions after the nonequilibrium process for the given potential energy are close to the minimum point. In other words, estimating the difference of the entropy can be an indicator how much the approximate solution differs from the actual minimum.

The detailed estimation by use of a simple instance is given in the reference [7]. Let us here emphasize another advantage in use of the entropy difference to measure the deviation from the true solution. As described in figure 1, if $\Delta E = \mathcal{E}_2 - \mathcal{E}_1$ takes a common value in both cases, the entropy gap ΔS can become quite different, since the entropy rapidly decreases when the system breaks through local minima of the potential energy. The energy cannot lead us to any information on the difference coming from the energy structure. Therefore the entropy difference can be a good measure of the deviation from the true solution in optimization problems.

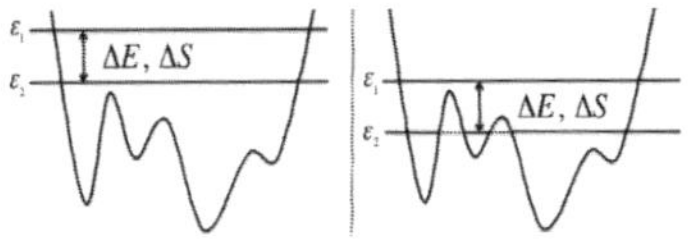

Figure 1. Case that equal ΔE but different ΔS.

4. Conclusion

We have extended the Adib identity inspired by the Jarzynski equality to an energy-controlled system, and shown the possibility as the measure of the deviation of the approximate solution from the true answer. However our study is still incomplete, since the efficient dynamics to search for the ground state have not been considered. In addition we have a trouble in numerical computation, since the value of the entropy of an approximate solution is compared with an infinite value in our protocol. To solve these problems, we might establish the quantum version of our identity. In the case of a quantum system, the entropy at the ground state is equal to 0 which is clearly more suitable for the quantitative estimation than $-\infty$ of classical systems. Furthermore, the tunneling effect can be useful, as quantum annealing, for searching for the global minimum of the complicated energy landscape as in spin glasses, with which most of the optimization problems are closely related [9].

We also remark that the quantum version of our formulation can also open the possibility of quantum computation. By use of the quantum degrees of freedom, we can implement our identity in the quantum computation [10]. The same techniques as in the literature would be available to suggest the algorithm to estimate the entropy as well as the free energy.

References

[1] Garey M R and Johnson D S 1979 *Computers and Intractability: A Guide to the Theory of NP-Completeness* (San Francisco: Freeman)
[2] Hartmann A K and Weigt M 2005 *Phase Transitions in Combinatorial Optimization Problems: Basics, Algorithms and Statistical Mechanics* (Weinheim: WileyVCH)
[3] Kirkpatrick S, Gelett C D and Vecchi M P 1983 *Science* **220** 671
[4] Kadowaki T and Nishimori H 1998 *Phys. Rev. E* **58** 5355
[5] Adib A B 2005 *Phys. Rev.* E **71** 056128
[6] Jarzynski C 1997 *Phys. Rev. Lett.* **78** 2690
[7] Katsuda H and Ohzeki M 2011 *Preprint* cond-mat/1101.3826
[8] Evans D J and Morris G P 1990 *Statistical Mechanics of Nonequilibrium Liquids* (London: Academic).
[9] Nishimori H 2001 *Statistical Physics of Spin Glasses and Information Processing: An Introduction* (New York: Oxford University Press) chapter 9
[10] Ohzeki M 2010 *Phys. Rev. Lett.* **105** 050401

Designer Atomic Nuclei an Emerging Tool for Science

B.M. Sherrill

Facility for Rare Isotope Beams, Michigan State University, East Lansing, MI 48824, USA

sherrill@frib.msu.edu

Abstract. The field of nuclear science is in the middle of several revolutions. One is centered on the availability of a much wider range of isotopes than was previously obtainable. A range of isotopes creates the possibility to devise experiments where specific features of the nuclear many-body problem are isolated and hence model assumptions tested and approximations refined or changed. The new isotope tools allow us to search for the fundamental physics underlying apparent regularities found in atomic nuclei. The revolution extends to astrophysical modeling where, finally, a vast majority of the key nuclear data can be measured and used in astrophysical models. This will allow observations, such element abundances in stars, to be used to infer conditions in stellar explosions, or in the environment in which the star formed. This paper discusses some of the advances and potential other benefits of a wider range of isotopes.

1. Introduction

The frontiers of nuclear science involve the quest to study many-body aspects of QCD in regimes of the highest temperatures at RHIC [1] and LHC [2], and the highest densities as occur in neutron stars [3]. The frontiers extend to the effort to understand the QCD underpinning of atomic nuclei and the force that binds nucleons together. Key open questions remain on the nature of quark confinement, the origin of the nucleon-nucleon force in terms of QCD, and how to predicatively model atomic nuclei and their interactions.

Central to meeting the challenge of modeling atomic nuclei is the experimental quest to synthesize atoms made of arbitrary combinations of neutrons and protons. The ability to specify a given proton and neutron number and then study that system is similar to the capability to produce designer materials for nanoscience; hence, nuclear science has entered an age of designer atomic nuclei. However, isotopes are a synthesis problem on the femtoscale, and the difficulty of producing nuclei with widely different numbers of neutrons and protons has meant that nuclear scientists normally work in a limited range of proton and neutron numbers. The interest in producing a wider range is that certain combinations highlight particular aspects of the nuclear many-body problem and allow specific parts of the nuclear interaction or specific forces to be studied. Insight will also come from the novel quantum structures of certain exotic isotopes with low binding energies that exhibit phenomena such as haloes and skins [4]. These features are illustrated schematically in figure 1.

In part, based on the current capabilities for creating new nuclides at facilities such as the RIBF accelerator at RIKEN [5], our understanding of atomic nuclei has changed dramatically. Many of the so called basic properties of atomic nuclei turn out to not be as universal as we thought. For example, we now understand the standard shell model of atomic nuclei must be modified to include three-body and tensor forces [6,7]. New theoretical approaches and computational techniques have also been important in the development of a more complete picture. Over that last decade we have developed new theoretical approaches inspired by QCD and benefiting from the emergence of computational tools that allow modeling the nuclear many-body system with far fewer approximations.

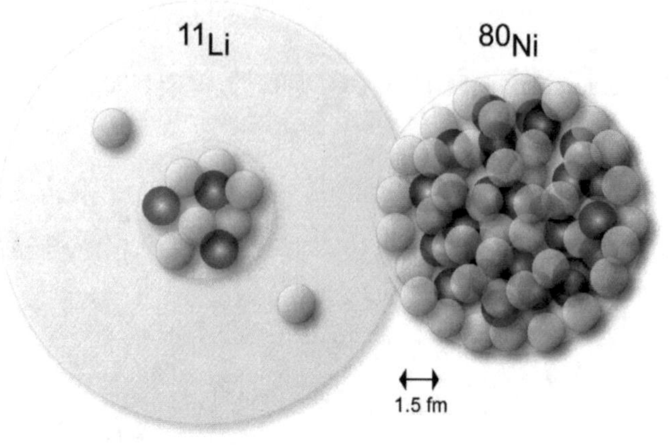

Figure 1. Schematic examples of halo (left) and skin nuclei (right). The weak binding of the last two neutrons of ^{11}Li allow their wave function to penetrate beyond the central ^{9}Li core to a size similar to that of a ^{208}Pb nucleus. Nuclei such as ^{80}Ni (28 protons and 52 neutrons) are predicted to have surface of neutron matter due to the large difference in Fermi energies of the neutrons and protons. The Coulomb force constraints the protons but the neutrons can extend to larger radii.

2. Understanding the nuclear many body system

An ultimate goal of nuclear physics would be to describe atomic nuclei from QCD. This is however an extremely difficult problem. Lattice QCD is now able to calculate the relative masses of hadrons and heavy mesons at the scale of 1 GeV [8]. However nuclear features, such as the rotational bands in ^{220}Rn are at a scale 20 times lower in energy. There is no fundamental theory of QCD that can explain these features.

2.1. Origins of the nuclear force

One approach to solving the nuclear many-body problem is to study nucleon-nucleon scattering and use this data to construct two-body potentials such as the Argonne V18 potential [9], add three-body potentials [10], and then use these along with modern computational power to calculate the lightest nuclei up to ^{12}C [10] using Greens Function Monte Carlo (GFMC) methods. These techniques, referred to as *ab initio*, are remarkably successful and can calculate the binding energy of ^{12}C to better than 1 MeV [10]. A review of *ab intio* techniques can be found in reference [11]. A particularly important insight from the GFMC work came from comparison of calculated and measured binding energies of very neutron-rich helium isotopes. This led to the refinement of the three body forces used and a significant overall improvement in the accuracy of the technique. This example demonstrates how rare isotopes are often the only way to test and refine certain aspects of nuclear models.

Other recent attempts to model the nuclear force start from QCD [12] or string theory [13] to construct the nuclear potentials. Lattice QCD may be able to constrain parameters for the nuclear three-body force and three-body systems [14], which can then be tested against what is learned from rare isotopes [7]. Another promising approach is the development of effective field theories based on the symmetries of QCD, such as chiral symmetry [15]. These approaches naturally introduce three body forces in third order and also hint that four-body forces will be important in nuclei at some level.

2.2. A roadmap to a predictive model of atomic nuclei and their reactions

The nuclear theory community has developed a road map for how to construct a comprehensive and predictive model of atomic nuclei [16]. This involves starting with *ab initio* models for the lightest nuclei (here, study of neutron-rich, light nuclei helps determine the force to use in models). Mid-mass nuclei are modeled by configuration-interaction theory such as the shell model [6] (here, study of key nuclei such as ^{54}Ca, ^{60}Ca, ^{122}Zr will be important to determine the correct degrees of freedom such as shell closures and effective interactions). The heaviest nuclei will be modeled with improved density functional theory (here new measurements to provide a broad view of the nuclear mass surface, deformation and vibrational modes as a function of Z and N, the nuclear fission barrier mass surface, and other basic quantities are important).

The improved nuclear models and computational tools will advance our ability to calculate the reactions of nuclei. This will provide the ability to, for example, reliably calculate all the reactions important for Big Bang nucleosynthesis and develop models for nuclear decay. Understanding the role of the continuum is a key to these developments. Halo nuclei (left side of figure 1) are also sensitive to proper treatment of the continuum and hence provide a testing ground for reaction theory. Next generation facilities will allow drip line nuclei up to around mass number 100 to be studied. A prediction to illustrate the extreme extent for one such case, ^{42}Mg, is shown in figure 2. The proton and neutron densities were calculated with HFB theory with the separation energy constrained to 100 keV (the predicted value).

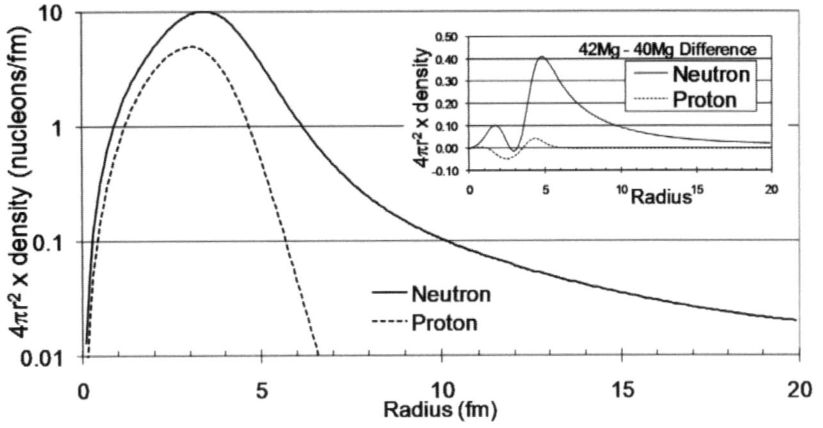

Figure 2. Radial density calculated for the nucleus ^{42}Mg, illustrating the extreme halo nuclei that can be produced at rare isotope facilities. The inset shows the predicted change in the core proton distribution caused the valence two neutrons.

3. Advances in nuclear astrophysics

3.1. Measurement of isotope abundances in stellar objects

The measurement of elemental abundances in stars is at the forefront of astronomical observation. Detailed abundance measurements are made possible by large area telescopes such as Subaru, Hubble Space Telescope, LBT, and the Keck telescopes, due to their high resolving power. A wealth of new data is available from detailed sky surveys that provide elemental abundance data in hundreds of thousands of stars [17]. A sample of the type of data available is shown in figure 3; taken from the Hamburg/ESO r-process star survey [18]. The figure shows the logarithmic ratios (defined by equation 1) of one element to another compared to the Solar value. In the figure barium to iron and iron to hydrogen ratios are shown. In general there are more than 30 elemental abundances heavier than iron that can be determined for stars in our galaxy. The study of rare isotopes is needed to provide the underlying nuclear physics to allow this evolution to be accurately modeled, or inversely to use models to infer the conditions in the universe over time. One of the goals would be to use the elemental ashes from the second and third generations of stars to let us infer the conditions that existed for the first stars in the universe.

$$[\mathrm{Fe/H}] = LOG\left[\frac{(Fe\ abundance\,/\,H\,abundance)_{Star}}{(Fe\ abundance\,/\,H\,abundance)_{Sun}}\right] \tag{1}$$

This is one of the main goals of future space telescope missions such as the James Webb Space Telescope - "is specifically designed for discovering and understanding the formation of the first stars and galaxies, measuring the geometry of the Universe and the distribution of dark matter, investigating the evolution of galaxies and the production of elements by stars, and the process of star and planet formation [19]."

3.2. Modeling of explosive astrophysical environments

Another key area of astrophysics where rare isotopes are important is in explosive environments and in modeling neutron stars. More than half of the elements heavier than iron are produced in a rapid neutron capture process, the r-process. In order to match observations with models nuclear data is needed on very neutron-rich isotopes [20] that up to now have been largely unavailable. Data are also needed on electron capture rates to model the process before and during supernovae explosions and to model process on the surface of neutron stars that lead to X-ray busts [21].

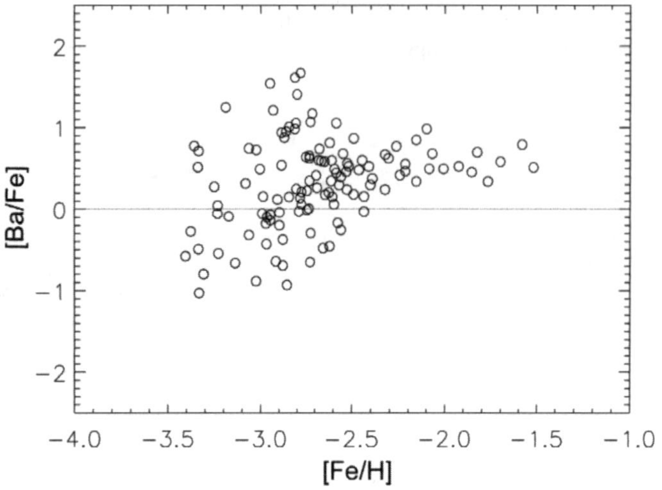

Figure 3. Relative log abundances for sample stars measured in the HERES survey [18]. The figure illustrates the chemical evolution that measurements with rare isotopes will help unravel. Because stars create Fe over time while H is close to constant, the horizontal axis is related to time. The variance among individual stars of Ba/Fe tell us about the history of those stars. Currently we do not understand what that is.

4. Other uses of rare isotopes

4.1. Tests of fundamental symmetries

Rare isotopes can be used to test the fundamental symmetries and search for the ingredients of the New Standard Model [for a review see reference 22]. Use of isotopes from facilities like FRIB will allow tests of parity and time reversal symmetries and searches for potential new contributions to the weak interactions. For example, the isotopes produced at FRIB would allow an improved measurement of parity-non-conserving (PNC) effects in atomic systems and a determination of the weak charge, Q_w, of the nucleus [23]. In atoms, PNC observables are proportional to Q_w times an atomic structure function that depends upon the distribution of neutrons in the nucleus. Determination of the latter limits the accuracy of the determination of Q_w. The uncertainty from atomic calculations can be greatly reduced by using ratios of PNC observables in a series of isotopes–but this method requires a precise knowledge of the isotopic dependence of the neutron skin, which may come from the improved density functional theory resulting from FRIB. Another area where facilities like FRIB may contribute is to provide samples of isotopes such as ^{229}Pa where the presentence sensitivity to and atomic EDM may be 10,000 greater than in current tests. An atomic EDM would be evidence for violation of time-reversal symmetry and provide tests for supersymmetric models and string theory [22].

4.2. Isotope applications in other sciences

Radioactive isotopes are used in a wide variety of applications ranging from medical diagnostics to the tracing of groundwater migration patterns. They serve as sensitive probes in materials science studies of nano-scale devices or in the study of mechanical wear of novel materials. In medical imaging, they form the basis of a multi-billion dollar industry. FRIB will have the capability to provide access to the full spectrum of known, important radioisotopes, such as ^{7}Be, ^{8}Li, ^{11}Be, ^{32}Si, ^{221}Rn and others. However, more broadly, FRIB can endow future advances in the applied sciences by providing scientists access to thousands of new isotopes that are not normally available.

Table 1 shows some examples that have been identified as examples of interesting isotopes that a facility such as FRIB would produce well and that could be harvested parasitically to normal operation [24].

Table 1. Sample applications of isotopes.

Nuclide	**Half-life**	**Use** [a]
Silicon-32	153 y	Oceanographic studies; climate change
Radon-221	25 m	Targeted alpha therapy
Protactinium-229	15 d	Electric dipole moment search
Radium-225	15 d	Electric dipole moment search
Krypton-85	11 y	High specific activity for s-process studies and homeland security
Titanium-44	60 y	Target and ion source material
Copper-67	62 h	Imaging and therapy for hypoxic tumours

[a] These are representative uses discussed at the Workshop on Isotope Harvesting September 2010 at Santa Fe [24], New Mexico. Other uses are likely as well.

5. Rare isotope production facilities and the next steps

The dramatic next steps are being taken at new facilities such as RIBF at RIKEN, and will be taken will be the construction of new facilities such as the Facility for Rare Isotope Beams, FRIB, in the United States [25], FAIR in Germany [26], and SPIRAL2 in France [27]. These facilities will be able to produce most of the astrophysically interesting isotopes needed to model element formation in the universe, neutron stars, and stellar explosions [25]. They will also allow specific, key measurements of nuclear properties needed for progress in nuclear theory. Finally, the new capability promises to provide a range of interesting isotopes of use for fields from condensed matter science to human health. [25].

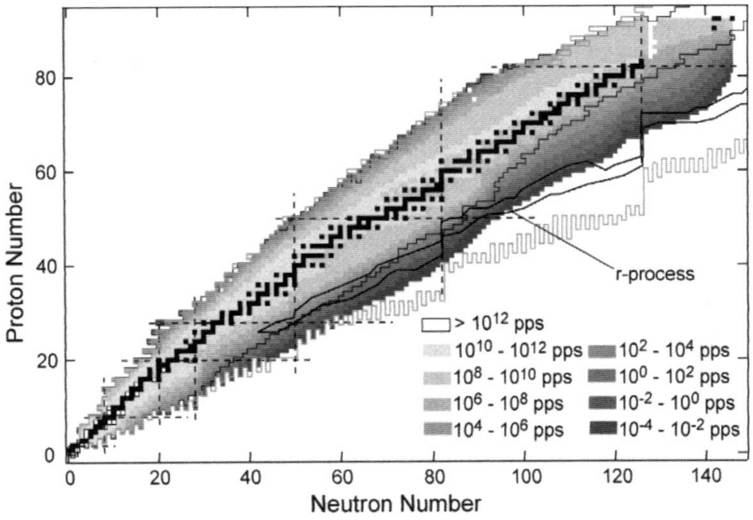

Figure 4. FRIB production rates indicated by shades of gray. Similar production yields will be possible at other modern facilities such as RIBF and FAIR.

The range of new isotopes that can be produced at one of these facilities, FRIB is illustrated in figure 4. The approximate limit of the current isotopes is marked by the jagged black line. FRIB (along with the other new facilities world-wide) will be able to produce more than 1000 new isotopes and allow detailed study of nearly 5000 isotopes compared to the 1700 or so studied in some detail today. It will allow the neutron drip line to be studied up to $A = 120$ (compared to $A = 24$ today) and production of most of the key nuclei for astrophysical modeling. The solid band shows a prediction for the range of isotopes that would be important for modeling the r-process, which is responsible for producing more than half of the elements heavier than iron, including making most of the gold atoms and all of the uranium we find on Earth.

6. Conclusions

Nuclear science has entered an age of designer isotopes, when arbitrary combinations of neutrons and protons can be assembled and used in experiments. The research with these isotopes is carried out at facilities such as RIBF at RIKEN and will be advanced at a number of facilities under construction world-wide such as FRIB. With the new tools of isotope production there is a pathway to develop a comprehensive and predictive model of atomic nuclei. This and the experiments with new, rare isotopes will allow researchers to understand the nuclear force, more accurately model astrophysical objects, test nature's symmetries and provide isotopes over a wide range for other sciences.

References

[1] Riordan M, Zajc WA 2006 Scientific American **294** 34
[2] Yurkewicz K 2011 "The skinny on the LHC's heavy ions" Symmetrybreaking Magazine **8,** http://www.symmetrymagazine.org/breaking/2010/11/05/the-skinny-on-the-lhcs-heavy-ions/
[3] Steiner AW, Lattimer JM, Brown EF (2010) Astrophysical Jour **722** 33-54
[4] Hansen G, Jensen AS and Jonson B 1995 Ann. Rev. Nucl. Part. Sci. 45 (1995) 591; and Tanihata I, Prog. Part. Nucl. Phys. **35** 505
[5] Yano Y 2007 Nucl. Instr. and Meth. B **261** 1009
[6] Otsuka T, *et al.* 2010 Phys. Rev. Lett. **104** 012501
[7] Otsuka T, Suzuki T, Holt JD 2010 Phys. Rev. Lett. **105** 032501
[8] Dürr S, Fodor Z, *et al.* 2008 Science **322** 1224-1227
[9] Wiringa R. B., Stoks V. G. J., and Schiavilla R. 1995 Phys. Rev. C **51** 38
[10] Pieper SC 2005 Nucl Phys A **751** 516c-532c and Pieper SC and Wiringa RB 2001 Annu. Rev. Nucl. Part. Sci. **51** 53
[11] Luu T, Savage MJ, Schwenk A, Vary JP (2010) Phys Rev C **82** 034003
[12] Ishii N., Aoki S., and Hatsuda T. 2007 Phys. Rev. Lett. **99** 022001
[13] Hashimoto K, Sakai T, Sugimoto S 2009 Prog Theor Phys **122** 427-476
[14] Kreuzer S, Hammer HW 2011 Phys Lett B **694** 424-429
[15] Epelbaum E., Hammer H.-W., and Meissner U. 2009 Modern theory of nuclear forces, Rev. Mod. Phys. **81** 1773
[16] RIA Theory Bluebook: A Road Map A Report from the RIA Theory Group; www.jinaweb.org/docs/Blue_Book_6.pdf
[17] *New Worlds, New Horizons in Astronomy and Astrophysics* 2010 (Washington DC:Nat Acad Press)
[18] Barklem PS, Christlieb N, Beers TC, Hill V Bessell MS, Holmberg J, marsteller B, Rossi S, Zickgraf F-J, Reimers D 2005 Astronomy and Astrophysics **439** 129-151
[19] James Webb Space Telescope web site: http://www.jwst.nasa.gov/
[20] Grawe H, Langanke K, Martinez-Pinedo G 2007 Rep. Prog. Phys. **70** 1525-1582
[21] Cyburt RH, Amthor AM, Ferguson R, et al. 2010 Astro Jour Supp **189** 240-252
[22] *Scientific Opportunities with a Rare Isotope Facility in the United States* 2007 (Washington DC:Nat Acad Press)
[23] Brown BA, Derevianko A Flambaum VV 2009 Phys Rev C 79 035501
[24] Harvesting Option for Rare Isotopes at FRIB, Santa Fe workshop 29 Sep – 1 Oct, 2010
[25] Geesaman DF, Gelbke CK, Janssens RVF, Sherrill BM 2006 Ann. Rev. Nucl. Part. Phys. **56** 53
[26] Henning W F 2008 FAIR: recent developments and status Nucl. Phys. A 805 502
[27] Gales S 2010 Nucl Phys A 834 717c-723c

Density functional approaches to atomic nuclei[1]

Takashi Nakatsukasa
RIKEN Nishina Center, Wako, 351-0198, Japan
E-mail: nakatsukasa@riken.jp

Abstract. Nuclear mean-field models are briefly reviewed to illustrate its foundation and necessity of state dependence in effective interactions. This state dependence is successfully taken into account by the density dependence, leading to the energy density functional. Recent results for photoabsorption cross sections in spherical and deformed Nd isotopes are shown.

1. Introduction

The nucleus is a self-bound quantum system which presents a rich variety of phenomena. It is composed of fermions of spin 1/2 and isospin 1/2, called nucleons (protons and neutrons), interacting with each other through a complex interaction with a short-range repulsive core [1]. Remarkable experimental progress in production and study of exotic nuclei requires us to construct a theoretical model with higher accuracy and reliability. Extensive studies have been made in the past, to introduce models and effective interactions to describe a variety of nuclear phenomena and to understand basic nuclear dynamics behind them [1, 2]. Simultaneously, significant efforts have been made in the microscopic foundation of those models. For light nuclei, the "first-principles" large-scale computation, starting from the bare nucleon-nucleon (two-body & three-body) forces, is becoming a current trend in theoretical nuclear physics.

Although the ab-initio-type approaches have recently shown a significant progress, they are still limited to nuclei with the small mass number. In contrast, the density functional model is a leading theory for describing nuclear properties of heavy nuclei and perhaps the only theory capable of describing all nuclei and nuclear matter with a single universal energy density functional. In the nuclear physics, it is often called the self-consistent mean-field model, because of a historical development based on the Brueckner-Hartree-Fock theory and introduction of the effective interaction. I briefly review basic properties of nuclei and discuss whether those properties can be understood by a simple independent-particle model. Then, I recapitulate developments in the microscopic many-body theory leading to the nuclear density functional model with a Skyrme energy functional. Here, the saturation property plays a key role to understand the nuclear force and the effective interaction.

Intensive studies in nuclear density functional models in recent years have produced numerous results and new insights into nuclear structure [3, 4]. However, it is impossible to review all of them in this short paper. Thus, I will present a result of our recent study with the time-dependent density functional approach, on the photoabsorption cross sections in the rare-earth nuclei [5].

[1] This work is supported by Grant-in-Aid for Scientific Research in Japan (Nos. 21340073 and 20105003).

2. Independent particle model for nuclei

Nuclei are known to be well characterized by the saturation property. Namely, they have an approximately constant density $\rho_0 \quad 0.17\ \mathrm{fm}^{-3}$, and a constant binding energy per particle $B/A \quad 16$ MeV.[2] In this section, I show that the nuclear saturation property has a great impact on nuclear models. Especially, it is inconsistent with the independent-particle model of nuclei with a "naive" average (mean-field) potential.

There are many evidences for the fact that the mean-free path of nucleons is larger than the size of nucleus. In fact, the mean free path depends on the nucleon's energy, and becomes larger for lower energy [1]. Therefore, it is natural to assume that the nucleus can be primarily approximated by the independent-particle model with an average one-body potential. The crudest approximation is the degenerate Fermi gas of the same number of protons and neutrons ($Z = N = A/2$). The observed saturation density of $\rho_0 \quad 0.17\ \mathrm{fm}^{-3}$ gives the Fermi momentum, $k_F \quad 1.36\ \mathrm{fm}^{-1}$, that leads to the Fermi energy (the maximum kinetic energy), $T_F = k_F^2/2M \quad 40$ MeV.

First, I show that the independent-particle model with a constant attractive potential $V < 0$ cannot describe the nuclear saturation property. It follows from the simple arguments. The constancy of B/A means that it is approximately equal to the separation energy of nucleons, S. In the independent-particle model, it is estimated as

$$S \quad B/A \quad -(T_F + V). \tag{1}$$

Since the binding energy is $B/A \quad 16$ MeV, the potential V is about -55 MeV. It should be noted that the relatively small separation energy is the consequence of the significant cancellation between kinetic and potential energies. The total (binding) energy is given by

$$-B = \sum_{i=1}^{A} \quad T_i + \frac{V}{2} \quad = A \quad \frac{3}{5}T_F + \frac{V}{2} \quad , \tag{2}$$

where we assume that the average potential results from a two-body interaction. The two kinds of expressions for B/A, Eqs. (1) and (2), lead to $T_F \quad -5V/4 \quad 70$ MeV, which is different from the previously estimated value (~ 40 MeV). Moreover, it contradicts the fact that the nucleus is bound ($T_F < |V|$).

To reconcile the independent-particle motion with the saturation property of the nucleus, the nuclear average potential should be state dependent. Allowing the potential V_i depend on the state i, the potential V should be replaced by that for the highest occupied orbital V_F in Eq. (1), and by its average value $\langle V \rangle$ in the right-hand side of Eq. (2). Then, we obtain the following relation:

$$V_F \quad \langle V \rangle + T_F/5 + B/A. \tag{3}$$

Therefore, the potential V_F is shallower than its average value.

Weisskopf suggested the momentum-dependent potential V, which can be expressed in terms of an effective mass m^* [6]:

$$V_i = U_0 + U_1 \frac{k_i^2}{k_F^2}. \tag{4}$$

Actually, if the mean-field potential is non-local, it can be expressed by the momentum dependence. Equation (4) leads to the effective mass, $m^*/m = (1 + U_1/T_F)^{-1}$. Using Eqs. (1), (3), and (4), we obtain the effective mass as

$$\frac{m^*}{m} = \left\{ \frac{3}{2} + \frac{5}{2}\frac{B}{A}\frac{1}{T_F} \right\}^{-1} \quad 0.4. \tag{5}$$

[2] This is the extrapolated value for the infinite nuclear matter without the surface and the Coulomb energy. The observed values for finite nuclei are $B/A \approx 8$ MeV.

Quantitatively, this value disagrees with the experimental data. The empirical values of the effective mass vary according to the energy of nucleons, $0.7 \lesssim m^*/m \lesssim 1$, however, they are almost twice larger than the value in Eq. (5). As far as we use a normal two-body interaction, this discrepancy should be present in the mean-field calculation with any interaction, because Eq. (5) is valid in general for a saturated self-bound system. Therefore, the conventional models cannot simultaneously reproduce the most basic properties of nuclei; the binding energy and the single-particle property. This suggest the importance of the state-dependent effective interaction, which will be discussed in Sec. 4.

3. Nucleon-nucleon interaction (nuclear force)

The saturation property of nuclear density indicates the balance between attractive and repulsive contributions to nuclear binding energy. One source of such repulsive effects is the nucleonic kinetic energy of the Fermi gas. However, its contribution per particle is proportional to $\rho^{2/3}$, which is not strong enough to resist against the collapse caused by the attractive force between nucleons. Therefore, the nucleonic interaction must contain a repulsive element. Indeed, the phase-shift analysis on the nucleon-nucleon scattering at high energy ($E > 250$ MeV) reveals a short-range strong repulsive core in the nucleonic force. The radius of the repulsive core is approximately c 0.5 fm. This strong repulsive core prevents the nucleons approaching closer than the distance c, which produces a strong two-body correlation, $\rho^{(2)}(\vec{r}_1, \vec{r}_2)$ 0 for $|\vec{r}_1 - \vec{r}_2| < c$. The attractive part of the interaction has a longer range, which can be characterized by the pion's Compton wave length λ , and is significantly weaker than the repulsion. Thus, a naive application of the mean-field calculation fails to bind the nucleus, since the mean-field approximation cannot take account of such strong two-body correlations.

At first sight, this seems inconsistent with the experimental observations. As I mentioned in Sec. 2, there are many experimental evidences for the independent-particle motion in nuclei. We may intuitively understand that it is due to the fact that the nucleonic density is significantly smaller than $1/c^3$. Therefore, the collisions by the repulsive core rarely occur and the system can be approximately described in terms of the independent-particle motion. Furthermore, the effects of the Pauli principle hinder the collisions, since the nucleons cannot be scattered into occupied states. Although the repulsive-core collisions are experienced by only a small fraction of nucleons ($\sim \rho_0 c^3$), each collision carries a large amount of energy. Therefore, the repulsive core provides an important contribution to the total energy and are responsible for the saturation.

Another important factor for the independent particle motion is the strong quantum nature due to the weakness of the attractive part of the nuclear force. The importance of the quantum nature can be measured by the magnitude of the zero-point kinetic energy compared to that of the interaction. If the attractive part of the nuclear force were much stronger than the unit of $\hbar^2/Mc^2$, the quantum effect would disappear and each nucleon would stay at the bottom of the interaction potential (cf. Fig. 2-36 in Ref. [1]). Then, the nucleus would crystallize at low temperature. In reality, the attraction of the nuclear force is so weak that it barely produces many-nucleon bound states at the relatively low density. In Sec. 2, I have shown that, in nuclear binding energy, there is a strong cancellation between the positive kinetic energy and the negative potential energy. The nucleonic kinetic energy plays an important role in many phenomena in nuclei, which can be described in terms of the independent-particle motion.

4. Density-dependent Hartree-Fock method and energy density functional

The nuclear matter theory pioneered by Brueckner gives a hint for a solution for the inconsistency between the nuclear saturation and the independent-particle model. Details of the theory can be found at Refs. [7, 2]. The independent-particle motion under the presence of the interaction with a repulsive core was qualitatively discussed in Sec. 3. The Brueckner theory may provide a first step toward the quantitative treatment to understand the saturation property and the independent-particle motion in nuclei.

The basic ingredient of the Brueckner theory is a two-body scattering matrix of particle 1 and 2 inside nucleus caused by the nuclear force v,

$$G(\omega) \equiv v + v \frac{Q}{\omega - Q(T_1 + T_2)Q} G(\omega), \tag{6}$$

where T_i is the kinetic energy of particle i, Q is the Pauli-exclusion operator to restrict the intermediate states, and ω is called a starting energy that depends on energies of particle 1 and 2. This is called G-matrix [8]. The G-matrix renormalizes high-momentum components in the bare nuclear force and becomes an effective interaction in nuclei under the independent-pair approximation. The G-matrix reflects an underlying structure of the independent many-nucleon system through the operator Q and the starting energy ω. Inevitably, the G-matrix becomes state (structure) dependent.

Since the short-range singularity is renormalized in the G-matrix, we can calculate the total energy in the independent-particle (mean-field) model, analogous to Eq. (2).

$$-B = \sum_{i=1}^{A} \left\{ T_i + \frac{1}{2} \sum_{j=1}^{A} \bar{G}_{ij,ij}(\omega_{ij}) \right\} \tag{7}$$

where $\omega_{ij} = \epsilon_i + \epsilon_j$, defines the self-consistency condition for the Brueckner's single-particle energies, and $\bar{G}_{ij,ij} \equiv G_{ij,ij} - G_{ij,ji}$. This is called Brueckner-Hartree-Fock (BHF) theory. The validity of the BHF theory is measured by the wound integral $\kappa = \langle \psi - \quad | \psi - \quad \rangle$, where $| \quad \rangle$ is an unperturbed two-particle wave function and $|\psi\rangle$ is a correlated two-particle wave function in nucleus. κ is known to be of the order of 15 %. The BHF calculation was successful to describe the nuclear saturation, however, could not reproduce simultaneously B/A and ρ_0, known as a problem of the Coester band [9]. Its applications to finite nuclei also quantitatively failed to reproduce the energy, radius, and density in the ground state.

These problems are somewhat miraculously solved by the density-dependent Hartree-Fock (DDHF) theory by Negele [10]. Starting from a realistic G-matrix, first, the local density approximation is introduced, using the expressions for the Pauli operator

$$\langle \vec{r}_1 \vec{r}_2 | Q | \vec{r}_1' \vec{r}_2' \rangle = \left\{ \delta(\vec{r}_1 - \vec{r}_1') - \rho(\vec{r}_1 - \vec{r}_1') \right\} \left\{ \delta(\vec{r}_2 - \vec{r}_2') - \rho(\vec{r}_2 - \vec{r}_2') \right\}, \tag{8}$$

and the average single-particle energy $\epsilon[\rho(\vec{r})]$. Then, a short-range part of the G-matrix, which is not fully understood, is phenomenologically added to the energy expression to quantitatively fit the saturation property, and finally, the total energy is treated variationally. This procedure is called the density matrix expansion (DME) [11]. The state dependence of the G-matrix is now replaced by the density dependence. The final result for the energy is of the form

$$E[\rho] = \int d\vec{R} H(\vec{R}), \quad H(\vec{R}) = H[\rho(\vec{R})] = H[\psi^*, \psi], \tag{9}$$

which is completely analogous to the Hamiltonian density of the Skyrme energy functional [12].

The essential aspect of the DDHF comes from the density dependence and the variational treatment. The variation of the total energy with respect to the density contains re-arrangement potential, $\partial V_{\text{eff}}[\rho]/\partial\rho$, which appear due to the density dependence of the effective force $V_{\text{eff}}[\rho]$. These terms turn out to be crucial to obtain the saturation condition. Now, the expression for the total energy, Eq. (2), should be modified to include the re-arrangement effect. This resolves the previous issue, then provides a consistent independent-particle description for the nuclear saturation.

In summary, the failure in the mean-field description of nuclei using phenomenological effective interactions can be traced back to the missing state (structure) dependence. The DDHF takes into account the state dependence in terms of the density dependence. The energy functional obtained by the DME is essentially identical to the Skyrme energy functional. This provides a foundation for the nuclear energy functional.

5. Applications to giant resonances

The giant resonance is a typical collective motion in nuclei, which exhausts a major part of the sum-rule value. They are also related to the basic properties of nuclear matter, such as incompressibility K_∞, effective mass m^*, etc. For instance, the nuclear matter incompressibility is extracted from properties of the giant monopole resonances [13],

$$K_\infty = 210 \pm 30 \text{ MeV}. \tag{10}$$

This gives a restriction on the density dependence of the phenomenological short-range repulsive part of the energy functional. The effective mass deduced from the analysis on the giant quadrupole resonances is [13]

$$m^*/m \quad 0.8 \sim 1. \tag{11}$$

These values are consistent with the DDHF in Sec. 4, but inconsistent with the naive mean-field value of Eq. (5).

The giant resonances can be reasonably described by a small-amplitude approximation of the time-dependent version of the DDHF. In this approach, again, the rearrangement terms, such as $\partial V_{\text{eff}}[\rho]/\partial\rho$ and $\partial^2 V_{\text{eff}}[\rho]/\partial\rho^2$ should be consistently taken into account. The theory can be regarded as the time-dependent density-functional theory, founded by Runge and Gross [14].

A modern energy functional for nuclei is a functional of many kinds of density, such as kinetic $\tau(\vec{r})$ and spin-orbit density $\vec{J}(\vec{r})$. In addition, to describe superfluid nuclei with pairing correlation, we need to add the pair (abnormal) density $\kappa(\vec{r})$. These densities are collectively denoted as $\tilde{\rho}$ in the followings. Variation of the total energy, $E[\tilde{\rho}]$, leads to the Hartree-Fock-Bogoliubov equation:

$$H[\Psi, \Psi^*]|\Psi_\mu\rangle = E_\mu|\Psi_\mu\rangle, \tag{12}$$

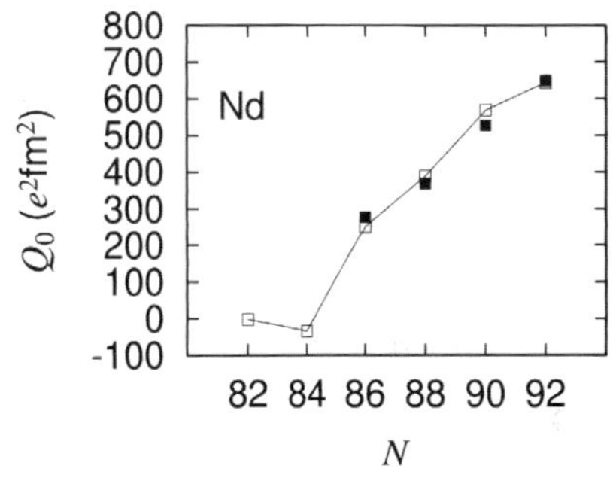

Figure 1. Calculated (open squares) and experimental (filled) intrinsic quadrupole moment for Nd isotopes [5].

where E_μ and $|\Psi_\mu\rangle$ are quasi-particle energies and states, respectively [2]. $|\Psi_\mu\rangle$ is composed of two components; the upper $|U_\mu\rangle$ and the lower one $|V_\mu\rangle$. The solution of Eq. (12) defines the normal density $\rho(\vec{r}) = \sum_\mu V_\mu(\vec{r})V_\mu^*(\vec{r})$, the pair density $\kappa(\vec{r}) = \sum_\mu U_\mu(\vec{r})V_\mu(\vec{r})$, and other densities at the ground state. Since $h[\Psi, \Psi^*]$ depends on these densities, Eq. (12) must be solved in a self-consistent way. Minimization of the energy density functional may lead to a spontaneous breaking of symmetry. An example is given in Fig. 1 for Nd isotopes [5]. The intrinsic quadrupole moment calculated with the Skyrme functional of SkM* is compared with the experimental data. At $N = 82$, the nucleus at the ground state is spherical $Q_0 = 0$, while for $N = 86 \sim 92$, the deformation gradually develops. The observed ground-state deformations deduced from the transition probability $B(E2; 2^+ \to 0^+)$ are nicely reproduced. Note that there are no adjustable parameters in this calculation.

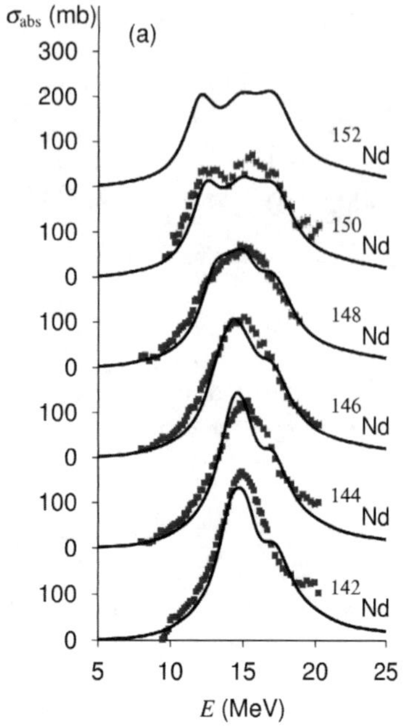

Figure 2. Calculated (lines) and experimental (symbols) photoabsorption cross sections for Nd isotopes [5].

For a study of the giant resonances, we need to extend the energy functional to include the time-odd densities, such as the spin density $\vec{s}(\vec{r})$ and the current density $\vec{j}(\vec{r})$. Now, these densities are time dependent. The time-dependent version of Eq. (12) is

$$i\frac{\partial}{\partial t}|\Psi_\mu(t)\rangle = H[\Psi(t), \Psi^*(t)]|\Psi_\mu(t)\rangle. \tag{13}$$

Assuming the oscillation with a fixed frequency, this equation is linearized with respect to the fluctuation of the densities around those at the ground state. This leads to the matrix form of the equation identical to the quasi-particle random-phase approximation (QRPA) [2].

$$\sum_{\gamma\delta} \begin{pmatrix} A_{\alpha\beta,\gamma\delta} & B_{\alpha\beta,\gamma\delta} \\ -B_{\alpha\beta,\gamma\delta} & -A_{\alpha\beta,\gamma\delta} \end{pmatrix} \begin{pmatrix} X_{\gamma\delta} \\ Y_{\gamma\delta} \end{pmatrix} = \hbar\omega \begin{pmatrix} X_{\alpha\beta} \\ Y_{\alpha\beta} \end{pmatrix}. \tag{14}$$

The QRPA matrix, A and B, are calculated in the quasi-particle basis, then the normal modes of excitation and their energies are obtained from Eq. (14).

The calculated photoabsorption cross sections for Nd isotopes are shown in Fig. 2. For spherical nuclei (N = 82 and 84), the photoabsorption has a single peak for the photon energy of E 15 MeV. The increase of the neutron number results in the broadening of the peak, which well agree with the experimental data. This is due to the increase of the ground-state deformation shown in Fig. 1. The calculated energy-weighted sum-rule values for these isotopes are about 40 % larger than the classical Thomas-Reiche-Kuhn sum-rule value, because of the momentum and isospin dependence of the mean field (Kohn-Sham) potential.

6. Summary

The nuclear density functional approaches were developed as the mean-field theory with the density-dependent effective interactions. The density-dependent Hartree-Fock theory succeeded to describe both the total energy and single-particle properties, simultaneously. In contrast, the normal mean-field calculation with a phenomenological two-body interaction fails. The density-dependence is a key ingredient to understand the nuclear saturation and the independent-particle motion. Numerical applications were shown for giant resonances in shape transitional nuclei.

References

[1] Bohr A and Mottelson B R 1969 *Nuclear Structure, Vol. I* (New York: W. A. Benjamin)
[2] Ring P and Schuck P 1980 *The nuclear many-body problems* (New York: Springer-Verlag)
[3] Bender M, Heenen P H and Reinhard P G 2003 *Rev. Mod. Phys.* **75** 121–180
[4] Lunney D, Pearson J M and Thibault C 2003 *Rev. Mod. Phys.* **75** 1021–1082
[5] Yoshida K and Nakatsukasa T 2011 *Phys. Rev. C* **83** 021304
[6] Weisskopf V F 1957 *Nucl. Phys.* **3** 423–432
[7] Day B D 1967 *Rev. Mod. Phys.* **39** 719; *Rev. Mod. Phys.* **50** 495
[8] Bethe H A and Goldstone J 1957 *Proc. Roy. Soc. (London)* **A238** 551
[9] Day B D 1981 *Phys. Rev. Lett.* **47** 226; *Phys. Rev. C.* **24** 1203
[10] Negele J W 1970 *Phys. Rev. C* **1** 1260–1321
[11] Negele J W and Vautherin D 1972 *Phys. Rev. C* **5** 1472–1493
[12] Vautherin D and Brink D M 1972 *Phys. Rev. C* **5** 626–647
[13] Blaizot J P 1980 *Phys. Rep.* **64** 171–248
[14] Runge E and Gross E K U 1984 *Phys. Rev. Lett.* **52** 997–1000

Search for a permanent EDM using laser cooled radioactive atom

Y.Sakemi, K.Harada, T.Hayamizu, M.Itoh, H.Kawamura,S.Liu, H.S.Nataraj, A.Oikawa, M.Saito, T.Sato, H.P.Yoshida

Cyclotron and Radioisotope Center, Tohoku University, Sendai, Miyagi 980-8578, Japan

T.Aoki

Graduate School of Arts and Sciences, University of Tokyo, Tokyo 153-8902, Japan

A.Hatakeyama

Department of Applied Physics, Tokyo University of Agriculture and Technology, Tokyo 184-8588, Japan

T.Murakami, K.Imai

Department of Physics, Kyoto University, Kyoto 606-8502, Japan

K.Hatanaka

Research Center for Nuclear Physics, Osaka University, Ibaraki, Osaka 567-0047, Japan

T.Wakasa

Department Physics, Kyushu University, Fukuoka 812-8581, Japan

Y.Shimizu

Department of Physics, Tohoku University, Sendai, Miyagi 980-8578, Japan

M.Uchida

Department of Physics, Tokyo Institute of Technology, Tokyo 152-855

E-mail: sakemi@cyric.tohoku.ac.jp

Abstract. An Electric Dipole Moment (EDM) of the elementary particle is a good prove to observe the phenomena beyond the Standard Model. A non-zero EDM shows the violation of the time reversal symmetry, and under the CPT invariance it means the CP violation. In paramagnetic atoms, an electron EDM results in an atomic EDM enhanced by the factor of the 3rd power of the charge of the nucleus due the relativistic effects. A heaviest alkali element francium (Fr), which is the radioactive atom, has the largest enhancement factor K ~ 895. Then, we are developing a high intensity laser cooled Fr factory at Cyclotron and Radioisotope Center (CYRIC), Tohoku University to perform the search for the EDM of Fr with the accuracy of 10^{-29} e · cm. The important points to overcome the current accuracy limit of the EDM are to realize the high intensity Fr source and to reduce the systematic error due to the motional magnetic field and inhomogeneous applied field. To reduce the dominant component of the systematic errors mentioned above, we will confine the Fr atoms in the small region with the Magneto-Optical Trap and optical lattice using the laser cooling and trapping techniques. The construction of the experimental apparatus is making progress, and the new thermal ionizer already produces the Fr of ~10^6 ions/s with the primary beam intensity 200 nA. The developments of the laser system and optical equipments are in progress, and the present status and future plan of the experimental project is reported.

1. Introduction

To explore the mechanism responsible for the generation of observed matter-antimatter asymmetry in the Universe, the research on fundamental symmetry violations and various fundamental interactions using the laser cooled and trapped atoms is being promoted. The understanding of how the symmetry between the matter and antimatter was broken during the evolution of the early universe requires laboratory experiments which search for symmetry violations in the elementary particles such as quarks and leptons; one such phenomenon of our interest is the intrinsic electric dipole moment (EDM) of either elementary or composite systems. The non-zero observation of EDM provides the direct and conclusive signatures of the violation of time-reversal symmetry. The study of EDM also paves a way for the continuing quest for the ultimate theory of the Universe and it has a great potential in uncovering many mysteries which have been puzzling the mankind for ages such as the very survival of ourselves amidst many cosmic catastrophes such as a complete annihilation of matter and antimatter in the Universe. Using the extreme quantum states of matter at the low-energy scales we study the high-energy physics related to the phenomena which are thought to have happened in various epochs in the very early Universe.

2. Electron EDM search with radioactive atom

The standard model (SM) predicts quite tiny value of the electron EDM with ~10^{-37} e · cm, which is appeared in the higher order diagram with gluon contribution. This means that the EDM is the good probe to search for the phenomena beyond the standard model with low background contributing from the SM process. Many theoretical works show the finite electron EDM ranging to the value which can be measured with the modern experimental technique [1]. The super symmetry (SUSY) show that the EDM depends on the mass of the SUSY particles and the new CP violating phases appeared in the SUSY, then the EDM becomes the powerful tool to explore the new symmetry such as the SUSY and its violating mechanism.

2.1. Sensitivity of the EDM measurement

Although many theoretical models predict the large EDM, but it is still tiny value to measure it with the current experimental technique. The electron EDM in paramagnetic atoms is appeared and enhanced by the relativistic effects with the 3rd power of the nuclear charge. The heaviest alkali element such as the francium (Fr) has the largest enhancement factor, and the detailed calculation with the relativistic coupled cluster model [2] shows the enhancement factor with ~895 [3]. The Fr atom has the function as a microscope to magnify the tiny electron EDM. Although all the Fr isotopes are radioactive with finite lifetimes, the ^{210}Fr has a lifetime with ~3 minutes sufficient to produce by the nuclear reaction, extract, transport, and trap in the cell to measure the EDM. The EDM itself can be determined from the difference of the Larmor frequencies by measuring between the states with the electric field applied parallel and anti-parallel to the magnetic field. To measure it, the Ramsey resonance technique will be utilized, where the application of two successive RF pulses results in resonance signals as a sinusoidal function of the phase of the second RF pulse. The sensitivity of the EDM measurement is limited by the quantum shot noise, and depends on the enhancement factor K, applied electric field E, the number of accumulated atoms N, the spin coherence time τ, the measurement time T with the equation $\delta d=h/\{(2e \cdot K \cdot E)(N \cdot \tau \cdot T)^{1/2}\}$. The high sensitivity search of the EDM is realized by the next generation experiment using laser cooled radioactive atoms, since the Fr has the largest enhancement factor, and in the high vacuum environment of the laser trapping cell, the spin coherence time becomes longer and the high electric field can be applied. The important point is that we need the sufficient number of the accumulated Fr atoms which will be shown in the next section.

2.2. EDM search with laser cooled atoms

Up to now, the EDM measurements have been done with the accumulated atoms in the cell and the collinear atomic beam using the Cs [4] and Tl which keeps the highest accuracy data at present [5]. However there are limitations to the systematic error as follows. The motional magnetic field $B=(v \times E)/c^2$, where the v is the velocity of the atomic beam, and the E is the applied electric field, causes the false signal of the EDM. Also magnetic field generated by the leakage and/or changing currents causes the false signal. The geometric phase shifts generated by the complicated field gradients and the magnetic Johnson noise generated by traditionally used metals in electric field plates, vacuum chamber, and magnetic shield are also contributed to the systematic error. This means that the motion of the atoms and the difficulty to realize the uniform external field in the large measurement area limit the systematic error. The laser cooled and trapped atoms have the unique feature that they are confined in the small space and their kinetic energies or temperatures become very low. Then, the motional magnetic field can be highly suppressed and inhomogeneous applied field can be lowered.

3. Development of the experimental apparatus

The construction of the high intensity laser cooled Fr factory has been started from is in progress at CYRIC, Tohoku University. The experimental apparatus consists of the beam swinger system, the thermal ionizer, the Fr beam transport system, the neutralizer, the laser cooling, the magneto-optical trap (MOT), and the laser trap system with an optical lattice to detect the EDM as shown in Figure 1.

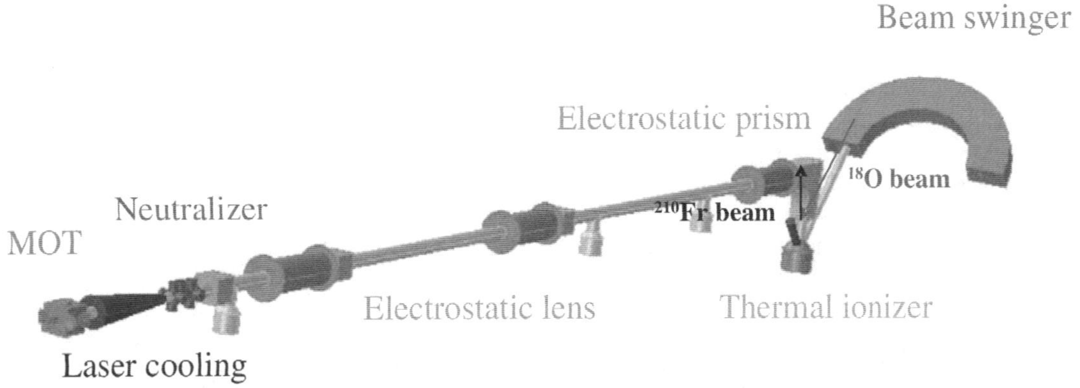

Figure 1. The overview of the experimental apparatus of the laser cooled Fr factory

3.1. Thermal ionizer

The Fr is produced in a heavy-ion fusion reaction using an oxygen beam and a gold target ($^{18}O+^{197}Au \rightarrow {}^{210}Fr$ +5n etc.) with the beam energy of 100 MeV just above the coulomb barrier. The target consists of a lump of gold melted and flattened onto the end of a tungsten rod with a thickness of 50 um. The target rod is surrounded by an oven which is heated by the coil heater, and the target material is heated by the radiation from the oven. The target temperature can be controlled precisely with the additional heater installed around the rod. The embedded Fr produced by fusion reaction diffuses rapidly to the surface and evaporates. The francium desorbs from the target surface as atoms and ions according to the Langmuir-Saha equation: $n_+/n_0 = (w_+/w_0) \cdot \exp\{(E_{WF}-E_{IP})/kT\}$, where n_+/n_0 is the ratio of ions to atoms desorbed, w_+/w_0 is the ratio of the statistical weights and equals 1/2 for alkali atoms, E_{WF} is the work function of the surface, and E_{IP} is the ionization potential of the desorbed atom. Since for gold E_{WF} is 5.1 eV and 4.08 eV for E_{IP}, we have $E_{WF}>E_{IP}$, and consequently the target emits primarily Fr ions. While Fr isotopes can be produces in other fusion reaction, the gold is used as a target since it is a noble metal, naturally monoisotopic, and provides an ionizing surface for alkali atoms. The structure of the thermal ionizer is shown in the Figure 2. The unique point of this system is to use the swinger system to inject the beam to the target from the upper direction with the target configured upward to be able to keep it with melting to realize the high extraction efficiency. The extracted Fr beam is bended to 90 degree with the electrostatic prism and transported with three sets of the lens systems along about 10 m from the vertex region to the MOT.

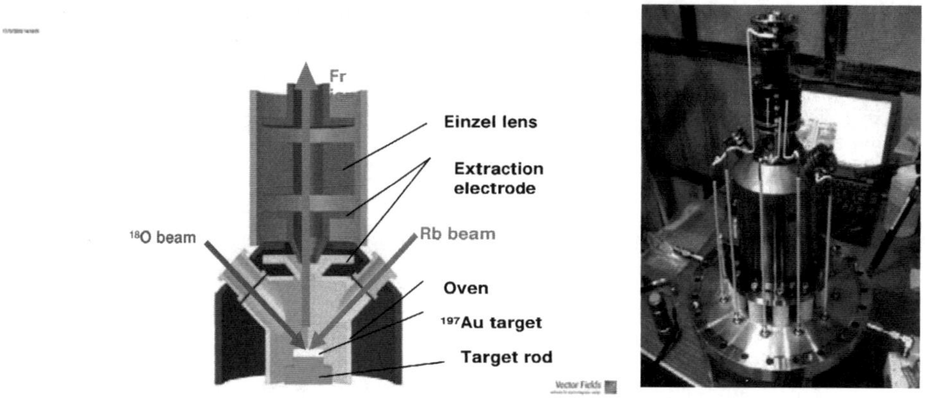

Figure 2. The structure of the newly developed thermal ionizer.

3.2. Magneto-optical trap system

The Fr ions will be neutralized by the neutralizer and injected to the Zeeman slower to slow down the Fr atomic beam to be trapped efficiently into the MOT. The neutralizer consists of the filament to emit the high density electrons and the electrode to control the position of the electron distribution. The Fr ion beam is decelerated up to the kinetic energy of 0.1 ~ 1 eV which is suitable for the efficient cooling with the Zeeman slower and injected into the high density electron distribution and neutralized with the electron recombination. The low speed Fr atomic beam is injected into Zeeman slower and loaded into the MOT. The Zeeman slower and the MOT require the high intensity laser light sources, which cool and trap not only the Fr but also the Rb, since the chemical properties of Fr and Rb are similar and all the parameters to operate the experimental apparatus can be adjusted with the high intensity Rb beam in the offline experiment. The Fr needs the laser with a wavelength of 718 nm for the trapping and the high power ~3W is realized by the Ti:S laser supplied by COHERENT with the combination of Verdi V18 and MBR-110. The repump light source 817 nm is prepared by the external cavity laser diode (ECLD). The laser with a wavelength of 780 nm is prepared by the ECLD and tapered amplifier for the Rb trapping and the output power of 1 W is achieved. The repump laser of a wavelength 795 nm is also prepared with the ECLD. All the light sources are ready at present and the MOT vacuum chamber will be installed soon.

4. Present status of the high intensity laser cooled Fr factory

4.1. Test experiment at CYRIC

The test experiments to check the performance of the new thermal ionizer has been performed at the newly constructed beam line for the Fr production in the CYRIC. The configuration of the experimental setup was minimized to study the Fr production rate, the transmission efficiency, and the neutralization efficiency efficiently, where only one electrostatic lens system consisting of three sets of the quadrapole lenses were installed to focus the Fr ion beam to the neutralizer. The two sets of the beam diagnosis system with the SSD were installed just after the thermal ionizer and the downstream of the beam line just before the neutralizer. The primary ^{18}O beam supplied from the AVF cyclotron was injected to the ^{197}Au target installed inside the thermal ionizer. The beam profile, observed by the ZnS beam viewer installed just before the thermal ionizer, was about 10mm diameter which was almost same size of the target. The beam intensity was ~200 nA on target, which was the highest intensity at present due to the low transmission efficiency between the ECR ion source and the AVF cyclotron although the beam intensity from ECR ion source was much higher. The temperature of the target was monitored with the thermocouple (W/Re) attached to the below of the target. The extraction voltage of the thermal ionizer was applied up to 2kV for the stable operation, although the design value was 5 kV but the leakage current became large due to the spattered materials from the target or other reasons. The operation parameters of the beam transport system were optimized using the Rb beam in advance based on the simulation of the ion optics.

4.2. Experimental results

The alpha decaying spectrum obtained by the SSD installed just after the thermal ionizer was shown in the Figure 3. The prominent peak from the alpha decay of ^{210}Fr can be seen at the energy of 6.543 MeV, and other peaks are from other produced isotopes ^{209}Fr and daughter nucleus decayed from Fr. The left peak is the reference alpha source of ^{241}Am with 5.486 MeV. It can be seen clearly that there are almost no backgrounds from other radioactive atoms produced in the fusion reaction with ^{18}O and ^{197}Au due to the condition on the surface ionization, and we can expect that the Fr can be transported to the final stage MOT with high purity.

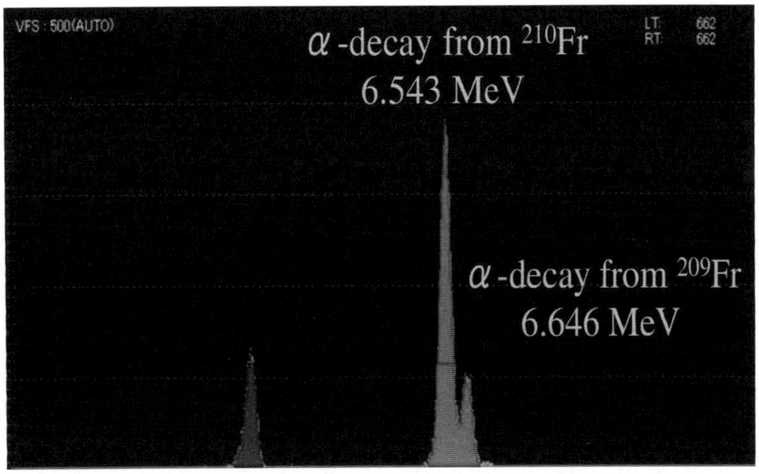

Figure 3. The horizontal axis show s the energy of the alpha particles decaying from ^{209}Fr, ^{210}Fr, and ^{241}Am obtained by the SSD. The vertical axis shows the number of the alpha particles. The ^{241}Am source is installed in the beam diagnosis system for the energy calibration.

The temperature dependence of the Fr yield is shown in the Figure 4. It should be noted that the vertical axis shows the number of alpha particles decaying from ^{210}Fr atoms which are caught by the catcher foil installed in the Fr beam line and detected by the SSD installed about 60 mm far from the catcher. The Fr yield has been evaluated with the correction of the detector acceptance in the final analysis. The yield is drastically increased at 970 degree, which is around the melting point of the gold 1064 degree. This similar feature was also observed at other institutes such as SUNY [6] and LNL [7], and also we observed that the extraction time of the Fr ions became short when the thermal ionizer was operated at the temperature higher than the melting point. The stable production yield of ^{210}Fr with ~10^6 Fr^+/sec was achieved successfully with the primary beam ^{18}O intensity of 200 nA. The evaluated extraction efficiency is more than 30 %. This yield is sufficient to search for the EDM as far as the design value of the efficiency of the neutralization and trapping is achieved.

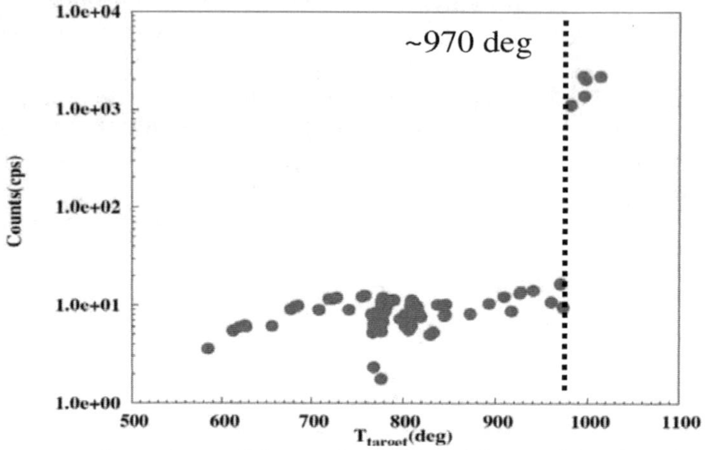

Figure 4. The temperature dependence of the number of the alpha particles decaying from ^{210}Fr. The horizontal axis is the target temperature. The vertical shows the number of the detected alpha particles.

5. Summary

The construction of the laser cooled Fr factory is in progress at CYRIC to search for the electron EDM with the sensitivity 10^{-29} e · cm. We developed the new thermal ionizer and succeeded to produce the 10^6 Fr^+/sec with the primary beam intensity of 200 nA. The expected Fr intensity will be 10^7 Fr^+/sec with the ^{18}O beam intensity 2 uA using the upgraded ECR ion source. We can realize the high intensity Fr source at a small scale accelerator facility compared with the TRIUMF ISAC and CERN ISOLDE, where the high energy proton beam is injected to the targets such as UO_2 / UCx which need the careful treatments to produce the Fr at the large scale facilities. This research was partially supported by Grant-in-Aid for Scientific Research on Innovative Areas "Extreme quantum world opened up by atoms" (No.21104005) and Grant-in-Aid for Scientific Research (A) (No.21244026) from the Ministry of Education, Culture, Sports, Science, and Technology.

References

[1] Maxim Pospelov and Adam Ritz 2005 *Annals Phys. 318* 119-169
[2] H.S.Nataraj, B.K.Sahoo, B.P.Das and D.Mukherjee 2008 *Phys. Rev. Lett. 101* 033002
[3] Debashis Mukherjee, B.K.Sahoo, H.S.Nataraj, and B.P.Das 2009 *Journal of Physical Chemistry A 113* 12549-12557
[4] S.A.Murthy, D.Krause, Jr., X.L.Li, and L.R.Hunter 1989 *Phys. Rev. Lett. 63* 965-968
[5] B.C.Regan, E.D.Commins, C.J.Schmidt, and D. Demille 2002 *Phys. Rev. Lett. 88* 071805
[6] J.E.Simsarian, A.Ghosh, G.Gwinner, L.A.Orozco, G.D.Sprouse, and P.A.Voytas 1996 *Phys. Rev. Lett. 76* 3522-3525
[7] G.Stancari, S.Veronesi, L.Corradi, S.N.Atutov, R.Calabrese, A.Dainelli, E.Mariotti, L.Moi, S.Sanguinetti and L.Tomassetti 2006 *Nucl. Instrum. Meth. A557* 390-396

Prospects of the Hadron Physics at J-PARC

Makoto Oka

Dept. of Physics, H-27, Tokyo Institute of Technology, Meguro, Tokyo 152-8551, Japan

E-mail: oka@th.phys.titech.ac.jp

Abstract. An overview of the leading subjects of hadron physics with an emphasis on those relevant to J-PARC is presented. J-PARC is a new high intensity proton beam accelerator facility built at Tokai, Japan. The subjects include (1) strangeness nuclear physics, (2) generalized nuclear force, and (3) exotic hadron spectroscopy.

1. Introduction

J-PARC, the acronym of the Japan Proton Accelerator Research Complex, located at Tokai, is a high-intensity proton accelerator jointly operated by KEK and JAEA[1]. It is composed of the proton Linac, the 3 GeV Synchrotron and the 50 GeV Main Ring. The main ring accelerates high intensity proton beam for nuclear, hadron and particle physics, where the beam is extracted for experiments at the Hadron Experimental Hall, and also for the Neutrino production beam line. In the Hadron Experimental Hall, the secondary meson beams, π and K of 1-2 GeV/c, are produced, which enable us to perform high precision hadron and particle physics experiments. The neutrino beam is ejected towards the Super-Kamiokande neutrino detector, located 295 km away from the J-PARC. The primary proton beam can also be used for high energy hadron physics experiments, such as studying parton distributions in the Drell-Yan processes. All the accelerators were commissioned before 2010, and the nuclear/hadron and neutrino experiments have just started.

The facility is the result of a long term effort of the Japanese nuclear physics community. Physics subjects for the hadron hall experiments range from the hadronic level to the quark level including the strange nuclear physics, hadron properties in nuclear medium, study of quark structure of hadrons and exotic hadron spectroscopy. The main research areas are summarized as follows:

- Strangeness Nuclear Physics

  Strong kaon (K) and anti-kaon ($\overline{K}$) beams (as well as pion beams) are utilized to produce various strangeness hadrons and nuclei. The main subjects there include production, structure and decays of hypernuclei, properties of generalized nuclear force, *i.e.,* YN and YY interactions, where Y stands for a hyperon (baryon with strangeness), and the kaon and anti-kaon interactions with the nucleon and nucleus. The generalized nuclear force seems to be the key to understand the hadronic interaction starting from the quark-gluon dynamics. J-PARC plans to carry out many hypernuclear experiments which are critically important to reveal the properties of the YN and YY forces. Searches for deep anti-kaon $\overline{K}$ bound states in nuclear medium is a hot subject recently[2]. Such a bound state will determine how mesons interact and behave in nuclear medium. There are experiments at

J-PARC to confirm the pentaquark resonance Θ^+. The $\Theta^+(1440)$ has been first reported from the photon beam facility at SPring-8 by the LEPS group[3]. However, its existence is not confirmed yet. It is urgent to make the situation clear and terminate the controversy by carrying out the searches with the hadronic beam experiments at the J-PARC.

- Hadron Properties in Nuclear Medium
  One of the main goals of QCD studies is to understand the phase structure and properties of hadronic matter at finite temperature and/or finite density. Many years of researches have revealed that the phase diagram at finite baryon chemical potential may be quite complicated and show many new structures of matter[4]. In particular, possibility of color superconductor phase at large baryon density with low temperature was predicted, which may be realized at the center core of the neutron stars in nature. It is, however, not possible to access such phases directly at laboratories. One possible way of probing finite density nuclear medium is by producing hadrons in nuclei. The density may not be large enough for exotic phases, but partial restoration of chiral symmetry may be manifested as mass shifts and width broadening of mesons produced inside nuclei[5].
- Exotic Hadron Spectroscopy
  Since the discovery of the pentaquark, many new approaches have been made for excited hadrons and exotic hadrons. The reaction dynamics and resonance structures are closely related and, in fact, some resonances are produced just by unitarizing scattering amplitudes of two hadrons in coupled channel systems. Such resonances may be regarded as molecular bound states of hadrons. It is quite important to identify the nature of individual excited hadrons, whether it is made of just two or three quarks, or is an exotic hadron made of extra quark or gluon components, or a hadronic molecular resonance state. In viewing the exotic multi-quark states, heavy quark may play important roles. In particular, several charmonium-like states found recently do not simply fit to the $c\bar{c}$ spectrum, and may contain tetra-quark or two-meson components.

In all these areas of researches, the third flavor, strangeness, plays unique and important roles. in the next section, we overview the roles of the strangeness in the strong interaction of quarks in hadrons.

2. Roles of Strangeness in Nuclear/Hadron Physics

The basic dynamics of quarks and gluons are given by a non-Abelian gauge theory called Quantum Chromodynamics (QCD). QCD is composed of gauge invariant interactions of quarks and gluons, which are characterized by the color charge, a new internal degrees of freedom. QCD has special unprecedented properties, such as color confinement and asymptotic freedom. Namely, QCD interactions are scale dependent so that at long distances the interaction is so strong that color charges are confined, while at short distances, the color interactions are weak and can be treated perturbatively.

The QCD lagrangian,

$$\mathcal{L} = \sum_f \bar{\psi}_f(x)(i\gamma^\mu D_\mu - m_f)\psi_f(x) - \frac{1}{4}\sum_a (G^a_{\mu\nu})^2 \tag{1}$$

has the exact SU(N_f) symmetry, if the quark masses of the N_f flavors are equal. In reality, the quark masses are distributed from a few MeV to 170 GeV as is shown in Fig. 1. The masses of the "light" quarks, u and d, are much smaller than the QCD energy scale Λ_{QCD}, and their interactions are constrained by chiral symmetry that is exact for massless quarks. On the other hand, the "heavy" quarks, c, b and t, are much heavier than Λ_{QCD} and therefore heavy quark symmetry can be applied. The strange quark, s, has the mass just about Λ_{QCD} and neither

Journal of Physics: Conference Series **302** (2011) 012052 doi:10.1088/1742-6596/302/1/012052

the chiral symmetry nor heavy quark symmetry is good enough to neglect symmetry breaking corrections. Thus, the dynamics of the strange quark is most sensitive to the dynamics of QCD.

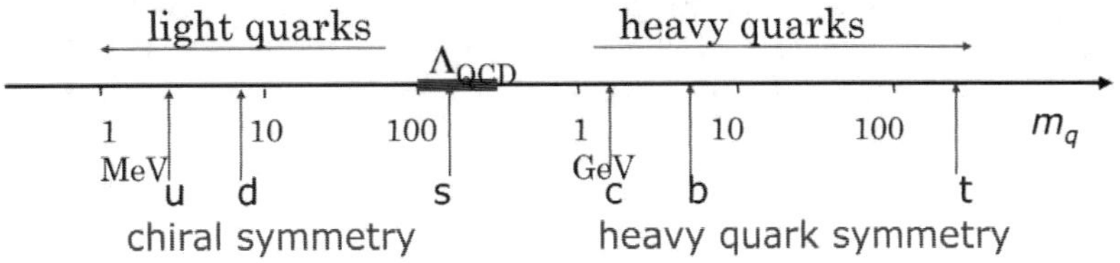

Figure 1. Quark masses and QCD scales.

The quark model with the SU(3) flavor or the SU(6) spin-flavor symmetry has played a major role in the history of hadron physics, but its limitation has become clear in the spectroscopy of excited and exotic hadrons. Indeed, accumulated evidences of new hadrons, such as $\Theta^+(1540)$, $X(3872)$, $Z^+(4430)$, and so on, have shown that there may exist "sharp" exotic resonances whose main component is not simply qqq or $q\bar{q}$. It is interesting and important to reveal the dynamics of these exotic states from the QCD point of view. Beyond the quark models, many refined effective models have been proposed to describe hadron spectra and hadron structures, but the predictability of the models is limited as they involve many unknown parameters. The validity of such models will be determined by examining whether they can explain experimental data involving strange and heavier quarks.

At high density, with $\mu(u,d) > m_s$, strange matter may become stable. The strange hadrons (K^-, Σ^-, Ξ^-) with negative charges may play significant roles in the core of the neutron star, as they may override the chemical potential of electron. Especially, a strong $\overline{K} - N$ attraction may drive the kaon condensation. At higher density, the quark matter with strangeness may appear in the neutron star matter, where the QCD ground state may be the color superconductor with strangeness. With $N_c = 3$ colors and $N_f = 3$ flavors, the color-flavor locking (CFL) phase with the condensate $\langle q_i^{Ta} C\gamma^5 q_j^b \rangle = \epsilon^{abc}\epsilon_{ijc}\Delta$ may become stable.

3. Generalized Nuclear Force: Origin of Short-range Repulsion

The hadronic interaction is a major subject in the nonperturbative properties of QCD. The nuclear force, among them, is well established and is tested very well by numerous experimental data. The nuclear force consists of the longest-range one-pion exchange potential, the medium-range multi-pion and boson exchange potential and the strong short-range repulsion. The scale of each part is of 100 MeV or more, while the attraction from the medium-range part and the short-range repulsion cancel out largely and result in the small binding energy of the deuteron, 2 MeV.

Recently, various suggestions of possible hadronic molecular states, which may appear in excited hadron spectra, have appeared. There reliable prediction of the hadronic interactions is critically important. In particular, it is interesting to examine whether two baryon bound states other than the deuteron may exist or not. The origin of the short-range repulsion between two nucleons is therefore very important to determine such possibility.

In 1977, it was suggested by Neudatchin et al.[6] that the antisymmetrization among valence quarks may give short-range repulsion between two nucleons. In 1980, we proposed the Quark Cluster Model (QCM) of two-baryon interactions[7], where it was shown that the quark exchange diagrams with the help of the color-magnetic interaction (CMI) in the one-gluon exchange between quarks explains the short-range repulsion. It was also shown that the generalization of the QCM to strange baryons gives a rich structure of two baryon interactions. Table 1 shows the classification of the S-wave ($L = 0$) two baryon interactions in terms of the spin-flavor symmetry of valence quarks[8]. From the analysis based on the quark model symmetry, we predict that the NN states are moderately repulsive, which is caused mainly by the CMI. On the other hand, there are some states which have very strong repulsion at short distances caused by the Pauli exclusion principle, such as ΣN ($I = 1/2$, $S = 0$) and ΣN ($I = 3/2$, $S = 1$). On the other hand, no repulsion is predicted for the flavor singlet $\Lambda\Lambda - N\Xi - \Sigma\Sigma$ system, where we may have a compact di-baryon bound/resonance state. (It was first predicted by Jaffe in 1977, called H dibaryon[9].) Thus, the QCM predicts a strong state dependence of the two baryon interactions, which should be confirmed experimentally at the J-PARC.

Table 1. SU(3) $\times$ SU(2)$_{\rm spin}$ $\rightarrow$ SU(6) classification of the $L = 0$ two baryon systems. The SU(6) representations are given by the Young partition numbers.

SU(3)	SU(2)	SU(6)	Baryon interactions
1	S=0	[33]	attractive in $\Lambda\Lambda, N\Xi, \Sigma\Sigma \rightarrow H$ dibaryon
8_s	S=0	[51]	strong repulsion in ΣN ($I = 1/2$, $S = 0$)
27	S=0	[33], [51]	repulsion due to CMI, NN $^1S_0$
8_a	S=1	[33], [51]	
10	S=1	[33], [51]	strong repulsion in ΣN ($I = 3/2$, $S = 1$)
$\overline{10}$	S=1	[33], [51]	repulsion due to CMI, NN $^3S_1$

The QCM predictions have been tested by comparing them with phenomenological potential models determined by the available experimental data of hypernuclei[10]. It has been confirmed that the following properties predicted by the quark model agree very well with the phenomenological analyses.

- Suppression of the ΛN spin-orbit interaction
- Strong short-range ΣN repulsion with spin-isospin dependence
- Attraction in the flavor singlet H dibaryon channel

Furthermore, technical development in computational QCD simulations, the lattice QCD (LQCD), now allows us to compute the baryon-baryon potential directly from the first principle. Recent LQCD calculations by the HAL QCD collaboration[11] have confirmed that the picture of the Pauli principle plus the CMI interaction for the short-range baryon-baryon interaction is consistent with QCD. In particular, Inoue et al. showed the classification given in Table 1 is indeed realized in the SU(3) limit of QCD.

Further investigations along this line, cooperation of J-PARC experiments in hypernuclear spectroscopies and lattice QCD calculations of the generalized nuclear force, should clarify the dynamical mechanism of the short-range baryon-baryon interactions and in general hadronic interactions in various hadron channels, including possible exotic hadronic molecular states.

4. Exotic hadrons: $\Lambda(1405)$

Since the report of $\Theta^+(1440)$ discovery, the exotic multi-quark hadrons have been re-examined extensively. During analyses of various channels of excited hadrons, it becomes clear that the strangeness and heavy quarks allow multi-quark hadrons and/or hadronic molecular resonances preferably. Two main reasons for this preference are (1) suppressed kinetic energy of heavy quarks and (2) strong channel couplings to the excited states.

In the strangeness sector, the lowest energy baryon resonance with negative parity, $\Lambda(1405)$, attracts much interest recently. Several first-principle calculations have been carried out. Among them, a new lattice QCD calculation[12] with three-quark interpolating field operators has revealed that there are two negative-parity Λ states around 1.65-1.8 GeV, while no trace of a state as low as 1.4 GeV is found. This result suggests that the physical $\Lambda(1405)$ resonance may not be a 3-quark bound state, but may contain more quarks, or is a hadronic molecule. Indeed, a QCD sum rule analysis with mixed 3- and 5-quark interpolating field operators indicates that the lowest Λ contains the 5-quark component dominantly[13].

On the other hand, the chiral unitary approach, which describes excited hadrons as poles in the scattering amplitudes of two hadrons, has successfully reproduced the $\Lambda(1405)$ peak structure using the dynamics constrained by chiral symmetry and unitarity of the amplitude[14]. Their results are consistent with and support the hadron molecule picture of $\Lambda(1405)$.

It is important to clarify whether the $\Lambda(1405)$ is a special case or there exist many other exotic states that are not accounted by ordinary 3-quark, or quark-antiquark pictures. In fact, several such multi-quark candidates have been proposed, which include tetra-quark resonances with heavy quarks, especially in charmonium and open charm hadrons, and dibaryon resonances. In order to study the origin and structure of such states, combinations of the first-principle QCD calculations and the hadron dynamics approaches are necessary. J-PARC may again play important roles in determining scattering amplitudes in various channels where theories suggest exotic resonances.

5. Conclusion

J-PARC is a unique facility for researches of the hadron physics and QCD. I have emphasized that the third flavor, strangeness, is a key to reveal the nonperturbative features of QCD at low energy. It is expected that new and exciting experimental data are coming out from the J-PARC experiments and our understanding of the hadron structures as well as the strong interaction dynamics will be largely developed.

References

[1] For overview of the J-PARC and its activities in particle and nuclear physics, see T. Sato, T. Takahashi and K. Yoshimura (Eds.), Particle and Nuclear Physics at J-PARC, Lecture Notes in Physics 781(Springer, Berlin Heidelberg, 2009). See also http://j-parc.jp/index-e.html.
[2] For proposed J-PARC experiments, see M. Iwasaki in ref.[1].
[3] T. Nakano et al., Phys. Rev. Lett. **91** (2003) 012002. For a recent status and proposed J-PARC experiment, see M. Naruki in ref.[1].
[4] For a recent review, see K. Fukushima, T. Hatsuda, Rept. Prog. Phys. **74** (2011) 014001.
[5] For proposed J-PARC experiments, see S. Yokkaichi in ref.[1].
[6] V. G. Neudatchin, Yu. F. Smirnov, R. Tamagaki, Prog. Theor. Phys. **58** (1977) 1072.
[7] M. Oka, K. Yazaki, Phys. Lett. **90B** (1980) 41; Prog. Theor. Phys. **66** (1981) 556 and 572.
[8] M. Oka, K. Shimizu, K. Yazaki, Nucl. Phys. **A464** (1987) 700
[9] R.L. Jaffe, Phys. Rev. Lett. **38** (1977) 195.
[10] T.A. Rijken, M.M. Nagels, Y. Yamamoto, Prog. Theor. Phys. **S185** (2010) 14, and the references therein.
[11] S. Aoki, T. Hatsuda, N. Ishii, Prog. Theor. Phys. **123** (2010) 89; T. Inoue et al., Prog. Theor. Phys. **124** (2010) 591.
[12] T.T. Takahashi, M. Oka, Phys. Rev. **D81** (2010) 034505.
[13] T. Nakamura, J. Sugiyama, T. Nishikawa, M. Oka, N. Ishii, Phys. Lett. **B662** (2008) 132.
[14] D. Jido et al., Nucl. Phys. **A725** (2003) 181; T. Hyodo, D. Jido, A. Hosaka, Phys. Rev. **C78** (2008) 025203.

Hadronic molecules in chiral dynamics

Tetsuo Hyodo

Department of Physics, Tokyo Institute of Technology, Meguro 152-8551, Japan

E-mail: hyodo@th.phys.titech.ac.jp

Daisuke Jido

Yukawa Institute for Theoretical Physics, Kyoto University, Kyoto 606-8502, Japan

Atsushi Hosaka

Research Center for Nuclear Physics (RCNP), Ibaraki, Osaka 567-0047, Japan

Abstract. Hadrons are composed of quarks and gluons through the strong interaction. When the hadrons themselves are strongly correlated by the inter-hadron forces, a composite system of hadrons can be formed. This is called hadronic molecule state. We present recent developments in the study of the structure of such hadrons in chiral dynamics. It is shown that the novel structure of hadronic molecules can be realized by the strong attraction in chiral low energy interaction.

1. Introduction

Strong interaction of quarks and gluons is governed by quantum chromodynamics (QCD). At low energies, due to color con nement, asymptotic degrees of freedom become hadrons (mesons and baryons) which are regarded as elementary excitations of the QCD vacuum. Recently much attention is paid to the discussion of hadronic molecule states, in which two or more hadrons are loosely bound by the inter-hadron interactions, keeping their identities as hadrons.

Well known example of the hadronic molecule is the *nucleus* which is the composite system of protons and neutrons. For instance, deuteron is considered as the bound state of a proton and a neutron. As far as the quantum number is concerned, however, one could form the deuteron as a compact six-quark state, so that it is a single hadron as a whole. It is clear that the former picture is the correct one, but it should be veri ed using a theoretical argument. It has been indeed shown in Ref. [1] that the deuteron is not a single hadron but a composite system, by introducing a compositeness condition through the eld renormalization constant.

At microscopic level, hadronic molecule structure can be interpreted as the clustering phenomena of quarks, given that the hadrons are composite by themselves. This quark-number picture is however not trivial when the antiquarks are taken into account. By writing the quark (antiquark) number as n_q (n_q), the net quark number $N = n_q \quad n_q$ is conserved in QCD by vector U(1) symmetry. On the other hand, neither n_q nor n_q is separately conserved, since the qq pair creation/annihilation can take place. In other words, the number of valence quarks is not a well-de ned classi cation scheme of the hadron structure.

Instead, the structure of hadrons can be classi ed by regarding them as the fundamental degrees of freedom. Even though the number of mesons are not conserved, one baryon state can be distinguished from the meson-baryon state. This is because the hadrons are the asymptotic states, and the two-body system forms continuous scattering states which are separated from the discrete levels of the one-body states. In contrast, multi-quark states in general do not form the continuous spectrum because of color con nement.

In this paper, we focus on the low energy chiral interaction as the driving force of hadronic molecules. We introduce a theoretical framework of chiral dynamics which describes the excited baryons as resonances in the meson-baryon scattering amplitude. We then review the recent attempts to pin down the structure of baryon resonances in chiral dynamics.

2. Chiral dynamics

QCD Lagrangian has approximate chiral symmetry for light avor quarks. Chiral symmetry and its spontaneous breaking in the underlying theory dictate the low energy interaction of hadrons. In addition, the scattering amplitude should satisfy the unitarity condition which ensures the probability conservation. Refs. [2, 3, 4, 5] have established the dynamical framework in which the scattering amplitude is constructed consistently with the coupled-channel unitarity and obeys the chiral theorem at low energy. In the N/D method [4], we write down the general form of the amplitude for an s-wave single-channel meson-baryon scattering at total energy $\sqrt{s}$ as

$$T(\sqrt{s}) = \frac{1}{V^{\ 1}(\sqrt{s}) \quad G(\sqrt{s};a)}, \tag{1}$$

where $V(\sqrt{s})$ is the kernel interaction speci ed below. The once-subtracted dispersion integral $G(\sqrt{s};a)$ can be identi ed as the meson-baryon loop function with dimensional regularization, and the subtraction constant a plays the role of the cuto of the loop integral. In the leading order of the low energy expansion, the interaction kernel is given by the Weinberg-Tomozawa term

$$V(\sqrt{s}) = V_{\mathrm{WT}}(\sqrt{s}) = \quad \frac{C}{2f^2}(\sqrt{s} \quad M_T), \tag{2}$$

where C, M_T and f are the group theoretical factor, the baryon mass, and the meson decay constant, respectively. Since Eq. (1) expresses nonperturbative resummation of the s-channel diagrams, resonances and bound states can be dynamically generated for su ciently attractive interactions. The coupling strength C is determined by group theoretical argument, and they are strong enough to generate resonances, especially in three avor sector [6, 7]. Various hadron resonances are well described in this approach, together with the successfully reproduced hadron scattering amplitude, thanks to the universality of the low energy chiral interaction [8, 9].

3. Origin of resonances

In the scattering theory, single particle contribution to the scattering amplitude is called the Castillejo-Dalitz-Dyson (CDD) poles. The N/D method describes the amplitude by including the CDD pole contribution in the interaction kernel V. In the previous studies, CDD poles were introduced in V as the explicit resonance propagator of the bare eld, or as the contracted resonance contribution in the higher order chiral Lagrangian. In Ref. [10], it has been shown that the loop function G can also contain the CDD pole contribution. This is understood from the viewpoint of the renormalization group; once the amplitude T is determined by experiments, then the change of the interaction V can be absorbed by modifying the renormalization parameter a in the loop function G. In this case, the CDD pole contribution in the kernel V may be e ectively transferred to the loop function G. Thus, in order to study the origin of resonances in this approach, we should make the CDD pole contribution in the model under control.

For this purpose, "natural renormalization scheme" is developed [10]. In this scheme, we constrain the loop function G to exclude the CDD pole contribution, by requiring G to be negative below threshold. In addition, the amplitude T is matched with the interaction kernel V at certain low energy scale, based on the chiral expansion. These requirements uniquely lead to the renormalization condition

$$G(\sqrt{s}; a_{\text{natural}}) = 0 \quad \text{at} \quad \sqrt{s} = M_T, \tag{3}$$

from which we determine the subtraction constant a_{natural}.

Once the natural renormalization scheme is established, the origin of the resonances in the phenomenological approach can be studied. Suppose that the experimental data is t by the leading order interaction V_{WT} and the subtraction constant a_{pheno}. The scattering amplitude can also be expressed in the natural renormalization scheme with the subtraction constant a_{natural} and the e ective interaction V_{natural}. To obtain the same amplitude, it follows from Eq. (1) that

$$V^{\ 1}_{\text{natural}}(\sqrt{s}) \quad G(\sqrt{s}; a_{\text{natural}}) = V^{\ 1}_{\text{WT}}(\sqrt{s}) \quad G(\sqrt{s}; a_{\text{pheno}}). \tag{4}$$

By this equation we obtain V_{natural} as

$$V_{\text{natural}}(\sqrt{s}) = \frac{C}{2f^2}(\sqrt{s} \quad M_T) + \frac{C}{2f^2}\frac{(\sqrt{s} \quad M_T)^2}{\sqrt{s} \quad M_{\text{e}}}, \tag{5}$$

with an e ective mass $M_{\text{e}} \quad M_T \quad (16 \ ^2 f^2)/CM_T \quad a)$ and $a = a_{\text{pheno}} \quad a_{\text{natural}}$.

The e ective interaction V_{natural} has a pole term which is interpreted as the CDD pole contribution. The relevance of the pole term depends on the scale of the e ective mass M_{e} ; for small a, the e ective pole mass M_{e} is large and the second term of Eq. (5) can be neglected in the resonance energy region $\sqrt{s} \quad M_T + m \quad M_{\text{e}}$. If the deviation a is large, the e ective mass M_{e} gets closer to the threshold and the pole contribution has substantial e ect on the observables. Thus, the e ect of the CDD pole contribution can be estimated from the values of the e ective mass M_{e} .

Applying this method to the physical baryon resonances, we nd that the (1405) resonance is predominantly formed by the meson-baryon molecule component, while the $N(1535)$ resonance requires a substantial CDD pole contribution.

4. Compositeness of bound states

Next we introduce a quantitative measure of "compositeness" through the eld renormalization constant [11, 12]. For simplicity, we consider the bound state with mass M_B. In the nonrelativistic quantum mechanics, the eld renormalization constant Z is given by the overlap of the bare state and the physical state [1]. Since Z expresses the elementarity of the bound state, the quantity 1 Z is interpreted as the compositeness.

In the relativistic eld theory, the eld renormalization constant is de ned as the residue of the bound state propagator. For an s-wave bound state, it is given by

$$(\sqrt{s}) = \frac{Z}{\sqrt{s} \quad M_B}. \tag{6}$$

Utilizing a eld theory with the Yukawa interaction, the constant Z is expressed in terms of the renormalized coupling constant of the bound state to the scattering state g and the derivative of the loop function G. Calculating the coupling constant g, we derive the compositeness of the bound state in chiral dynamics as [12]

$$1 \quad Z = \quad 1 + \frac{G(M_B; a)}{(M_B \quad M_T)G'(M_B)} \quad ^{1}. \tag{7}$$

It is shown that the natural renormalization condition (3) gives a bound state which is dominated by the composite structure. This result reinforces the quantitative aspect of the discussion on the origin of resonances with the natural renormalization scheme. For instance, based on the result in Ref. [10], it is shown that more than 80 % of (1405) can be interpreted as a meson-baryon composite structure [12].

5. Summary

We have illustrated a description of hadron resonances in a dynamical approach with chiral symmetry, and have shown the recent progress in the study of hadron structure. With the help of the natural renormalization scheme, it is possible to extract the CDD pole contribution hidden in the loop function. This scheme, together with the phenomenological amplitude, enables us to clarify the origin of resonances in chiral dynamics. The compositeness of the bound state is then introduced to elaborate the quantitative aspect. These techniques reveal that the (1405) resonance is a representative of the hadronic molecule state. It is important to clarify the property and the binding mechanism of the hadronic molecules for the understanding of the nonperturbative aspects of the strong interaction.

Acknowledgments

T.H. thanks the support from the Global Center of Excellence Program by MEXT, Japan through the Nanoscience and Quantum Physics Project of the Tokyo Institute of Technology. This work was partly supported by the Grant-in-Aid for Scienti c Research from MEXT and JSPS (Nos. 22740161, 22105507 and 21840026).

References

[1] Weinberg S 1965 *Phys. Rev.* **137** B672–B678
[2] Kaiser N, Siegel P B and Weise W 1995 *Nucl. Phys.* **A594** 325–345 (*Preprint* nucl-th/9505043)
[3] Oset E and Ramos A 1998 *Nucl. Phys.* **A635** 99–120 (*Preprint* nucl-th/9711022)
[4] Oller J A and Meissner U G 2001 *Phys. Lett.* **B500** 263–272 (*Preprint* hep-ph/0011146)
[5] Lutz M F M and Kolomeitsev E E 2002 *Nucl. Phys.* **A700** 193–308 (*Preprint* nucl-th/0105042)
[6] Jido D, Oller J A, Oset E, Ramos A and Meissner U G 2003 *Nucl. Phys.* **A725** 181–200 (*Preprint* nucl-th/0303062)
[7] Hyodo T and Weise W 2008 *Phys. Rev.* **C77** 035204 (*Preprint* 0712.1613 [nucl-th])
[8] Hyodo T, Jido D and Hosaka A 2006 *Phys. Rev. Lett.* **97** 192002 (*Preprint* hep-ph/0609014)
[9] Hyodo T, Jido D and Hosaka A 2007 *Phys. Rev.* **D75** 034002 (*Preprint* hep-ph/0611004)
[10] Hyodo T, Jido D and Hosaka A 2008 *Phys. Rev.* **C78** 025203 (*Preprint* 0803.2550 [nucl-th])
[11] Hyodo T, Jido D and Hosaka A 2010 *AIP Conf. Proc.* **1322** 374–378 (*Preprint* 1009.5754 [nucl-th])
[12] Hyodo T, Jido D and Hosaka A *in preparation*

Photon detection from stopped ^{87}Rb atoms injected into superfluid helium for a new nuclear laser spectroscopy of rare radioisotopes

T Furukawa[1], T Wakui[2], A Sasaki[2], S Izumi[2], Y Ichikawa[3], A Yoshimi[3], K Tajiri[4], Y Ishii[1], N Yoshida[1], Y Matsuura[5], Y Kato[5], Y Yamaguchi[5], K Imamura[5], M Makuta[6], A Hatakeyama[6], M Wada[3], T Sonoda[3], Y Ito[7], T Nanao1, T Kobayashi[8], S Nishimura[3], M Nishimura[3], Y Kondo[1], N Aoi[3], K Yoneda[3], S Kubono[9], Y Ohshiro[9], H Ueno[3], T Shimoda[4], T Shinozuka[2], K Asahi[1] and Y Matsuo[3,1,4]

[1] Department of Physics, Tokyo Institute of Technology, 2-12-1 Ookayama, Meguro-ku, Tokyo 152-8551, Japan
[2] Cyclotron and Radioisotope Center, Tohoku University, 6-3 Aoba, Aramaki, Aoba-ku, Sendai, Miyagi, 980-8578, Japan
[3] RIKEN Nishina Center, 2-1, Hirosawa, Wako, Saitama 351-0198, Japan
[4] Department of Physics, Osaka University, 1-1 Machikaneyama, Toyonaka, Osaka 560-0043, Japan
[5] Department of Physics, Meiji University, 1-1-1 Higashi-Mita, Tama-ku, Kawasaki-shi, Kanagawa 214-8571, Japan
[6] Department of Applied Physics, Tokyo University of Agriculture and Technology, 2-24-16 Naka-cho, Koganei-shi, Tokyo 184-8588, Japan
[7] Department of Physics, University of Tsukuba, 1-1-1 Tennodai, Tsukuba, Ibaraki 305-8577, Japan
[8] RIKEN Advanced Science Institute, 2-1, Hirosawa, Wako, Saitama 351-0198, Japan
[9] Center for Nuclear Study, University of Tokyo, 2-1, Hirosawa, Wako, Saitama 351-0198, Japan

E-mail: takeshi@yap.nucl.ap.titech.ac.jp

Abstract. We have been developing a new nuclear laser spectroscopic method "OROCHI" for determining spins and moments of exotic RIs. This is a method based on the combination of superfluid helium as a stopper of radioactive isotope (RI) beam and in-situ laser spectroscopy of RI atoms. This unique method will enable us to measure nuclear spins and electromagnetic moments of extremely low yield RI (less than 1 particle per second). To confirm the feasibility of this method for rare radioisotopes, we carried out the first on-line experiment with a ^{87}Rb beam, aiming at evaluating the minimum beam intensity required for the measurement. The LIF (Laser Induced Fluorescence) photons from the ^{87}Rb atoms stopped in He II are successfully observed in the on-line experiment. The obtained LIF intensity is nearly consistent with the intensity calculated from the photo-absorption cross section of atoms in He II.

1. Introduction

Nuclear spins and electromagnetic moments are often key observables for studying nuclear structures because they are directly correlated with the quantum states and configurations of valence nucleons [1]. So far, the studies of nuclear spins and moments have revealed precisely the nuclear structures from the stable to the unstable. Further attempts have been made to apply such studies to short-lived unstable nuclei far from the β-stability line [2]. To investigate the exotic structures of short-lived exotic nuclei, intensive studies have been attempted in the field of nuclear physics in decades. However, they have not always been successful mainly due to the low-yield of such the short-lived exotic nuclei produced by the nuclear reaction.

To measure the spins and moments of low yield exotic radioisotopes (RI) which are difficult to measure with existing methods, we have been developing a new nuclear laser spectroscopy method using superfluid helium (He II). We call this method as "OROCHI (Optical Radioisotope atom Observation in Condensed Helium as Ion-catcher)." In the OROCHI, He II liquid is adopted as a stopping material of RI beams and also as host material of in-situ laser spectroscopy for the introduced RI atoms. With characteristic features of the atoms immersed in He II [3], such as the blue-shifted and broadened absorption spectra due to the pressure of surrounding helium liquid, it will become feasible to measure the nuclear spins and moments of the extremely low yield RIs.

By using the OROCHI method, we demonstrated that we could deduce successfully the nuclear spins and moments with stable Rb, Cs, Ag and Au isotopes which were introduced into He II by laser sputtering technique [4]. Recently, we carried out the first on-line experiment with a ^{87}Rb beam to observe the atoms from an accelerated ion beam injected and stopped in He II. In this on-line experiment, we aim at evaluating the minimum beam intensity required for the observation. In this paper, we briefly report the result of the recent on-line experiment using the OROCHI setup.

2. Experiment

The experiment was performed at RIPS (RIKEN Projectile fragment Separator) in RIKEN RI Beam Factory. Figure 1 shows the schematics of our experimental setup. The ^{87}Rb beam (energy: 66 *A*MeV, diameter: 3 mm) was injected into He II. The typical beam intensity of ^{87}Rb beam was approximately 10^5 particles per second. The introduced Rb ions were counted one by one with a plastic scintillator (BC408, thickness: 100 μm) placed in front of the cryostat. Aluminum foils of thickness ranging from 0 μm to 800 μm ("Energy degraders" in Fig. 1) reduced the injected beam energy to adjust the range of ^{87}Rb in He II to the center of cryostat (7 mm down from the injection window). The vacuum of both beamline and cryostat was sealed by thin Kapton foils (75 μm thickness), respectively. The injection window was made of a thin havar foil (21 μm thickness). The stopped ^{87}Rb atoms were subjected to irradiation of the cw pumping laser light (cw Ti:Sapphire laser, 899-01, Coherent Co.Ltd., power: 200 mW, diameter: 5 mm). The wavelength of the pumping laser light was tuned to the D1 absorption line of Rb atoms in He II (780 nm). The LIF photons from laser-excited ^{87}Rb atoms were collected, wavelength-separated, and then detected with the photo-detection system [5]. Figure 1-b) shows the schematic diagram of the fluorescence detection system. It consist of three Fresnel lenses, a pair of slits (parallel and perpendicular to the laser light), wavelength filters, and a Peltier cooled photomultiplier tube (PMT, Hamamatsu Co.Ltd., R633-10P + C9143). The light emitted from atoms was collected and focused on the slits by the first Fresnel lens (Lens 1). Light which has passed through the slits was collimated with the second lens (Lens 2) so that the light enters at normal incidence to the wavelength filters for perfect separation of wavelength. The light was then focused on the photoelectric surface of the PMT by the third lens (Lens 3).

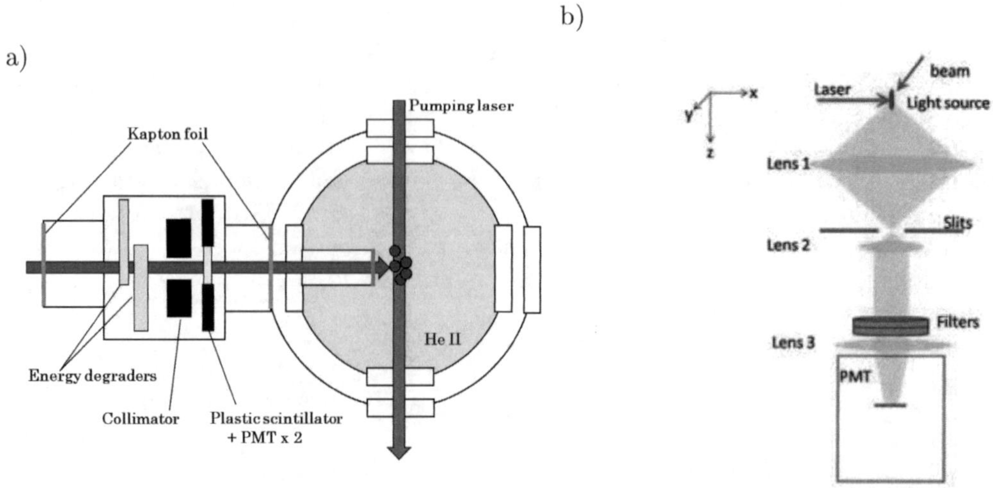

Figure 1. Schematic view of the experimental setup of a) the beam line and cryostat and b) the photo-detection system. The ^{87}Rb beams (66 AMeV) were injected into the cryostat through two energy degraders and a plastic scintillator. The LIF intensity from the introduced atoms were collected with three large lenses, wavelength-separated from the scattered laser light with two interference filters, and then detected with the Peltier-cooled photomultiplier tube.

3. Observed LIF intensity in the on-line experiment

In the on-line experiment we measure the LIF intensity with the injection of the ^{87}Rb beam on and off. Figure 2 shows the typical time evolution of the detected photon intensity with the ^{87}Rb beam on and off. The observed intensity is increased with injecting ^{87}Rb beam and decreased with stopping the beam injection. The decrease of LIF intensity with the stop of Rb beam injection indicates that the atoms introduced into He II flow out from the observed region of He II. The resident time of introduced ^{87}Rb atoms are approximately 0.5 sec, which is as long as that of laser sputtered atoms in the off-line experiments. Such long resident time suggests that this injection method with accelerated ion beam achieved equivalent performance to that with laser sputtering method, and hence it is feasible to introduce this method to study atomic properties of all the elements in He II.

From the difference in counting rates, we concluded that the observed LIF intensity is approximately 0.8×10^5 photons/sec, with the injected ^{87}Rb of 1.7×10^5 ions/sec. The obtained LIF intensity is nearly consistent with the intensity calculated from the photo-absorption cross section of atoms in He II with 780 nm wavelength photons [6].

4. Conclusion

In summary, we have been developing the new laser spectroscopy method "OROCHI" for determining nuclear spins and moments of exotic RI far from the β-stability. In the OROCHI, we use superfluid helium (He II) as an effective stopper of RI beam and a host matrix of laser spectroscopy. The blue-shifted and broadened absorption spectra of atoms in He II enable us to measure the extremely low yield RI whose yield is less than 1 particle per second. So far, we have demonstrated successfully the determination of nuclear spins and moments of stable Rb, Cs, Ag and Au isotopes from their atomic hyperfine and Zeeman splittings.

Here, we have reported the first on-line experiment with a ^{87}Rb beam to observe the atoms from an accelerated ion beam stopped in He II. In this on-line experiment, we aim at evaluating the minimum beam intensity required for the observation. In the on-line experiment we measure

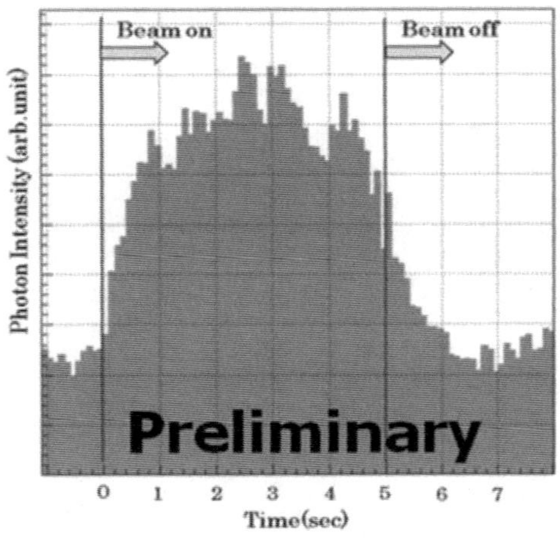

Figure 2. Time evolution of LIF intensity with turning on and off the beam injection. The LIF intensity is increased as the ^{87}Rb beam is injected.

the LIF intensity in turning on and off the injection of the ^{87}Rb beam. The decrease of LIF intensity with the stop of Rb beam injection indicates that the atoms introduced in He II flow out from the observation region of He II. The resident time of introduced ^{87}Rb atoms are approximately 0.5 sec, which is as long as that of laser sputtered atoms in off-line experiments. The observed LIF intensity is approximately 0.8×10^5 photons/sec with the injected ^{87}Rb of 1.7×10^5 ions/sec. The observed Intensity is nearly consistent with the intensity calculated from the photo-absorption cross section of atoms in He II with 780 nm wavelength photons. In the near future, we will measure the hyperfine structure of unstable ^{84,86}Rb isotopes in He II with the OROCHI method to demonstrate the determination of the spins and moments of radioisotopes, and will apply it to measure the nuclear spins and moments of rare isotopes.

Acknowledgment
This work was supported in part by the Grant-in-Aid for Scientific Research (No.19740170 and No.19204029) of the Ministry of Education, Culture, Sports, Science, and Technology, Japan. One of the author (T.F.) also acknowledges the financial support from the Global Center of Excellence Program by MEXT, Japan through the "Nanoscience and Quantum Physics" Project of Tokyo Institute of Technology.

References

[1] A. Bohr and B.R. Mottelson, *Nuclear Structure* (World Scientific, 1998)
[2] E.W. Otten, *Treatise on Heavy-Ion Science*, vol.8, 515 (Prenum Press, New York, 1987)
[3] P. Moroshkin et al.,: Low. Temp. Phys. **32**, 981 (2006).
[4] T. Furukawa et al.: Hyp. Int. **196**, 191 (2010).
[5] A. Sasaki et al.: RIKEN. Acc. Prog. Rep. **42**, 193(2009)
[6] T. Furukawa et al.: RIKEN. Acc. Prog. Rep. **41**, 153 (2008)

International Symposium "Nanoscience and Quantum Physics 2011" IOP Publishing
Journal of Physics: Conference Series **302** (2011) 012055 doi:10.1088/1742-6596/302/1/012055

Drell-Yan Experiment: Studying Anti-Quarks in the Proton

Kenichi Nakano and Toshi-Aki Shibata

Tokyo Institute of Technology, 2-12-1, Ookayama, Meguro-ku, Tokyo 152-8550, Japan

E-mail: knakano@nucl.phys.titech.ac.jp

Abstract. The internal structure of the proton is one of the most vital topic in the present hadron physics. At a high-energy scale the proton reveals its complicated parton structure which is beyond the scope of the Quark Parton Model (QPM), i.e. quarks (q), anti-quarks ($\bar{q}$) and gluons (g). Gluons and anti-quarks are dynamically created via $q \to q + g$ and $g \to q + \bar{q}$, respectively. The dynamics of the proton structure is being investigated in both experiment and theory.

The Drell-Yan process in proton+proton collisions is the reaction in which a quark in one proton and an anti-quark in the other proton annihilate into a virtual photon and then create a muon pair: $q + \bar{q} \to \gamma^* \to \mu^+ + \mu^-$. This process is suited to study the anti-quarks in the proton, and many experiments measuring the Drell-Yan process have been carried out in the past 20 years. The E906/SeaQuest experiment at Fermi National Accelerator Lab (FNAL) in USA is a Drell-Yan experiment being prepared now to explore a new kinematic region with very high statistics. The main purpose of this experiment is to precisely measure the asymmetry of $\bar{u}$ and $\bar{d}$ distributions in the proton at the x_{Bj} range from 0.25 to 0.45.

In this paper, the anti-quark flavor asymmetry in the proton and the past Drell-Yan experiments are summarized, and the status of the E906 experiment is reported.

1. Introduction

The proton in a low-energy (static) condition can be expressed with three valence quarks, namely two up quarks and one down quark. But the proton in a high-energy condition consists of a huge number of quarks (q), anti-quarks ($\bar{q}$) and gluons (g), where gluons and anti-quarks are dynamically created via $q \to q + g$ and $g \to q + \bar{q}$, respectively, by the strong force. For example, at a momentum scale of $Q^2 \gtrsim 100$ GeV^2, a half of the proton momentum is carried by quarks and the remaining half by gluons. The Q^2 dependence of the parton distributions are well described with the DGLAP evolution equation [1, 2, 3].

Starting from the naive parton model [4], recent developments in both measurement and theory are revealing rich features of the proton internal structure. A flavor asymmetry between the light anti-quarks, i.e. $\bar{u}$ and $\bar{d}$, has been observed by many experiments, and is the main topic of this paper. Another example is the proton spin problem initiated by the EMC Collaboration in 1988 [5, 6]. The contribution of the quark spin to the proton spin has been measured to be only 30% [7, 8]. The remaining component should be carried by the gluon spin and the orbital angular momenta of quarks and gluons, although it has not been determined well by neither measurement nor theory. The state of partons in the proton is the key information to understand the detailed characteristics of Quantum ChromoDynamics (QCD).

2. Anti-Quark Flavor Asymmetry, $\bar{d}/\bar{u}$, in Proton

Because the masses of $\bar{u}$ and $\bar{d}$ are approximately equal to each other, the distributions of $\bar{u}$ and $\bar{d}$ in the proton were initially assumed to be flavor symmetric, namely $\bar{d} = \bar{u}$. Based on this assumption the Gottfried Sum Rule [9] was proposed in 1967,

$$S_G = \int_0^1 \frac{dx_{Bj}}{x_{Bj}} \{F_{2p}(x_{Bj}) - F_{2n}(x_{Bj})\} = \frac{1}{3}\{(u-\bar{u})-(d-\bar{d})\} - \frac{2}{3}(\bar{d}-\bar{u}) = \frac{1}{3} \quad (1)$$

where $F_{2p}(x_{Bj})$ and $F_{2n}(x_{Bj})$ are the structure functions of the proton and the neutron, respectively. In 1990 the New Muon Collaboration (NMC) at CERN measured the strcture functions and examined the Gottfried sum rule [10, 11]. The result is shown in Fig. 1 and demonstrates $S_G = 0.2281(65)$ at $x_{Bj} \in (0.004, 0.8)$ & $Q^2 = 4$ GeV^2. This result was the first indication that the flavor symmetry is broken.

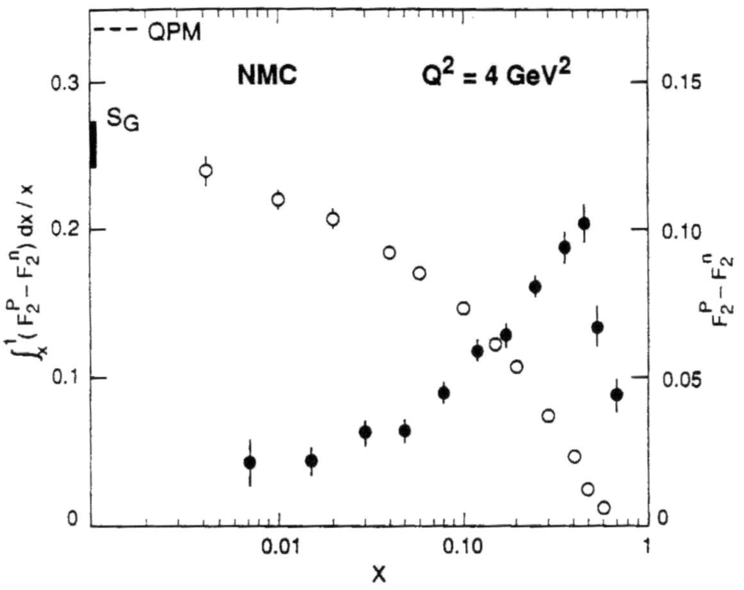

Figure 1. Structure functions and Gottfried Sum measured by NMC [10, 11]. The filled and open circles are $F_2^p(x_{Bj}) - F_2^n(x_{Bj})$ and $\int_{x_{Bj}}^1 (F_2^p(x_{Bj}) - F_2^n(x_{Bj}))dx_{Bj}/x_{Bj}$ as functions of x_{Bj}, respectively. The vertical bar at $x_{Bj} = 0$ is S_G extrapolated from the open circles. The expected S_G value, 1/3, based on the flavor-symmetry assumption is indicated by the dashed line labeled "QPM" (Quark Parton Model).

The NA51 experiment at CERN in 1994 [12] and the E866/NuSea experiment at FNAL in 1998 [13, 14, 15] directly measured the flavor asymmetry with the Drell-Yan process. As shown in Fig. 2, the measured asymmetry is as large as $\bar{d}/\bar{u} = 170\%$ at $x_{Bj} \sim 0.2$. After this observation, many theoretical models [16, 17, 18, 19, 20] have been proposed to explain the large asymmetry. Although all models reproduces the basic shape of the measured asymmetry, no model can show a trend that the measured asymmetry reverses ($\bar{d} \gtrless \bar{u}$) at $x_{Bj} \sim 0.3$. To further investigate the consistency between the measurement and the models, more precise data particularly at high x_{Bj} is needed.

The HERMES experiment also measured the flavor asymmetry in the lepton-nucleon deep inelastic scattering and obtained a consistent result [21].

3. Drell-Yan Process

The Drell-Yan process [23, 24] is, as shown in Fig. 3, a process of hadron+hadron reactions in which a quark in one hadron and an anti-quark in the other hadron annihilate into a virtual photon and then create a muon pair: $q + \bar{q} \to \gamma^* \to \mu^+ + \mu^-$. The Drell-Yan cross section is written as

$$\frac{d^2\sigma}{dx^{\mathrm{T}}dx^{\mathrm{B}}} = \frac{4\pi\alpha^2}{9x^{\mathrm{T}}x^{\mathrm{B}}}\frac{1}{s}\sum_i e_i{}^2\left\{q_i^{\mathrm{T}}(x^{\mathrm{T}})\bar{q}_i^{\mathrm{B}}(x^{\mathrm{B}}) + \bar{q}_i^{\mathrm{T}}(x^{\mathrm{T}})q_i^{\mathrm{B}}(x^{\mathrm{B}})\right\}, \tag{2}$$

where s is the square center-of-mass energy of two interacting hadrons, $q_i(x)$ is the parton distribution function of a flavor i, and the superscripts "T" and "B" denote partons in one hadron (target) and the other hadron (beam), respectively.

The distribution of invariant mass (M) of $\mu^+\mu^-$ pairs in proton+proton collisions is shown in Fig. 4, which was measured by the E866/NuSea experiment. The continuum is of the Drell-Yan process. It is contaminated by backgrounds, namely accidental $\mu^+\mu^-$ pairs and the tail of the meson resonances, and should be experimentally corrected. From this distribution the Drell-Yan cross section at the invariant-mass region outside the meson resonances can be derived in measurement. The parton distribution functions can be extracted from the measured cross

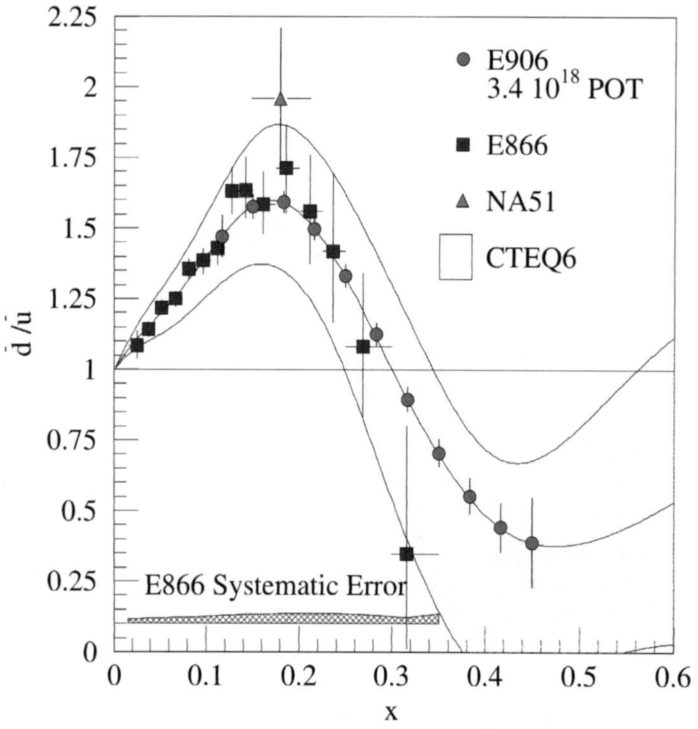

Figure 2. Anti-quark flavor asymmetry, $\bar{d}/\bar{u}$, as a function of x_{Bj} ($= p_{\mathrm{parton}}/p_{\mathrm{proton}}$). The pink triangle is the result of CERN NA51 in 1994 [12] and the blue squares are the result of FNAL E866/NuSea in 1998 [13, 14, 15]. The solid line with the yellow band is the CTEQ6 fit result with one-standard-deviation interval [22], which includes the E866/NuSea data. The red circles with the vertical bars are the anticipated x_{Bj} range and statistical accuracy measured by FNAL E906/SeaQuest.

Journal of Physics: Conference Series **302** (2011) 012055 doi:10.1088/1742-6596/302/1/012055

section via Eq. (2).

Since the Drell-Yan process always involve anti-quark, it is the ideal process for the measurement of the anti-quark flavor asymmetry. In addition, the final state and the kinematics are so simple that x_{Bj} of two scattered partons can be reconstructed event-by-event to resolve the x_{Bj} dependence of the flavor asymmetry.

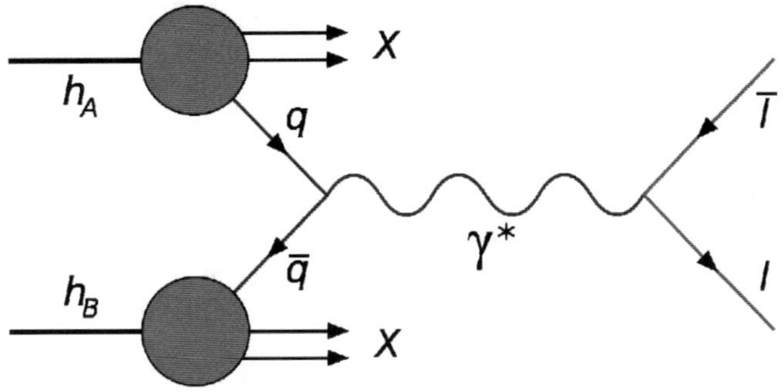

Figure 3. Diagram of Drell-Yan reaction.

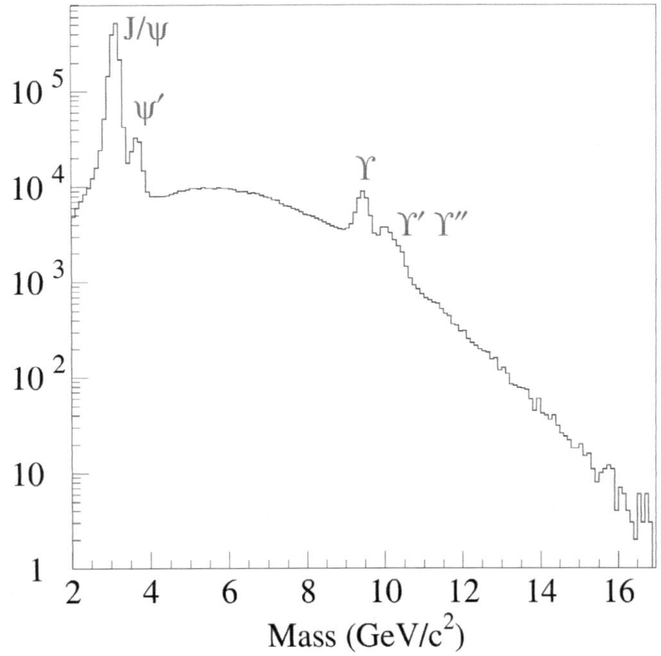

Figure 4. Distribution of invariant mass (M) of $\mu^+\mu^-$ pairs in proton+proton collisions measured by E866/NuSea.

4. E906/SeaQuest Experiment at FNAL

E906/SeaQuest at FNAL is a fixed-target experiment to measure the Drell-Yan process with a 120-GeV proton beam ($\sqrt{s} = 15$ GeV). It measures the Drell-Yan cross sections in $p + p$ and $p + d$ reactions with hydrogen and deuterium targets, and derives the flavor asymmetry from the ratio of the cross sections:

$$\frac{\sigma_{pd}}{2\sigma_{pp}} \approx \frac{1}{2}\left(1 + \frac{\bar{d}}{\bar{u}}\right), \tag{3}$$

where taking this ratio cancels out most systematic errors. E906/SeaQuest aims to achieve high precision at high x_{Bj} by

- using a beam from the Fermilab Main Injector, which is of high intensity and low energy (where, as $\sqrt{s}$ decreases, the cross section of Drell-Yan process increases in proportion to $1/s$ and that of background decreases in proportion to s), and
- designing the spectrometer so that muon pairs with large invariant mass are efficiently measured (where invariant mass correlates with x_{Bj} as $M^2 = x^{\mathrm{T}} x^{\mathrm{B}} s$).

The anticipated accuracy is 10 times better at $x_{Bj} \sim 0.3$ than E866/NuSea, as shown in Fig. 2.

Figure 5 is a schematic drawing of the E906/SeaQuest spectrometer. The beam and a huge number of background particles are absorbed with the solid iron just after the target. Charged particles that penetrate the hadron absorber and hit the proportional tubes at Station 4 are identified as muons. The momentum vector of muons is reconstructed with the wire chambers at Station 1, 2 and 3 and the KTeV tracking magnet. All stations are equipped with a hodoscope array for trigger.

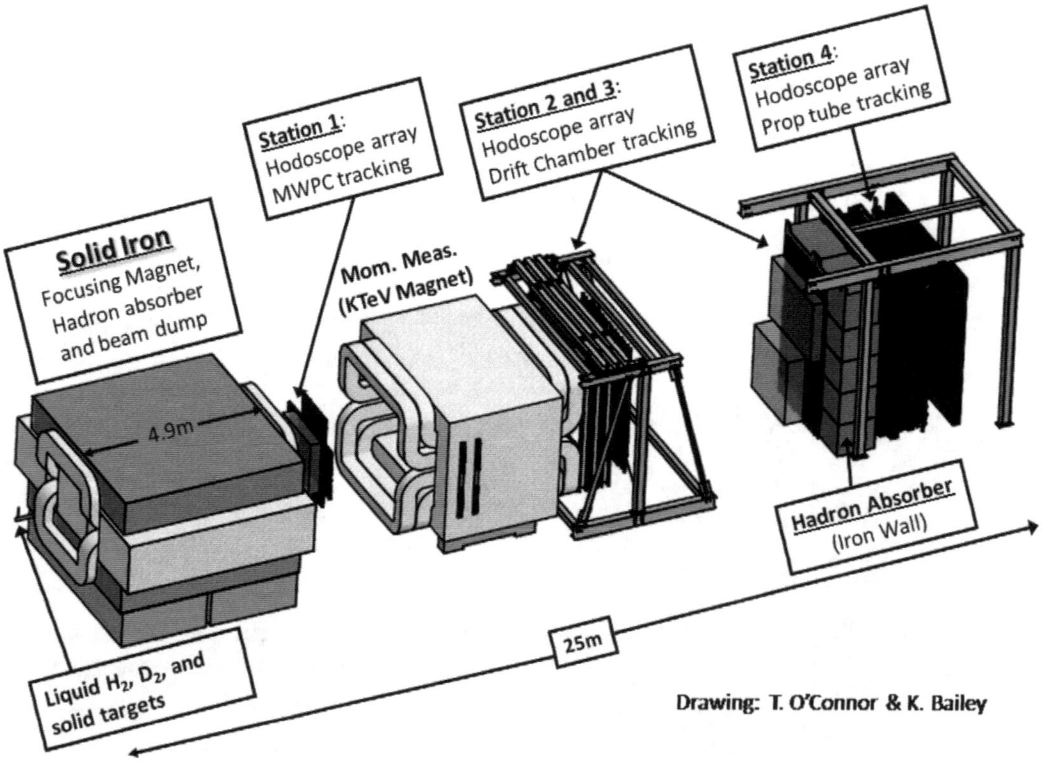

Figure 5. Schematic drawing of E906/SeaQuest spectrometer.

The spectrometer is being set up for data taking. Four sets of trigger hodoscopes have successfully been set up by February 2011 and are ready to measure the rates of signals and backgrounds with the first beam coming. The construction of a target system will be finalized in March 2011. The first beam for commissioning will be available in May 2011. The beam experiment will be adjourned halfway for one year due to an upgrade of the Main Injector, and will finish in 2014.

References

[1] Gribov V N and Lipatov L N 1972 *Sov. J. Nucl. Phys.* **15** 438–450
[2] Altarelli G and Parisi G 1977 *Nucl. Phys.* **B126** 298
[3] Dokshitzer Y L 1977 *Sov. Phys. JETP* **46** 641–653
[4] Bjorken J D and Paschos E A 1969 *Phys. Rev.* **185** 1975–1982
[5] Ashman J *et al.* (European Muon) 1988 *Phys. Lett.* **B206** 364
[6] Ashman J *et al.* (European Muon) 1989 *Nucl. Phys.* **B328** 1
[7] Airapetian A *et al.* (HERMES Collaboration) 2007 *Phys. Rev. D* **75** 012007
[8] Alexakhin V Y *et al.* (COMPASS Collaboration) 2007 *Phys. Lett.* **B647** 8–17 (*Preprint* hep-ex/0609038)
[9] Gottfried K 1967 *Phys. Rev. Lett.* **18** 1174
[10] Amaudruz P *et al.* (New Muon) 1991 *Phys. Rev. Lett.* **66** 2712–2715
[11] Arneodo M *et al.* (New Muon) 1994 *Phys. Rev.* **D50** 1–3
[12] Baldit A *et al.* (NA51) 1994 *Phys. Lett.* **B332** 244–250
[13] Hawker E A *et al.* (FNAL E866/NuSea) 1998 *Phys. Rev. Lett.* **80** 3715–3718 (*Preprint* hep-ex/9803011)
[14] Towell R S *et al.* (FNAL E866/NuSea) 2001 *Phys. Rev.* **D64** 052002 (*Preprint* hep-ex/0103030)
[15] Peng J C *et al.* (E866/NuSea) 1998 *Phys. Rev.* **D58** 092004 (*Preprint* hep-ph/9804288)
[16] Alberg M A, Henley E M and Miller G A 2000 *Phys. Lett.* **B471** 396–399 (*Preprint* hep-ph/9907417)
[17] Nikolaev N N, Schafer W, Szczurek A and Speth J 1999 *Phys. Rev.* **D60** 014004 (*Preprint* hep-ph/9812266)
[18] Szczurek A, Speth J and Garvey G T 1994 *Nucl. Phys.* **A570** 765–781
[19] Pobylitsa P V, Polyakov M V, Goeke K, Watabe T and Weiss C 1999 *Phys. Rev.* **D59** 034024 (*Preprint* hep-ph/9804436)
[20] Dorokhov A E and Kochelev N I 1993 *Phys. Lett.* **B304** 167–175
[21] Ackerstaff K *et al.* (HERMES) 1998 *Phys. Rev. Lett.* **81** 5519–5523 (*Preprint* hep-ex/9807013)
[22] Pumplin J *et al.* 2002 *JHEP* **07** 012 (*Preprint* hep-ph/0201195)
[23] Drell S D and Yan T M 1970 *Phys. Rev. Lett.* **25** 316–320
[24] Christenson J H *et al.* 1970 *Phys. Rev. Lett.* **25** 1523–1526

Study of Fragmentation from A Quark to Hadrons

N. Kobayashi

Department of Physics, Tokyo Institute of Technology, 2-12-1, Oh-okayama, Meguro-ku, Tokyo 152-8551, Japan

E-mail: nkobayashi@nucl.phys.titech.ac.jp

Abstract. Hadrons are known to be strongly interacting particles. A single quark is not detected in experiments but is always bound in a hadron because of color confinement. We observe hadrons in the final states as a result of quark reactions. We now consider a process of a quark to hadrons in e^+e^- annihilation. An electron collides with a positron, and a quark-antiquark pair is generated. The quark and antiquark have opposite momenta. The quark radiates gluons and the gluons produce further quark-antiquark pairs. These quarks and gluons finally form hadrons. This process is called 'fragmentation'. Such a process is formulated by the fragmentation function. I will discribe the process of the fragmentation from a quark to hadrons by using several models and also fragmentation function, based on lepton-nucleon deep inelastic scatterings and e^+e^- reactions where I am involved in data analyses.

1. Introduction

Parton fragmentation refers to the process of converting high-energy, colored partons (quark and gluon) out of a hard scattering into hadron jets observed in detectors. Dramatic demonstrations of quark substructure are obtained in e^+e^- annihilation to hadrons at very high energy (Figure 1.). The hadrons appear in the form of two 'jets' collimated around the $q\bar{q}$-axis. The first observation of jet was in 1975 at Stanford Linear Accelerator Center on $e^+e^- \to q + \bar{q}$ reaction [1]. After that, jet was observed on deep inelastic lepton scattering and proton-antiproton reaction. Now fragmentation function, which associate hadron with parton, is a powerful probe for the study of particle physics. For example, 'spin structure' of the proton can be investigated when the experimental information from deep inelastic scattering is combined with the fragmentation function from e^+e^- collider. However, we need to think about some models to associate the jet with parton because nonperturbative treatment is necessary for this. We now mention two of the famous fragmentation function models.

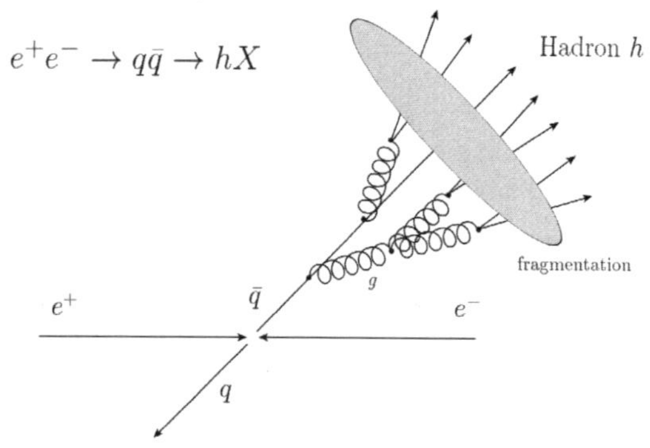

Figure 1. Fragmentation process in e^+e^- annihilation

2. Fragmentation Function

Fragmentation functions represent the probability for a parton to fragment into a particular hadron carrying a certain fraction of the parton's energy. A parton fragmentation $D_i^h(z)$ encodes the probability that the parton i fragments into a hadron h carrying a energy fraction z of the parton ' s momentum. energy fraction z is written as :

$$z = \frac{p_h}{|p|} \tag{1}$$

where p_h is the momentum of a hadron h, p is momentum of parton i. In an actual experiment, z is decided like this.

$$z = \frac{p_h}{|p|} \simeq \frac{2E_h}{E_{\rm CM}} \tag{2}$$

Figure 2. shows a energy fraction z of a pion generated from e^+e^- annihilation experiment of $\rm E_{CM}$ (center-mass-energy) = 10.58 GeV.

The following relation consists in $D_q^h(z)$ and cross sections :

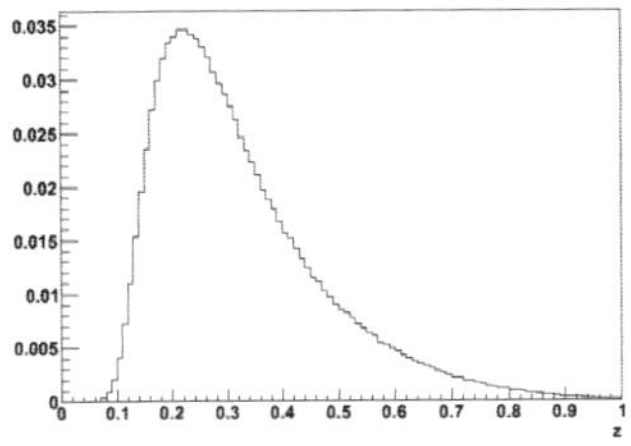

Figure 2. A energy fraction z of a pion generated from e^+e^-annihilation experiment of $\rm E_{CM}$ = 10.52 GeV. All bins are normalized.

$$\frac{1}{\sigma}\frac{d\sigma(e^+e^- \to hX)}{dz} = \frac{\sum_q e_q^2[D_q^h(z) + D_{\bar{q}}^h(z)]}{\sum_q e_q^2} \tag{3}$$

where e_q is a charge of quarks. The right-hand value doesn't depend on center-mass-energy $\rm E_{CM}$ and only depends on z. And that is scaling consists. Figure 3. shows a distribution varying $\rm E_{CM}$ in e^+e^- annihilation experiment[2]. This data uses the fact $\sigma \propto {\rm E_{CM}}^{-1}$. The distribution shows it's independence of $\rm E_{CM}$.

Fragmentation Function describes parton characteristics, thus it doesn ' t depend on the method of making parton. At inclusive ep scattering ($ep \to hX$), cross section σ is represented like this:

$$\frac{1}{\sigma}\frac{d\sigma(ep \to hX)}{dz} = \frac{\sum_q e_q^2 f_q(x) D_q^h(z)}{\sum_q e_q^2 f_q(x)} \tag{4}$$

where f_q is parton distribution function of proton. Sums are extends to all quarks and anti-quarks which can become mother of hadron h. A function $f_q(x)$ represents momentum distribution of a parton in proton. Variable x is called Bjorken x and defined as $x = \frac{Q^2}{2M\nu}$. Q^2 is momentum transfer from electron to proton and μ is energy transfer. M is mass of the proton.

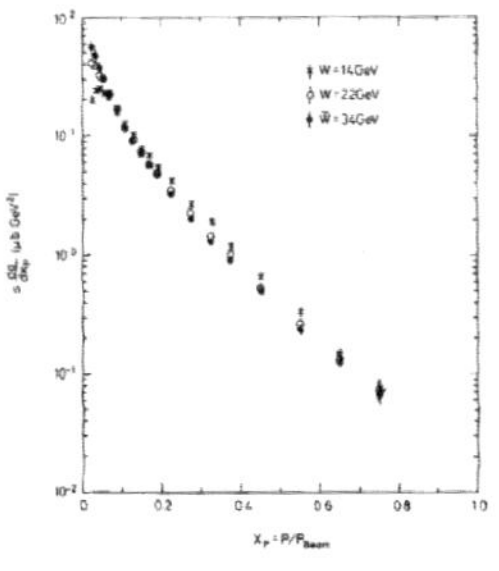

Figure 3. ${\rm E_{CM}}^{-1} d\sigma/dz$ distributions with $E_{CM} = 14, 22, 34$ GeV in e^+e^- annihilation experiment[2]

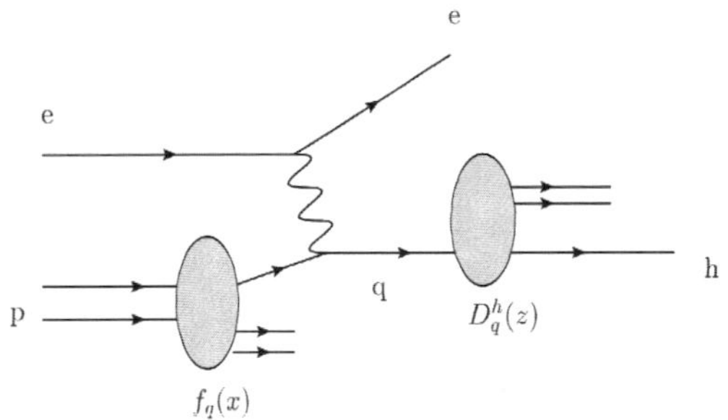

Figure 4. Hadron Production from Deep inelastic lepton scattering: $ep \to hX$

3. Fragmentation Models

For comparing a experimental data to the theory, it is necessary to understand how a parton is converted into hadrons. The rigorous method doesn't exist mathematically now but there are various approaches. Two of the very popular models are the independent fragmentation and the string fragmentation .

3.1. Independent fragmentation model

Independent fragmentation model is based on a rule that scaling consists about parton's longitudinal momentum and distribution of transverse momentum doesn't depend on it's energy[3]. We assume that partion mass is zero and transverse momentum is ignored. Then partion q_i with longitudinal momentum $q_{\parallel}$ makes meson $M(q_i\bar{q}_{i+1})$ with longitudinal momentum $p_{mz} = zp_{q\parallel}$ choosing $\bar{q}_{i+1}$ from $q_{i+1}\bar{q}_{i+1}$ in vacuum at the probability of $f(z)dz$. The other partion q_{i+1} is going through at the probability of $1 - f(z)dz$ (Figure 5.).

When hadron mass is not ignored, we use light cone variable $(E + p_{\parallel})$ instead of momentum. Then energy fraction z is defined as

$$z = \frac{(E_h + p_{h\parallel})}{(E_q + p_{q\parallel})} \tag{5}$$

If parton q_i and q_j hadronizes independently at each process, we can use Monte Carlo calculations to this iteration.

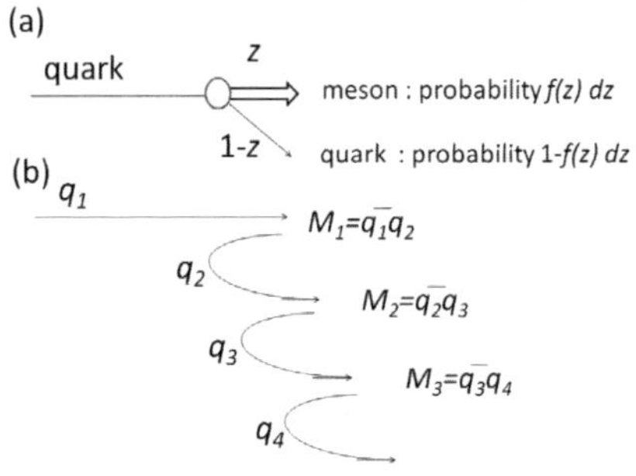

Figure 5. Independent fragmentation model.
(a) A quark which has a longitudinal momentum makes meson at the probability of $f(z)dz$. The other quark goes through at the probability of $1 - f(z)dz$.
(b) A quark q_1 makes meson $M_1 = q_1\bar{q}_2$ choosing $\bar{q}_2$ from $q_2\bar{q}_2$ in vacuum. The other quark q_2 also picks up $q_3\bar{q}_3$ from vacuum. Each quarks generate mesons independently but in the same way.

The total probability $D(z)dz$ of first partion to hadronize consists of the probability of the parton to mesons $f(z)dz$ at first hadronization and that of other partons.

$$D(z) = f(z) + \int_z^1 f(1-\eta)D(\frac{z}{\eta})\frac{d\eta}{\eta} \tag{6}$$

$$\int f(z)dz = 1 \tag{7}$$

Then rapidity η is

$$\eta = \frac{1}{2}\ln\frac{(E_h + p_{h\parallel})}{(E_q + p_{q\parallel})} = \ln(E_h + p_{h\parallel})m_T \sim \ln z + \text{const.} \tag{8}$$

In light quarks (u, d, s), $f(z)$ and $D(z)$ are

$$f(z) = (1+\alpha)(1-z)^{\alpha} \tag{9}$$

$$D(z) = (1+\alpha)\frac{(1-z)^{\alpha}}{z} \tag{10}$$

where α= 0.6 is a free parameter to set calculations.

3.2. String fragmentation model

String fragmentation model innovates a string concept of QCD[4][5][6]. After $q\bar{q}$ arises at $t = x = 0$, they go through along a light-cone $t^2 - x^2 = 0$. Then they store a potential energy proportional to their distance. So if parton has a small energy compare to this, the parton loses it's kinetic energy at the end point of x-axis. They move inverse direction and ther trajectory gets to zigzag (Figure 6.(a)). If parton has a large energy, they break up string and make another $q\bar{q}$ pairs (Figure 6.(b)). The $q\bar{q}$ pairs are created according to the probability of a tunnelling process $\exp(-\pi \frac{m_{q,T}^2}{\kappa})$ which depends on the transverse mass squared $m_T^2 \equiv m_q^2 + p_T^2$ and the string tension $\kappa \sim 1$ GeV/fm. The transverse momentum p_T is locally compensated between quark and antiquark. Due to the dependence on the parton mass m_q and hadron mass m_h, the production of strange and heavy-quark hadrons is suppressed. The light-cone momentum fraction $z = (E + p_\parallel)_h/(E + p)_q$, where $p_\parallel$ is the momentum of the formed hadron h along the direction of the quark q, is given by the string-fragmentation function

$$f(z) \sim \frac{(1-z)^a}{z} \exp(-b\frac{m_{h,T}^2}{z}) \tag{11}$$

where a and b free parameters. These parameters are obtained by measured data $a = 0.11$ and $b = 0.52$ GeV^{-2} as determined in [6]

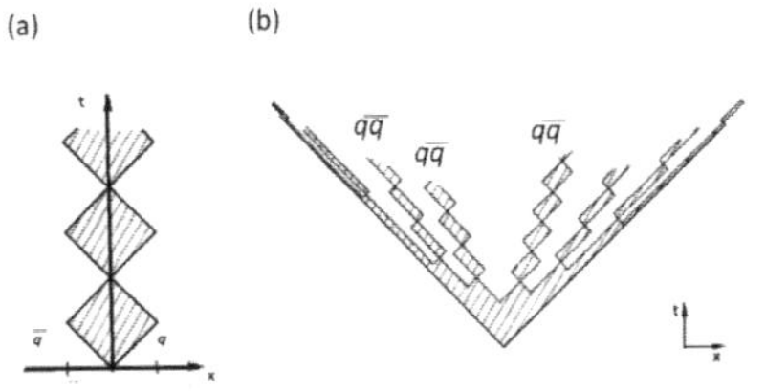

Figure 6. (a) The motion of q and $\bar{q}$ in the CM frame. The hatched areas show where the field is nonvanishing.
(b) The final picture when q and $\bar{q}$ move with large energies in opposite directions. The field has broken at many places by the production of $q\bar{q}$ pairs. Hatched areas indicate nonvanishing field.

4. Summary

Fragmentation function described the process from a quark to Hadrons . It gives detailed information about the parton with observed hadron jets. A parton fragmentation function D_i^h depends on energy fraction z. To compare a experimental data to the theory, independent fragmentation model and string fragmentation models are effective.

Acknowledgments

I thank my supervisor T.-A. Shibata and K. Nakano. I also appreciate R. Seidl and Belle FF analysis group members. Financial support by GCOE Nanoscience and Quantum Physics is acknowledged.

References

[1] Schwitters R F et al. 1975 *Phys. Rev. Lett.* **35** 1320
[2] TASSO Collab. Brandelik R et al. 1982 *Phys. Lett.* B**114** 65.
[3] Field R D and Feynman R P 1978 *Nucl. Phys.* B**136** 1
[4] Field R D and Feynman R P 1977 *Phys. Rev.* D**15** 2590
[5] Artru X and Menessier G 1974 *Nucl. Phys.* B**70** 93
[6] Andersson B et al. 1983 *Phys. Reports.* **97** 31
[7] Schmelling M 1995 *Phys. Scripta* **51** 683

A Drift Chamber for a Muon Spectrometer and Measurement of Sea Quark Flavor Asymmetry in the Proton at E906/SeaQuest Drell-Yan Experiment

Florian Sanftl and Toshi-Aki Shibata

Department of Physics, Tokyo Institute of Technology, 2-12-1, Ohokayama, Meguro-ku, Tokyo, 152-8551, Japan

E-mail: `florian.sanftl@nucl.phys.titech.ac.jp`

Abstract. Fermilab E-906/SeaQuest is a particle physics experiment which will use Drell-Yan process to measure the contributions of antiquarks to the structure of the proton or neutron and how this structure is modified when the proton or neutron is included within an atomic nucleus. It is part of a series of fixed target Drell-Yan experiments. The Drell-Yan process occurs in high energy hadron–hadron scattering. It takes place when a quark of one hadron and an antiquark of another hadron annihilate, creating a virtual photon which then decays into a pair of oppositely-charged muons. The kinematics of these myons allow a direct relation to the antiquark structure of the proton. In addition the E-906/SeaQuest experiment can also examine the modifications to the antiquark structure of the proton from nuclear binding or access transverse momentum dependent parton distribution functions. The E906/SeaQuest experiment will use a 120 GeV proton beam extracted from the Fermilab Main Injector. This beam energy provides a tremendously increasing cross section compared to past Drell-Yan experiments. Taking the reduction of background events due to J/Ψ decays into account a gain in statistics by a factor of 50 compared to the latest Drell-Yan experiments is expected. The spectrometer is currently being assembled at Fermilab. It is expected to be commissioned in spring 2011 and collect data for two years. The group of Tokyo Institute of Technology is among Japanese collaborators in charge of the tracking stations in the E-906/SeaQuest experiment. In this report we will give an overview and a status report of the ongoing commissioning of the Drift Chambers (DC).

1. Introduction

E906/SeaQuest is a fixed-target experiment that is currently being installed at Fermilab in USA. For its measurements it is using the 120 GeV proton beam extracted from the Fermilab Main Injector. The E906/SeaQuest detector is a forward spectrometer and is designed to be sensitive to the Drell-Yan process, $q\bar{q} \to \gamma^* \to \mu^+\mu^-$, in proton-nucleon reactions by detecting the final-state muon pair. The main goal of this experiment is to investigate the flavor asymmetry of light sea quarks ($\bar{u}$ and $\bar{d}$) in the proton[1]. Particularly it focuses on the asymmetry at larger x_{Bj} of sea quarks (≥ 0.25), where x_{Bj} is the momentum fraction carried by the scattered quark with respect to the target nucleon.

2. The E906/SeaQuest Detector

The E906/SeaQuest spectrometer is designed to be capable of not only detecting muon pairs from the Drell-Yan process but also eliminating a huge number of low-momentum charged particles, particularly muons that come mainly from J/Ψ decays.

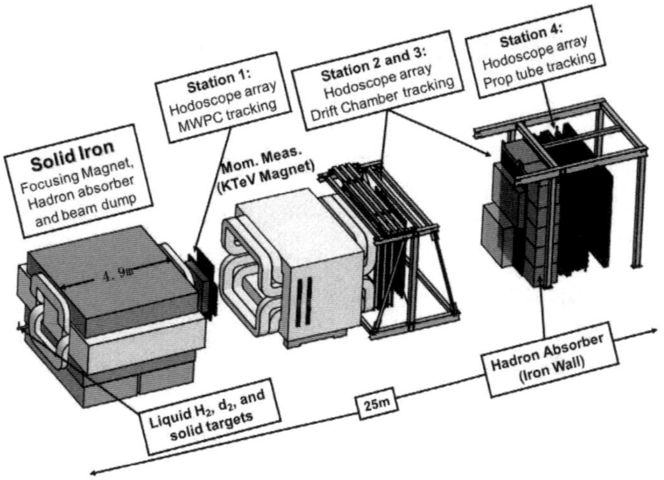

Figure 1. Schematic drawing of detector setup in side-top view.

Figure 1 shows the setup of the E906/SeaQuest detector, where the proton beam comes from the left side. The target materials will be liquid-hydrogen, liquid-deuterium and several solid targets, like Carbon, Calcium or Tungsten, stored in a cryogenic target cell. A focusing magnet labeled as "Solid Iron" in Fig. 1 condenses especially high-p_T muons into the detector acceptance and sweeps out low-momentum background muons. A large iron block is placed inside the focusing magnet, and serves both as a hadron absorber and also as a beam dump. Station 1, 2 and 3 are multi-wire drift chambers (MWDC). Station 3 consists of two different chambers, the Top and the Bottom chamber. They measure hit positions of passing muons in order to reconstruct their tracks and momenta. The momenta are determined with an analyzing magnet labeled as "KTeV Magnet" in Fig. 1 that bends muons in the horizontal direction. Station 4 consists of alternating layers of drift tubes and iron absorbers. It identifies high-momentum muons by only being sensitive to particles that penetrate the complete spectrometer including its absorbers due to its far back position in the overall detector setup. Each station is sandwiched by hodoscope arrays to trigger di-muon events.

The proton beam becomes available in spring 2011. Data will be taken intermittently for the next three years. The E906 Japanese group designed and constructed the Station 3 Top DC and is in charge of the operation of all Station 3 drift chambers. This article reports on the commissioning status of the new Station 3 Top DC at Fermilab. Throughout this report we refer to the new Station 3 Top DC by Station 3 DC.

3. The new Station 3 Drift Chamber

The dimensions of the Station 3 DC are 3.2 m in height $\times$ 2.2 m in width. In order to maximize the acceptance its position in the overall spectrometer has been determined to be 19 m from the target. The required position resolution at Station 3 is supposed to be less than 400μm. It is required to provide a very high (above 98%) muon pair detection efficiency at background rates which correspond to 5 kHz/cm^2 at the spectrometer location of Station 3. The background rate has been evaluated using the GEANT detector simulator[2].

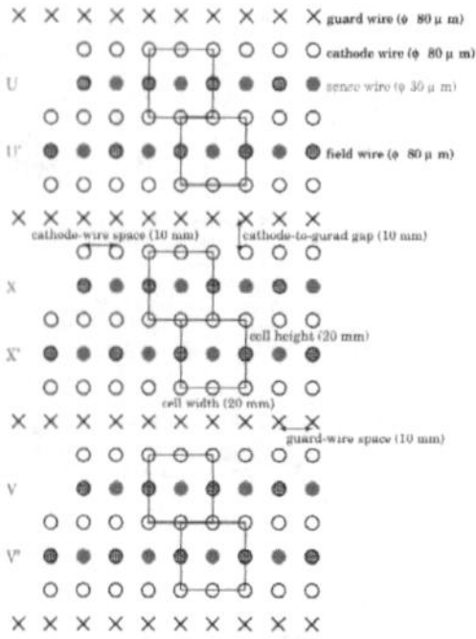

Figure 2. Cell structure (unit cell in blue boxes) of the Station 3 DC.

Figure 2 shows the cell structure of Station 3 DC. It consists of six layers labeled as U, U', X, X', V and V'. All wires are vertically arranged, whereas the wires of the U'- and V'-layers are tilted by 14°. In total, the number of sense wires, and thus of readout channels, is 768, and the number of field and guard wires is 4368.

The construction of the Station 3 DC was finalized in February 2010. After basic operational tests at RIKEN during spring and summer 2010 it was shipped to Fermilab in July 2010.

4. Commissioning - The first Steps

We present the first performance studies of the Station 3 DC. All measurements have been performed since August 2010 at Fermilab. For all studies, only eight neighboring sense wires were read out in an Argon:CO_2 (80%:20%) gas mixture.

As a first step, the gas amplification gain for different high voltage (HV) values was measured. Raw signals caused by cosmic rays were recorded and the corresponding gas gain was evaluated. A Garfield simulation [3] predicted the gas amplification gain to be $\mathcal{O}(10^5)$. The gas gain was obtained by dividing the charge output by the predicted ionization deposit. As seen in Fig. 3, the simulation and measurement differ by about a factor of 2. Further studies have to investigate the ionization process in a more detailed way.

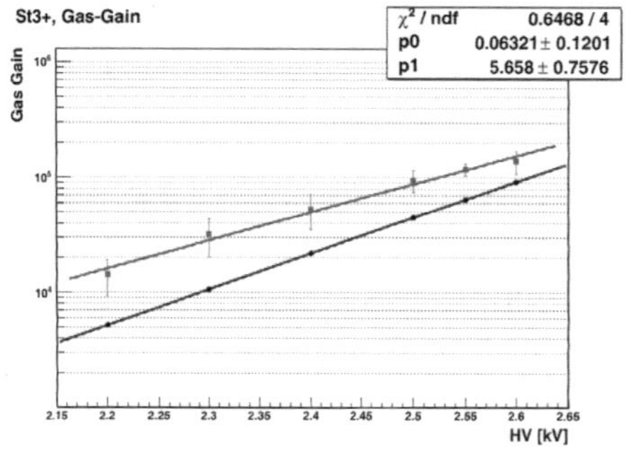

Figure 3. Measured (red line) and simulated (blue line) gas gain for Argon:CO_2 (80%:20%).

Furthermore, the one-layer efficiency curve was determined. Two plastic scintillators served as cosmic ray triggers for the measurement. Figure 4 shows the efficiency curve for two different discriminator thresholds. For the high threshold $98 \pm 1\%$-detection efficiency is reached at a HV of $\approx 2.5\,\mathrm{kV}$. This turns out to be sufficient considering that the future nominal operational HV of the Station 3 DC lies at 2.8 kV.

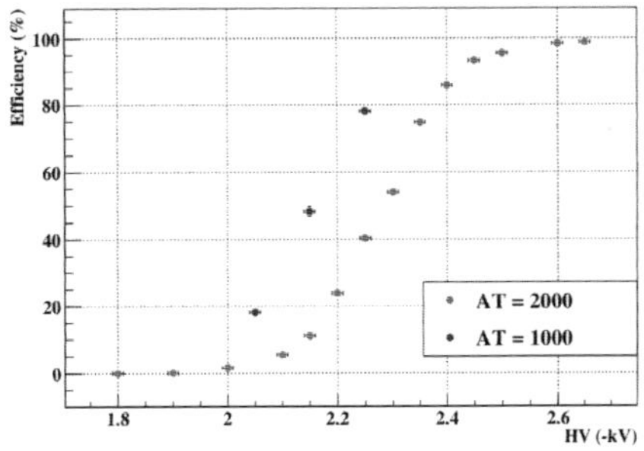

Figure 4. Efficieny curve for low (blue points) and high (red points) discriminator thresholds.

Finally, a preliminary TDC distribution was measured with a common-stop trigger signal. Considering only single hit events (red line in Fig. 5) the measurement is in good agreement with our Garfield simulations. The width of the TDC distribution is 150 TDC channels, which corresponds to 300 ns. The right side of the TDC distribution corresponds to a shorter drift time and has more hits by a factor of 3 compared to the left part. According to our simulation, this can be explained by the non-linear drift velocity being faster near the sense wire than at the cell edge by a factor between $2 - 3$.

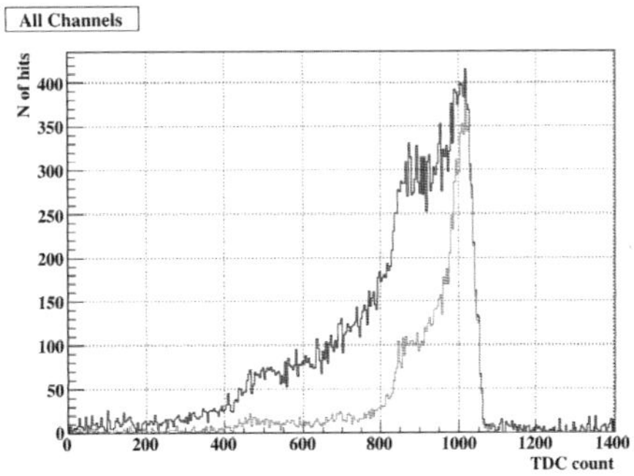

Figure 5. Integrated TDC distributions for multiple (black line) and single hits (red line).

5. Conclusion, Outlook and Acknowledgments

We reported on preliminary performance measurements of the Station 3 DC. The operation and behaviour of the chamber are well understood and match our expectations. At the moment we are optimizing the future read-out electronics. More detailed performance measurements will be possible as soon as the setup of the full read-out electronics is finished.

The collaboration of the E906/SeaQuest members is acknowledged. In particular, I would like to thank the group of Tokyo Institute of Technology and all Japanese members of E906/SeaQuest who built the Station 3 DC with me. The support of the Global Center of Excellence "Nanoscience and Quantum Physics" at Tokyo Institute of Technology is highly acknowledged.

References

[1] J. Arrington *et al.*, Proposal of E906 Experiment, http://www.phy.anl.gov/mep/drell-yan/proposals/
[2] CERNLIB Webpage, http://cernlib.web.cern.ch/cernlib/
[3] Garfield Webpage, http://garfield.web.cern.ch/garfield/

International Symposium "Nanoscience and Quantum Physics 2011" IOP Publishing
Journal of Physics: Conference Series **302** (2011) 012058 doi:10.1088/1742-6596/302/1/012058

Λ^* hypernuclei with chiral dynamics

Toshitaka Uchino, Tetsuo Hyodo and Makoto Oka

Department of Physics, Tokyo Institute of Technology, Meguro 152-8551, Japan

E-mail: uchino@th.phys.titech.ac.jp

Abstract. As a strangeness $S = -1$ and baryon number $B = 2$ system, the two-body bound state of $\Lambda^* = \Lambda(1405)$ and a nucleon is studied. To solve the $\Lambda^* N$ system, we construct the $\Lambda^* N$ potential by extending the Jülich model with couplings estimated in the chiral unitary approach. We have the $\Lambda^* N$ quasi-bound state with the mass, $M_{\Lambda^* N} \sim 2364$MeV which is shallowly bound about 9.5 MeV from the $\bar{K}NN$ threshold. Decay width of the fall apart process, where the $\Lambda^* N$ resonance decays to $\pi\Sigma N$ with a nucleon being as a spectator, is estimated to be $\Gamma_{F.A} \sim 49$MeV.

One of the most interesting topics of strange nuclear physics is possible existence of the $\bar{K}$ bound state in nuclei. The interaction of $\bar{K}$ and a nucleon is known to be attractive to support a quasi-bound state, which is called $\Lambda^* = \Lambda(1405)$, slightly below the $\bar{K}N$ threshold. By the $\bar{K}N$ attraction, the $\bar{K}NN$ three-body system is also considered to make a bound state. From a theoretical viewpoint, it is difficult to solve the $\bar{K}NN$ system due to the existence of the Λ^* resonance and the $\bar{K}$ absorption by two nucleons. Meanwhile, based on several theoretical arguments, we can take the viewpoint that Λ^* is regarded as a fundamental particle in the $\bar{K}NN$ bound system. This picture has an advantage that the inclusion of the $YN(Y = \Sigma, \Lambda)$ channels, which represent the two-nucleon absorption processes, is relatively easy, compared with the $\bar{K}NN$ picture. To search for a possible bound state of the $\Lambda^* N$ system, we construct the $\Lambda^* N$ potential. It is the most fundamental interaction in the "Λ^*-hypernuclei", which we call the nuclei with a Λ^*[1]. The $\Lambda^* N$ system is labeled by two quantum numbers, the total spin S and the orbital angular momentum L, and we consider $S = 0, 1$ and $L = 0$ cases as candidates of the ground state.

We construct the $\Lambda^* N$ potential by extending the Jülich model [2, 3], which is one of the hyperon-nucleon one-boson-exchange potential models. Since the isospin of Λ^* is zero, isoscalar mesons $X(X = \sigma, \omega)$ are exchanged, while the pseudoscalar η exchange is omitted as its coupling to the nucleon is small. In addition, considering the $\bar{K}$ exchange which contributes with an interchange of Λ^* and nucleon, the $\Lambda^* N$ potential can be written by

$$V = V_\sigma + V_\omega + V_{\bar{K}}, \tag{1}$$

where V_σ, V_ω and $V_{\bar{K}}$ represent σ, ω and $\bar{K}$ meson exchange contributions respectively.

For the lack of the information of Λ^*, vertex properties concerning Λ^* are not clear. To estimate relevant parameters, coupling constants, we adopt the microscopic structure of Λ^* obtained by the chiral unitary approach. In the chiral unitary model, Λ^* is described as a resonance in a coupled-channel meson and baryon $(\pi\Sigma, \bar{K}N, \eta\Lambda, K\Xi)$ multiple scattering, and we take the Λ^* meson baryon coupling constants from Refs. [4, 5]. The $\Lambda^*\bar{K}N$ coupling constant is directly obtained in the chiral unitary model, while the $\Lambda^*\Lambda^* X(X = \sigma, \omega)$ couplings are to

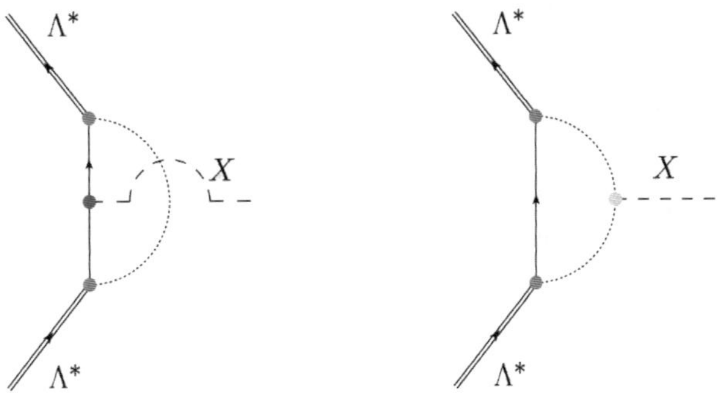

Figure 1. Microscopic description of the $\Lambda^*\Lambda^*X(X=\sigma,\omega)$ coupling. Left(Right) panel of the figure represents the meson X coupled diagram to the intermediate baryon(meson) in Λ^*.

be estimated. Since the exchanged meson couples to the meson and baryon in the multiple scattering, the $\Lambda^*\Lambda^*X$ coupling can be estimated by evaluating microscopic contributions which correspond to one-loop diagrams shown in Fig. 1. In the estimation, we take only dominant contributions, $\pi\Sigma$ and $\bar{K}N$ for making mechanism more visible.

Moreover, from the analysis of Λ^* with the chiral unitary approach, Λ^* is a superposition of two states. Therefore, the Λ^*N system consists of two states, Λ_1^*N and Λ_2^*N. Here, we call higher(lower) energy state $\Lambda_1^*(\Lambda_2^*)$. Then, we solve the two-channel coupled Schrödinger equation given by

$$H\psi=E\psi,\ H=T+V, \tag{2}$$

with

$$T=\begin{pmatrix} T_1 & 0 \\ 0 & T_2 \end{pmatrix},\ V=\begin{pmatrix} V_{11} & V_{12} \\ V_{21} & V_{22} \end{pmatrix},\ \psi=\begin{pmatrix} \psi_1 \\ \psi_2 \end{pmatrix}, \tag{3}$$

where T_a, V_{aa} and ψ_a stand for the kinetic energy term, potential term and wave function of Λ_a^*, while, off-diagonal components, V_{12} and V_{21}, contribute to the transition of the Λ_1^*N system and the Λ_2^*N system.

In order to convert the $\bar{K}$ exchange amplitude into the potential, an exchange factor should be applied, which introduces the spin dependence. Since the $\Lambda^*\bar{K}N$ vertex is a scalar type and the scalar exchange in the NN potential is an attractive force, the $\bar{K}$ exchange contribution is attractive(repulsive) for total spin $S=0(S=1)$. Nonzero energy transfer, due to the difference of the mass between Λ^* and nucleon, is included by the use of an effective $\bar{K}$ mass given by

$$\tilde{m}_{\bar{K}}=\sqrt{m_{\bar{K}}^2-(M_{\Lambda^*}-M_N)^2}, \tag{4}$$

where $m_{\bar{K}}$, M_{Λ^*} and M_N stand for $\bar{K}$, Λ^* and nucleon masses respectively. Since the effective $\bar{K}$ mass become lighter as the mass of Λ^* gets closer to the $\bar{K}N$ threshold, the $\bar{K}$ exchange is stronger in the Λ_1^*N system than in the Λ_2^*N system.

In the Λ^*N potential, isoscalar meson exchange contributions, the attractive σ exchange and the repulsive ω exchange, are largely cancelled out each other, except for the repulsive core at the very short range region. On the other hand, the $\bar{K}$ exchange contribution is a long range interaction by the light effective $\bar{K}$ mass. Therefore, total spin, which controls the sign of the $\bar{K}$ exchange contribution, determine whether the Λ^*N potential has attraction or not. In Fig. 2, we show the diagonal components of the obtained Λ^*N potential in $S=0$. In both case, the

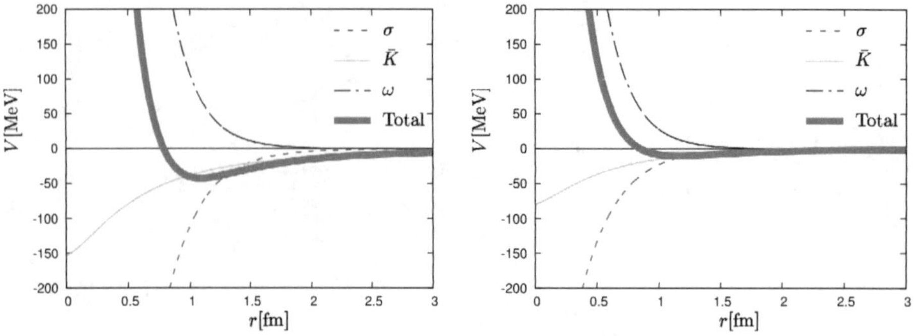

Figure 2. Λ^*N potential for $S = 0$. Left(Right) panel of the figure corresponds to $V_{11}(V_{22})$.

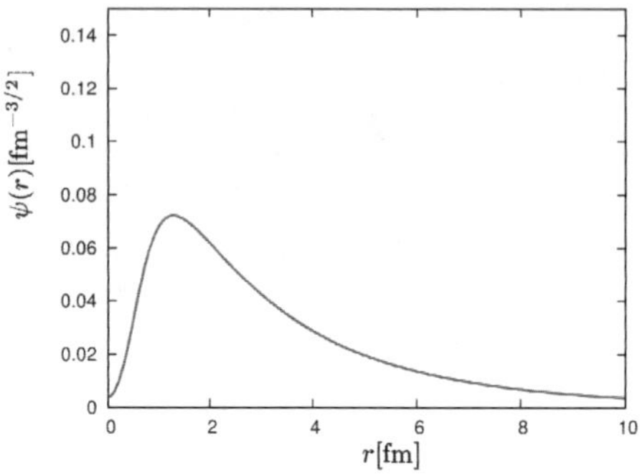

Figure 3. Wave function of the Λ_1^*N bound state.

potential has an attractive pocket at the intermediate range region due to the attraction in the $\bar{K}$ exchange, which is not the case in $S = 1$. It can be seen that the Λ_1^*N potential is more attractive because of the lighter effective $\bar{K}$ mass.

We solve the full coupled channel Schrödinger equation with the obtained Λ^*N potential, and study the bound state of the Λ^*N system. As a first step, to study the property of each Λ_a^*N system, we search for a possible bound state of each Λ_a^*N channel separately by switching off the off-diagonal components of the Λ^*N potential, $V_{12} = V_{21} = 0$. In the total spin $S = 1$, the Λ^*N potential has no attraction to make any bound states of the Λ^*N system. While, in $S = 0$, we find the bound state of the Λ_1^*N system only, with the binding energy $B = 9.5$ MeV measured from the $\bar{K}NN$ threshold. The wave function in coordinate space of the Λ_1^*N bound state is shown in Fig. 3. We can see that the wave function has a long tail due to small binding energy.

In the next place, performing the full coupled channel calculation, the Λ_1^*N bound state aquires a finite width through the coupling to the Λ_2^*N channel. We find the resonance state of the Λ^*N system by using the real scaling method, with the two-body mass of the resonance,

$$M_{\Lambda^*N} \sim 2364 \text{ MeV}. \tag{5}$$

Since the resonance wave function is not normalized, it is not straightforward to extract the properties of the Λ^*N quasi-bound state from the resonance solution. However, the coupling to the Λ_2^*N channel does not change the mass of the (quasi-)bound state very much, so we expect that the Λ_1^*N component dominates the Λ^*N quasi-bound state. We thus use the wave function

Table 1. Comparison with other chiral models, [4, 5], [6], [7] and [8, 9].

Refs.	$M_{\Lambda^* N}$(MeV)	$\sqrt{\langle r^2 \rangle}$(fm)	$\Gamma_{F.A}$ (MeV)
[4, 5]	2364	5.1	49
[6]	2364	6.3	49
[7]	2371	5.2	28
[8, 9]	2364	3.2	59

of the bound $\Lambda_1^* N$ system to estimate the property of the quasi-bound state. The mean-distance of the $\Lambda^* N$ system is estimated as, $\sqrt{\langle r^2 \rangle_{\Lambda^* N}} \sim \left[\int d^3 r r^2 |\psi_1|^2\right]^{1/2} \sim 5.1$fm, which indicates the loosely bound of the $\Lambda^* N$ system. The decay width can also be obtained using the wave function, together with some appropriate matrix element for the decay process. The $\Lambda^* N$ quasi-bound state can decay into the three-body $\pi Y N$ states and two-body YN states. Among them, here we estimate the decay width, $\Gamma_{F.A}$, for the fall apart process where Λ^* in the quasi-bound state of the $\Lambda^* N$ system decays to $\pi\Sigma$ with a nucleon being as a spectator, with $\Gamma_{F.A} \sim 49$ MeV.

We construct the $\Lambda^* N$ potential based on the particular chiral unitary model, [4, 5]. To study ambiguities due to the properties of Λ^*, we apply the Λ^* parameters, pole positions of $\Lambda_a^* (a = 1, 2)$ with the Λ_a^* and meson-baryon couplings, taken from different models in Refs. [6], [7] and [8, 9], to our potential model. Solving the Schrödinger equation, we have results which are not so different between the models qualitatively. Namely, the two-body system of Λ^* and a nucleon develops a shallow and loose quasi-bound state, and the decay width is narrow in the fall apart process for its large phase space. The latter property may be related to the small coupling of the Λ_1^* state to the $\pi\Sigma$ channel. Numerical results are listed in Table. 1. Note that the $\Lambda_1^* N$ threshold is different between the chiral unitary models, due to the different Λ_a^* mass. Thus, the same two-body masses with the models [4, 5] and [8, 9] come from the different $\Lambda_1^* N$ binding energy, which is related to each mean-distance. For the same reason, the model [7] gives the smallest decay width for its largest phase space, because of its smallest $\Lambda_1^* \pi\Sigma$ coupling.

Acknowledgments
We thank Drs. W. Weise, E. Hiyama and Y. Ikeda for useful discussion. This work is partially supported by a grant for the Tokyo Institute of Technology Global COE program, "Nanoscience and Quantum Physics", from the Ministry of Education, Culture, Sports, Science and Technology of Japan, and by the Grant-in-Aid for Scientific Research from MEXT and JSPS (No. 21840026).

References
[1] A. Arai, M. Oka and S. Yasui, Prog. Theor. Phys. **119**, 103 (2008).
[2] B. Holzenkamp, K. Holinde and J. Speth, Nucl. Phys. A **500**, 485 (1989).
[3] A. Reuber, K. Holinde and J. Speth, Czech. J. Phys. **42**, 1115 (1992).
[4] T. Hyodo, S. I. Nam, D. Jido and A. Hosaka, Phys. Rev. C **68**, 018201 (2003).
[5] T. Hyodo, S. I. Nam, D. Jido and A. Hosaka, Prog. Theor. Phys. **112**, 73 (2004).
[6] E. Oset, A. Ramos and C. Bennhold, Phys. Lett. B **527**, 99 (2002) [Erratum-ibid. B **530**, 260 (2002)].
[7] J. A. Oller and U. G. Meissner, Phys. Lett. B **500**, 263 (2001).
[8] B. Borasoy, R. Nissler and W. Weise, Phys. Rev. Lett. **94**, 213401 (2005).
[9] B. Borasoy, R. Nissler and W. Weise, Eur. Phys. J. A **25**, 79 (2005).

Cosmic-Ray accelerators in Milky Way studied with the Fermi Gamma-ray Space Telescope

Tuneyoshi Kamae[1]
SLAC and KIPAC, Stanford University,
2575 Sand Hill Road, Menlo Park, CA 94025, USA
E-mail: kamae@slac.stanford.edu

Abstract. High-energy gamma-ray astrophysics is now situated at a confluence of particle physics, plasma physics and traditional astrophysics. Fermi Gamma-ray Space Telescope (FGST) and upgraded Imaging Atmospheric Cherenkov Telescopes (IACTs) have been invigorating this interdisciplinary area of research. Among many new developments, I focus on two types of cosmic accelerators in the Milky-Way galaxy (pulsar, pulsar wind nebula, and supernova remnants) and explain discoveries related to cosmic-ray acceleration.

1. Introduction

Fermi Gamma-ray Space Telescope (*Fermi*) [1,2] and Imaging Atmospheric Cherenkov Telescopes (IACTs) [3] have been invigorating the interdisciplinary research area at the confluence of particle physics, plasma physics and traditional astrophysics. Among them are particle acceleration mechanisms at work in cosmic environment. Most cosmic accelerators have been discovered in gamma-ray observations so far including Supernova remnants (SNR), Pulsars (PSR), and Pulsar wind nebulae (PWN) in the Galactic sky, and, Active Galactic Nuclei (AGN) and Gamma-Ray Bursts (GRB) in the Extra-Galactic Sky [4].

Particles are accelerated in similar mechanisms both in cosmic and man-made accelerators. Gamma-rays are emitted via synchrotron radiation (e^-/e^+ in B-field), inverse Compton scattering (between e^-/e^+ and seed photons), pion production (interaction of p/α with gas), and bremsstrahlung (interaction between e^-/e^+ and gas).

We explain how and where particles (e-/e+ in this case) are accelerated and how and where gamma-

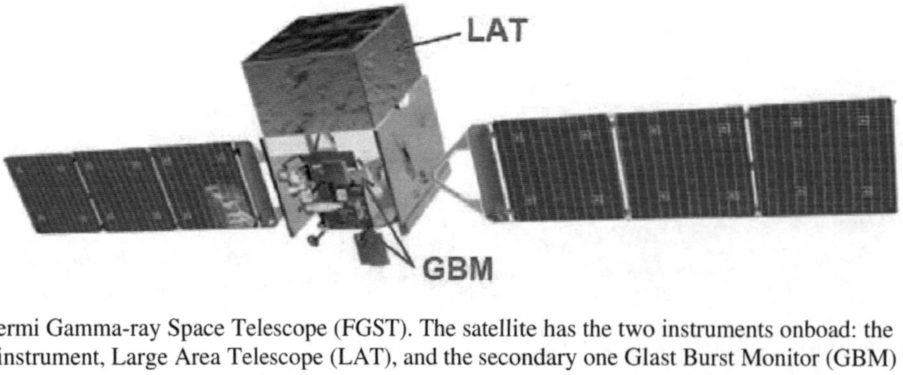

Fig. 1. Fermi Gamma-ray Space Telescope (FGST). The satellite has the two instruments onboad: the primary instrument, Large Area Telescope (LAT), and the secondary one Glast Burst Monitor (GBM) [1].

rays are emitted, taking the pulsars and the pulsar wind nebulae as examples.

Here, the pulsar is a fast-rotating (a few to ~500 times per second), highly-magnetized (10^8-10^{14} Gauss on the surface), neutron star with a radius of ~10km and mass of ~1.5 solar-mass). Particles are accelerated by magnetic induction to a few 10^{12} eV (TeV) in the pulsar. The pulsar wind nebula (PWN) is a plasma cloud filled with energetic e^-/e^+ extending around a pulsar where particles coming out of the pulsar are accelerated further to $>10^{15}$eV (PeV) [4].

2. Fermi Gamma-ray Space Telescope and Imaging Atmospheric Cherenkov Telescopes

In this presentation we use gamma-ray data taken by the Large Area telescope (LAT) onboard *Fermi Gamma-ray Space Telescope* (FGST) for 10^9 eV (GeV) energy range [1,2] and Imaging Atmospheric Cherenkov Telescopes (IACTs) for the TeV energy range [3]. Performance of the former and that of a typical IACT, *High Energy Stereoscopic System* (H.E.S.S.) [5] are given below and in Table 1.

2.1 *Large Area Telescope on Fermi Gamma-ray Space Telescope*

FGST-LAT was built by an international space program funded by USA (NASA, DOE), Japan, Italy (INFN, ASI), France, and Sweden. The collaboration was formed in 1995 and Japan officially joined in 1997. The LAT collaboration's proposal was selected in the NASA Announcement of Opportunity in 2001. The instrument was assembled and tested in 2006 at SLAC, integrated with the satellite in 2007, and launched to a Low-Earth Orbit on June 11, 2008 (Fig.1). Science operation started on August 4, 2008 after a month verification period. A catalog of ~1400 point sources detected by the instrument was published in March 2010 [6]. The instrument has been surveying the entire sky since the August 4, 2008 making 144k orbits and recording 474M gamma-rays by Jan 20, 2011. The key performance parameters are given in Table 1.

Table 1. Performance of FGST-Large Area Telescope[a] and HESS[b]

Parameters	Fermi-LAT	H.E.S.S.
Energy range	100MeV-300GeV	100GeV-3TeV
Effective area	~7000cm^2 (E>1GeV)	~10^5m^2 per mirror
Field of view	~2.4 sr	~3 deg in radius
Angular resolution	$0.8(E[GeV])^{-0.8}$deg	1-5 arcmin
Sensitivity (E^2dN/dE)	~10^{-12} erg/cm^2/s (2yrs)	~3x10^{-13} erg/cm^2/s
Sky coverage	~all sky (every ~3hrs)	selected pointing

[a] http://www-glast.slac.stanford.edu/software/IS/glast_lat_performance.htm and [1,2]
[b] [3,5]

2.2 High Energy Stereoscopic System (HESS)

HESS has been a most productive Imaging Atmospheric Cherenkov Telescope. The current version (HESS-I) has been completed in 2003 in Namibia and has been operational ever since [3,5]. It consists of 4 nearly identical telescopes and has been observing astronomical objects accessible from the southern hemisphere. The key performance parameters are summarized in Table 1.

3. Acceleration Mechanisms: Man-made accelerators vs Cosmic Accelerators

There are 4 acceleration mechanisms known to operate in cosmos: they have a counter part in the man-made accelerator or in the plasma experiment as shown in Figs.2a-2d.

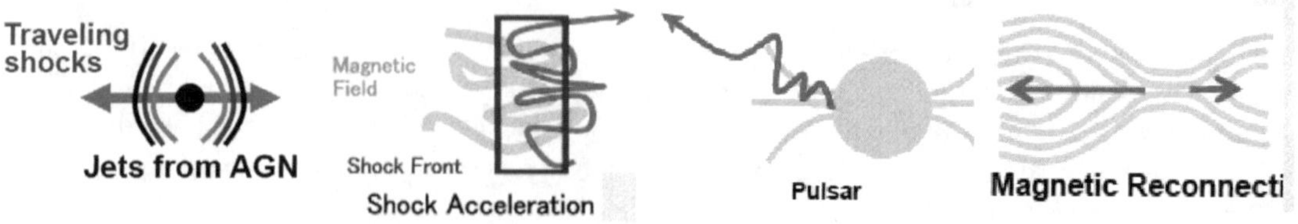

$$\varepsilon_s = \sqrt{\frac{3}{2}\frac{m^2}{q^{3/2}}}B^{-1/2} \qquad \varepsilon_d \simeq \frac{3}{2}\frac{m^4}{q^4}B^{-2}R^{-1} \qquad \varepsilon_c = \frac{3^{1/4}}{2}\frac{m}{q^{1/4}}B^{1/4}R^{1/2}$$

Fig.2a. Shock fronts formed within Lorentz-boosted plasma in jets of AGNs and GRBs.

Fig.2b. Particles recycle through the shock front made of turbulent B-field and reach to high energy.

Fig.2c. Electric field induced by the rapidly rotating dipole B-field of a neutron star.

Fig. 2d. Energy stored in B-field is released as kinetic energy of particles.

These acceleration mechanisms are believed to be accelerating particles to TeV to PeV energy range. As particle energies increase synchrotron radiation loss becomes non-negligible [4,7]. Even if the accelerator has capability to reach beyond PeV energy range, B-field inherent to the accelerator site limit the maximum energy as listed in the second row of Fig.2. Here m=mass, q=charge, B=mag field.

Fig.3a. A cosmic one-shot linear accelerator in a near-by AGN, Cen A. Faint linear structure is extending to up-left and down-right.

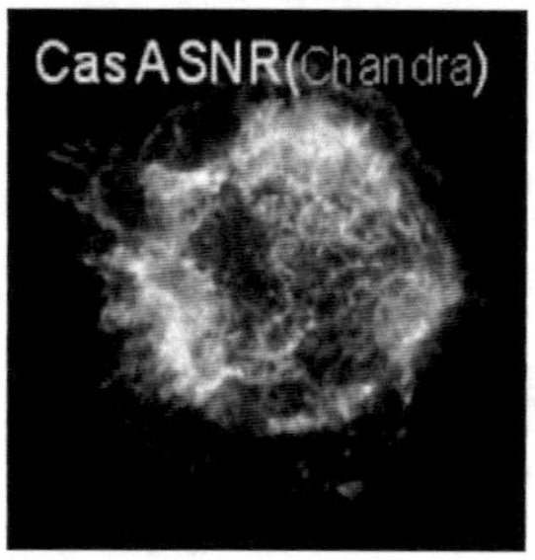

Fig.3b. A cosmic recycling accelerator in an SNR, Cas A. The shock accelerator is at the thin blue broken ring in the outer periphery.

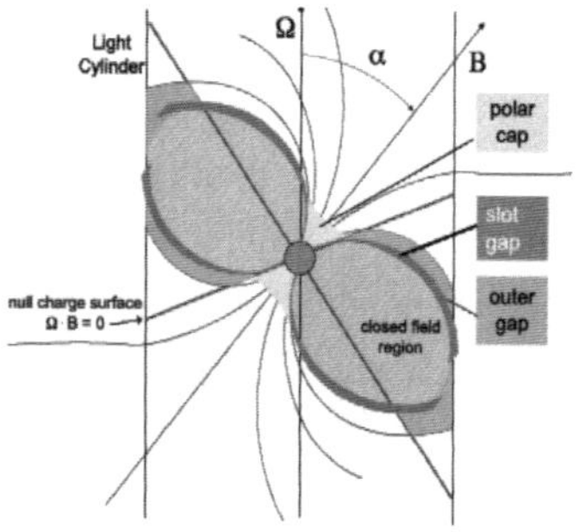

Fig. 3c. A cartoon of Crab Pulsar. The neutron star has a magnetic dipole pointing at B and rotates about an axis Ω at ~30Hz. Particles are accelerated in the blue stripes.

Shown in Fig.3 are 3 types of cosmic accelerators: (a) galaxies hosting a super-massive black-hole, Active Galactic Nuclei (AGN), where energetic particles stream out along jets; (b) supernova remnants (SNR) with a ring-like shock front expands into interstellar space; (c) a fast-rotating magnetized neutron star where induction generate high potential drop near the boundary of its magnetosphere [4].

4. Gamma-ray Emission from the Pulsar and the Pulsar Wind Nebula

We present new discoveries by Fermi about particle acceleration in the pulsar and the PWN taking the Vela pulsar [8] and the Crab nebula [9] as examples, respectively.

4.1 Fermi-LAT narrows down the γ-ray emitting region in the pulsar

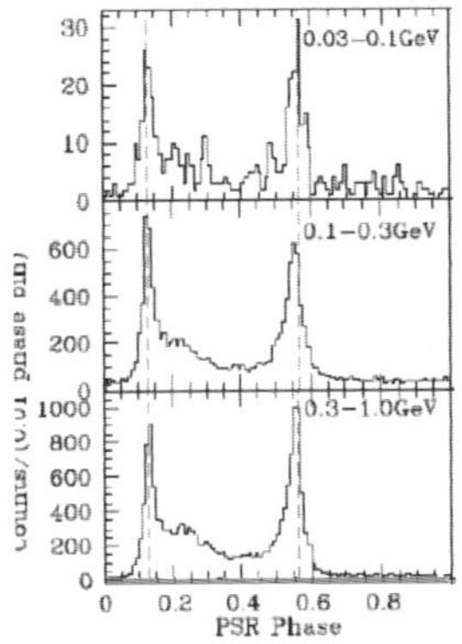

Fig.4a. The Vela pulsar pulse profile in 3 energy bands by *Fermi* [8].

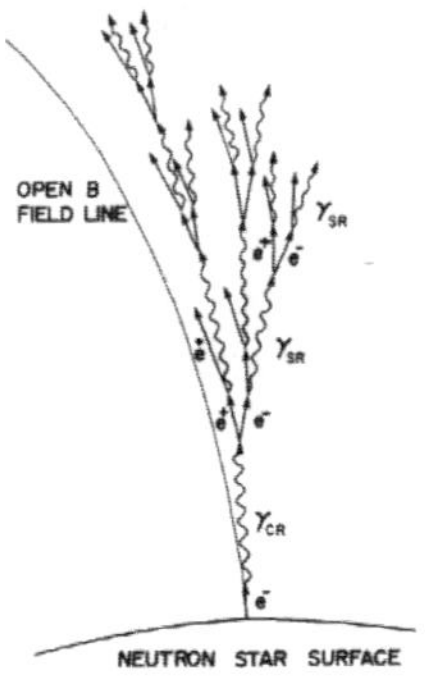

Fig.4b. γ-rays develop into a shower in strong B-field [10].

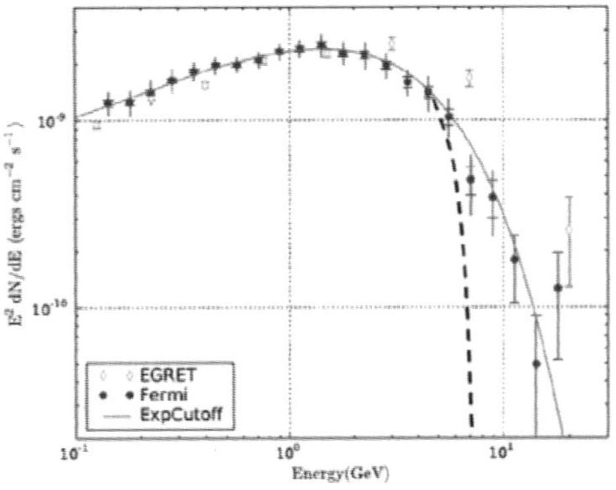

Fig.4c. The γ-ray spectrum observed by *Fermi* (data) and predicted for strong B-filed (dashed line) [8].

Spectrum of pulsed gamma-ray emission has been measured for several pulsars including that from the Vela pulsar and the Crab pulsar. We show the gamma-ray pulse profile of the Vela pulsar in 3 energy bands in Fig.4a. If the pulsed gamma-rays are emitted near the neutron star surface (the polar cap model) where B-field is extremely strong, they will be converted to an e^-/e^+ shower and will not reach to us [10]. The predicted spectrum is shown by the dashed line in Fig.4c contradicting the *Fermi* observation. We conclude that the emission region is away from the neutron star by least 3 times its radius as has been predicted by the outer gap model [11].

4.1 Discovery of giant flares in the Crab nebula by *Fermi*

Crab nebula (see Figs.5a and b) has been the best studied astronomical object and considered to be a stable standard candle on which almost all X/γ-ray telescopes have been calibrated. Two unexpected surprises came in 2009 and 2010: it is neither stable in the X-ray band [12] nor in the γ-ray band [13].

The nebula consists of the Crab Pulsar at the center, a ring-like termination shock where the wind of high energy particles stalls outside of the pulsar light cylinder, the cocoon-like structure housing the particles diffusing out of the pulsar and termination shock (Fig.5b). The finger-like structure in Fig.5a is believed to be created by the Rayleigh-Taylor instability as shown in the cartoon (Fig.5b).

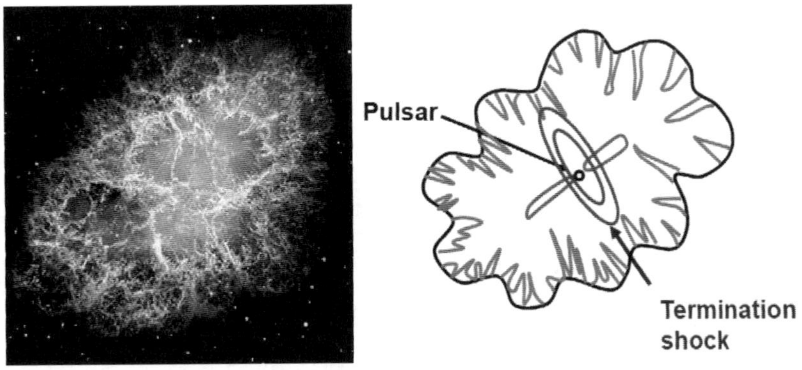

Fig.5. The Crab nebula seen by Hubble Telescope (left) and a cartoon showing major components (right).

Crab Nebula has been found, in the X-ray band, to be "flickering" at ~0.5-1 year time-scale at ~5% flux-level in the past ~10 years [12]. Even more surprising is two big flares (Fig.6a) *Fermi* found in 100MeV-GeV gamma-rays [13]. The γ-ray spectrum during the two flares is consistent with the synchrotron emission (Fig.6b). The flare #2 is particularly interesting because the maximum electron

Journal of Physics: Conference Series **302** (2011) 012059 doi:10.1088/1742-6596/302/1/012059

energy inferred from the spectrum is ~PeV. The rise time of flare #2 is less than 1 day or less than the time the accelerated particle making one circle: the diffusive shock acceleration would not work and non-standard mechanism has to be invoked [13].

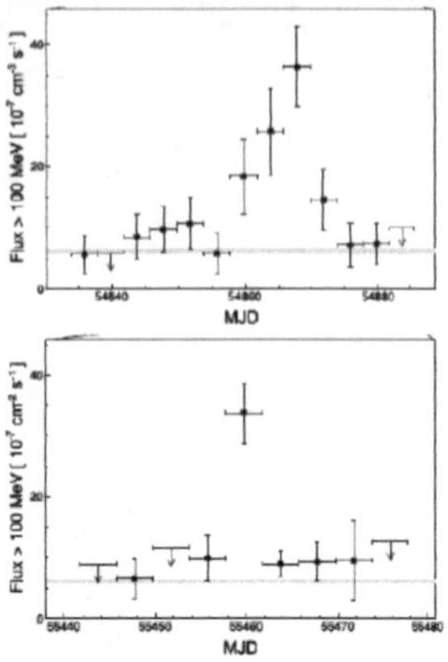

Fig.6a. Time profiles of flares #1 (top) and #2 (bottom) seen by Fermi [13]

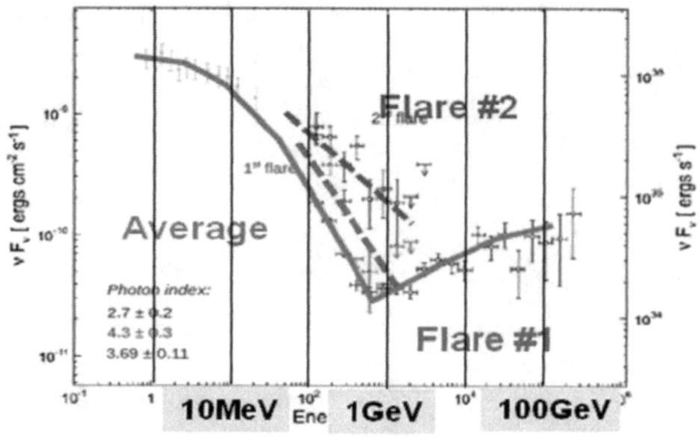

Fig.6b. The γ-ray spectra for flares #1 (lower dashed) and #2 (upper dashed) compared with that during the non-flaring period (solid line) [13].

5. Gamma-ray Emission from Supernova Remnants

1.1. Fermi finds middle-aged SNRs interacting with dense interstellar clouds

Fermi LAT observed nearly 10 SNRs of which about a half are found to be very bright in the GeV γ-ray band. They are found to be middle-aged ones where the shock front is impinging into dense clouds [14,15]. The standard diffusive shock acceleration theory does not apply to such environment and some new development is called for.

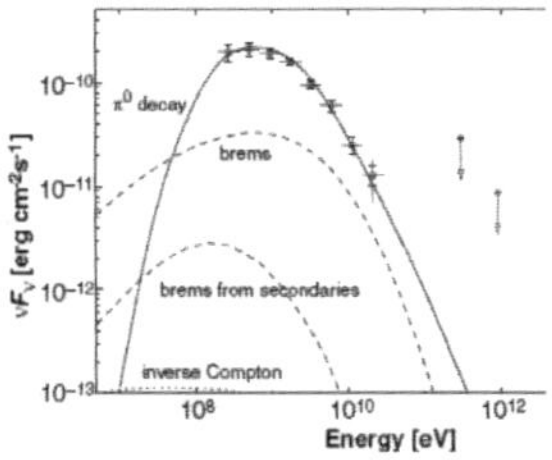

Fig.7a γ-ray spectrum from SNR W44 by Fermi and IACT [14]

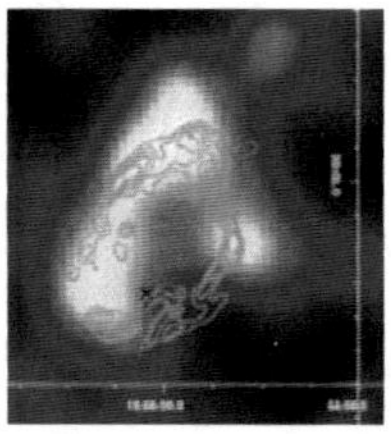

Fig.7b.Distr of γ-rays (color) and gas (green) in W44 [14]

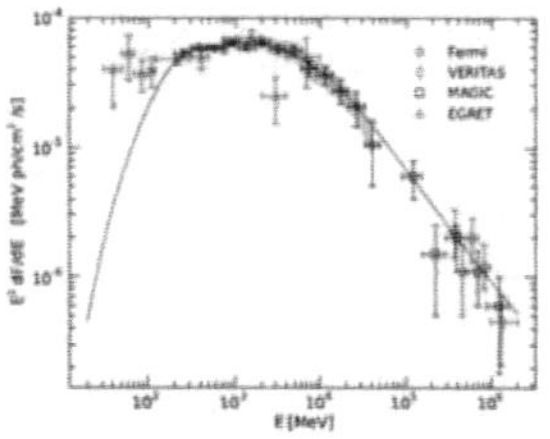

Fig.7c γ-ray spectrum from SNR IC443 by Fermi and IACTs [15]

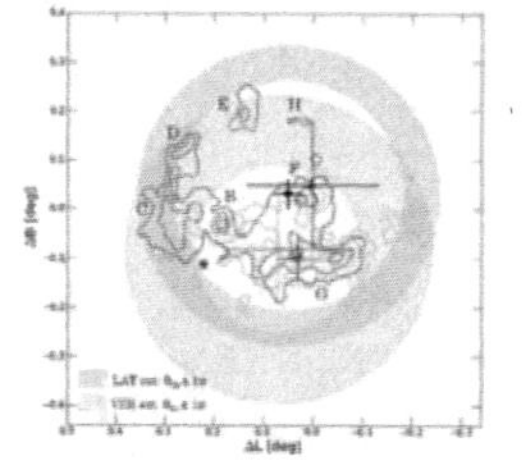

Fig.7d. 68% containment circles (circles) and centers (crosses) of γ-ray distr by Fermi and IACTs and gas (contours) for IC443 [15]

The broad-band γ-ray spectra obtained by combining *Fermi* and IACTs observations are compatible that the γ-rays are decay products of π^0 produced in nuclear interactions of p/α accelerated in the SNRs, presumably at the shock front [14,15]. The spectral shapes of p/α deduced from γ-rays spectra are power-law with a break ranging from ~1GeV to 70GeV as tabulated in Table 2.

Table 2

SNR	Age	Distance	Gas dens × CR [a]	PL or Broken-PL [b]		
Name	kyr	kpc	$n \times W_p$ [erg cm^{-3}]	E_{br} [GeV]	α_{LE}	α_{HE}
W49B	4.0	7.5	1.1×10^{52}	4.0	2.0	2.7
W44	20	3.0	6×10^{51}	8.0	1.74	3.7
W51C	30	6.0	5.2×10^{51}	15	1.5	2.9
IC 443	30	1.5	6.7×10^{50}	69	2.09	2.87
W28 (North)	40	2.0	1.3×10^{51}	1.0	2.09	2.74

6. Conclusions

The new generation of astronomical γ-ray experiments are discovering many cosmic accelerators with their much-improved sensitivity and sky-coverage. Some of them seems to behave as expected but some show unexpected behaviour. These new behaviours call for physics processes which are new to astrophysics but have been well studied in other disciplines of physics. High energy γ-ray astrophysics is located at a confluence of disciplines and congruous with the spirit of this symposium. To be specific, collaboration with plasma physicists simulating the particle-turbulence interaction will be most appreciated.

The γ-ray data obtained by Fermi are publically available at Fermi Science Support Centre with science analysis tools [16]. The support centre offers assistance to users and posts many FAQs.

Acknowledgements

I would like to thank Prof. N. Kawai and the organizing committee for inviting me to this symposium. The results presented here come from the *Fermi* collaboration and the IACT collaborations, HESS, MAGIC and VERITAS.

References

[1] http://fermi.gsfc.nasa.gov/ and http://www-glast.stanford.edu/
[2] Atwood, W. B. et al. 2009 The Large Area Telescope on the Fermi Gamma-ray Space Telescope Mission *ApJ, 697, 1071*
[3] Aharonian, F.; Buckley, J.; Kifune, T.; Sinnis, G. 2008 High energy astrophysics with ground-based gamma ray detectors *Reports on Progress in Physics, 71, 096901*
[4] Textbooks such as, Longair, M.S. 1994 High Energy Astrophysics I and II (2nd edition) *Cambridge Univ. Press*; Aharonian, F. 2004 Very High Energy Cosmic Gamma Radiation *World Scientific*
[5] Hinton, J.A. 2004 *New Astronomy Reviews 48, 331*
[6] Abdo, A. A. et al. 2010 Fermi Large Area Telescope First Source Catalog *ApJS, 188, 405*
[7] Maximum energy formulae taken from http://cdsweb.cern.ch/record/1249755/files/p533.pdf
[8] Abdo, A. A. et al. 2009 Fermi Large Area Telescope Observations of the Vela Pulsar *ApJ, 696, 1084*
[9] Abdo, A. A. et al. 2010 Fermi Large Area Telescope Observations of the Crab Pulsar and Nebula *ApJ, 708, 1254*
[10] Daugherty,J.K., Harding, A.K. 1996 Electromagnetic Cascade in Pulsars *ApJ, 252, 337*
[11] Cheng, K. S., Ho, C., Ruderman, M. 1986 Energetic radiation from rapidly spinning pulsars. I - Outer magnetosphere gaps *ApJ, 300, 500* and II. VELA and Crab ApJ, 300, 522; Romani, R. 1996 Gamma-Ray Pulsars: Radiation Processes in the Outer Magnetosphere *ApJ, 470, 469*;
[12] Wilson-Hodge, C.A., et al. 2011 When a Standard Candle Flickers *ApJ, 727, L40*
[13] Tavani, M. et al. 2011 Discovery of Powerful Gamma-Ray Flares from the Crab Nebula *Science, 331, 736*; Abdo, A. A. et al. 2011 Detection of gamma-ray flares from the Crab Nebula *Science, 331, 739*
[14] Abdo, A. A. et al. 2010 Gamma-ray Emission from the Shell of Supernova Remnant W44 Revealed by the Fermi LAT *Science, 327, 1103*
[15] Abdo, A. A. et al. 2010 Observation of Supernova Remnant IC 443 with Fermi-Large Area Telescope *ApJ, 712, 459*
[16] http://fermi.gsfc.nasa.gov/ssc/

A numerical challenge on the core-collapse supernovae: physics of neutrino and matter at extreme conditions

Kohsuke Sumiyoshi
Numazu College of Technology, Ooka 3600, Numazu, Shizuoka 411-0943, Japan
E-mail: sumi@numazu-ct.ac.jp

Abstract. Core-collapse supernovae are the explosive phenomena, which occur at the end of the life of massive stars. Despite the importance of these astrophysical events, the mechanism of supernova explosion has been a mystery even after the extensive studies for decades. The unsolved problem involves nuclear and neutrino physics at extreme conditions as well as the hydrodynamical aspects in astrophysics. The physics in femto-meter scale may change drastically the gigantic outcome of the core bounce after the gravitational collapse of the massive stars. In this contribution, I overview recent topics on the numerical challenges to clarify the supernova phenomena.

Recent advance of nuclear physics for unstable nuclei and exotic hadrons helps us to provide the nuclear data inside the supernova core. The influence of the dense matter on the explosion and supernova neutrinos has been clarified using the newly constructed data tables based on the theoretical and experimental developments. Recent studies demonstrated that the short neutrino bursts from the black hole formation in more massive stars can be a probe of exotic matter including hyperons and quarks by observations at the neutrino detector facilities. Despite the progress in nuclear and neutrino physics, no explosion is found in the numerical simulations under the spherical symmetry and hence novel effects in multi-dimensions are argued to be essential for the explosion mechanism. Toward the final goal to clarify the supernova mechanism, it is necessary to combine the best knowledge of nuclear, particle physics and astrophysics with computing science on supercomputing facilities.

1. Physics of core-collapse supernovae

Core-collapse supernova is a bright display frequently observed among astronomical events. A star becomes suddenly very bright, shines for months and fades away afterwards as in the case of a famous supernova observed in 1987 (SN1987A). These phenomena are actually the stellar explosions at the end of life of massive stars of $\sim 20 M_{\odot}$ [1, 2]. They are playing important roles in the Universe as the birth place of compact objects (neutron stars and black holes), the source of cosmic rays (including gravitational waves and neutrino bursts) and the driving force of the evolution of galaxies. Pulsars, which are rapidly rotating neutron stars, are found in the remnant of supernova explosion as in the case of the Crab nebula. Neutron stars have the typical mass of $\sim 1.4 M_{\odot}$ in the radius of ~ 10 km, being the laboratory of extremely dense conditions. The burst of neutrinos on the occasion of SN1987A was observed at the neutrino detector facilities (ex. Kamiokande in Japan) and has led to the Nobel prize in Physics in 2002. Supernovae are also the production sites of heavy elements (for example, precious metals such as gold, platinum and uranium) through the explosive nucleosynthesis.

The explosion mechanism starts from the gravitational collapse of the Fe core of the massive star. Electron capture processes in the central core reduce the pressure support and trigger the collapse. Neutrinos are created and trapped inside the core due to high densities during the collapse. Further collapse leads to the core bounce due to the hard repulsive core of the nuclear force above the nuclear matter density, 3×10^{14} g/cm^3. The shockwave is launched and its successful propagation leads to the supernova explosion, leaving a neutron star, heavy elements and supernova neutrinos. During this episode, the gravitational energy of $\sim10^{53}$ erg is released by the contraction of the core radius down to $\sim$10 km. Most of this energy is carried away by neutrinos and accords with the total energy of the neutrino burst observed at the terrestrial detector in 1987 [2]. Note that the explosion energy of ejected material amounts to $\sim10^{51}$ erg, which is a tiny fraction ($\sim10^{-2}$) of the gravitational energy. Therefore, the energy transfer via neutrino-matter interactions is essential for the clarification of the explosion mechanism.

To attack this problem, one needs quantitative and careful studies by covering ingredients from nuclear physics in $\sim$fm scale to astrophysics in $\sim$1000 km scale. It is necessary to provide the equation of state, neutrino reactions and nuclear data from nuclear physics in order to solve hydrodynamics and neutrino transfer with stellar modeling of massive stars. One has to put all these ingredients to the numerical simulations of core-collapse supernovae on supercomputers. Therefore, this is a challenging problem of nuclear physics, particle physics, astrophysics and computational science. I focus here on the challenges on the properties of dense matter at high density and temperature. I explain that the role of neutrinos in the supernova core is important and the observation of neutrino bursts is helpful to get the information of dense core.

2. Nuclear physics at extreme conditions

The properties of dense matter at extreme conditions are necessary inputs to understand quantitatively the supernova mechanism. For example, pressure-density relation is essential to determine stellar structure and hydrodynamics. It crucially determines the maximum mass of neutron stars, which is the border to the black hole. Thermodynamical quantities such as entropy and composition (neutron, proton and nuclei) are important to determine neutrino reactions and neutrino distributions. Hence, one has to provide the data set of the equation of state (EOS) for numerical simulations of supernovae. The challenge here is to cover the wide range of conditions. Densities in the central core become higher than that inside nuclei, which is the nuclear matter density, 3×10^{14} g/cm^3. Dense matter becomes eventually very neutron rich as the neutron star contains a small fraction ($\sim$0.1) of free protons in contrast to the proton fraction of $0.46=\frac{26}{56}$ for ^{56}Fe. Temperature becomes high above 10 MeV ($\sim10^{11}$ K) during the evolution.

In order to cover consistently the wide range of environment, one has to work with the unified framework, which is checked by experimental data as much as possible. Such efforts have been made recently by the progress of nuclear physics. Together with the advance of relativistic nuclear many body theories [3] in 1990s, the experimental data of unstable nuclei provided by the radioactive nuclear beam facilities such as RIKEN are helpful for us. It is important to have new data on neutron rich nuclei in the nuclear chart in order to constrain the neutron-rich matter of neutron stars and supernovae. For example, the radii of Na isotopes have been measured systematically [4]. The radii of neutron distribution in very neutron-rich Na nuclei are found systematically larger than those of proton distribution (neutron skin). Such systematic studies are important to constrain the equation of state in astrophysics.

Based on the relativistic Brückner Hartree-Fock theory [5], the relativistic equation of state for supernovae has been obtained [6] by utilizing the experimental data of unstable nuclei to constrain the nuclear interaction [7]. The framework is based on the relativistic mean field theory with the local density approximation, which is necessary to describe the non-uniform matter distribution. The data table of the supernova EOS is constructed to cover the wide range of density, proton fraction and temperature [8]. This set of supernova EOS (Shen EOS) is available on the web and becomes very popular in the wide area of astrophysics. The Shen EOS table has been extended to the hyperon EOS with strange baryons [9] and to the quark EOS with the MIT bag model [10]. It is vital to have the experimental data of hyper-nuclei for this. Our reference is the conventional EOS by Lattimer and Swesty (LS EOS) based on the extension of the compressible liquid drop model [11]. I make comparisons by using the three EOS sets (Shen, Hyperon and LS) in the following discussion. They have different properties of dense matter such as stiffness and symmetry energy while they fulfill the basic requirement at the saturation point of nuclear matter. They give different values for the maximum mass of neutron stars due to different stiffness at high densities, for example. Having the sets of EOS newly developed, it is interesting to see the influence on the phenomena of core-collapse supernovae.

In addition to the properties of dense matter, it is mandatory to provide the rates of neutrino reactions with the dense matter in the supernova core [12]. The neutrino reactions change the number and energy of neutrinos, therefore, they contribute to the heating and cooling of matter as well as the change of composition. Numerical simulations need the implementation of neutrino reactions through the emission and absorption, the scattering with nucleons and nuclei as well as the thermal pair processes for three neutrino flavors [13]. Since the neutrino reactions proceed via the weak interaction, experiments to measure the reaction rates are difficult. In addition, the reaction rates depend on the energy and the structure of nuclei. Therefore, implementing the neutrino reaction rates is also a challenging task for the numerical simulation of supernovae.

3. Influence of equation of state of dense matter

Supernova neutrinos carry the information of dense matter in the supernova core [2]. Starting from the Fe core of a massive star of $\sim 20 M_\odot$, a proto-neutron star is born after the core bounce and the following explosion of outer layer. Since this hot object contains plenty of neutrinos, it cools down by emitting neutrinos. Due to the neutrino emission through the electron capture on protons, the object becomes neutron-rich, leading to a cold neutron star. During this cooling, a bunch of neutrinos ($\sim 10^{58}$) is emitted. They are the supernova neutrinos and were detected at the Kamiokande neutrino detector facility. In a case of the next galactic supernova, neutrino events of $\sim 10^4$ will be detected. By examining these neutrinos, one can extract the information of dense core since neutrinos are emitted from the hot and dense matter.

Numerical studies of the cooling of proto-neutron stars have been done extensively [2, 14, 15]. The neutrino burst lasts for the duration of ~ 20 s, which is determined by the diffusion time scale of neutrinos inside the central core. The average energy and luminosity decrease gradually, reflecting the decrease of temperature of the proto-neutron star. These behaviors depend on the equation of state to determine the profile of density and temperature [16]. The time evolution of average energy is different in the comparison adopting two EOSs (Shen EOS and LS EOS, for example). Therefore, it should be possible to probe the EOS by examining the properties of the supernova neutrinos in detail (See also [15, 17] regarding exotic matter).

For more massive stars, the character of neutrino signal is different since they lead to the black hole formation promptly [18]. Starting from the Fe core of a $\sim 40 M_\odot$ star, there is no explosion intrinsically because of a large mass of Fe core [19, 20]. However, the proto-neutron star is born at center even in this case. Since the material from the outer layer accretes onto the proto-neutron star, its mass increases quite rapidly. When the mass reaches the critical mass, which is determined by the EOS, the central object dynamically collapses to the black hole. During this evolution in ~ 1 s, neutrinos are emitted from the proto-neutron star until the black hole formation [21]. In this event, there is no optical display but the burst of neutrinos.

There will be 10^4 events of neutrinos, which are comparable to the case of ordinary supernova neutrinos, expected at the Super-Kamiokande for this kind of failed supernovae [22]. Hence, this is also a chance to probe the EOS at high density and temperature. Moreover, this is a new way to find the black hole formation. A black hole candidate corresponding to this scenario was discovered [23] and a systematic survey of failed supernovae is planned [24].

The properties of neutrino burst until the black hole formation are different from those of the ordinary supernovae. The duration of the burst is short within 1 s in contrast to 20 s for ordinary supernovae. The energy and luminosity increase rapidly due to fast increase of temperature of the intensively accreting proto-neutron star. For example, Figure 1 shows the time evolution of the average energies of neutrinos obtained from the numerical simulation of neutrino-radiation hydrodynamics [21, 25]. The average energy becomes very high after the core bounce for all cases. The neutrino bursts stop at 1.3 and 0.6 s for Shen EOS and LS EOS, respectively. The softer EOS leads to the smaller critical mass for the collapse, therefore, it leads to the earlier collapse and the shorter burst [18, 21]. Therefore, the neutrino burst from the failed supernovae can be used to probe the softness of EOS in a clear manner.

Since the proto-neutron star becomes massive, the temperature and density become high quickly. This leads to the appearance of exotic particles of hadrons and quarks. By performing the numerical simulation with the hyperon EOS, it is found that the hyperons appear abundantly at ~0.5 s after the bounce [25]. The appearance of hyperons triggers the dynamical collapse to the black hole from the proto-neutron star because of additional softening of the EOS. The duration of the neutrino burst from the failed supernovae becomes shorter with the hyperon EOS (0.7 s) as compared to the case with the nucleonic Shen EOS (1.3 s) as in Fig. 1. Therefore, it should be possible to probe the emergence of hyperons by examining the duration of the neutrino burst and the rise of energies for different flavors by the detection of neutrinos at the Super-Kamiokande [26].

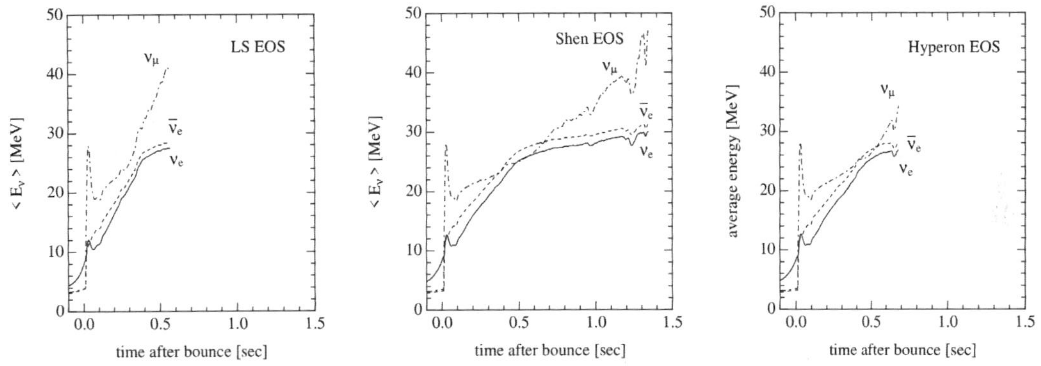

Figure 1. Time evolution of average energies of neutrinos (three species) emitted from the collapse of a $40M_\odot$ star with LS EOS (left), Shen EOS (middle) and Hyperon EOS (right).

4. Role of neutrinos in the explosion mechanism

Despite the extensive studies of the core-collapse supernovae, the explosion mechanism is still elusive. One has to examine carefully a delicate balance of counter-effects in a quantitative manner. As long as the numerical simulation of supernovae is performed in spherical (1D) geometry, no explosion is obtained in the neutrino-radiation hydrodynamics (for example, [27]). Note that the numerical simulations in 1D are the first principle calculations with the up-to-date EOS and neutrino physics. The shock wave is launched by the core bounce, however, it stalls on the way above 100 km for both LS EOS and Shen EOS [27]. This is mainly because the initial shock energy is used up by the dissociation of Fe during the propagation of the shock wave through the outer layer of Fe core. The initial shock energy by the release of gravitational energy amounts to several 10^{51} erg. However, the energy loss due to the Fe dissociation amounts to 1.6×10^{51} erg per $0.1M_{\odot}$, using up the initial energy.

Since the 1D calculations do not bring the explosion, one may need the multi-dimensional effects such as convection and rotation [28, 29]. The asymmetry away from the spherical geometry is also supported by the supernova observation and the analysis of spectra. The convection contributes to the effective heating of material and may help the revival of shock wave. It has been shown recently that the hydrodynamical instability of the standing accretion shock wave (SASI) may help further the launch of the shock wave to a larger radius and longer time scale for the neutrino heating. There are a handful cases of successful explosion in the state-of-the-art 2D simulations. However, there are several scenarios competing each other and the problem is not settled yet. This is partly because the numerical simulations by different groups adopt different models and methods with approximate treatment of neutrino transfer.

One of the important factors is the neutrino heating mechanism. When the shock wave stalls on the way above 100 km, the material is heated up by the absorption process of the neutrinos, which are emitted from the central core (proto-neutron star). This is the famous mechanism to have the delayed explosion [1]. Through the transfer of energy from neutrinos to matter, the size of heating amounts to typically $\sim2.2\times10^{51}$ erg, which is again comparable to the other effects. Apparently, the neutrino heating depends on the properties of supernova neutrinos (flux and energy), the available amount of targets and duration. Therefore, one has to evaluate this effect very carefully.

To examine this effect in a precise manner, one has to solve the neutrino transfer. Neutrinos diffuse outward in the central part and stream freely with few reactions through the outer layer. The neutrino heating takes place in the intermediate regime between them, therefore, one has to solve the equation of neutrino transfer together with all of neutrino reactions with the EOS data. The Boltzmann equation in 3 dimensional space and a phase space of 3D neutrino momentum is actually a 6 dimension problem, which is a challenging subject in astronomy, astrophysics and engineering. As for the supernova problem in 1D, the first principle calculations have been done to examine the microphysics and the systematics. In 2D, the approximate treatment has been adopted for the state-of-the-art simulations to explore the explosion mechanism. In 3D, simple prescriptions of neutrinos are adopted to explore the hydrodynamical instabilities. In order to establish the supernova mechanism, one has to perform the full 3D simulation of hydrodynamics and neutrino transfer with the EOS data and the neutrino reactions. Since the hydrodynamical instabilities can launch the position of shock wave to larger radii than that in the spherical calculations, the material is hovering around the heating region for longer time during the accretion, therefore, there would be more time for the neutrino heating.

In order to attack this problem, I am working on the project of the 3D neutrino-radiation hydrodynamics as a collaboration. I have developed a new numerical code of the Boltzmann equation in 3D to follow the time evolution of the neutrino distribution in 6D (3 space, 2 angles and 1 energy of neutrinos). The numerical code includes the progress of the physics of dense matter and neutrino reactions. This project is a computational challenge, which requires a large memory to store the neutrino distribution and a computational power to solve the large matrix for the implicit method for time steps. I have recently made the validation of the numerical code and started applications to the realistic supernova cores. Understanding the mechanism of supernova explosion is one of targets in the selected scientific areas for the K-computer, which is a 10 Pflops supercomputer being built at Kobe in Japan.

5. Summary

In summary, the physics of core-collapse supernovae is fascinating and puzzling. The supernovae from the gravitational collapse of massive stars lead to the explosion or collapse; the birth of neutron stars and black holes. The burst of neutrinos can be used to probe the state of matter deep inside the supernova core. The clarification of the supernova mechanism requires the wide knowledge of nuclear physics, particle physics and astrophysics. One needs the reliable information of matter and neutrino interaction at extreme conditions. The development of the EOS tables has made us possible to clarify its effect on the supernova neutrinos and the explosion mechanism. However, the full understanding of the explosion mechanism requires the large-scale numerical simulations in 3D on the next generation supercomputers with the collaboration of computational scientists.

Acknowledgements

This contribution is based on the collaboration with S. Yamada, H. Suzuki, K. Nakazato. The author is grateful for the development of the EOS tables with C. Ishizuka, A. Ohnishi, K. Oyamatsu, H. Shen and H. Toki. The author would like to express thanks for the discussion on supercomputing with H. Matsufuru. Numerical computations have been performed on the supercomputers at SX9/RCNP/Osaka, KEK, SR/UT, YITP, JAEA and NAOJ. This work is supported by the Grant-in-Aid for Scientific Research on Innovative Areas (20105004) and the Grant-in-Aid for Scientific Research (22540296).

References

[1] Bethe H A 1990 *Rev. Mod. Phys.* **62** 801
[2] Suzuki H 1994 *Physics and Astrophysics of Neutrinos* ed Fukugita M and Suzuki A (Tokyo: Springer-Verlag) p 763
[3] Serot B D and Walecka J D 1986 *Advances in Nuclear Physics* vol 16 ed Negele J W and Vogt E (New York: Plenum Press) p 1
[4] Suzuki T *et al.* 1995 *Phys. Rev. Lett.* **75** 3241
[5] Brockmann R and Machleidt R 1990 *Phys. Rev.* **C42** 1965
[6] Shen H, Toki H, Oyamatsu K and Sumiyoshi K 1998 *Nucl. Phys.* **A637** 435
[7] Sugahara Y and Toki H 1994 *Nucl. Phys.* **A579** 557
[8] Shen H, Toki H, Oyamatsu K and Sumiyoshi K 1998 *Prog. Thoer. Phys.* **100** 1013
[9] Ishizuka C, Ohnishi A, Tsubakihara K, Sumiyoshi K and Yamada S 2008 *J. Phys. G* **35** 085201
[10] Nakazato K, Sumiyoshi K and Yamada S 2008 *Phys. Rev.* **D77** 103006
[11] Lattimer J M and Swesty F D 1991 *Nucl. Phys.* **A535** 331
[12] Burrows A, Reddy S and Thompson T A 2006 *Nucl. Phys.* **A777** 356
[13] Bruenn S W 1985 *Astrophys. J. Suppl.* **58** 771
[14] Burrows A 1988 *Astrophys. J.* **334** 891
[15] Pons J A, Reddy S, Prakash M, Lattimer J M and Miralles J A 1999 *Astrophys. J.* **513** 780
[16] Suzuki H 2005 *Proceedings of the 5th International Workshop on Neutrino Oscillations and their Origin* ed Suzuki Y, Nakahata M, Moriyama S and Koshio Y (Singapore: World Scientific) p 332
[17] Pons J A, Miralles J A, Prakash M and Lattimer J M 2001 *Astrophys. J.* **553** 382
[18] Sumiyoshi K, Yamada S, Suzuki H and Chiba S 2006 *Phys. Rev. Lett.* **97** 091101
[19] Heger A, Fryer C L, Woosley S E, Langer N and Hartmann D H 2003 *Astrophys. J.* **591** 288
[20] Maeda K and Nomoto K 2003 *Astrophys. J.* **598** 1163
[21] Sumiyoshi K, Yamada S and Suzuki H 2007 *Astrophys. J.* **667** 382
[22] Nakazato K, Sumiyoshi K, Suzuki H and Yamada S 2008 *Phys. Rev.* **D78** 083014
[23] Prestwich A H *et al.* 2007 *Astrophys. J.* **669** L21
[24] Kochanek C S *et al.* 2008 *Astrophys. J.* **684** 1336
[25] Sumiyoshi K, Ishizuka C, Ohnishi A, Yamada S and Suzuki H 2009 *Astrophys. J.* **690** L43
[26] Nakazato K, Sumiyoshi K, Suzuki H and Yamada S 2010 *Phys. Rev.* **D81** 083009
[27] Sumiyoshi K, Yamada S, Suzuki H, Shen H, Chiba S and Toki H 2005 *Astrophys. J.* **629** 922
[28] Kotake K, Sato K and Takahashi K 2006 *Rep. Prog. Phys.* **69** 971
[29] Janka H T, Langanke K, Marek A, Martínez-Pinedo G and Muller B 2007 *Phys. Rep.* **442** 38

Journal of Physics: Conference Series **302** (2011) 012061 doi:10.1088/1742-6596/302/1/012061

Outburst of LS V+44 17 observed by MAXI and RXTE discovery of dip structure in pulse profile

R Usui[1], N Kawai[1], M Morii[1], K Sugimori[1], T Mihara[2], T Yamamoto[2,3] and M Matsuoka[2]

[1] Department of Physics, Tokyo Institute of Technology, 2-12-1 Ookayama, Meguro-ku, Tokyo, Japan

[2] MAXI team, RIKEN, 2-1 Hirosawa, Wako, Saitama, Japan

[3] Department of Physics, Nihon University, 1-8-14 Surugadai, Chiyoda, Tokyo, Japan

E-mail: usui@hp.phys.titech.ac.jp

Abstract. We report the discovery of a dip structure in the pulse pro le of Be/X-ray binary LS V +44 17 observed by RXTE during an outburst detected by MAXI. The outburst was rst transient activity of LS V +44 17 since 1997 when the source was identi ed as a Be/X-ray binary. After the discovery with MAXI, the RXTE observations provided a pulse pro le with a narrow dip structure in the soft X-ray band (3 10 keV). Our pulse-phase-resolved spectroscopy showed that the absorption column density at the dip phase is much higher than those at the other intervals. Thus we can conclude that the partial eclipse of the emission region by the accretion column of the neutron star causes the dip structure in the pulse pro le of LS V +44 17 during the rst outburst.

1. Introduction

The accreting pulsar LS V +44 17/RX J0440.9+4431 is one of Be/X-ray binaries consisting of a neutron star (a pulsar) and a Be companion star. Be/X-ray binaries typically show periodic and/or non-periodic outbursts which are thought due to a mass accretion from the companion star to the neutron star through the circumsteller disk of the Be star. LS V +44 17 was discovered during the ROSAT all-sky survey [1]. Reig and Roche [2] performed the RXTE observation and obtained the sinusoidal pulse pro le with 202.5 0.5 sec period. They classi ed LS V +44 17 into one of persistent Be/X-ray binaries which show no outburst and stay in low luminosity ($10^{34\ \ 35}$ergs s ^{1}).

Monitor of All-sky X-ray Image (MAXI) [3] is a highly sensitive all-sky monitor mounted on the Exposed Facility of the Japanese Experimental Module "Kibo" of the International Space Station. MAXI has two X-ray cameras, GSC (Gas Slit Camera) and SSC (Solid-state Slit Camera) and is monitoring the ux of more than two hundred sources. Fig. 1 shows the2 20 keV light curve of LS V +44 17 obtained with GSC from 2009 August 15 to 2010 December 30. On 2010 March 31, MAXI detected an outburst of LS V +44 17 and reported it as the rst transient activity for the source [4]. The X-ray ux has increased for a week and nally reached 150 mCrab in the 2 20 keV band.

Following the detection with MAXI, RXTE con rmed the outburst on 2010 April 6 (see [5]) and carried out further observations on April 12 and 15 (the epochs of these observations, Obs.A, B and C, are shown in the small panel of Figure 1). We analyzed the RXTE data in the 2 60

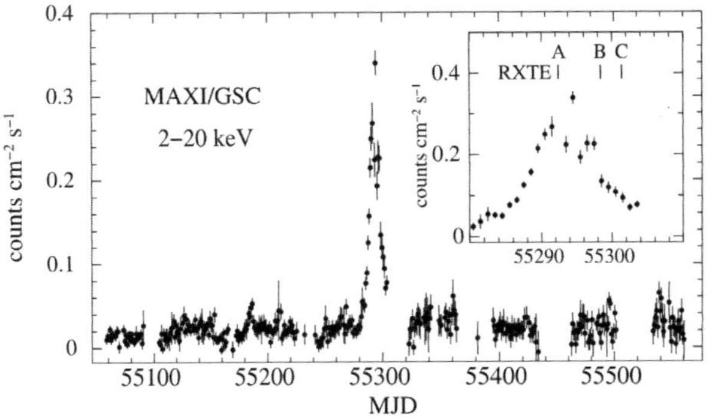

Figure 1. MAXI/GSC light curve of LS V +44 17. The count rate corresponding to the ux of Crab pulsar is 2.5 count cm ^{2} s ^{1}.

keV band with Proportional Counter Array (PCA) and 15 250 keV with High-Energy X-ray Timing Experiment (HEXTE).

2. Analysis and Results

2.1. Timing Analysis

We estimated pulse periods from the RXTE/PCA light curves after applying the barycentric correction. Performing the epoch folding search, we obtained 205 sec pulse periods of LS V +44 17 during the outburst. Considering a statistical error (almost several seconds) and a long-term spin period change, the estimated period is consistent with the past RXTE result. Folding the light curve with this period, we obtained energy-resolved pulse pro le of the PCA observation as shown in Figure 2. Phase 0.0 was chosen at the minimum in the 3-20 keV PCA pulse pro le for each observation. Notable point is that a narrow dip structure can be seen at phase 0.7 in Figure 2. The dip appeared only in the soft energy bands (3 6 and 6 10 keV), but not in the hard band (10 20 keV). Similar structure was reported for two Be/X-ray binaries (A0535+262 and RX J0812.4 3114) and can originate from a local absorption with higher density matter, such as an accretion stream on the neutron star.

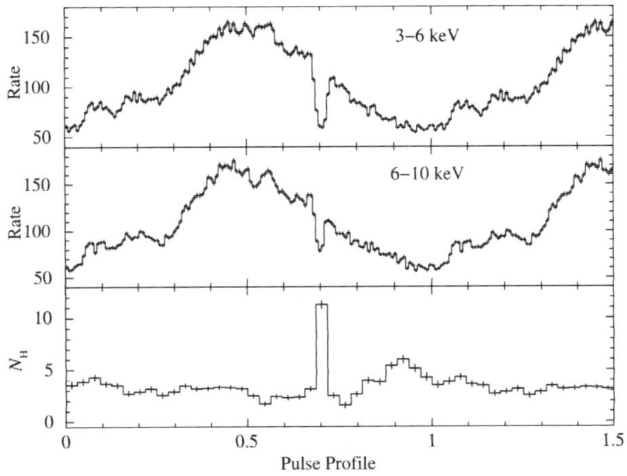

Figure 2. Pulse pro les of LS V +44 17 in the 3 6 and 6 10 keV bands in Obs.A (top and middle panel) and result of the phase-resolved spectroscopy (bottom panel). The unit of absorption column density $N_{\rm H}$ is 10^{22} cm ^{2}.

2.2. Pulse-phase-average Spectroscopy

We carried out pulse-phase-averaged spectroscopy for the three RXTE observations. We used a power-law model with an exponential cut-o plus a blackbody and a Gaussian function corresponding to an Fe emission line (the width of this component is xed to be 0.1 keV). The best tting values are shown in Table 1, where the errors correspond to 1sigma.

Table 1. Pulse phase-averaged spectral ts results.

Parameters	value
$N_{\rm H}$ (10^{22} cm ^{2})	3.3 0.3
	1.15 0.04
cut-o energy $E_{\rm fold}$ (keV)	26 1
blackbody temperature kT (keV)	$2.17^{+0.06}_{0.05}$
Gaussian peak energy $E_{\rm Fe}$ (keV)	$6.44^{+0.10}_{0.09}$
Reduced chi-square / degree of freedom	1.10/116

2.3. Pulse-phase-resolved Spectroscopy

To investigate a phase dependence of the absorption column density we performed pulse-phase-resolved spectral analysis for Obs.A. We used the same spectral model of the phase-average spectrum, where the photon index, the cut-o energy, the blackbody temperature and the peak energy of an iron line were xed to the values obtained with the pulse-phase-averaged analysis. The ratio between the power-law and blackbody normalization was also xed the phase-averaged value. Figure 2 shows the phase dependence of the absorption column density and the highest absorption density at the dip phase (phase 0.7) is clearly seen. Further, we present a ratio between the phase-average and dip-phase spectra as Figure 3.

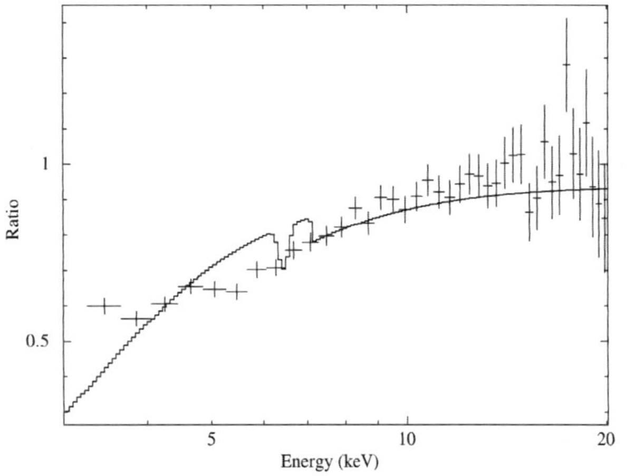

Figure 3. Ratio between the phase-average and dip-phase PCA (3 20 keV) spectrum of LS V +44 17 observed in Obs.A. The solid line presents a ratio between the models.

3. Discussion

According to Figure 2, the pulse pro le obtained during Obs.A had the dip structure. Similar sharp dips have been seen in the pulse pro les of the some Be/X-ray binaries (e.g. A 0535+262 and RX J0812.4 3114). The origin of the dip has been considered as an accretion ow eclipses [6] [7]. In this picture the X-ray emission originating from the surface of the neutron star is reprocessed by the accretion ow when it passes through the line of sight as the stars revolve around each other. Cemeljic and Bulik [6] modeled the emission from a neutron star considering the accretion ow eclipse and reconstructed the pulse pro les of A0535+262. Galloway et al. [7] showed through pulse-phase-resolved spectroscopy that had the absorption column density took maxima at dip phases of GX 1+4 and RX J0812.4 3114 . They argued that the results con rmed the dip origin which was the partial eclipse by the accretion column. Our result shows that the accretion column density is the highest at the dip phase and a ects the dip-phase spectrum (Figure 2, 3). This result may be consistent with the model proposed for the three binary systems, indicating that the accretion column eclipse a ects the pulse pro le of LS V +44 17 and generates the dip structure. In gure 3, the residuals remaining below 5 keV can propose another low-temperature component. This component may be emitted from the accretion disk or the accretion column. We will report the analysis for the dip-phase spectrum in our PASJ paper soon.

Acknowledgments
The authors acknowledge the MAXI team for their support and the RXTE team for providing the observational data. This work has been nancially supported from the Global Center of Excellence Program by MEXT, Japan through the "Nanoscience and Quantum Physics" Project of the Tokyo Institute of Technology.

References

[1] Motch, C., Haberl, F., Dennerl, K., Pakull, M., and Janot-Pacheco, E. 1997, A&A, 323, 853
[2] Reig, P., and Roche, P., 1999, MNRAS, 306, 100
[3] Matsuoka, M., et al., 2009, PASJ, 61, 999
[4] Morii, M., et al., 2010, Astron Telegram, 2527
[5] Finger, M. H., and Camero-Arranz, A., 2010, Astron Telegram, 2537
[6] Cemeljic, M., and Bulik, T., 1998, Acta Actron, 48, 65
[7] Galloway, D. K., Giles, A. B., Wu, K., and Greenhill, J. G., 2001, MNRAS, 325, 419

MAXI Monitoring of Crab Pulsar during the GeV Gamma-ray Flare on September 2010

Mikio Morii[1], Nobuyuki Kawai[1], Ryuichi Usui[1], Kousuke Sugimori[1], Mutsumi Sugizaki[2], Tatehiro Mihara[2], Takayuki Yamamoto[2], Masaru Matsuoka[2], and the MAXI team

[1]Department of Physics, Tokyo Institute of Technology, Ookayama 2-12-1, Meguro-ku, Tokyo 152-8551, Japan

[2]Coordinated Space Observation and Experiment Research Group, Institute of Physical and Chemical Research (RIKEN), 2-1 Hirosawa, Wako, Saitama 351-0198, Japan

E-mail: `morii@hp.phys.titech.ac.jp`

Abstract.
We report on the MAXI GSC X-ray monitoring of the Crab nebula and pulsar during the GeV gamma-ray flare on 2010 September detected by AGILE and Fermi-LAT. There were no significant variations on the pulse-phase-averaged and pulsed fluxes during the gamma-ray flare. The pulse profile also showed no significant change during the period. The upper limits on the variation of the pulse-phase-averaged and pulsed fluxes on MJD 55457.5−55462.5 in the 4-10 keV band are 1% and 19% at the 90% confidence limit of the statistical uncertainty, respectively. Here, the measured fluxes include a 2% systematic uncertainty at the 1-σ limit due to the error on the instrument calibration. The lack of variations in the pulsed component in multi-wavelength range (radio, soft X-ray, hard X-ray, and gamma-ray) supports the nebula origin for the gamma-ray flare.

1. Introduction

The Crab nebula is a remnant of the supernova on 1054. There is a neutron star (pulsar) with a spin period of 33 ms at the center of the nebula. The Crab nebula has been the standard candle of X-ray and gamma-ray astronomy, because the flux and spectrum have been expected to be steady for a long time scale of years. Surprisingly, AGILE and Fermi-LAT reported a gamma-ray flare on 2010 September 18−23 in more than 100 MeV energy range [1, 2]. RXTE-ASM, Swift-BAT, INTEGRAL did not detect variation for the pulse-phase-averaged flux [3]. Radio observation showed no evidence for a glitch, and no changes of pulsed flux and pulse profile shape [4]. Swift-XRT and RXTE-PCA follow-up observations did not detect any changes of flux, spectrum and pulse profile [5, 6]. Chandra and HST follow-up observations found some activated region around the nebula [7]. Recently, two more GeV flare episodes, happened on October 2007 and February 2009, were reported by AGILE and Fermi, respectively [2, 7]. Swift, Super-AGILE and Fermi have not seen significant variations in the pulsed emission in hard X-ray to gamma-ray during the three flaring periods.

2. Observation

MAXI (Monitor of All-sky X-ray Image) is an all-sky X-ray monitor, mounted on the Japanese Experiment Module - Exposed Facility (JEM-EF or Kibo-EF) on the International Space Station (ISS) [8]. It carries two types of X-ray cameras: the Gas Slit Camera (GSC; [9, 10]) for 2–30 keV band and Solid-state Slit Camera (SSC; [11, 12]) for 0.5–12 keV band, using gas proportional counters and X-ray CCDs, respectively. GSC scans almost all sky every 92 minutes with the field of view (FoV) with width of 1.5 degree (FWHM). MAXI GSC scans almost all sky every 92 mins by a transit duration about 1 min. The time resolution of GSC is 50 μs. The absolute time assignment is confirmed to be accurate within the stability of the time ($1\sigma = 0.2$ ms) [10]. MAXI has been monitoring the Crab nebula since the start of the operation on August 15, 2009 (UT), including the whole duration of the gamma-ray flare on September 2010. Here, we report on the MAXI monitoring of the Crab nebula during the GeV gamma-ray flare. The details will be shown in Morii *et al.* (2011)[13] in the near future.

3. Light curve and pulse profile

To measure the pulse-phase-averaged flux, we extracted the circular region on the sky coordinate and subtracted the background counts from the near-by region as done in the usual photometry method. Figure 1 (top) shows the light curves of the pulse-phase-averaged flux in 4–10 keV band. On the other hand, to measure the pulsed flux, we proceeded the following steps. At first, the times of events were corrected to the solar system barycenter. The time sequence was folded in the pulse period of Crab pulsar provided at the web site of the Jodrell Bank radio observatory [14]. In this folding, the time variation of effective area and exposure times were corrected [15, 16]. We fitted the pulse profiles with a model of $a \times T + b$, where T was a template pulse profile made from the data from December 13, 2009 to January 11, 2010, and the baseline offset of un-pulsed component was subtracted from the all phase bins. From this fit, we obtained the pulsed flux by $F_{\rm pulse} = a \sum_i T_i/N$. Figure 1 (bottom) shows the light curve of pulsed fluxes in 4–10 keV band. There is no significant flux change on the gamma-ray flare period. The 90% confidence level upper limits on the variation of the pulse-phase-averaged and pulsed fluxes on the flare period (MJD 55457.5–55462.5) were 1% and 19% in the 4–10 keV band. The pulse profile during the flare period is shown in Figure 2. There is no significant change from the template pulse profile.

4. Summary

We report on the MAXI GSC observation of the Crab pulsar during the gamma-ray flare. It is the only report showing the pulse profile and pulsed flux of the Crab pulsar on the whole duration of the gamma-ray flare in the X-ray band. The lack of the changes on the pulsed component in the gamma-ray energy range ($>$ 100 MeV) suggests the nebula origin for the gamma-ray flare. The same result obtained in our MAXI observation as well as that obtained in radio and hard X-ray energy ranges support this scenario.

Acknowledgments

Mikio Morii acknowledges support by a Grant-in-Aid for Young Scientists (B)(21740140) and the Global COE Program, "Nanoscience and Quantum Physics Project of the Tokyo Institute of Technology."

References

[1] Tavani, M. *et al.*, 2010, Astronomer's Telegram #2855
[2] Abdo, A. A. *et al.*, 2011, Science, **331**, 739
[3] Markwardt, C. B. *et al.*, 2010, Astronomer's Telegram #2858
[4] Espinoza, C. M. *et al.*, 2010, Astronomer's Telegram #2889

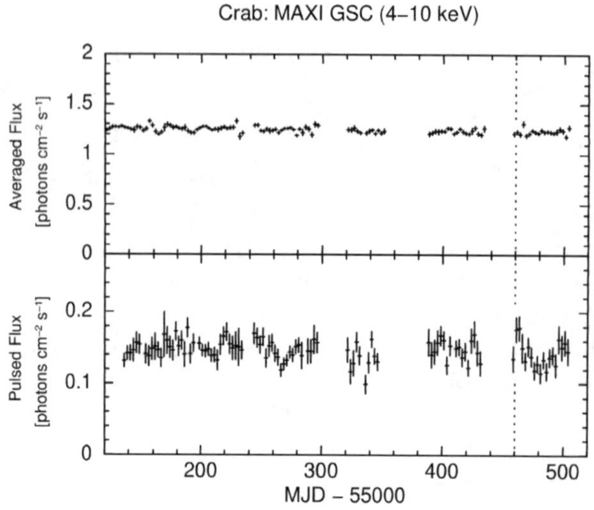

Figure 1. GSC light curves of the pulse-phase-averaged flux (top panel) and pulsed flux (bottom panel) of Crab nebula in the energy range of 4–10 keV in 2.5-day time bin. The horizontal and vertical axes are shown in units of MJD (modified Julian day) minus 55000 (2009 June 18) and photons cm^{-2} s^{-1}, respectively. The vertical error bars correspond to 1 σ statistical errors. The time center (MJD 55460.0) of the GeV gamma-ray flare period (MJD 55457.5–55462.5) is denoted by a vertical dotted line.

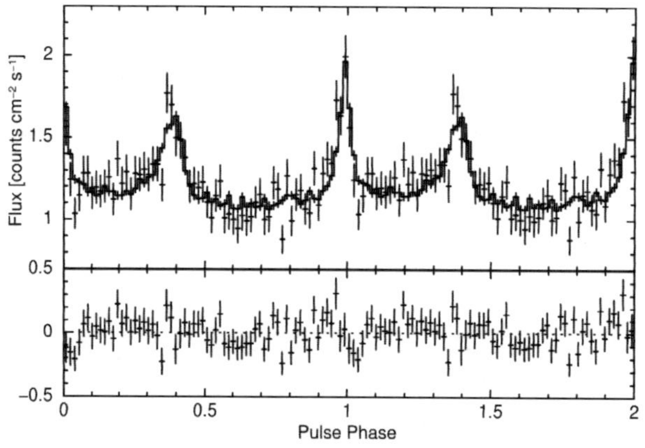

Figure 2. Top panel: Pulse profile of Crab pulsar in the 4–10 keV during the gamma-ray flare on MJD 55457.5–55462.5 obtained by GSC. The vertical axis is the flux shown in an unit of counts cm^{-2} s^{-1}. The solid histogram is the template pulse profile with the best fit. The bottom panel is the residual of the data from the best fitting model. In both panels, the vertical error bars correspond to 1 σ statistical errors.

[5] Evangelista, Y. *et al.*, 2010, Astronomer's Telegram #2866
[6] Shaposhnikov, N. *et al.* 2010, Astronomer's Telegram #2872
[7] Tavani, M. *et al.*, 2011, Science, **331**, 6018, 736
[8] Matsuoka, M. *et al.*, 2009, PASJ, **61**, 999
[9] Mihara, T. *et al.*, 2011, PASJ, accepted
[10] Sugizaki, M. *et al.*, 2011, PASJ, accepted

[11] Tsunemi, H. *et al.*, 2010, PASJ, **62**, 1371
[12] Tomida, H. *et al.*, 2011, PASJ, accepted
[13] Morii, M. *et al.*, 2011, PASJ, in preparation
[14] Lyne, A. G., Pritchard, R. S., & Smith, F. G., 1993, MNRAS, **265**, 1003. http://www.jb.man.ac.uk/research/pulsar/crab.html.
[15] Morii, M. *et al.* 2006, Proc. of SPIE **6266**, 6263U
[16] Morii, M., Sugimori, K., and Kawai, N., 2010, Physica E: Low-dimensional Systems and Nanostructures, **43**, 692.

On detecting the quantum correlations in the early universe

Yutaka Shikano[1,2,*], **Yusuke Hayashi**[3,†], **and Masaaki Hashimoto**[3,♠]

[1]Department of Physics, Tokyo Institute of Technology, 2-12-1 Oh-Okayama, Meguro, Tokyo 152-8551 Japan

[2]Department of Mechanical Engineering, Massachusetts Institute of Technology, 77 Massachusetts Avenue, Cambridge, MA 02139 USA

[3]Department of Physics, Kyushu University, 6-10-1 Hakozaki, Higashi-ku, Fukuoka, Fukuoka 812-8581 Japan

E-mail: *shikano@th.phys.titech.ac.jp, †hayashi@phys.kyushu-u.ac.jp, ♠hashimoto@phys.kyushu-u.ac.jp

Abstract. The superhorizon region has been supported by the Wilkinson Microwave Anisotropy Probe experimental data. In this paper, we point out that the conventional entanglement criterion cannot be applied in the superhorizon region.

1. Introduction

In the standard scenario of the early universe, the inflation theory was individually proposed by Sato [1] and Guth [2] to resolve the fundamental problems on the early universe in the follows.

The horizon problem While the density distribution of matter in the big bang cosmology is in principle inhomogeneous, this prediction is mismatched to the experimental data.

The flatness problem According to the experimental observation, the current curvature density Ω is nearly flat. This is mismatched to $\Omega \propto 1/H_0^2$ with the Hubble horizon H_0 due to the small horizon in the early universe.

The relic problem The unresolved issue in standard cosmology in which various theories of particle physics would invariably produce massive particles which are not observed.

The fine tuning problem The cosmological constant in the Einstein equation is mismatched to the current experimental data.

Many versions of the inflation theory to adjust the experimental cosmological data were proposed, *e.g.*, see the review papers [3, 4, 5]. However, nobody experimentally has yet confirmed this theory. However, recent observations of the cosmic microwave background (CMB) anisotropies [6] strongly support the *superhorizon* region in the early universe. The *superhorizon* region is called the time region that the physical scale of the quantum correlation exceeds one of the event horizon. It is remarked that the superhorizon region is dealt in many alternative approaches for the inflation theory, for example, ekpyrotic universe [7], cyclic universe theory [8, 9], and brane world theory [10]. The ultimate aim of this work is to establish how to detect the quantum correlations in the superhorizon region. This may contribute to

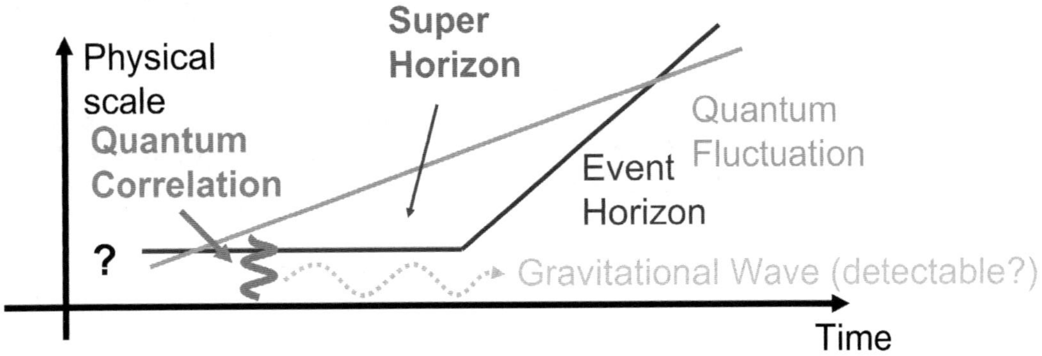

Figure 1. The schematic figure of the inflation theory to be focused on the superhorizon region. Information of the quantum correlations at that time may be stored in the gravitational wave.

understanding the role of the quantum field after the big bang. In this paper, we point out that the conventional entanglement criterion cannot be applied in this region.

2. Quantum correlation in the superhorizon region

Before the main discussion, we briefly recapitulate the relationship between the quantum correlation and the entanglement, which is defined such that the quantum state in the combined system, $|\psi\rangle_{AE} \in \mathcal{H}_A \otimes \mathcal{H}_E$, cannot be described as the separable state;

$$|\psi\rangle_{AE} \neq |\phi_1\rangle_A \otimes |\phi_2\rangle_E, \ \forall|\phi_1\rangle_A \in \mathcal{H}_A, \ \forall|\phi_2\rangle_E \in \mathcal{H}_E. \tag{1}$$

Here, the Hilbert spaces, $\mathcal{H}_A$ and $\mathcal{H}_E$, express the considered and the other systems, respectively. If the state has the quantum correlation, even the state is in general not entangled. However, if the state is entangled, the state must have the quantum correlation.

While we would like to detect the quantum correlations between beyond and not beyond the physical scale of the event horizon to be illustrated in Fig. 1, there are crucial problems:

(i) the lack of the concept of the entanglement for the quantum field theory,

(ii) how to confirm the quantum correlations between beyond and not beyond the physical scale of the event horizon.

On the statement (i), the entanglement criterion for the quantum field seems not to be well defined from the quantum mechanical sense since the quantum field is defined as the global state. There are some approaches in this context. Fuentes *et al.* showed that the logarithmic negative criterion for the entanglement [11] was applied to the particles created by the Bogoliubov transformation. However, this criterion cannot distinguish which particles are entangled [12]. Reznik claimed that the two Unruh-Dewitt detectors couple to the quantum field. By the entanglement between the coupled detectors, the entanglement for the inflaton field is defined [13]. This idea is also used in Ref. [14]. On the statement (ii), the entanglement criterion is essentially used in the quantum informational method. In the conventional method of quantum information science, the detection of the quantum correlation is used in the local operation and classical communication (LOCC) [15]. However, in this situation, in principle, classical communication between beyond and not beyond the event horizon cannot be used due to the causality. To propose the judgment of the quantum correlation, we have known the alternative method using dynamics of the purity, the trace of the square of the density operator, in one-side subsystem [16]. According to Ref. [16], they showed that

$$\frac{d}{dt}\left(\mathrm{Tr}\rho_A^2\right) \neq 0 \Rightarrow \rho \text{ is entangled}, \tag{2}$$

where the density operator in the considered subsystem $\rho_A := \mathrm{Tr}_E \rho$. Intuitively, when the considered field is coupled to the other field, obviously the state in the considered system must be changed. However, this criterion also has the serious problem how to define the local Hilbert space, that is, $\mathcal{H}_A$ and $\mathcal{H}_E$. On studying the quantum correlations in the superhorizon region, the local Hilbert space has not yet well defined. Therefore, the conventional well-known entanglement criterion cannot be applied.

3. Conclusion and Outlooks

We strikingly pointed out that the conventional quantum informational methods to judge the entanglement cannot be applied to the early universe context due to the causality.

From considering this problem, many open problems on quantum information science are shown up. As mentioned in the statement (i), first, the definition, the operational meaning, and the applications of the entanglement in quantum field theory are not known very well. Since many-body quantum mechanical systems can be characterized by the quantum field theory, this problem may be crucial to build up the large-scaled quantum computer. This is essentially different from the concept of the entanglement in a few body quantum system. Second, our proposal seems to allow the superluminal communication since we can extract information. However, the similar situation can be occurred in the context of the information loss paradox [18, 19] motivated by quantum teleportation. Therefore, our result has the potential to open the discussion on the role of the entanglement in the context of the cosmological situation. For example, related to the quantum teleportation, the first author, who is the co-author of Refs. [20, 21], discussed the quantum description of the closed timelike curve. However, Refs. [20, 21] proposals are only taken as the closed timelike curve consistently but were not and were not simulated to this since the arrow of time only has one direction, that is, does not have the reverse direction and the loop structure. To summarize these, quantum circuit representation may be not related to the spacetime. If so, the classification of the entanglement by LOCC is not physical. To resolve it, we need the quantum information theory to be expanded to the context in relativistic quantum field theory. Finally, what is meaning of the local operation in quantum field theory? However, in this context, the causality can be well defined. Since the locality is disconnected to the cluster principle, the local operation has no relationship to causality.

Acknowledgments

One of the authors (YS) thanks Akio Hosoya, Masahide Yamaguchi, and Si-Hui Tan for useful discussion. YS acknowledges the great hospitality and great support, especially from Margaret Sullivan, at Massachusetts Institute of Technology hosted by Seth Lloyd partially supported by JSPS Excellent Young Researcher Overseas Visit Program. YS is supported by JSPS Research Fellowships for Young Scientists (No. 21008624) and Global Center of Excellence Program "Nanoscience and Quantum Physics" at Tokyo Institute of Technology. MH is supported by Grant-in-Aid for Scientific Researches (19104006, 21540272) of the Ministry of Education, Culture, Sports, Science and Technology of Japan.

References

[1] Sato K 1981 *Mon. Not. R. Astron. Soc.* **195** 467
[2] Guth A 1981 *Phys. Rev. D* **23** 347
[3] Bartolo N, Komatsu E, Matarresse S, and Riotto A 2004 *Phys. Rep.* **402** 103
[4] Baumann D 2009 arXiv:0907.5424 as Lecture Note from the 2009 Theoretical Advanced Study Institute at University of Colorado, Boulder
[5] Yamaguchi M 2011 *Class. Quantum Grav.* **28** 103001
[6] Komatsu E, Smith K M, Dunkley J, Bennett C L, Gold B, Hinshaw G, Jarosik N, Larson D, Nolta N R, Page L, Spergel D N, Halpern M, Hill R S, Kogut A, Limon M, Meyer S S, Odegard N, Tucker G S, Weiland J L, Wollack E, and Wright E L 2011 *Astrophys. J. Suppl.* **192** 18

[7] Khoury J, Ovrut B A, Steinhardt P J, and Turok N 2001 *Phys. Rev. D* **64** 123522
[8] Steinhardt P J and Turok N 2002 *Phys. Rev. D* **65** 1260003
[9] Baum L and Frampton P H 2008 *Mod. Phys. Lett. A* **23** 33
[10] Lisa R and Sundrum R 1999 *Phys. Rev. Lett.* **83** 3370
[11] Vidal G and Werner R F 2002 *Phys. Rev. A* **65** 032314
[12] Fuentes I, Mann R B, Martin-Martinez E, and Moradi S 2010 *Phys. Rev. D* **82**, 045030
[13] Reznik B 2003 *Found. Phys.* **33** 167
[14] Nambu Y and Ohsumi Y 2009 *Phys. Rev. D* **80** 124031
[15] Horodecki R, Horodecki P, Horodecki M, and Horodecki K 2009 *Rev. Mod. Phys.* **81** 865
[16] Kimura G, Ohno H, and Mosonyi M 2010 *Open Sys. Infor. Dyn.* **17** 233
[17] Lin S-Y and Hu B L 2006 *Phys. Rev. D* **73** 124018
[18] Horowitz G T and Maldacena J 2004 *JHEP* **0402** 8
[19] Gottesmann D and Preskill J 2004 *JHEP* **0403** 26
[20] Lloyd S, Maccone L, Garcia-Patron R, Giovannetti V, Shikano Y, Pirandola S, Rozema L A, Darabi A, Soudagar Y, Shalm L K, and Steinberg A M 2011 *Phys. Rev. Lett.* **106** 040403
[21] Lloyd S, Maccone L, Garcia-Patron R, Giovannetti V, and Shikano Y 2011 arXiv:1007.2615

AUTHOR INDEX

AUTHOR INDEX

AUTHOR INDEX